T0202486

Lecture Notes in Computer Science 11137

Commenced Publication in 1973
Founding and Former Series Editors:
Gerhard Goos, Juris Hartmanis, and Jan van Leeuwen

More information about this series at http://www.springer.com/series/7409

Denny Vrandečić · Kalina Bontcheva
Mari Carmen Suárez-Figueroa · Valentina Presutti
Irene Celino · Marta Sabou
Lucie-Aimée Kaffee · Elena Simperl (Eds.)

The Semantic Web – ISWC 2018

17th International Semantic Web Conference
Monterey, CA, USA, October 8–12, 2018
Proceedings, Part II

Editors
Denny Vrandečić
Google
San Francisco, CA
USA

Kalina Bontcheva
University of Sheffield
Sheffield
UK

Mari Carmen Suárez-Figueroa
Universidad Politécnica de Madrid (UPM)
Madrid, Madrid
Spain

Valentina Presutti
National Research Council
Rome, Roma
Italy

Irene Celino
Cefriel - Politecnico di Milano
Milan
Italy

Marta Sabou
TU Wien
Vienna
Austria

Lucie-Aimée Kaffee
University of Southampton
Southampton
UK

Elena Simperl
University of Southampton
Southampton
UK

ISSN 0302-9743 ISSN 1611-3349 (electronic)
Lecture Notes in Computer Science
ISBN 978-3-030-00667-9 ISBN 978-3-030-00668-6 (eBook)
https://doi.org/10.1007/978-3-030-00668-6

Library of Congress Control Number: 2018954489

LNCS Sublibrary: SL3 – Information Systems and Applications, incl. Internet/Web, and HCI

This Springer imprint is published by the registered company Springer Nature Switzerland AG
The registered company address is: Gewerbestrasse 11, 6330 Cham, Switzerland

Preface

Now in its 17th year, the ISWC continues to be a focal point of the Semantic Web community. Year after year, it brings together researchers and practitioners from all over the world to present new approaches and findings, share ideas, and discuss experiences. It features a balanced mix of fundamental research, innovative technology, scientific artefacts such as ontologies or benchmarks, and applications that showcase the power of semantics, data, and the Web.

The Web, and all the ideas, technologies, and values that surround it, are at a crossroads. After several decades of growth and prosperity, it is increasingly seen as a means to lock-in customers and their data, spread misinformation, and increase polarization in society. At the same time, there is a palpable sense of excitement as we witness new voices and developments from the community that are fighting this trend in various ways – from more open and transparent forms of scholarly publishing and peer review in some of the workshops featured at the conference to cutting-edge research and applications on topics such as fake news, semantic coherence, and fact checking. Against this background, this year we decided to revive the Blue Sky Ideas track, chaired by Carolina Fortuna and supported by the Computing Community Consortium, to seek visionary ideas and opportunities for research and innovation, which are outside the mainstream topics of the conference.

A child of its times, the 17th ISWC featured a stellar, all-female keynote lineup: Jennifer Golbeck from the University of Maryland talked about human factors in semantic technologies; Vanessa Evers, University of Twente, introduced us to social robotics, an area with interesting applications for the models and technologies developed in our community; while Natasha Noy of Google discussed how we could use semantics to make structured data on the web more accessible and useful for everyone.

This volume contains the proceedings of ISWC 2018, i.e. papers that were peer reviewed and accepted into the main conference program, which covered three tracks: research, resources, and in-use. Altogether, a total of 254 submissions were received, which were evaluated by 486 reviewers. A total of 62 papers were accepted – 39 for the research track, 17 for the resources track, and six for the in-use track. The substantial number of papers in the resources category attests the commitment of the community to sharing and collaboration and to repeatable, reproducible research.

ISWC has an excellent scientific profile – as such, the research track continues to be the most popular venue for submissions. This year the track received overall 167 valid full-paper submissions, which turned into 39 acceptances, leading to an acceptance rate of 23%. We recruited 272 PC members and 67 sub-reviewers, guided by 17 senior PC members. Each paper received at least four reviews, including one from a senior PC member. The papers were assessed for originality, novelty, relevance, and impact of the research contributions, soundness, rigour and reproducibility, clarity and quality of presentation, and grounding in the literature. Each paper was then discussed by the PC chairs and the senior PC members, who helped us reach a consensus.

The resources track promotes the sharing of high-quality information artifacts that have contributed to the generation of novel scientific work. Resources can be datasets, ontologies, vocabularies, ontology design patterns, benchmarks, crowdsourcing designs, software frameworks, workflows, protocols, metrics, among others. The track is becoming demonstratively more and more important to our community as the sharing of reusable resources is key to allowing other researchers to compare new results, reproduce experimental research, and explore new lines of research, in accordance with the FAIR principles for scientific data management. All published resources must address a set of requirements: persistent URI, canonical citation, license specification, to mention a few. This year the track received 55 submissions, of which 17 were accepted (31% acceptance rate), covering a wide range of resource types such as benchmarks, ontologies, datasets, software frameworks, and crowdsourcing designs; a variety of domains such as music, health, education, drama, and audio; and addressing multiple problems such as RDF querying, ontology alignment, linked data analytics, and recommending systems. The reviewing process involved 70 PC members and 9 subreviewers, supported by 8 senior PC members. The average number of reviews per paper were 3.7 (at least three per paper), plus a meta-review provided by a senior PC member. Papers were evaluated based on the availability of the resource, its design and technical quality, impact, and reusability. The review process also included a rebuttal phase and further discussions among reviewers and senior PC members, who provided recommendations. Final decisions were taken following a detailed analysis and discussion of each paper conducted by the program chairs and the senior PC.

The in-use track at ISWC 2018 continued the tradition of demonstrating and learning from the increasing adoption of Semantic Web technologies outside the boundaries of research institutions, by providing a forum for the community to explore the benefits and challenges of applying these technologies in concrete, practical applications, in contexts ranging from industry to government and science. This year, the 32 submissions were reviewed by at least three PC members each and assessed in terms of novelty of the proposed use case or solution, uptake by the target user group, demonstrated or potential impact, as well as the overall soundness and quality. The PC consisted of 43 members. It helped us select 6 papers for acceptance, covering different domains (e.g., healthcare, cultural heritage, industry) and addressing a multitude of research problems (e.g., data integration, collaborative knowledge management, recommendations).

The industry track provides an opportunity for industry adopters to highlight and share the key learnings and new research challenges posed by real-world implementations. This year we had many exciting submissions from small to large companies that are making revealing leaps forward in science and engineering by using and adopting semantic technologies, web of data sources, and knowledge graphs. Each short submission was reviewed by at least three PC members. We accepted 14 out 27 abstracts that showcased a wide range of real-world industrial strength applications. The submissions were assessed in terms of the impact of semantics as a competitive differentiator in industry and discussions on the business value, experiences, insights, as well obstacles that stand in the way of large-scale adoption of semantic technologies.

The main conference program was complemented by presentations from the journal, industry, and posters and demos tracks, as well as the Semantic Web Challenge and a panel on future trends in knowledge graphs.

The conference included a variety of events appreciated by the community, which created more opportunities to present and discuss emerging ideas, network, learn, and mentor. Thanks to Amrapali Zaveri and Elena Demidova, the workshops and tutorials program includes a mix of established topics such as ontology matching and ontology design patterns alongside newer ones that reflect the commitment of the community to innovate and help create systems and technologies that people want and deserve, including re-decentralizing the Semantic Web, augmenting intelligence with humans in the loop, and a perspective workshop discussing open issues and trends. Application-centric workshops range from statistics to science to healthcare. The tutorials covered topics such as ontology modeling, crowdsourcing methods and metrics, RDF data validation and visualization, as well as knowledge graph machine learning and applications.

The conference also included a Doctoral Consortium track, which was chaired by Lalana Kagal and Sabrina Kirrane. The DC afforded PhD students from the Semantic Web community the opportunity to share their research ideas in a critical but supportive environment, where they received feedback from senior members of the community. This year the Program Committee accepted 12 papers for presentation at the event, while a total of 18 students were selected to participate in the DC poster and demo session. All student participants were paired with mentors from the PC who provided guidance on improving their research, producing slides, and giving presentations.

The program was complemented by activities put together by Bo Fu and Anisa Rula as student coordinators, who secured funding for travel grants, managed the grant application process, and organized the mentoring lunch alongside other informal opportunities for students and other newcomers to get to know the community.

Posters and demos are one of the most vibrant parts of every ISWC. This year, the track was chaired by Marieke van Erp and Medha Atre. It included 40 demos and 39 posters selected from a total of 95 submissions. A minute madness session offered time to those who wanted to take to the stage to present a brief preview of their poster or demo to generate interest in the work.

The Semantic Web Challenge has now been a part of ISWC for 15 years. Started as an open challenge to provide a forum for new and prestigious applications of Semantic Web technologies, and seconded by a challenge for scalability with the Billion Triple Challenge since 2003, the challenge was reanimated in 2017 with a new direction, with fixed datasets, and objective measures allowing for direct comparison of challenge entries. The 2018 challenge used a partly public, partly private knowledge graph about company networks owned by Thomson Reuters, and participants were asked to predict supply chain relations between those companies, using both knowledge in the graph itself as well as external sources. The best solutions were presented and discussed at the conference, both in a dedicated plenary session as well as during the poster session.

Delivering a conference is so much more than assembling a program. An event of the scale and complexity of ISWC requires the help, resources, and time of hundreds of people, organizers of satellite events, reviewers, volunteers, and sponsors. We are very grateful to our local team at Stanford University, who have expertly managed conference facilities, accommodation, registrations, the website, and countless other details. They made the conference a place we want to be every year and helped us grow this exciting scientific community.

Our thanks also go to Maribel Acosta, our tireless publicity chair – she played a critical role in ensuring that all conference activities and updates were communicated and promoted across mailing lists and on social media. Oana Inel was the metadata chair this year – her work made sure that all relevant information about the conference was available in a format that could be used across applications, continuing a tradition established at this conference many years ago. We are especially thankful to our proceedings chair, Lucie-Aimée Kaffee, who oversaw the publication of this volume alongside a number of CEUR proceedings for other tracks.

Sponsorship is crucial to the realization of the conference in its current form. We had a highly committed trio of sponsorship chairs, Annalisa Gentile, Maria Maleshkova, and Laura Koesten, who went above and beyond to find new ways to engage with sponsors and promote the conference to them. Thanks to them, the conference now features a social program that is almost as exciting as the scientific one – including a jam session accompanying the posters and demos presented on the second day of the conference and a bike ride from San Jose to Asilomar, the venue of the conference. Our special thanks go to the Semantic Web Science Association (SWSA) for their continuing support and guidance and to the organizers of the conference from 2017 and 2016, who were a constant inspiration, role model, and source of practical knowledge.

August 2018

Denny Vrandečić
Kalina Bontcheva
Mari Carmen Suárez-Figueroa
Valentina Presutti
Irene Celino
Marta Sabou
Lucie-Aimée Kaffee
Elena Simperl

Organization

Organizing Committee

General Chair

Elena Simperl — University of Southampton, UK

Local Chair

Rafael Gonçalves — Stanford University, USA

Research Track Chairs

Denny Vrandečić — Google, Mountain View, USA
Kalina Bontcheva — University of Sheffield, UK

Resources Track Chairs

Mari Carmen Suárez-Figueroa — Universidad Politécnica de Madrid, Spain
Valentina Presutti — Italian National Research Council, Italy

In-Use Track Chairs

Irene Celino — Cefriel, Italy
Marta Sabou — Vienna University of Technology, Austria

Workshop and Tutorial Chairs

Amrapali Zaveri — Maastricht University, Netherlands
Elena Demidova — L3S Research Center in Hannover, Germany

Poster and Demo Track Chairs

Medha Atre — Indian Institute of Technology, India
Marieke van Erp — Knaw Humanities Cluster, Netherlands

Journal Track Chairs

Abraham Bernstein — University of Zurich, Switzerland
Pascal Hitzler — Wright State University, Dayton, USA
Steffen Staab — University of Koblenz-Landau, Germany

Industry Track Chairs

Vanessa Lopez — IBM Research, Dublin, Ireland
Kavitha Srinivas — Rivet Labs, USA

Doctoral Consortium Chairs

Sabrina Kirrane Vienna University of Economics and Business, Austria
Lalana Kagal MIT CSAIL, USA

Semantic Web Challenge Chairs

Heiko Paulheimt University of Mannheim, Germany
Axel Ngonga University of Paderborn, Germany
Dan Bennett Enterprise Data Services at Thomson Reuters, USA

Proceedings Chair

Lucie-Aimée Kaffee University of Southampton, UK

Metadata Chair

Oana Inel Vrije Universiteit Amsterdam, Netherlands

Sponsorship Chairs

Laura Koesten Open Data Institute, UK
Maria Maleshkova Karlsruhe Institute of Technology, Germany
Annalisa Gentile IBM Research, San Jose, USA

Student Coordinators

Bo Fu California State University, Long Beach, USA
Anisa Rula University of Milano-Bicocca, Italy

Publicity Chair

Maribel Acosta Karlsruhe Institute of Technology, Germany

Program Committee

Senior Program Committee – Research Track

Christian Bizer University of Mannheim
Kai-Uwe Sattler TU Ilmenau
Paul Groth Elsevier Labs
Stefan Schlobach Vrije Universiteit Amsterdam
Thomas Lukasiewicz University of Oxford
Steffen Staab Institut WeST, University Koblenz-Landau and WAIS, University of Southampton
Pascal Hitzler Wright State University
Harith Alani The Open University
Alessandro Moschitti Qatar Computing Research Institute
Claudia d'Amato University of Bari
Vojtěch Svátek University of Economics, Prague

Sören Auer	TIB Leibniz Information Center for Science and Technology and University of Hannover
Oscar Corcho	Universidad Politécnica de Madrid
Abraham Bernstein	University of Zurich
Lalana Kagal	MIT
David Martin	Nuance Communications
Jose Manuel Gomez Perez	Expert System Iberia

Program Committee – Research Track

Maribel Acosta	Karlsruhe Institute of Technology
Alessandro Adamou	The Open University
Nitish Aggarwal	IBM
Harith Alani	The Open University
Panos Alexopoulos	Textkernel B.V.
Jose Julio Alferes	Universidade NOVA de Lisboa
Marjan Alirezaie	Orebro University
José Luis Ambite	University of Southern California
Renzo Angles	Universidad de Talca
Grigoris Antoniou	University of Huddersfield
Manuel Atencia	Univ. Grenoble Alpes and Inria
Ioannis N. Athanasiadis	Wageningen University
Sören Auer	TIB Leibniz Information Center for Science and Technology and University of Hannover
Nathalie Aussenac-Gilles	IRIT CNRS
Franz Baader	TU Dresden
Payam Barnaghi	University of Surrey
Valerio Basile	University of Turin
Zohra Bellahsene	LIRMM
Michael K. Bergman	Cognonto Corporation
Abraham Bernstein	University of Zurich
Elisa Bertino	Purdue University
Leopoldo Bertossi	Carleton University
Christian Bizer	University of Mannheim
Fernando Bobillo	University of Zaragoza
Kalina Bontcheva	The University of Sheffield
Alex Borgida	Rutgers University
Mihaela Bornea	IBM
Loris Bozzato	Fondazione Bruno Kessler
Adrian M. P. Brasoveanu	MODUL Technology GmbH
Charalampos Bratsas	Aristotle University of Thessaloniki
John Breslin	NUI Galway
Carlos Buil Aranda	Universidad Técnica Federico Santa María
Gregoire Burel	The Open University
Elena Cabrio	Université Côte d'Azur, CNRS, Inria, I3S, France
Andrea Calì	University of London, Birkbeck College

David Carral	TU Dresden
Gerard Casamayor	Universitat Pompeu Fabra
Davide Ceolin	Vrije Universiteit Amsterdam
Ismail Ilkan Ceylan	University of Oxford
Pierre-Antoine Champin	LIRIS, Université Claude Bernard Lyon1
Gong Cheng	Nanjing University
Christian Chiarcos	Universität Frankfurt am Main
Michael Cochez	Fraunhofer
Pieter Colpaert	Ghent University
Simona Colucci	Politecnico di Bari
Olivier Corby	Inria
Oscar Corcho	Universidad Politécnica de Madrid
Luca Costabello	Accenture Labs
Fabio Cozman	University of São Paulo
Isabel Cruz	University of Illinois at Chicago
Philippe Cudre-Mauroux	U. of Fribourg
Bernardo Cuenca Grau	University of Oxford
Claudia d'Amato	University of Bari
Aba-Sah Dadzie	The Open University
Enrico Daga	The Open University
Florian Daniel	Politecnico di Milano
Laura M. Daniele	TNO - Netherlands Organization for Applied Scientific Research
Victor de Boer	Vrije Universiteit Amsterdam
Jeremy Debattista	Trinity College Dublin
Thierry Declerck	DFKI GmbH
Jaime Delgado	Universitat Politècnica de Catalunya
Daniele Dell'Aglio	University of Zurich
Emanuele Della Valle	Politecnico di Milano
Elena Demidova	L3S Research Center
Ronald Denaux	University of Leeds, UK
Dennis Diefenbach	Université Jean Monet
Stefan Dietze	GESIS - Leibniz Institute for the Social Sciences
Ying Ding	Indiana University Bloomington
Mauro Dragoni	Fondazione Bruno Kessler - FBK-IRST
Michel Dumontier	Maastricht University
Henrik Eriksson	Linköping University
Vadim Ermolayev	Zaporozhye National University
Jérôme Euzenat	Inria and Univ. Grenoble Alpes
James Fan	HelloVera.ai
Nicola Fanizzi	Università degli studi di Bari "Aldo Moro"
Anna Fensel	Semantic Technology Institute (STI) Innsbruck, University of Innsbruck
Alberto Fernandez	CETINIA, University Rey Juan Carlos
Javier D. Fernández	Vienna University of Economics and Business
Sebastien Ferre	Université de Rennes 1

Besnik Fetahu	L3S Research Center
Tim Finin	University of Maryland, Baltimore County
Lorenz Fischer	Sentient Machines
Fabian Flöck	GESIS Cologne
Antske Fokkens	Vrije Universiteit Amsterdam
Muriel Foulonneau	Luxembourg Institute of Science and Technology
Flavius Frasincar	Erasmus University Rotterdam
Fred Freitas	Universidade Federal de Pernambuco (UFPE)
Adam Funk	University of Sheffield
Aldo Gangemi	Università di Bologna and CNR-ISTC
Daniel Garijo	Information Sciences Institute
Anna Lisa Gentile	IBM
Jose Manuel Gomez Perez	Expert System Iberia
Rafael S. Gonçalves	Stanford University
Gregory Grefenstette	IHMC and Biggerpan, Inc.
Paul Groth	Elsevier Labs
Tudor Groza	The Garvan Institute of Medical Research
Cathal Gurrin	Dublin City University
Christophe Guéret	Accenture
Peter Haase	metaphacts
Armin Haller	Australian National University
Harry Halpin	World Wide Web Consortium
Karl Hammar	Jönköping University
Andreas Harth	University Erlangen-Nuremberg
Oktie Hassanzadeh	IBM
Tom Heath	Open Data Institute
Johannes Heinecke	Orange Labs
Andreas Herzig	IRIT-CNRS
Pascal Hitzler	Wright State University
Aidan Hogan	DCC, Universidad de Chile
Laura Hollink	Vrije Universiteit Amsterdam
Matthew Horridge	Stanford University
Katja Hose	Aalborg University
Andreas Hotho	University of Wuerzburg
Geert-Jan Houben	Delft University of Technology
Wei Hu	Nanjing University
Eero Hyvönen	Aalto University
Yazmin A. Ibanez-Garcia	Institute of Information Systems, TU Wien
Luis Ibanez-Gonzalez	University of Southampton
Oana Inel	Vrije Universiteit Amsterdam
Mustafa Jarrar	Birzeit University
Ernesto Jimenez-Ruiz	The Alan Turing Institute
Clement Jonquet	University of Montpellier - LIRMM
Lucie-Aimée Kaffee	University of Southampton
Lalana Kagal	MIT
Martin Kaltenboeck	Semantic Web Company

Mark Kaminski	University of Oxford
Pavan Kapanipathi	IBM T.J. Watson Research Center
Md. Rezaul Karim	Fraunhofer FIT, Germany
Tomi Kauppinen	Aalto University School of Science
Takahiro Kawamura	Japan Science and Technology Agency
Mayank Kejriwal	Information Sciences Institute
Carsten Keßler	Aalborg University Copenhagen
Prashant Khare	Knowledge Media Institute, Open University, UK
Haklae Kim	Samsung Electronics
Sabrina Kirrane	Vienna University of Economics and Business - WU Wien
Matthias Klusch	DFKI
Matthias Knorr	Universidade NOVA de Lisboa
Stasinos Konstantopoulos	NCSR Demokritos
Roman Kontchakov	Birkbeck, University of London
Jacek Kopecky	University of Portsmouth
Adila A. Krisnadhi	Wright State University and Universitas Indonesia
Udo Kruschwitz	University of Essex
Tobias Kuhn	Vrije Universiteit Amsterdam
Benedikt Kämpgen	Empolis Information Management GmbH
Patrick Lambrix	Linköping University
Steffen Lamparter	Siemens AG, Corporate Technology
Agnieszka Lawrynowicz	Poznan University of Technology
Danh Le Phuoc	TU Berlin
Chengkai Li	University of Texas at Arlington
Juanzi Li	Tsinghua University
Nuno Lopes	TopQuadrant, Inc.
Chun Lu	Université Paris-Sorbonne and Sépage
Markus Luczak-Roesch	Victoria University of Wellington
Thomas Lukasiewicz	University of Oxford
Carsten Lutz	Universität Bremen
Alexander Löser	Beuth Hochschule für Technik Berlin
Frederick Maier	Institute for Artificial Intelligence
David Martin	Nuance Communications
Trevor Martin	University of Bristol
Mercedes Martinez-Gonzalez	University of Valladolid
Miguel A. Martinez-Prieto	University of Valladolid
Wolfgang May	Universitaet Goettingen
Diana Maynard	The University of Sheffield
Franck Michel	Université Côte d'Azur, CNRS, I3S
Nandana Mihindukulasooriya	Universidad Politécnica de Madrid
Riichiro Mizoguchi	Japan Advanced Institute of Science and Technology
Marie-Francine Moens	Katholieke Universiteit Leuven
Pascal Molli	University of Nantes - LS2N

Gabriela Montoya	Aalborg University
Federico Morando	Nexa Center for Internet and Society at Politecnico di Torino
Alessandro Moschitti	Qatar Computing Research Institute
Paul Mulholland	The Open University
Raghava Mutharaju	GE Global Research
Lionel Médini	LIRIS lab., University of Lyon
Ralf Möller	University of Luebeck
Hubert Naacke	Sorbonne Université, UPMC, LIP6
Axel-Cyrille Ngonga Ngomo	Paderborn University
Andriy Nikolov	metaphacts GmbH
Lyndon Nixon	MODUL Technology GmbH
Leo Obrst	MITRE
Francesco Osborne	The Open University
Raul Palma	Poznan Supercomputing and Networking Center
Matteo Palmonari	University of Milano-Bicocca
Jeff Z. Pan	University of Aberdeen
Rahul Parundekar	Toyota Info-Technology Center
Bibek Paudel	University of Zurich
Heiko Paulheim	University of Mannheim
Tassilo Pellegrini	University of Applied Sciences St. Pölten
Silvio Peroni	University of Bologna
Catia Pesquita	LaSIGE, Universidade de Lisboa
Reinhard Pichler	Vienna University of Technology
Emmanuel Pietriga	Inria
Giuseppe Pirrò	Institute for High Performance Computing and Networking (ICAR-CNR)
Dimitris Plexousakis	Institute of Computer Science, FORTH
Mike Pool	Goldman Sachs Group
Livia Predoiu	University of Oxford
Cédric Pruski	Luxembourg Institute of Science and Technology
Yuzhong Qu	Nanjing University
Dnyanesh Rajpathak	General Motors
Dietrich Rebholz-Schuhmann	Insight Centre for Data Analytics
Georg Rehm	DFKI
Achim Rettinger	Karlsruhe Institute of Technology
Martin Rezk	Rakuten
German Rigau	IXA Group, UPV/EHU
Carlos R. Rivero	Rochester Institute of Technology
Giuseppe Rizzo	ISMB
Marco Rospocher	Fondazione Bruno Kessler
Camille Roth	Sciences Po
Marie-Christine Rousset	University of Grenoble Alpes
Ana Roxin	University of Burgundy, UMR CNRS 6306

Harald Sack	FIZ Karlsruhe, Leibniz Institute for Information Infrastructure & KIT Karlsruhe
Sherif Sakr	The University of New South Wales
Angelo Antonio Salatino	The Open University
Muhammad Saleem	AKSW, University of Leizpig
Cristina Sarasua	University of Zurich
Felix Sasaki	Lambdawerk
Bahar Sateli	Concordia University
Kai-Uwe Sattler	TU Ilmenau
Vadim Savenkov	Vienna University of Economics and Business (WU)
Marco Luca Sbodio	IBM
Johann Schaible	GESIS - Leibniz Institute for the Social Sciences
Bernhard Schandl	mySugr GmbH
Ansgar Scherp	Kiel University and ZBW – Leibniz Information Center for Economics, Kiel, Germany
Marvin Schiller	Ulm University
Stefan Schlobach	Vrije Universiteit Amsterdam
Claudia Schon	Universität Koblenz-Landau
Marco Schorlemmer	Artificial Intelligence Research Institute, IIIA-CSIC
Lutz Schröder	Friedrich-Alexander-Universität Erlangen-Nürnberg
Daniel Schwabe	PUC-Rio
Erich Schweighofer	University of Vienna
Giovanni Semeraro	University of Bari
Juan F. Sequeda	Capsenta Labs
Luciano Serafini	Fondazione Bruno Kessler
Saeedeh Shekarpour	University of Dayton
Gerardo Simari	Universidad Nacional del Sur and CONICET
Elena Simperl	University of Southampton
Hala Skaf-Molli	Nantes University
Sebastian Skritek	Vienna University of Technology
Monika Solanki	University of Oxford
Dezhao Song	Thomson Reuters
Steffen Staab	Institut WeST, University Koblenz-Landau and WAIS, University of Southampton
Yannis Stavrakas	Institute for the Management of Information Systems
Armando Stellato	University of Rome, Tor Vergata
Audun Stolpe	Norwegian Defence Research Establishment (FFI)
Umberto Straccia	ISTI-CNR
Markus Strohmaier	RWTH Aachen University and GESIS
Heiner Stuckenschmidt	University of Mannheim
Jing Sun	The University of Auckland
York Sure-Vetter	Karlsruhe Institute of Technology
Vojtěch Svátek	University of Economics, Prague
Marcin Sydow	PJIIT and ICS PAS, Warsaw
Mohsen Taheriyan	Google
Hideaki Takeda	National Institute of Informatics

Kerry Taylor	Australian National University and University of Surrey
Annette Ten Teije	Vrije Universiteit Amsterdam
Kia Teymourian	Boston University
Dhavalkumar Thakker	University of Bradford
Allan Third	The Open University
Krishnaprasad Thirunarayan	Wright State University
Ilaria Tiddi	The Open University
Thanassis Tiropanis	University of Southampton
Konstantin Todorov	LIRMM, University of Montpellier
David Toman	University of Waterloo
Nicolas Torzec	Yahoo
Yannick Toussaint	LORIA
Sebastian Tramp	eccenca GmbH
Cassia Trojahn	UT2J & IRIT
Raphaël Troncy	EURECOM
Jürgen Umbrich	Vienna University of Economy and Business (WU)
Joerg Unbehauen	University of Leipzig
Jacopo Urbani	Vrije Universiteit Amsterdam
Herbert Van De Sompel	Los Alamos National Laboratory, Research Library
Jacco van Ossenbruggen	CWI & VU University Amsterdam
Ruben Verborgh	Ghent University – imec
Serena Villata	CNRS - Laboratoire d'Informatique, Signaux et Systèmes de Sophia-Antipolis
Denny Vrandečić	Google
Domagoj Vrgoc	Pontificia Universidad Católica de Chile
Simon Walk	Graz University of Technology
Kewen Wang	Griffith University
Zhichun Wang	Beijing Normal University
Paul Warren	Knowledge Media Institute, Open University, UK
Grant Weddell	University of Waterloo
Erik Wilde	CA Technologies
Cord Wiljes	CITEC, Bielefeld University
Gregory Todd Williams	Hulu
Jiewen Wu	Institute for InfoComm Research, A*STAR
Yong Yu	Shanghai Jiao Tong University
Fouad Zablith	American University of Beirut
Ondřej Zamazal	University of Economics, Prague
Benjamin Zapilko	GESIS - Leibniz Institute for the Social Sciences
Sergej Zerr	L3S Research Center
Qingpeng Zhang	City University of Hong Kong
Ziqi Zhang	Sheffield University
Jun Zhao	University of Oxford

Additional Reviewers – Research Track

Abdel-Qader, Mohammad
Acar, Erman
Akrami, Farahnaz
Allavena, Davide
Angelidis, Iosif
Annane, Amina
Badenes-Olmedo, Carlos
Bakhshandegan Moghaddam, Farshad
Batsakis, Sotiris
Bhardwaj, Akansha
Bilgin, Aysenur
Biswas, Russa
Borgwardt, Stefan
Braun, Tanya
Calleja, Pablo
Chaloux, Julianne
Charalambidis, Angelos
Chaves-Fraga, David
Cheatham, Michelle
Chen, Jiaoyan
Daniel, Ron
Deng, Shumin
Ding, Jiwei
Dudas, Marek
Espinoza, Paola
Frommhold, Marvin
Hildebrandt, Marcel
Hogan, Aidan
Jimenez, Damian
Jouanot, Fabrice
Khosla, Megha
Kilias, Torsten
Kondylakis, Haris
Koopmann, Patrick
Kostylev, Egor V.
Koutraki, Maria
Krieg-Brückner, Bernd
Krishnamurthy, Gangeshwar
Kritikos, Kyriakos
Mireles, Victor
Mogadala, Aditya
Moodley, Kody
Musto, Cataldo
Padhee, Swati
Palumbo, Enrico
Piao, Guangyuan
Priyatna, Freddy
Revenko, Artem
Ribeiro De Azevedo, Ryan
Ringsquandl, Martin
Rodrigues, Cleyton
Rodriguez Muro, Mariano
Rosso, Paolo
Schneider, Rudolf
Shimizu, Cogan
Siciliani, Lucia
Silvestre Vilches, Jorge
Sun, Zequn
Tachmazidis, Ilias
Thoma, Steffen
Umbrich, Jürgen
Volk, Martin
Wielemaker, Jan
Ziebelin, Danielle
Özcep, Özgür Lütfü

Senior Program Committee – Resources Track

Maria Esther Vidal	Universidad Simon Bolivar
Valentina Tamma	University of Liverpool
Anna Lisa Gentile	IBM
Steffen Lohmann	Fraunhofer
Aidan Hogan	DCC, Universidad de Chile
Serena Villata	CNRS - Laboratoire d'Informatique, Signaux et Systèmes de Sophia-Antipolis
Jorge Gracia	University of Zaragoza
Stefan Dietze	L3S Research Center

Program Committee – Resources Track

Muhammad Intizar Ali	Insight Centre for Data Analytics, National University of Ireland, Galway
Ghislain Auguste Atemezing	Mondeca
Mattia Atzeni	University of Cagliari
Maurizio Atzori	University of Cagliari
Elena Cabrio	Université Côte d'Azur, CNRS, Inria, I3S
Timothy Clark	University of Virginia
Francesco Corcoglioniti	University of Trento
Daniele Dell'Aglio	University of Zurich
Emanuele Della Valle	Politecnico di Milano
Stefan Dietze	GESIS - Leibniz Institute for the Social Sciences
Ying Ding	Indiana University Bloomington
Mauro Dragoni	Fondazione Bruno Kessler - FBK-IRST
Mohnish Dubey	University of Bonn
Fajar J. Ekaputra	Vienna University of Technology
Diego Esteves	University of Bonn
Stefano Faralli	University of Mannheim
Mariano Fernández López	Universidad San Pablo CEU
Jesualdo Tomás Fernández-Breis	Universidad de Murcia
Aldo Gangemi	Università di Bologna and CNR-ISTC
Anna Lisa Gentile	IBM
Claudio Giuliano	Fondazione Bruno Kessler
Jose Manuel Gomez-Perez	ExpertSystem
Alejandra Gonzalez-Beltran	University of Oxford
Rafael S Gonçalves	Stanford University
Jorge Gracia	University of Zaragoza
Alasdair Gray	Heriot-Watt University
Tudor Groza	The Garvan Institute of Medical Research
Amelie Gyrard	Kno.e.sis - Ohio Center of Excellence in Knowledge-enabled Computing
Pascal Hitzler	Wright State University
Robert Hoehndorf	King Abdullah University of Science and Technology
Rinke Hoekstra	University of Amsterdam
Aidan Hogan	DCC, Universidad de Chile
Antoine Isaac	Europeana and VU University Amsterdam
Ernesto Jimenez-Ruiz	The Alan Turing Institute
Simon Jupp	European Bioinformatics Institute
Tomi Kauppinen	Aalto University School of Science
Christoph Lange	University of Bonn and Fraunhofer IAIS
Agnieszka Lawrynowicz	Poznan University of Technology
Alejandro Llaves	Fujitsu Laboratories of Europe
Steffen Lohmann	Fraunhofer

Phillip Lord	Newcastle University
Markus Luczak-Roesch	Victoria University of Wellington
Maria Maleshkova	Karlsruhe Institute of Technology
Fiona McNeill	Heriot Watt University
Nandana Mihindukulasooriya	Universidad Politécnica de Madrid
Raghava Mutharaju	GE Global Research
Giulio Napolitano	Fraunhofer Institute and University of Bonn
Vinh Nguyen	National Library of Medicine, NIH
Andrea Giovanni Nuzzolese	University of Bologna
Alessandro Oltramari	Bosch Research and Technology Center
Raul Palma	Poznan Supercomputing and Networking Center
Bijan Parsia	The University of Manchester
Silvio Peroni	University of Bologna
María Poveda-Villalón	Universidad Politécnica de Madrid
Valentina Presutti	CNR - Institute of Cognitive Sciences and Tecnologies
Mariano Rico	Universidad Politécnica de Madrid
German Rigau	IXA Group, UPV/EHU
Giuseppe Rizzo	ISMB
Edna Ruckhaus	Universidad Politécnica de Madrid
Anisa Rula	University of Milano-Bicocca
Michele Ruta	Politecnico di Bari
Satya Sahoo	Case Western Reserve University
Cristina Sarasua	University of Zurich
Stefan Schlobach	Vrije Universiteit Amsterdam
Jodi Schneider	University of Illinois Urbana Champaign
Stefan Schulte	Vienna University of Technology
Hamed Shariat Yazdi	University of Bonn
Elena Simperl	University of Southampton
Mari Carmen Suárez-Figueroa	Universidad Politécnica de Madrid
Valentina Tamma	University of Liverpool
Krishnaprasad Thirunarayan	Wright State University
Cassia Trojahn	UT2J & IRIT
Raphaël Troncy	EURECOM
Maria Esther Vidal	Universidad Simon Bolivar
Natalia Villanueva-Rosales	University of Texas at El Paso
Serena Villata	CNRS - Laboratoire d'Informatique, Signaux et Systèmes de Sophia-Antipolis
Fouad Zablith	American University of Beirut
Amrapali Zaveri	Maastricht University
Jun Zhao	University of Oxford

Additional Reviewers – Resources Track

Bader, Sebastian
Daquino, Marilena
Ebrahimi, Monireh
García-Silva, Andrés
Heling, Lars
Kismihók, Gábor
Nayyeri, Mojtaba
Nechaev, Yaroslav
Pham, Thu Le
Sadeghi, Afshin
Shimizu, Cogan
Weller, Tobias
Zhou, Lu

Program Committee – In-Use Track

Irene Celino	Cefriel
Marco Comerio	Cefriel
Oscar Corcho	Universidad Politécnica de Madrid
Philippe Cudre-Mauroux	University of Fribourg
Mathieu D'Aquin	Insight Centre for Data Analytics, National University of Ireland, Galway
Brian Davis	Insight Centre for Data Analytics, Galway
Stefan Dietze	GESIS - Leibniz Institute for the Social Sciences
Mauro Dragoni	Fondazione Bruno Kessler - FBK-IRST
Achille Fokoue	IBM
Daniel Garijo	Information Sciences Institute
Anna Lisa Gentile	IBM
Jose Manuel Gomez-Perez	ExpertSystem
Rafael S Gonçalves	Stanford University
Paul Groth	Elsevier Labs
Tudor Groza	The Garvan Institute of Medical Research
Christophe Guéret	Accenture
Peter Haase	metaphacts
Lucie-Aimée Kaffee	University of Southampton
Tomi Kauppinen	Aalto University School of Science
Elmar Kiesling	Vienna University of Technology
Craig Knoblock	University of Southern California
Freddy Lecue	Accenture Labs
Vanessa Lopez	IBM
Vassil Momtchev	Ontotext AD
Andriy Nikolov	metaphacts GmbH
Francesco Osborne	The Open University
Jeff Z. Pan	University of Aberdeen
Artem Revenko	Semantic Web Company GmbH
Giuseppe Rizzo	ISMB
Dumitru Roman	SINTEF
Marta Sabou	Vienna University of Technology
Harald Sack	FIZ Karlsruhe, Leibniz Institute for Information Infrastructure and KIT Karlsruhe

Juan F. Sequeda	Capsenta Labs
Elena Simperl	University of Southampton
Dezhao Song	Thomson Reuters
Thomas Steiner	Google
Simon Steyskal	Siemens AG Austria
Anna Tordai	Elsevier B.V.
Raphaël Troncy	EURECOM
Josiane Xavier Parreira	Siemens AG Österreich

Additional Reviewers – In-Use Track

Bakhshandegan Moghaddam, Farshad
Chen, Jiaoyan
Fawei, Biralatei
García-Silva, Andrés
Li, Chenxi
Lutov, Artem
Mireles, Victor
Nikolov, Nikolay
Smirnova, Alisa
Tietz, Tabea
Türker, Rima
Zernichow, Bjørn Marius Von
Zhao, Yuting

Sponsors

Platinum Sponsors

https://www.elsevier.com/

IBM **Research**

http://www.ibm.com/

https://franz.com/

Gold Sponsors

https://data.world/

http://www.metaphacts.com/

https://ontotext.com/

http://www.oracle.com/

https://www.thomsonreuters.com/

http://videolectures.net/

iNovex

https://inovexcorp.com/

Bronze Sponsors

Google

https://www.google.com/

Contributors

https://www.springer.com/

https://capsenta.com/

https://www.iospress.nl/

Student Travel Award Sponsors

https://www.nsf.gov/

http://swsa.semanticweb.org/

Doctoral Consortium

https://www.journals.elsevier.com/artificial-intelligence

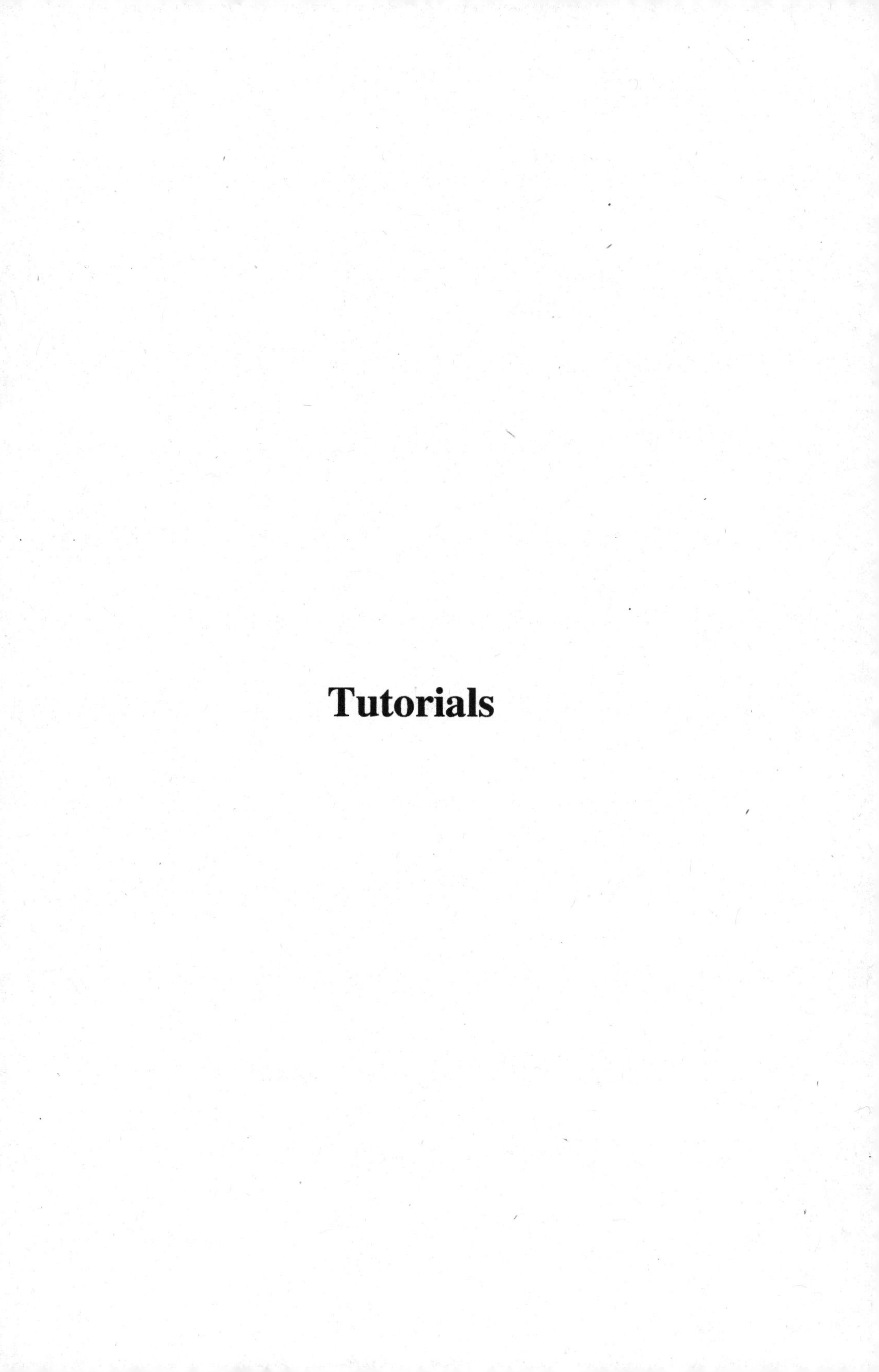

Tutorials

ISWC 2018 Workshop and Tutorial Chairs' Welcome

Besides the main technical program, ISWC 2018 hosts a selection of workshops and tutorials on a range of emerging and established topics. The key areas addressed by the workshop and tutorial programme include core Semantic Web technologies such as knowledge graphs and scalable knowledge base systems, ontology design and modelling, semantic deep learning and statistics, and well as novel applications of semantic technologies to audio and music, IoT, robotics, healthcare, social media and social good topics. Furthermore, several events address the topics on the interface of Semantic Web technologies and humans, including visualization and interaction paradigms for Web Data as well as crowdsourcing applications. The workshops and tutorials provide a setting for focused, intensive scientific exchange among researchers and practitioners in a variety of formats.

The decision on acceptance of workshops and tutorial proposals was made on the basis of their overall quality and their appeal to a reasonable fraction of the Semantic Web community while also targeting diversity of the programme. Overall, we received 31 workshop and tutorial proposals, of which 8 were accepted as full-day events and 17 as half-day events. The full workshop and tutorials programme is available at: http://iswc2018.semanticweb.org/workshops-tutorials.

We would like to take this opportunity to thank the workshop and tutorial organizers for their invaluable and inspiring contributions to the ISWC 2018 programme. We look forward to seeing you in Monterey!

March 2018

Elena Demidova
Amrapali Zaveri
Workshop & Tutorial Chairs

Methods and Tools for Modular Ontology Modeling

Karl Hammar[1], Pascal Hitzler[2], Cogan Shimizu[2], and Md Kamruzzaman Sarker[2]

[1] Department of Computer Science and Informatics, Jönköping University, Sweden
karl.hammar@ju.se
[2] Data Semantics Lab, Wright State University, USA
{pascal.hitzler,shimizu.5,sarker.3}@wright.edu

Ontology design patterns and other methods for modular ontology engineering have recently experienced a revival, and several new promising tools and techniques have been presented. The use of methods for modular ontology development and these newly developed tools and technologies promise simpler ontology development and management, in turn furthering increased adoption of ontologies and ontology-based tech, both within and outside of the semantic web academic environment. This workshop intends to spread the word about these method and tooling improvements beyond "the usual crowd"of pattern developers and researchers, for the benefit of the Semantic Web research community as a whole.

This full-day tutorial targets ontology designers, data publishers, and software developers interested in employing semantic technologies and ontologies. We present the state-of-the-art in terms of methods and tools, exemplifying their usage in several real-world cases. We then tutor the attendees on the use of three sets of related tooling for modular ontology development, allowing them to try out leading-edge software that they might otherwise have missed, under the supervision of the tools' main developers. We expect that at the end of the day, the attendees will have developed the ability to independently and with confidence develop ontologies in a modular fashion, using the tools and techniques showcased in this tutorial.

Validating RDF Data Tutorial

Jose Emilio Labra Gayo[1] and Iovka Boneva[2]

[1] University of Oviedo, Spain
labra@uniovi.es
[2] Univ. Lille - CRIStAL, F-59000 Lille, France
iovka.boneva@univ-lille1.fr

RDF promises a distributed database of repurposable, machine-readable data. Although the benefits of RDF for data representation and integration are indisputable, it has not been embraced by everyday programmers and software architects who care about safely creating and accessing well-structured data. Semantic web projects still lack some common tools and methodologies that are available in more conventional settings to describe and validate data. In particular, relational databases and XML have popular technologies for defining data schemas and validating data which had no analog in RDF.

Two technologies have been proposed for RDF validation: Shape Expressions (ShEx) and Shapes Constraint Language (SHACL).

ShEx was designed as an intuitive and human-friendly high level language for RDF validation in 2014 [4]. ShEx 2.0 has recently been proposed by the W3C ShEx community group [3].

SHACL was proposed by the Data Shapes Working Group and accepted as a W3C Recommendation in July 2017 [1].

In this tutorial we will present both ShEx and SHACL using examples, presenting the rationales for their designs, a comparison of the two, and some example applications. The contents of the tutorial will be complemented by the *Validating RDF Data* book [2] written by the presenters.

References

1. Knublauch, H., Kontokostas, D.: Shapes Constraint Language (SHACL). W3C Proposed Recommendation, June 2017
2. Labra Gayo, J.E., Prud'hommeaux, E., Boneva, I., Kontokostas, D.: Validating RDF Data. Morgan & Claypool (2017)
3. Prud'hommeaux, E., Boneva, I., Labra Gayo, J.E., Kellog, G.: Shape Expressions Language 2.0. W3C Community Group Report, Apr 2017
4. Prud'hommeaux, E., Labra, J.E., Solbrig, H.: Shape expressions: an RDF validation and transformation language. In: 10th International Conference on Semantic Systems, Sept 2014

Hybrid Techniques for Knowledge-Based NLP - Knowledge Graphs Meet Machine Learning and All Their Friends

Jose Manuel Gomez-Perez and Ronald Denaux

Expert System, Madrid, Spain
{jmgomez,rdenaux}@expertsystem.com

Many different artificial intelligence techniques can be used to explore and exploit large document corpora that are available inside organizations and on the Web. While natural language is symbolic in nature and the first approaches in the field were based on symbolic and rule-based methods, like ontologies, semantic networks and knowledge bases, many of the most widely used methods are currently based on statistical approaches. Each of these two main schools of thought in natural language processing, knowledge-based and statistical, have their limitations and strengths and there is an increasing trend that seeks to combine them in complementary ways to get the best of both worlds. This tutorial covers the foundations and modern practical applications of knowledge-based and statistical methods and techniques as well as their combination for the exploitation of large document corpora. Following a practical and hands-on approach, the tutorial tries to address a number of fundamental questions to achieve this goal, including: (i) how can machine learning extend previously captured knowledge explicitly represented as knowledge graphs in cost-efficient and practical ways, (ii) what are the main building blocks and techniques enabling such hybrid approach to natural language processing, (iii) how can structured and statistical knowledge representations be seamlessly integrated, (iv) how can the quality of the resulting hybrid representations be inspected and evaluated, and (v) how can this improve the overall quality and coverage of our knowledge graphs. The tutorial will first focus on the foundations that can be used to this purpose, including knowledge graphs and word embeddings, and will then show how these techniques can be effectively combined in NLP tasks (and other data modalities in addition text) related to research and commercial projects where the instructors currently participate.

Building Enterprise-Ready Knowledge Graph Applications in the Cloud

Peter Haase[1] and Michael Schmidt[2]

[1] metaphacts GmbH, 69190 Walldorf, Germany
ph@metaphacts.com
[2] Amazon Web Services, Seattle, WA, USA
schmdtm@amazon.com

Knowledge Graphs are a powerful tool that changes the way we do data integration, search, analytics, and context-sensitive recommendations. Consisting of large networks of entities and their semantic relationships, they have been successfully utilized by the large tech companies, with prominent examples like the Google Knowledge Graph and Wikidata, which makes community-created knowledge freely accessible. Cloud computing has fundamentally changed the way that organizations build and consume IT resources, enabling services to be provisioned on-demand in a pay-as-you-go model. Building Knowledge Graphs in the cloud makes it easy to leverage their powerful capabilities quickly and cost effectively.

In this tutorial, we cover the fundamentals of building Knowledge Graphs in the cloud. In comprehensive hands-on exercises we will cover the end-to-end process of building and utilizing an open Knowledge Graph based on high-quality Linked Open Data sets, covering all aspects of the Knowledge Graph life cycle including enterprise-ready data management, integration and interlinking of sources, authoring, exploration, querying, and search. The hands-on examples will be performed using prepared individual student accounts set up in the AWS cloud, backed by an RDF/SPARQL graph database service with an enterprise Knowledge Graph application platform deployed on top.

Crowdsourcing with CrowdTruth
Harnessing Disagreement in Human Interpretation for Ambiguity-Aware Machine Intelligence

Lora Aroyo[1], Anca Dumitrache[1], Oana Inel[1], and Chris Welty[2]

[1] Vrije Universiteit Amsterdam, Netherlands
lora.aroyo@gmail.com, anca.dumitrache@gmail.com,
oana.inel@gmail.com
[2] Google Research, New York, USA
cawelty@gmail.com
http://crowdtruth.org

In this tutorial, we introduce the *CrowdTruth methodology* for crowdsourcing ground truth by harnessing and interpreting inter-annotator disagreement. CrowdTruth is a widely used crowdsourcing methodology[1] adopted by industrial partners and public organizations, e.g. Google, IBM, New York Times, The Cleveland Clinic, Crowdynews, The Netherlands Institute for Sound and Vision, Rijksmuseum, and in a multitude of domains, e.g. AI, news, medicine, social media, cultural heritage, social sciences. The central characteristic of CrowdTruth is *harnessing the diversity in human interpretation* to capture the wide range of opinions and perspectives, and thus, provide more reliable and realistic real-world annotated data for training and evaluating machine learning components. Unlike other methods, we do not discard dissenting votes, but incorporate them into a *richer and more continuous representation of truth*. The goal of this tutorial is to introduce the Semantic Web audience to a *novel approach to crowdsourcing* that takes advantage of the diversity of opinions (human semantics) inherent to the Web. We believe it is quite timely, as methods that deal with disagreement and diversity in crowdsourcing have become increasingly popular. Creating this more *complex notion of truth* contributes directly to the larger discussion on *how to make the Web more reliable, diverse and inclusive*.

[1] http://crowdtruth.org.

Challenges and Opportunities with Big Linked Data Visualization

Laura Po

"Enzo Ferrari" Engineering Department, University of Modena and Reggio Emilia, Italy
laura.po@unimore.it
http://www.dbgroup.unimo.it/laurapo

The Linked Data Principles defined by Tim-Berners Lee promise that a large portion of Web Data will be usable as one Big interlinked RDF database. Today, we are assisting at a staggering growth in the production and consumption of Linked Open Data (LOD). In this scenario, it is crucial to provide intuitive tools for researchers, domain experts, but also businessmen and citizens to view and interact with increasingly large datasets. Visual analytics integrates the analytic capabilities of the computer and the abilities of the human analyst, allowing novel discoveries and empowering individuals to take control of the analytical process.

This tutorial aims to identify the challenges and opportunities in the representation of Big Linked Data by reviewing some current approaches for exploring and visualizing LOD sources. First, we introduce the problem of finding relevant sources in catalogues of thousands of datasets, we present the issues related to the understanding and exploration of unknown sources. We list the difficulties to visualize large datasets in static or dynamic form. We focus on the practical use of LOD/ RDF browsers and visualization toolkits and examine the support at big scale. In particular, we experience the exploration of some LOD datasets by performing searches of growing complexity. At last, we sketch the main open research challenges with Big Linked Data visualization. By the end of the tutorial, the audience will be able to get started with their own experiments on the LOD Cloud, to select the most appropriate tool for a defined type of analysis and they will be aware of the open issues that remain unsolved in the scenario of the exploration of Big Linked Data.

Contents–Part II

Resources Track

In-Use Track

Contents–Part I

Resources Track

DOREMUS: A Graph of Linked Musical Works

Manel Achichi[1], Pasquale Lisena[2], Konstantin Todorov[1](✉), Raphaël Troncy[2], and Jean Delahousse[3]

[1] LIRMM, University of Montpellier, CNRS, Montpellier, France
{achichi,todorov}@lirmm.fr
[2] EURECOM, Sophia Antipolis, France
{pasquale.lisena,raphael.troncy}@eurecom.fr
[3] OUROUK, Paris, France
delahousse.jean@gmail.com

Abstract. Three major French cultural institutions—the French National Library (BnF), Radio France and the Philharmonie de Paris—have come together in order to develop shared methods to describe semantically their catalogs of music works and events. This process comprises the construction of knowledge graphs representing the data contained in these catalogs following a novel agreed upon ontology that extends CIDOC-CRM and FRBRoo, the linking of these graphs and their open publication on the web. A number of specialized tools that allow for the reproduction of this process are developed, as well as web applications for easy access and navigation through the data. The paper presents one of the main outcomes of this project—the DOREMUS knowledge graph, consisting of three linked datasets describing classical music works and their associated events (e.g., performances in concerts). This resource fills an important gap between library content description and music metadata. We present the DOREMUS pipeline for lifting and linking the data, the tools developed for these purposes, as well as a search application allowing to explore the data.

1 Introduction

The Linked Open Data (LOD) paradigm for data representation, sharing and publishing has been more and more appealing to the world of museums and libraries over the past years. The LOD project and the semantic web in general offer technological means for data reuse, increased visibility and data sharing on the web, data federation and facilitated exchange of metadata by the creation of links across resources. Attracted by these possibilities, many major actors from the library world, such as the Library of Congress (LOC) or the French National Library (BnF), have embraced semantic web technologies with the goal to open their archives and catalogs to the web. This process has resulted in a number of openly available and explorable RDF graphs reflecting the rich content of numerous libraries and cultural institutions from all over the world [1].

D. Vrandečić et al. (Eds.): ISWC 2018, LNCS 11137, pp. 3–19, 2018.
https://doi.org/10.1007/978-3-030-00668-6_1

The DOREMUS project follows this line of research and practice, with a particular interest in classical and traditional music, so far relatively under-represented on the LOD.[1] Three major French cultural institutions—the BnF, Radio France (RF) and the Philharmonie de Paris (PP)—have joint efforts with data and social science academics in order to develop shared methods to describe semantically their catalogs of music works and events and open them to the web community. A major contribution of the project is the development of the DOREMUS ontology[2] which extends the well-known CIDOC-CRM and FRBRoo models for representing bibliographic information[3], adapting it to the domain of music, thus filling an important representational gap. A number of shared vocabularies about music-specific concepts (such as musical genres or keys) have been collected or developed, linked and published using the SKOS standard. The data from the catalogs of the three partner institutions comes in MARC or XML formats. Specific tools for data conversion to RDF following the DOREMUS model have been developed. This process results in the construction of several knowledge graphs about music works and events, which have been linked using a specifically developed for this purpose data linking tool. For evaluation purposes, a benchmark has been created manually by the library experts and shared to the semantic web community as part of the Ontology Alignment Evaluation Initiative (OAEI). The data fusion process results in the construction of a pivot graph of shared and unique musical works. Finally, an exploratory search engine is developed that allows to browse the knowledge graphs.

This paper covers the components of the DOREMUS workflow described above, which altogether form a paradigm for lifting, linking and publishing music library metadata. We present in detail the DOREMUS knowledge graphs with a focus on the (re-)used models and vocabularies and the processes that allow for their (re-)production and fusion. The contributions of this work are:

- A model for describing musical works and events extending FRBRoo together with a number of shared and linked music-specific controlled vocabularies.
- Three knowledge graphs about music works that represent the catalogs of three major French cultural institutions.
- An approach to interlink these graphs resulting in the construction of a pivot graph, containing all unique works and links to the original graphs.
- A set of benchmark datasets for data linking evaluation.
- A set of tools for data generation, vocabulary alignment and validation, data linking, pivot graph construction, and data search and exploration.

The remainder of the paper is structured as follows. In the next section, we provide general information about the graphs, their form and content, the (re-) used ontologies and controlled vocabularies, as well as statistics. In Sect. 3, we detail the different components of the DOREMUS data production pipeline, and in particular, the data conversion and linking approaches. In Sect. 4, we

[1] http://www.doremus.org.

[2] http://data.doremus.org/ontology/.

[3] http://new.cidoc-crm.org/frbroo/.

demonstrate how this resource has already been used and we discuss its wider expected impact. We present related initiatives in Sect. 5 before we conclude and discuss future work in Sect. 6.

2 The DOREMUS Datasets of Linked Musical Works

The DOREMUS knowledge graph consists of several datasets, each containing the information coming from a specific database of an institution. In that, a given real-world entity (e.g., a music work) is represented at most once in each graph. Currently, three stable datasets have been published: **(1)** `bnf`: Works and Artists, originally described in MARC records of the BnF; **(2)** `philharmonie`: Works and Concerts, originally described in MARC records of the PP; **(3)** `itema3`: Concerts and Recordings, originally described in XML records of RF. Each dataset has two access points: (1) A specific named graph in the DOREMUS triplestore, accessible through a public SPARQL endpoint. Each graph follows the pattern http://data.doremus.org/<dataset_name>. (2) A set of RDF files in Turtle format, available to public download. All datasets are licensed for free distribution, following a Creative Commons Attribution 4.0 license[4] and have a DCAT description in the triplestore itself. All links to DOREMUS datasets or tools are given in Table 1.

2.1 Content and Form of the Resource

The DOREMUS knowledge graphs contain information about works (referred to as *expressions*) and related entities from the field of classical and traditional music. Each entity is identified by an univocal persistent URI, which follows the pattern http://www.data.doremus.org/<group>/<uuid>, where the group is determined by the class of the entity (e.g. `expression`) and the UUID (Universal Unique Identifier) is generated at conversion time in a deterministic way using the dataset name, the class and the identifier of the source record as seed.[5] Currently, the resource is shaped by three knowledge graphs of music works (one per institution), that are linked together in a pivot graph, which is the union of all unique expressions across the three bases, together with `owl:sameAs` links to the original graphs (cf. Sect. 3). We find information pertaining to the instruments, genre and key of a music work (e.g., "piano", "sonata", "A-flat major"), its composer and title(s), date of creation, catalog numbers, opus numbers, etc. As an example, we can find all this information linked to http://data.doremus.org/expression/d72301f0-0aba-3ba6-93e5-c4efbee9c6ea, representing Beethoven's *Moonlight Sonata*. Jazz music would contain a different kind of information, as in http://data.doremus.org/expression/cc7fc9a6-124d-3cc1-95e7-5644ecb394a6, representing Coltrane's *Naima*.

---

[4] https://creativecommons.org/licenses/by/4.0/.
[5] https://github.com/DOREMUS-ANR/marc2rdf/blob/master/URI.patterns.md.

Table 1. Links to DOREMUS resources and tools.

Name of resource or tool and description, URL
Data
`bnf`: Works and Artists from the BnF, http://data.doremus.org/bnf
`philharmonie`: Works and Concerts from the PP, http://data.doremus.org/philharmonie
`itema3`: Concerts and Recordings from RF, http://data.doremus.org/itema3
DOREMUS triplestore, https://github.com/DOREMUS-ANR/knowledge-base/tree/master/data
DOREMUS sparql endpoint, http://data.doremus.org/sparql
Example queries, http://data.doremus.org/queries.html
DOREMUS ontology, http://data.doremus.org/ontology
DOREMUS vocabularies, https://github.com/DOREMUS-ANR/knowledge-base/tree/master/vocabularies
Vocab. Alignments, https://github.com/DOREMUS-ANR/knowledge-base/tree/master/vocabularies/alignments
DOREMUS linked data, https://github.com/DOREMUS-ANR/knowledge-base/tree/master/linked-data
DOREMUS benchmarks 2016, http://islab.di.unimi.it/content/im_oaei/2016/#doremus
DOREMUS benchmarks 2017, http://islab.di.unimi.it/content/im_oaei/2017/#doremus
Tools
`marc2rdf` converter, https://github.com/DOREMUS-ANR/marc2rdf
`itema3` converter, https://github.com/DOREMUS-ANR/itema3converter
`euterpe` converter, https://github.com/DOREMUS-ANR/euterpe-converter
Legato: instance matcher, https://github.com/DOREMUS-ANR/legato
DOREMUS pivot graph constructor, https://github.com/DOREMUS-ANR/pivot-graph-constructor
OVERTURE search engine, http://overture.doremus.org
YAM++ vocabulary mapping and validation, http://yamplusplus.lirmm.fr
Learning materials, https://github.com/DOREMUS-ANR/training

2.2 (Re-)used Ontologies and Vocabularies

The DOREMUS model is an ontology for the description of music catalogs. It is an extension for the music domain of the FRBRoo model for describing librarian information, which has in turn been born as a dialog of the librarian FRBR model and the CIDOC-CRM ontology for representing museum information, putting togheter the distinction between Work, Expression, Manifestation, Item of the former with the centrality of creation events for describing the cultural object lifecycle coming from the latter [2]. On top of the FRBRoo original classes and properties, specific ones have been added in order to describe aspects of a work that are related to music, such as the musical key, the genre, the tempo, the medium of performance (MoP), etc. [3]. The model is ready to be used for describing the interconnection of different arts: it is the case of the soundtrack of a movie, or a song that uses the text of a poem.

DOREMUS imports the Work-Expression-Event triplet[6] pattern of FRBRoo: the abstract intention of the author (Work) exists only through an Event (i.e. the composition) that realizes it in a distinct series of choices called Expression(s). This pattern ensures that each step of the life of a musical work can be modeled separately, following the same triplet structure. Thinking about a classic work, we will have a triplet for the composition, one for any performance event, one for every manifestation (e.g., the score), all connected in the graph. This means also that each part of the music production process is considered as an Event that gives birth to a new Work and a new Expression: this leads to the creation of classes like Performance Work or Recording Expression. Each triplet contains information that at the same time can live autonomously and be linked to the other entities. This provides the freedom of representing, for example, a jazz improvisation as extemporaneous performance not connected to a particular pre-existing work, or to collect all the recordings of a piece of world music. The result is a model, which is quite complex and hard to adopt if we look at the levels of distribution of information: from an Expression, one has to pass through Event and Activity to reach a composer, or through Casting and Casting detail to get the MoP. On the other hand, the model has a very detailed expressiveness that allows, for instance, to describe different kinds of contributors (not only authors or performers), to detail the casting of a composition (with number, roles, notes for each instrument/voice), to specify performers at level of single performance inside a whole concert. As an extension of FRBRoo, the model appears familiar to librarian catalogers (documentation: http://data.doremus.org/ontology).

For the description of music-specific concepts like the key, the genre or the MoP, we publish controlled vocabularies (using SKOS and MODS standards), realized and enriched by an editorial process that involved also librarians, in order to overcome multilingualism and alternative names issues. Some of these vocabularies were already available and in use by the community: in this case our contribution consists in their collection, conversion to SKOS (if needed) and alignment. As a result, we collected, implemented and published 17 controlled vocabularies belonging to 7 different categories (musical keys, types of derivation, modes, thematic catalogs, functions, musical genres and MoP) (Table 1) [4]. The vocabularies are all available in the DOREMUS triplestore server via its public SPARQL endpoint. Alternatively, they can be explored by a web browser starting from http://data.doremus.org/vocabularies/. Each vocabulary is licensed for free distribution, following a Creative Commons Attribution 4.0 license.

The categories of genres and MoPs contain each 6 different vocabularies, including well-established reference thesauri, as well as institution-specific lists. The vocabularies of these two categories have been aligned by establishing `skos:`

[6] Not to be confused with an RDF triple.

exactMatch relations between their elements in a pairwise manner using an automatic ontology and thesaurus matching system and these alignments have been manually validated and enriched by the experts of the three institutions. This process has been assisted by a dedicated generic web-application for ontology matching and mapping validation, YAM++ *online* [5], developed in part for the purposes of the project (link available in Table 1).

Statistics. Currently, the DOREMUS dataset includes more than 16 million triples, which describe over 3 million distinct entities. The classes and properties used come mostly from the DOREMUS ontology, FRBRoo and CIDOC-CRM, counting in total 57 distinct classes and 120 distinct properties. Table 2 summarizes the number of entities for the most representative classes and reports details about the presence of specific information.

Table 2. Number of entities of given classes for each dataset.

class	bnf	philharmonie	itema3	total
Expression	135818	9005	8319	153142
-with casting detail	123219	4621	0	127840
- with key	19645	1973	0	21618
- with genre	128497	3820	8071	140388
- with composer	133371	7741	8231	149343
- with composition date	91566	5712	4856	102134
- with catalogue	20796	2908	0	23704
- with order number	11598	1612	0	13210
- with opus number	21836	1985	0	23821
Performance	15784	784	1531	18099
- with more than 1 performed work	0	713	1277	1990
Track	0	6538	18273	24811

3 Resource Development and Reconstruction

The general workflow of DOREMUS is depicted in Fig. 1. The data from the three partner institutions is first converted to RDF following the DOREMUS model, resulting in three independent knowledge graphs (one per institution), which are then linked. After a manual validation of a set of uncertain links, a pivot graph is built containing identifiers of the union of all works found in the three graphs, together with identity links to the resources in each of the three institutional graphs. We detail on these stages of the workflow in this section.

3.1 Data Conversion

The data collected from the BnF and the PP describing music works is represented in the UNI– and INTERMARC variants of the MARC format. A MARC file is a succession of fields, each carrying a 3-digit label, and subfields, delimited by the $ symbol (e.g., "50011$313908188$qSonates$rPiano$sOp.27, no 2$uDo mineur").[7]. We have developed an open source prototype, named `marc2rdf` to automatically convert UNI- or INTERMARC bibliographic records to RDF, implementing the DOREMUS model (link to the tool given in Table 1). The conversion process relies on explicit expert-defined transfer rules (or mappings), which provide the corresponding property path in the model as well as useful examples [6]. We have used the DOREMUS properties to name the extracted relations (e.g., `mus:U12_has_genre` is the property describing the genre of a work). Beyond being a documentation for the MARC records, these rules embed information on specific and distinct librarian practices in the formalization of the content (format of dates, syntax of textual fields, default values for missing information), making `marc2rdf` a robust generic converter for MARC files.

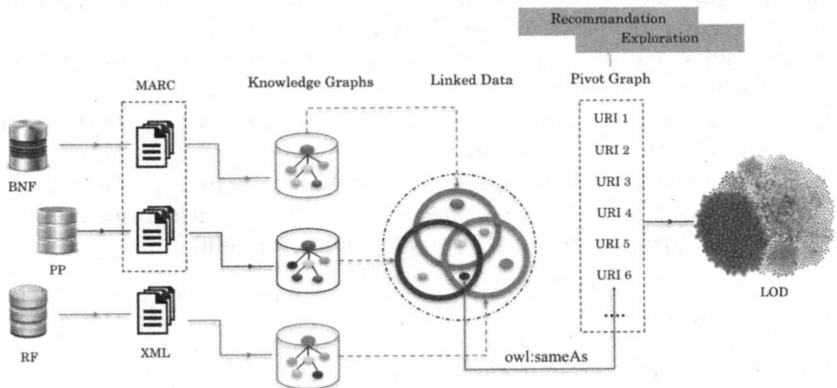

Fig. 1. The DOREMUS data lifecycle.

The converter is composed of different modules that work in succession (Fig. 2). First, a *file parser* reads the MARC file and makes the content accessible by field and subfield number. We implemented a converting module for both the INTERMARC and UNIMARC variants. Then, it builds the RDF graph reading the fields and assigning their content to the DOREMUS property suggested in the transfer rules. The *free-text interpreter* extracts further information from the plain text fields, that includes editorial notes. This amounts to do a

[7] For detailed information, we refer to the documentation released by The International Federation of Library Associations and Institutions (IFLA) http://www.ifla.org/publications/ifla-series-on-bibliographic-control-36.

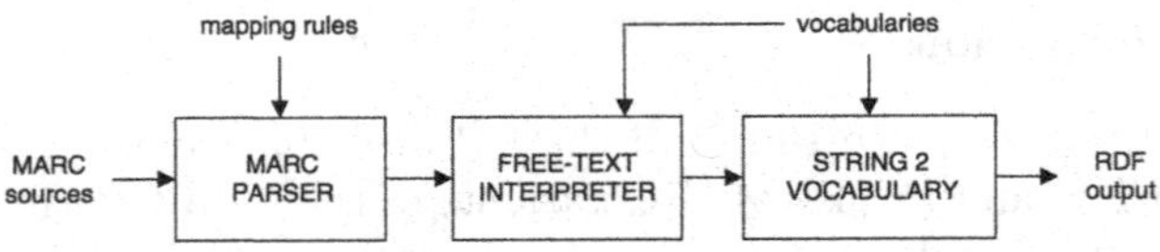

Fig. 2. The application flow of `marc2rdf`.

knowledge-aware parsing, since we search in the string exactly the information we want to instantiate from the model (i.e., the MoP from the casting notes, or the date and the publisher from the first publication note). The parsing is realized through empirically defined regular expressions, that are going to be supported by Named Entity Recognition techniques as future work. Finally, the *string2vocabulary* component performs an automatic mapping of string literals to URIs coming from controlled vocabularies. All variants for a concept label are considered in order to deal with potential differences in naming terms. As additional feature, this component is able to recognize and correct noise that is present in the MARC file: this is the case of musical keys declared as genre, or fields for the opus number that actually contain a catalog number and vice-versa.

The `marc2rdf` tool allows to reproduce deterministically the conversion process at any moment in time, providing the opportunity to seamlessly take into account possible updates of the ontology (e.g., the addition of a new property) and/or the data entries (a new record entering the catalog of one of the institutions), ensuring in that way the currentness and dynamics of the graphs.

The works from Radio France, described in XML, are managed by an *ad hoc* software that parses the input file, collects the required information, creates the RDF graph structure and runs the *string2vocabulary* module.

3.2 Data Linking

The three datasets that are currently subject to interlinking are highly heterogeneous: a given entity (e.g., a musical work) can be described quite differently across the three institutions. In addition to well-known data discrepancies such as lexical, semantic (polysemy, synonymy) and orthographic mismatches of string literals, the use of acronyms and abbreviations or differences in formats and types of numerical values, we have encountered several commonly occurring issues that are specific to our data. We outline some of them below.

- *Differences in coverage* and particularly lack of information in one of the graphs as compared to a richer description in another. In our case, the works coming from RF are systematically described by a considerably smaller set of attributes, than those found in the catalogs of the BnF and the PP (see Table 2).
- *Different depths in the graphs*, at which we find the value of interest—e.g., the birthplace of a composer can be directly assigned to the entity in one graph, or via a longer property chain in another.

- Presence of *comments in the form of free text* (given by the property `ecrm:P3_has_note`) that are difficult to compare, as well as presence of *institution-specific resource identifiers* (bibliographical records ID's) given under the same property name across different datasets, although not comparable.
- Presence of *blocks of highly similar in their descriptions, but yet distinct instances* in each of the graphs—e.g., the set of all piano sonatas by Beethoven, differing from one another in only one or two property values, which makes their disambiguation difficult and is likely to produce false positives.

In a first attempt to interconnect these graphs, we relied on state-of-the-art linking systems [7,8] that adopt a property-based philosophy where a set of attributes is selected in order to compare instances across two datasets based on (an aggregation of) similarity measures computed on their literal values. The results obtained proved to be not satisfactory.[8]. Consequently, we develop our own linking tool, named *Legato* [10]—a generic data linking system motivated by the DOREMUS use-case scenario and data linking challenges.

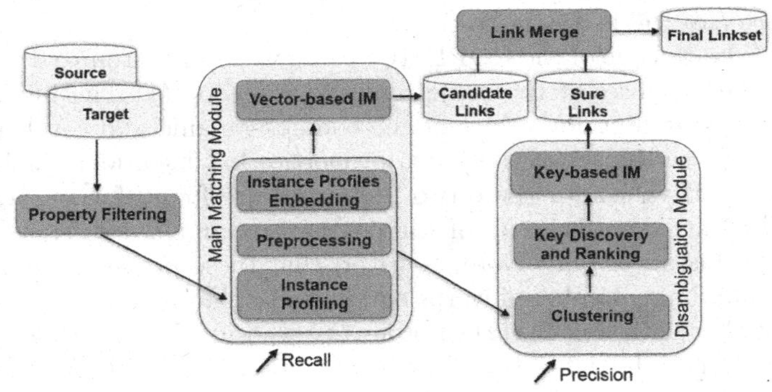

Fig. 3. Processing pipeline of *Legato*.

Legato is designed to match entities from highly heterogeneous graphs, effectively disambiguating highly similar yet distinct resources. Figure 3 shows the generic workflow of the system. The data cleaning module ensures to only keep properties that are comparable across the datasets (hence, comments in the form of free text, as well as institution-specific instance identifiers are removed). The instance profiling module represents instances by a subgraph corresponding to the union of the Concise Bounded Descriptions (CBD)[9] of each resource

[8] Evaluation results of *Legato* on OAEI benchmarks can be found at https://github.com/DOREMUS-ANR/legato/blob/master/Legato-Results.png. Data and configuration files for SILK are available at https://github.com/manoach/SILK-Evaluation. Note that SILK is configured by using the best keys selected by the algorithm in [9].

[9] https://www.w3.org/Submission/CBD/.

and its direct neighbors. In that, contrarily to SILK or Limes, Legato (in its default version) does not compare property values, but considers all extractable literal values as a bag-of-words. This representation addresses in its mechanism a number of data heterogeneities without requiring user input, in particular, the description differences and property depths discrepancies outlined above. The literals of these subgraphs are then used to project each instance in a vector space and the matching consists in comparing the resulting vectors. A deliberately low threshold is used for the vector similarity in order to ensure high recall. Then, highly similar instances are grouped together by the help of a standard hierarchical clustering algorithm [11]. An RDF key discovery [12] and a key ranking [9] algorithms are applied on each pair of similar clusters (identified by comparing cluster centroids) across the two graphs, in order to identify the set of properties that best allow to discriminate between the resources contained in each cluster. A new linkset (called "sure links") results from this process and is then compared to the links produced at the matching step (called "candidate links") in order to eliminate errors and increase precision, leading to the production of the final linkset. The outcome of *Legato* is presented in the EDOAL format,[10] allowing to keep track of the associated confidence scores, or as `owl:sameAs` triples. We provide an open source implementation of the system together with a simple user interface (see Table 1 for a link).

Given our knowledge of the DOREMUS data, we have customized the linking process for the purposes of the project in two respects. (1) The linking workflow begins by searching for values of the composer name and catalog number properties, because the set of these two properties has been identified as a key by our experts. If values for these two properties are found for a given pair of instances, they are directly used for the comparison and *Legato* is executed on the remaining instances only. Note, however, that these properties do not have values for a very large number of works and in particular, no entry of `itema3` has a catalog number (cf. Table 2). (2) In order to speed-up the execution of *Legato*, we have partitioned the datasets per composer and linked pairs of subsets across two graphs that gather works by the same composer.

To evaluate *Legato*, we have constructed benchmarks of music works from the BnF and the PP, by asking the librarian experts to manually select pairs of identical resources from the their respective catalogs. We have ensured that our benchmarks are representative and provide a fair account of the heterogeneity issues outlined above. This results in the generation of two benchmarks that have been released by the Instance Matching track of OAEI 2016 and 2017 (cf. Table 1). *Legato* has participated to the 2017 edition of the campaign, ranking first on the DOREMUS-FPT task. NjuLink surpasses *Legato* by 0.025 points (F1) on the DOREMUS-HT track, but performs worse by 0.044 points on the FPT track. Our data exhibits characteristics of the two, therefore we decided to go for *Legato*, in addition to the customizability argument given above.

[10] http://alignapi.gforge.inria.fr/edoal.html.

3.3 Link Validation and Pivot Graph Construction

As a result of the pair-wise alignments of the three graphs, we end up with three sets of links. We exploit the topology of the connectivity of the entities of the three graphs in order to define subsets of links to provide to data experts for manual validation, aiming to ensure a final set of links of high quality. We identify four connectivity patterns, shown in Fig. 4, according to which we classify the produced links to three categories: *certain links*, *invalid links* and *validation candidates*. The classification of a link as "certain" depends both on its confidence value and on the connectivity pattern in which it falls. The certain links are retained and included in the pivot graph constructed from our data (see below). If a link is approved by an expert during the user validation process, its confidence value is set to 1, which automatically classifies it as"certain", else it is declared as "invalid". We consider the following link patterns.

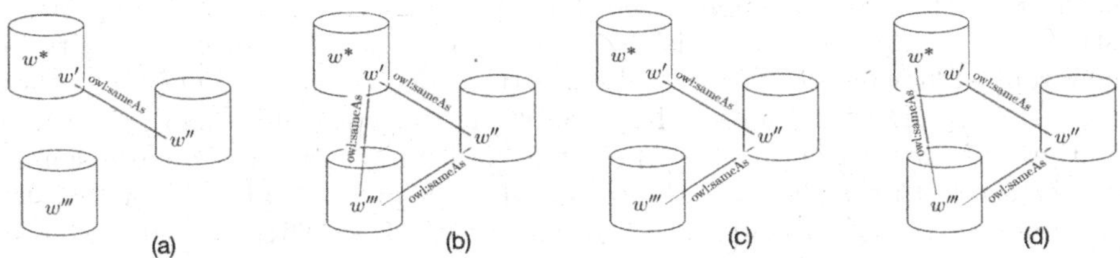

Fig. 4. Links across the three graphs: four connectivity patterns.

(1) Single link. This is the case when two works are connected via an identity link across their corresponding graphs as a result of the automatic linking process (Fig. 4(a)). According to the confidence value of the link, it is either classified as certain, passed over to the experts for validation, or discarded.
(2) Triangle. This is the case when three works from the three graphs are linked via three `owl:sameAs` relations. In this case, the three links are considered as certain and the expert is not solicited (Fig. 4(b)).
(3) Missing link. This is the case when an instance w' from one graph is linked to an instance w'' from the second graph, which in turn is linked to an instance w''' from the third graph, but no link has been created between w''' and w' (Fig. 4(c)). Instead of inferring that link, independently on the links confidence values, we pass the two link candidates $< w', \texttt{owl:sameAs}, w'' >$, $< w'', \texttt{owl:sameAs}\ w''' >$ to the experts for validation. If the validation process results in classifying these two links as certain, the link $< w'''$, $\texttt{owl:sameAs}, w' >$ is inferred and classified as certain. Note that the third link inference mechanism is activated only in case we have two certain links.
(4) Conflict. This is the case when an instance w' from one graph is linked to an instance w'' from the second graph, which in turn is linked to an instance w''' from the third graph, and w''' is linked to an instance w^* from the first

graph, where $w' \neq w^*$ (Fig. 4(d)). All three links are passed to the experts for validation. This necessarily leads to invalidating at least one of the three links, in which we fall into one of the three cases described above.

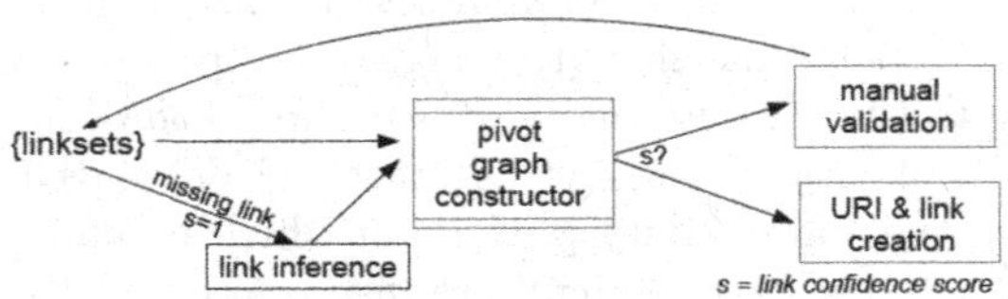

Fig. 5. Link validation and pivot graph construction workflow.

Pivot Graph Construction. We construct a referential pivot graph of music works that is the mathematical union of the three sets of works from the three partner institutions. As an editorial decision, a novel URI is created for every entity in that graph, following the URI creation pattern described previously, together with a `owl:sameAs` link to the URIs identifying this entity in each of the three input graphs (at least one such URI exists). For example, if a given expression is described in both the BnF and the PP graphs, the pivot graph will contain the following two triples: <PIVOT_URI> `owl:sameAs` <BNF_URI>, <PP_URI>. If the work exists in one single graph only (e.g., the one of BnF), one single triple will be declared: <PIVOT_URI> `owl:sameAs` <BNF_URI>. To reconstitute these links, we rely on the linksets produced in the data linking phase and on the manual link validation task. As explained above, as a result of these processes, we end up with three sets of "certain" links. Only links from that category will appear in the pivot graph. As Fig. 5 shows, the process of pivot graph construction and that of the manual validation of links are tangled up in a single workflow. The code of the algorithm for (re-)generating the pivot graph is released as open source (cf. Table 1).

Currently, the manual validation process is in progress. Therefore, the published pivot graph contains the links that have been identified automatically (i.e., corresponding to the patterns in Figs. 4(a) and (b)) by using the non-conservative thresholds of *Legato* tuned by the help of our benchmarks (0.2 for `bnf-philharmonie` and 0.5 for the other two pairs of datasets). The graph contains also the links of all unique works (those that have no matches found in any of the two other bases) to their original URIs. The results of the automatic link discovery process on the three bases together with the resulting pivot graph in its current shape are available at https://github.com/DOREMUS-ANR/knowledge-base/tree/master/linked-data.

Links Statistics. We have currently a total of 7495 links created automatically across the three graphs, among which we have 2520 links of type *single link*, 396 links of type *triangle*, 3378 links of type *missing link* and 261 links of type *conflict* (as labeled in Fig. 4), plus additional 940 links of type 1: many that are currently subject to post-processing. Updates in the datasets will not decrease the number of links since the source databases monotonically grow.

4 The DOREMUS Resource in Use

We proceed to discuss aspects related to the use of the resources, starting with their exploration and search.

`Overture`: *an Exploratory Search Engine.* We develop OVERTURE (Ontology-driVen Exploration and Recommendation of mUsical REcords), a prototype of an exploratory search engine for DOREMUS data, available at http://overture.doremus.org. The application makes requests directly to the SPARQL endpoint and provides information in a web user interface (UI).

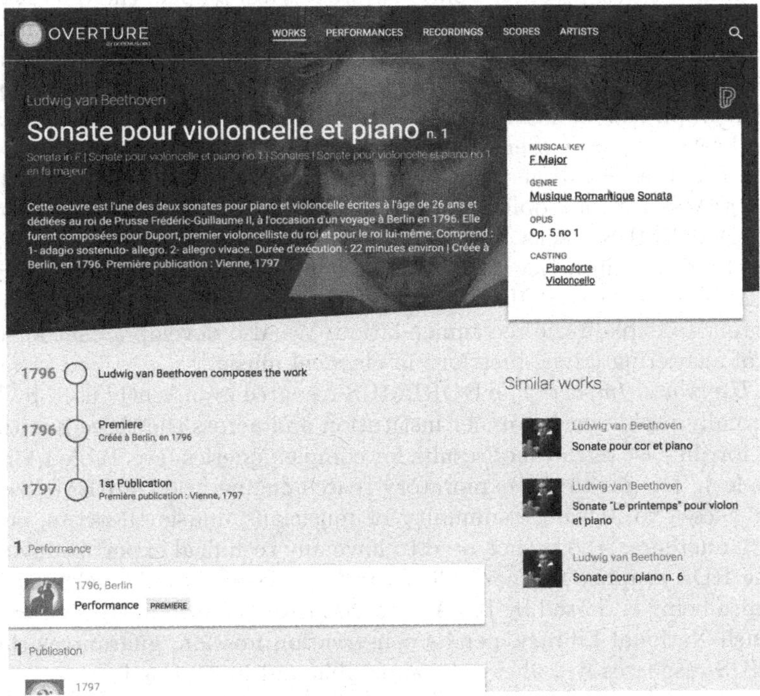

Fig. 6. The detail of an expression in OVERTURE

At the top of the UI, the menu bar allows the user to navigate between the main concepts of the DOREMUS model: expression, performance, score, recording, artist. Figure 6 represents Beethoven's *Sonata for piano and cello n.1* as seen in OVERTURE. Aside from the different versions of the title, the composer and a textual description, the page provides details on the information we have about the work, like the musical key, the genres, the intended MoP, the opus number. When these values come from a controlled vocabulary, a link is present in order to search for expressions that share the same value, for example, the same genre or the same musical key. A timeline shows the most important events in the story of the work (the composition, the premiere, the first publication). Other performances and publications can be represented below.

The richness of the DOREMUS model offers to the end-user the chance to perform a detailed advanced search. All expressions (works) are searchable by facets, that include the title and the composer, but also keys, genres, detailed castings, making it possible to select very precise subsets of data, like all sonatas (genre) that involve a clarinet and a piano (MoPs). The hierarchical properties in the controlled vocabulary allow the smart retrieval not only of the entity that match exactly the chosen value (i.e. *Strings*), but also any of its narrower concepts (i.e. *violin*, *cello*, etc.).

A work-in-progress recommendation system is also implemented in `Overture` in order to suggest to the final user different works to discover. The recommended works have similar properties to the current one, like the genre, the composer and the foreseen instruments. The recommendation is realized by computing knowledge graph embeddings using *node2vec* [13] on the DOREMUS knowledge graph and selecting the closer works using the euclidean distance [14].

Other client applications that also make use of the DOREMUS dataset include CityMus [15], a mobile application that generates Spotify playlist composed of DOREMUS tracks based on the surrounding important buildings of a geo-localized user in a city. More precisely, interesting paths in the DBpedia knowledge base between POIs and composer are sought and shown to the end user in order to explain the recommendation. We also develop a chatbot that is capable of answering trivia questions in classical music.[11]

Current Users and Impact. The DOREMUS resource is currently used by librarians internally within each partner institution and across the three institutions, allowing for the fast retrieval of results for complex queries (see Table 1 for a link to examples). Thanks to the exploratory search engine, the DOREMUS data is open for access to a wide community of musicians, music theorists, connoisseurs and amateurs, who do not need to have any technical expertise in order to query the RDF graphs. The controlled vocabularies and the DOREMUS ontology are also being endorsed by IFLA, as a de-facto standard for this community. The French National Library, per its conservation mission, guarantees that the DOREMUS resources will always be accessible and maintained.

Our goal is also to use the resources for both pedagogical and editorial purposes. The recommendation system that is currently under development will assist the creation of playlists for radios, allowing to group works together by very specific criteria, or to uncover rare works and provide insights about possible relations between composers, genres, events, etc.

We contribute to the semantic web community at large by providing open source implementations of novel and generic tools for data linking and fusion. We

[11] https://chatbot.doremus.org.

foster the adoption of semantic web technologies via the publication of numerous pedagogical materials, aiming to guide and encourage other cultural institutions to reuse the DOREMUS model and vocabularies and reproduce our data production framework (see Table 1), as similar initiatives exist in other fields [16].

5 Related Work and Graphs

There has been a significant effort in the last years to open and publish data from the field of cultural heritage [17]. An overview of related projects is given in [1], where the authors provide an evaluation of the various initiatives with regard to the well-known five-stars open data rating, applied to the cultural domain.

Regarding the more specific problem of producing linked data out of library records, addressed by the DOREMUS project, a number of related initiatives have recently been introduced. We refer to the multiple contributions of the Europeana project,[12], unifying and making accessible the catalogs of numerous libraries, museums and archives across Europe. One of the early efforts in that respect is made by the Library of congress,[13] which has become a dataset of reference in the field. In the same spirit, related projects include the German National Library linked data service,[14] the British National Bibliography Linked Data Platform,[15] the open data project of the French National Library BnF[16] or, more recently, the Virtual Library Miguel de Cervantes project [18].

In the majority of the cases, data comes in a given MARC variant and has to be converted to RDF. In certain cases the migration process goes through an intermediate phase of translation to relational database [18], or data is being directly converted to RDF based on the standards of bibliographical description, such as FRBR. DOREMUS follows this line of work by implementing its own expert-defined mappings-based conversion mechanism, enriching FRBRoo with more than 40 classes and 100 properties. The resulting (DOREMUS) model fills the important gap between library content description and music metadata.

As compared to music-related datasets, we outline that the BBC open datasets have tracks only, the Dutch Library (part of Europeana) has only publications, CPDL[17] is specialized for chorus (with scores and midi), while DOREMUS is general and can glue these datasets. MusicBrainz [19], one of the most popular knowledge bases about music metadata, started a few years ago its process of exposing its data as semantic triples through the platform LinkedBrainz [20]. In contrast to DOREMUS, which follows a librarian structure, MusicBrainz follows a more commercial practice giving a central role to tracks, albums and artists (un-distinguishing the composer from the performer), at the expense of all the information connected to the work concept (genre, casting, key, etc).

[12] http://www.europeana.eu.
[13] http://id.loc.gov.
[14] http://www.dnb.de/EN/lds.
[15] http://bnb.data.bl.uk/docs.
[16] http://data.bnf.fr.
[17] https://www.cpdl.org/.

6 Conclusion and Future Work

We have presented the DOREMUS resource—a collection of linked RDF datasets representing the catalogs of music works of three major French cultural institutions. The construction of this resource implies the implementation of a processing pipeline that allows for the conversion of the original data to RDF following the DOREMUS ontology, the development, SKOS-ification and alignment of a number of music-specific vocabularies and the interlinking of the datasets, which results in the construction of a reference pivot graph of musical works shared by or unique to the three institutions. This pipeline defines the data production paradigm of DOREMUS that is applicable to other music-library data—the described process is deterministic, extensible, reproducible and documented in numerous pedagogical materials published online. A number of tools acting at different layers of this pipeline have been introduced: the `marc2rdf` data converter, the *Legato* data linking system, the web-interface for SKOS thesauri alignment and mapping validation and enrichment YAM++ *on line*, as well as the exploratory search engine OVERTURE. We have relied on existing tools where appropriate (matching strings to URI), but the heterogeneity of the input data and the specificity of the librarian practices made this impossible in many cases. In terms of datasets, DOREMUS currently has published (1) three RDF graphs of musical works coming from the BnF, the Philharmonie de Paris and Radio France, (2) a pivot graph currently containing the certain links established automatically between the graphs of musical works, together with the results of the pairwise linking of these graphs, (3) expert curated benchmarks for evaluation of data linking systems, (4) a rich set of music-specific SKOS vocabularies together with their alignments.

We are currently in the process of applying the data conversion and linking workflow to two additional databases from Radio France. Natural Language Processing techniques are being included in the conversion process in order to parse the numerous free-text fields. OVERTURE will soon host all the links between the interlinked works, giving access at the same time to the joined knowledge and to the different information provenances. We have developed a web interface to assist the process of manual validation of links reducing the human effort, which is currently being deployed online. Alignments of our data to established datasets (in particular MusicBrainz) are currently being generated.

Acknowledgments. This work has been partially supported by the French National Research Agency within the DOREMUS project, under grant ANR-14-CE24-0020.

References

1. Marden, J., Li-Madeo, C., Whysel, N., Edelstein, J.: Linked open data for cultural heritage: evolution of an information technology. In: ICDC (2013)
2. Doerr, M., Bekiari, C., LeBoeuf, P.: FRBRoo: a conceptual model for performing arts. In: CIDOC Annual Conference, pp. 6–18 (2008)

3. Choffé, P., Leresche, F.: DOREMUS: connecting sources, enriching catalogues and user experience. In: IFLA World Library and Information Congress (2016)
4. Lisena, P., Todorov, K., Cecconi, C., Leresche, F., Canno, I., Puyrenier, F., Voisin, M., Troncy, R.: Controlled vocabularies for music metadata. In: ISMIR (2018)
5. Bellahsene, Z., Emonet, V., Ngo, D., Todorov, K.: YAM++ *Online*: a web platform for ontology and thesaurus matching and mapping validation. In: Blomqvist, E., et al. (eds.) ESWC 2017. LNCS, vol. 10577, pp. 137–142. Springer, Cham (2017). https://doi.org/10.1007/978-3-319-70407-4_26
6. Lisena, P., Achichi, M., Fernández, E., Todorov, K., Troncy, R.: Exploring linked classical music catalogs with OVERTURE. In: ISWC (Posters & Demos) (2016)
7. Jentzsch, A., Isele, R., Bizer, C.: Silk-generating RDF links while publishing or consuming linked data. In: ISWC (2010)
8. Ngomo, A.N., Auer, S.: LIMES - a time-efficient approach for large-scale link discovery on the web of data. In: IJCAI, pp. 2312–2317 (2011)
9. Achichi, M., Ben Ellefi, M., Symeonidou, D., Todorov, K.: Automatic key selection for data linking. In: EKAW, pp. 3–18 (2016)
10. Achichi, M., Bellahsene, Z., Todorov, K.: Legato: Results for OAEI 2017. In: Ontology Matching (2017)
11. Rokach, L., Maimon, O.: Clustering methods. In: Maimon, O., Rokach, L. (eds.) Data Mining and Knowledge Discovery Handbook, pp. 321–352. Springer, Boston (2005). https://doi.org/10.1007/0-387-25465-X_15
12. Symeonidou, D., Armant, V., Pernelle, N., Saïs, F.: SAKey: scalable almost key discovery in RDF data. In: Mika, P., Tudorache, T., Bernstein, A., Welty, C., Knoblock, C., Vrandečić, D., Groth, P., Noy, N., Janowicz, K., Goble, C. (eds.) ISWC 2014. LNCS, vol. 8796, pp. 33–49. Springer, Cham (2014). https://doi.org/10.1007/978-3-319-11964-9_3
13. Grover, A., Leskovec, J.: node2vec: scalable feature learning for networks. In: SIGKDD (2016)
14. Lisena, P., Troncy, R.: Combining music specific embeddings for computing artist similarity. In: ISMIR, Late-Breaking Demo Track (2017)
15. Lisena, P., Canale, L., Ellena, F., Troncy, R.: CityMus: music recommendation when exploring a City. In: ISWC, Poster Track (2017)
16. Villazón-Terrazas, B., Vilches-Blázquez, L.M., Corcho, O., Gómez-Pérez, A.: Methodological Guidelines for Publishing Government Linked Data. In: Wood, D. (ed.) Linking Government Data, pp. 27–49. Springer, New York (2011). https://doi.org/10.1007/978-1-4614-1767-5_2
17. Dijkshoorn, C., Jongma, L., Aroyo, L., van Ossenbruggen, J., Schreiber, G., ter Weele, W., Wielemaker, J.: The rijksmuseum collection as linked data. In: Semantic Web, pp. 1–10 (2014)
18. Candela, G., Escobar, P., Carrasco, R.C., Marco-Such, M.: Migration of a library catalogue into rda linked open data. Semantic Web, pp. 1–11 (2017)
19. Swartz, A.: Musicbrainz: a semantic web service. IEEE Intell. Syst. **17**(1), 76–77 (2002)
20. Jacobson, K., Dixon, S., Sandler, M.: Linkedbrainz: providing the musicbrainz next generation schema as linked data. In: Demo Session ISMIR (2010)

Audio Commons Ontology: A Data Model for an Audio Content Ecosystem

Miguel Ceriani(✉) and György Fazekas(✉)

Centre for Digital Music, Queen Mary University of London, London, UK
{m.ceriani,g.fazekas}@qmul.ac.uk

Abstract. Multiple online services host repositories of audio clips of different kinds, ranging from music tracks, albums, playlists, to instrument samples and loops, to a variety of recorded or synthesized sounds. Programmatic access to these resources maybe used by client applications for tasks ranging from customized musical listening and exploration, to music/sounds creation from existing sounds and samples, to audio-based user interaction in apps and games. We designed an ontology to facilitate interoperability between repositories and clients in this domain. There was no previous comprehensive data model for our domain, however the new ontology relates to existing ontologies, such as the Functional Requirements for Bibliographic Records for the authoring and publication process of creative works, the Music Ontology for the authoring and publication of music, the EBU Core ontology to describe media files and formats and the Creative Commons Licensing ontology to describe licences. This paper documents the design of the ontology and its evaluation with respect to specific requirements gathered from stakeholders.

1 Introduction

Over the last decade there has been an explosion in the amount of multimedia content available online. This is due in part to the advent of Web 2.0, i.e. the availability of online tools that facilitate creating and sharing user-generated content. The change can also be attributed to growth in internet connectivity and bandwidth that permitted the progressive increase of quality in streamed multimedia content. Audio content is a fundamental part of the multimedia content consumed, as shown by the popularity of audio streaming services such as Spotify and SoundCloud.

Most of the online audio content is available also to software agents through web-based Application Programming Interfaces (APIs) developed by the maintainers of online repositories primarily to provide access to their content. This enables scenarios, still mostly unexplored, that go beyond simple consumption through maintainer-provided apps. Applications include highly customized user interfaces, seamless exploration of multiple repositories, advanced analysis of content-based on audio features, integration in creative workflows for transformation and reuse of sounds and music.

D. Vrandečić et al. (Eds.): ISWC 2018, LNCS 11137, pp. 20–35, 2018.
https://doi.org/10.1007/978-3-030-00668-6_2

In order to facilitate the integration of the multiple existing repositories as wel as content consumption by software agents, we propose a common data model called the Audio Commons (AC) ontology. This paper describes the design process of the AC ontology and its first (and current) version, i.e. 1.0.0. The ontology is available online[1] with licence CC0.

Our ontology design is novel in many of its aspects: 1. it represents audio media in the broader context of audio production and sharing, going beyond media object-centric models like the EBU Core ontology or the W3C Ontology for Media Resources (e.g., including audio categories and collections as "first-class citizens"); 2. it employs a layered approach in which information can be represented at multiple levels of granularity (e.g., including optional details on how/when a content was recorded) associated with different perspectives (e.g., using a genre classification for music content or a sound effects taxonomy for sound effects); 3. it includes as a requirement support for an API from the developers' perspective, considering that this is a central aspect in the adoption of models nowadays.

The ontology is described following the MIRO (minimum information for the reporting of an ontology) guidelines [6]. All required information items are provided in the text. For reference, we will use the MIRO designations (e.g., C.3 for Communication), where the specific information item is provided[2]. Ontology need (B.1), its name (A.1), and its licence (A.3) have been already introduced.

Section 2 of the paper introduces the methodology and scope. Section 3 describes existing related models while Sect. 4 describes knowledge acquisition through a user survey. Section 5 details use cases and requirements, and Sect. 6 shows how existing ontologies are reused. The resulting ontology is described in Sect. 7 and evaluated in Sect. 8. Conclusions are discussed in Sect. 9.

2 Methodology and Scope

In order to frame the ontology design, this section describe more in detail its audience (B.3) and scope (C.1).

2.1 Methodology for Ontology Development

The ontology is developed and maintained by the Audio Commons consortium[3], composed of leading research institutes in sound and music computing and key players in the creative industries (A.2 and C.2). The development happens in an online public git repository on GitHub[4] (A.5). The GitHub issue tracking system associated with the repository will be used as communication channel for maintenance and future development of the ontology (C.3).

[1] https://w3id.org/ac-ontology/aco.

[2] The role of the information is conveyed anyway in the text, so the reader does not need necessarily to check the MIRO codes.

[3] http://www.audiocommons.org/team/.

[4] https://github.com/AudioCommons/ac-ontology.

Ontology design and development broadly follows the METHONTOLOGY [2] methodological framework (A.6) that identifies six phases: 1. the *specification* i.e., the identification of the audience, scope, scenarios of use, and requirements (Sects. 2 and 5); 2. the *conceptualization* of an informal model (first paragraph of Sect. 7); 3. the *formalization* of the ontology in OWL [9] (Sect. 7); 4. the *integration* of existing ontologies (Sect. 6); 5. the *implementation* of the ontology with a JSON-LD OWL serialization.

METHONTOLOGY identifies also two *activities* carried on during the whole design process and orthogonal to the six phases: 1. *acquiring knowledge* through research of related ontologies and models (Sect. 3) and gathering data from potential users (Sect. 4), to inform multiple phases of the design process, mainly conceptualization and integration; 2. *documentation* of the process phases (internal) and the ontology specification (public).

2.2 Audience and Scope: Audio Commons Ecosystem

The role of the AC ontology is to offer a common data model enabling an ecosystem that integrates multiple online repositories and tools and allows agents to seamlessly explore, access, transform, and redistribute audio content, the Audio Commons Ecosystem (ACE) [3].

As a first step towards the ACE, a web API, the Audio Commons API, has been designed. This provides integrated access to a set of existing repositories (currently Freesound, Jamendo, and Europeana Sounds). Tailor made clients were developed that integrate with common tools in standard production workflows and use the AC API to access the repositories. This process validated the general idea of the ecosystem and informed the design of the ontology.

3 Related Ontologies and Data Models

This section describes related ontologies and data models (B.2). They have been gathered through the research of literature and online resources (D.1 and D.2) and evaluated as part of the design process (D.3).

In the 1990s, the International Federation of Library Associations and Institutions (IFLA) developed a conceptual model called **Functional Requirements for Bibliographic Records (FRBR)** [7]. FRBR defines four main entities to represent the products of intellectual or artistic endeavour: "*Work* (a distinct intellectual or artistic creation) and *expression* (the intellectual or artistic realization of a work) reflect intellectual or artistic content. *Manifestation* (the physical embodiment of an expression of a work) and *item* (a single exemplar of a manifestation) reflect physical form." [7] The entities of the FRBR model and their relationships have been later represented as an OWL ontology[5]. The model is relevant to the audio publishing domain, but the concepts are too generic by themselves. They need to be specialised to clarify the usage.

[5] http://purl.org/vocab/frbr/core#.

The **Music Ontology**[6] [14] aims to provide a comprehensive, yet easy to use and easily extended domain specific knowledge representation for describing music related information. It relies on, and extends the FRBR model, and provides an event based conceptualisation of music production workflows. The Music Ontology describes a domain that is very close to the one we model. Its terms are bound however to the music production workflow [1], without considering the broader, non-musical audio domain that includes e.g. natural sounds or field recordings with their own unique production model.

The European Broadcasting Union (EBU) developed the **EBU Core ontology**[7], which, among other things, specify how to describe properties of media files and formats. The EBU Core ontology has much broader scope, modelling other aspects of media handling. But its approach is centred on broadcast, hence most other entities cannot be easily reused outside of that domain.

The World Wide Web Consortium (W3C) developed an **Ontology for Media Resources**, which aims to integrate multiple metadata vocabularies in the context of media resources. This work is very interesting but the model presented is flat: it is a set of properties that can be attached to a single type of individual, the *MediaResource*. It is hence, as EBU Core, too limited to completely fulfil the requirements of the Audio Commons ontology.

In order to retrieve and explore repositories of audio content, it is useful to have some structured classification of audio. Several classifications have been developed both for manual and automatic categorisation of audio content. Some of them may be applied to audio [4,11] without restrictions, while others are specifically tailored for relevant subsets. Given the importance of musical audio, several classification deal with music, for example organising content by genre or musical instruments. These are usually taxonomies (i.e., simple hierarchical classifications) which may be represented in RDF through the Simple Knowledge Organization System (SKOS) [8].

4 Knowledge Acquisition

Given the use cases, audience, and basic requirements of the ontology, the Audio Commons consortium designed a survey to gather specific requirements from potential users.

The survey contained 24 questions (15 questions with predefined answers and 9 open ended questions) asking people working with audio content about various subjects. Besides demographics, we enquired about the workflows they use and metadata they would like to acess when searching for new audio content on the Web[8] (D.1). The first 8 survey questions assessed the context of audio content usage in the participants' work. Twelve questions (including 7 open ended questions) asked them to describe their ideal query interface, the attributes

[6] http://purl.org/ontology/mo/.
[7] https://www.ebu.ch/metadata/ontologies/ebucore/.
[8] The questionnaire is available online at https://goo.gl/forms/gWdzeHJuPIZhUwzD3.

they would use to query/filter audio content, and the frustrations faced using current tools. The last 4 questions gathered basic demographic information on the participants. The Audio Commons industry partners were given the task to ask their user base to fill in the survey (D.2).

4.1 Survey Results and Analysis

The survey had 661 responses[9]. Participants are split almost in half between professionals (45.5%) and amateurs (54.5%). 42.1% of the participants have more than 10 years of experience with audio technologies and 84.6% of the participants have at least 2 years of experience. The two main contexts of use are music production/performance (63.5%) and audio creation for either film and TV programmes or games (56.3%). There is a significant overlap between these two usages (26.6%).

Most of the participants work with an Internet connected device (84.4%) and get at least *sometimes* audio content from the web (84.8%). For the majority (52.7%), finding the right file is the most time consuming activity in their workflow. This is in strong contrast with the fact that most of the participants consider audio processing as the creative part of their work (65.6%). As for the types of audio content they look for, it is mostly sound effects (82.9%), but also audio loops (36.8%) and full songs (22.3%).

As an ideal way of retrieving audio content, most participants would use a web browser (67.7%), while a substntial part of them would prefer not to leave their digital audio workstation (DAW) software (41.1%). Regarding the query interface, most participants desire to search textually using keywords (86.5%). Half of them would find it useful to have keywords suggested through drop-down lists or similar (46.7%). Some of them would be interested in writing queries using natural languages (26.1%). A relatively small fraction of participants would like to use a full-fledged query language (16.5%) or a graphical interface (12.4%).

Most participants would like to use audio perceptual attributes like "Punchy", "Bright", or "Powerful" (71.4%) while many would also use musical attributes like key, tempo, or instrumentation (47.7%). The analysis of the open ended questions reveals a wide range of attributes used for audio search, ranging from musical properties (e.g., rhythm), to used hardware (e.g., equipment), to moods (e.g., happy).

The answers to the question on frustration show problems related with licensing (not clear enough, hard to understand the rules), syntax (problematic labelling of audio content), sparseness of metadata, lack of workflow integrations (easily retrieving the data into some part of the workflow), bad recording quality of audio sources, various interface problems (bad design, pop-ups, redirections, etc.) and lack of quality curation/recommendation.

[9] All the responses, along with the list of questions are publicly available online at https://doi.org/10.5281/zenodo.832644.

4.2 Conclusions

Answers to those questions allowed us to get the insight into how users would like to search for specific files and how such strategy would impact the design of the user interfaces. Some answers can guide the general proposed approach of the AC ontology while others inform the specific development strategy and content (D.3).

The expressed need for keyword-based search using descriptive text and a variety of attributes, alongside the perceived heterogeneity and sparseness of metadata, support the need of a common data model that unify how metadata is represented and facilitates spotting missing information in data sets.

The declared limits and idiosyncrasies of user interfaces or tools, make a case for having a common API (based on the ontology) that fosters the development of an ecosystem of tools while decoupling the tools from the audio repositories. The fact that most users work on internet connected devices in order, among other reasons, to access potentially "unlimited" audio content, mitigate the most-obvious drawback of a web API based architecture.

Regarding the structure of the ontology, the main takeaway is that audio categorisation should be flexible. The expressed desire to have a text-based search interface, possibly augmented with keyword suggestion/selection, and the variety of attributes used for search/filter would not be compatible with a simple monolithic centralized universal categorisation of audio content. To answer the desiderata the AC ontology need to support multiple categorisations instead.

Another important result of the survey is that there is a need for supporting specific subdomains of musical content associated with musical attributes, on top of supporting the more general domain of *audio content*, not necessarily musical. The significant overlap among the contexts in which musical and non-musical content is used confirms the argument for a comprehensive ontology.

5 Specification

Based on the scope and the survey results, the ontology design is framed by developing use cases and requirements.

5.1 Use Cases

Three user stories have been identified as highy relevant.

- As a café owner I would like to search for whole songs, which are free of any licensing fees. As an example, I would like to search via a browser for "Slow funk track without vocals". Once I found something I like, I would like to find tracks that play well together.
- As an audio producer, I would like to have access to high-quality audio loops from within my digital audio workstation (DAW). I want to search by instrument type, genre, key, tempo.
- As a game sound designer, I would like to have access to high-quality audio files from within my DAW. I want to search by effect type, mood, and perceptual features like "warm","bright", etc.

5.2 Requirements

Using the analysis of the scope and the use cases the ontology designers identified a set of requirements. They are represented as a list of example questions that the ontology should be able to support answering, and a list of formal requirements.

Competency Questions. The following sample questions are meant to be asked with respect to a set of source repositories of audio content.

1. Which are the songs that are slow (tempo) funk (genre) tracks without vocals (instrumentation)?
2. What other tracks "play well" together with a given song in a playlist (e.g., are in the same category according to some classification)?
3. Retrieve high-quality (sample rate, bits per sample) audio loops (type of audio content) for a given instrument type, genre, key, tempo.
4. Retrieve high-quality (sample rate, bits per sample) sound effects (type of audio content) for a given effect type, mood, and a set of perceptual features (e.g., warm and bright).

Formal Requirements. The AC ontology should be able to ...

1. represent the concept of an audio clip as a piece of audio content published in a repository, alongside basic metadata (e.g., title, duration, licence);
2. describe attributes of the digital signal related to an audio clip (e.g., number of channels, sample rate);
3. describe attributes of the media file(s) related to an audio clip (e.g., media format, bit rate);
4. permit the classification of audio content along multiple axes (e.g., musical genre, mood, effect type);
5. represent the organisation of audio clips in collections (e.g., music albums, sound packs, search result sets);
6. optionally, describe additional details of the audio production/publishing process (e.g., where and when an audio clip was recorded).

6 Integration of Existing Ontologies

Following what is considered good practice, this section describes how external vocabularies and ontologies are reused for the AC ontology (E.4). In some cases, owing to discrepancies in the exact meaning or usage context, certain related terms from other vocabularies could not be used directly. In these cases, in order to promote interoperability, we tried to formally express the relationship between new terms and existing ones in the new ontology. This is often expressed by defining the novel classes (properties) as superclasses (superproperties) or subclasses (subproperties) of existing classes (properties). The structure of existing data models also informed our own modelling decisions.

The FRBR concepts, while generic, are relevant to the present case, hence the AC ontology specialize them to the audio production and publication domain. The Music Ontology model is also relevant, namely when dealing with musical content. So the classes (or properties) of the AC ontology are often defined as subclasses (subproperties) of the corresponding classes (properties) in the FRBR ontology (version 2005-08-10) and as superclasses (superproperties) of their counterparts in the Music Ontology (revision 2.1.5). The FRBR model fulfils the role of an upper ontology for the AC ontology, so no general-purpose upper ontologies are used (E.8).

The EBU Core ontology (version 1.8) is used for the detailed formalization of media resources, their metadata (e.g., file size) and their formats (e.g., encoding format). As there is a formal mapping from part of the EBU Core ontology to the attributes defined in the W3C Ontology for Media Resources, the W3C vocabulary is indirectly supported too.

For the generic metadata items (title, description, depiction, ...) the Dublin Core Metadata Initiative (DCMI Metadata Terms[10], version 2012-06-14) and schema.org[11] (Version 3.3) vocabularies are used.

To manage the life-cycle of creative works, e.g. most published audio content, it is especially important to track licensing information, in order to know how a content may be used and redistributed. Dublin Core defines a simple model to attach licence information to a resource. This simple model however does not establish semantics for this licence information and hence does not support comparison and reasoning about properties (permissions, prohibitions, etc.) of licences. The AC ontology reuses the more detailed model specified in the Creative Commons Licensing ontology[12] (version 2017-11-17).

The production of certain entities in AC, for instance, the recording of a track or a sound, are temporal in nature and thus best described as events. The Event ontology[13] [13] (version 1.0) is used for this purpose. This ontology describes different aspects of temporal events, which could either be instantaneous or have a duration.

7 Ontology Description

This section introduces the Audio Commons ontology. Rather than providing a formal specification in this paper, we focus on practical and theoretical considerations in the design of the ontology. We contrast the Audio Commons ontology with other related ontologies and provide rationale for design decisions. The formal specification is provided as an on-line document using OWL (E.1). It can be accessed at https://w3id.org/ac-ontology/aco (A.4). The design is based on a layered approach in which entities are organised in three main groups (see Fig. 1): 1. the *content* of a repository, i.e. the physical *sounds*, the (digital) *signals*, the

[10] http://purl.org/dc/terms/.
[11] http://schema.org/.
[12] https://creativecommons.org/ns.
[13] http://purl.org/NET/c4dm/event.owl.

(published) *audio clips*, and the *audio files*; 2. the *events* associated with the entities and their transitions, i.e. *recording* or *synthesis* producing a signal and the *publication* of a signal as an audio clip; 3. the multiple *categorisations* that can be used to classify content.

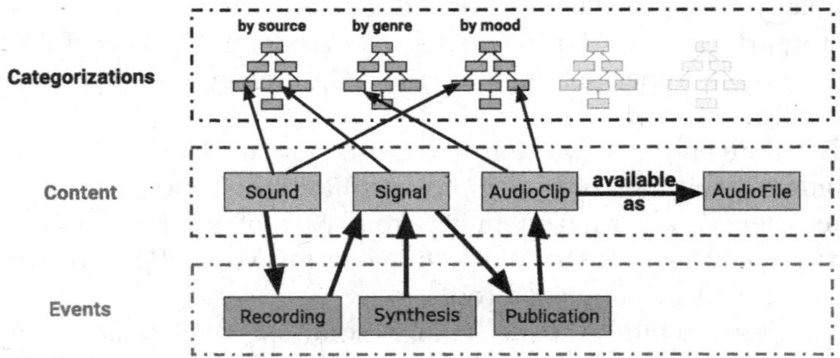

Fig. 1. A layered view of the Audio Commons ontology

Figure 2 shows the most general classes and properties of Audio Commons ontology and their relationship with elements of the FRBR and the Music ontologies. Following the FRBR model, the following three base classes have been defined: `ac:AudioExpression`, the specific intellectual or artistic form that a work takes each time it is "realized", in the audio domain (e.g., the recording or synthesis of music or sounds); `ac:AudioManifestation`, the physical embodiment of an audio expression (e.g., a musical track, a sound, an album); `ac:AudioItem`, a single exemplar of an audio manifestation (a copy of a CD or a specific media file).

The FRBR class Work, representing a distinct intellectual or artistic creation on a more conceptual level, has not been specialized in Audio Commons because this does not generalise sufficiently to all types of sounds relevant in the Audio Commons ecosystem. This concept is used to represent the common creation act in FRBR between different expressions, for instance, different drafts of a symphony, or its existence in the composer's mind at its most abstract level. For musical resources, the `mo:MusicalWork` class can still be used instead. An interesting crossing point is artistic conceptualisation for instance in *sound design* which we consider musical at this stage.

In the Music Ontology some specific properties (e.g., `mo:genre` and `mo:instrument`) are used orthogonally to classify both musical works, expressions, manifestations, and items, attaching them to specific instances of some classification schema (e.g., instances of `mo:Genre` and `mo:Instrument`). In the Audio Commons ontology these properties are generalised by the `ac:audioCategory` property that associates any audio expression or manifestation (or item, but the practical use of the latter case seems limited) to some generic `ac:AudioCategory`. Using this formalisation, different taxonomies specific to a domain of interest or a content provider can be "plugged in" and

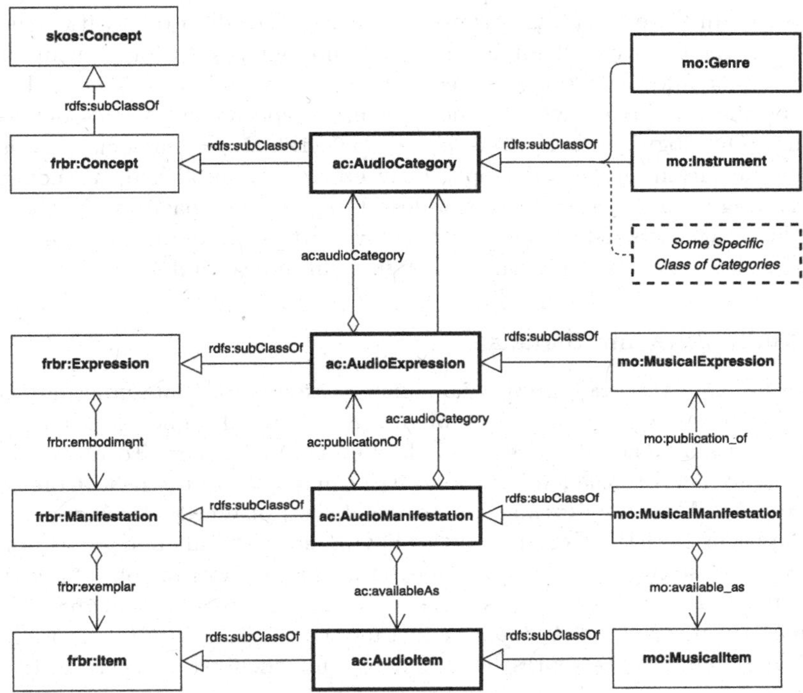

Fig. 2. Audio Commons ontology: a UML-like diagram of the top-level entities

matched to our core concepts enabling interoperability for generic tools, but retaining specificity required for expert users. Specific subclasses and properties related to audio expressions, manifestations, and items will be described in the rest of this section.

7.1 Audio Clips and Audio Collections

The class `ac:AudioManifestation` is a generalisation of (i.e., superclass of) a central entity in the Audio Commons ecosystem, `ac:AudioClip`. An instance of `ac:AudioClip` is any audio segment that has been published in some form or uploaded for consumption, for example, a track in a music label's repository or a sound in an audio repository, library or archive.

In order to represent collections of audio clips, the Audio Commons ontology offers an abstraction termed `ac:AudioCollection`, which is itself a subclass of `ac:AudioManifestation`. The content of each node of a collection is not limited to an `ac:AudioClip`, but may contain any `ac:AudioManifestation`. Collection can thus contain other collections to support specific cases, e.g. a mapping to the Music Ontology model where an `mo:Release` can contain multiple `mo:Record`(s) that can in turn contain multiple `mo:Track`(s).

The Dublin Core vocabulary is used for basic meta-data (e.g., title, description), while the Creative Commons licensing ontology is used for licensing information. Other information more specific to the domain is represented through audio-specific properties, which generalise music-specific properties defined in the Music Ontology: `ac:compiled` and `ac:published`, that associate an agent with manifestations he/she/it respectively created or published; `ac:homepage` and `ac:image`, that associate a manifestation with its page on a site (e.g., Jamendo) or with its depiction (e.g., the cover art of an album); `ac:duration`, that associate an audio clip with its duration (in milliseconds).

7.2 Audio Files and Signals

The class `ac:AudioItem` represents a concrete exemplar of an audio manifestation. In our domain, the main exemplars are the actual audio files. The corresponding class `ac:AudioFile` is a subclass of `ac:AudioItem`. To represent the information related to the audio file and its format, part of the EBU Core ontology is reused. The class `ac:AudioFile` is subclass of `ebu:MediaResource` too and the properties EBU Core properties having `ebu:MediaResource` as domain are used to describe the file (e.g., `ebu:hasEncodingFormat`, `ebu:fileSize`). The property `ac:availableAs` associates an `ac:AudioManifestation`with one or more corresponding `ac:AudioItem` instances.

While `ac:AudioFile` represents a concrete file encoded in a certain format, `ac:DigitalSignal` is the representation of the corresponding digital signal. `ac:DigitalSignal` is a subclass of `ac:MusicalExpression`. This conceptualisation was chosen because it pertains to the weakest ontological commitment with respect to how the signal is represented or encoded and where it is situated in a specific workflow. The data properties `ac:sampleRate`, `ac:bitsPerSample`, and `ac:channels`, associate a signal with its basic features specific to digital representations. The property `ac:publicationOf` can be used to associate an `ac:AudioClip` with the corresponding digital signal. The property `ac:encodes` instead, associates an `ac:AudioFile` with the encoded digital signal. The latter property works as a short-cut of traversing the inverse of `ac:availableAs` and `ac:publicationOf` which is introduced for representational convenience.

7.3 Audio Processes and Events

The description of temporal events is crucial to describe transitions in the workflow of audio production and publication. We thus extend the Event Ontology, offering subclasses of `event:Event` for specific actions that are interesting for the audio domain: `ac:SignalProduction`, the act of producing a `ac:Signal` (that could be either a `ac:AnalogSignal` or a `ac:DigitalSignal`), which is specialized by `ac:Recording` that represents the recording of a `ac:Sound` (e.g., the sound created by a musical band that is playing) and by `ac:Synthesis`; `ac:AudioPublication`, the event representing the public release of a piece of work (e.g., the release of a new album by a band). Using `event:Event`, details of the event such as its location in time and space, its factor, and its products may be explicitly described. Moreover, the events can be composed using the property `event:sub_event`, to build complex events.

7.4 Audio Collections as Lists

The `ac:AudioCollection` entity provides a mechanism to describe collections of audio content in a way that is coherent and integrated with the rest of the Audio Commons ontology. However, the full serialisation of an instance of `ac:AudioCollection` is an explicit representation of a linked list and tends to be quite convoluted no matter what specific RDF syntax is used. For the Audio Commons ecosystem it is important to support usability by conveying information about instances concisely, so a simpler representation should be supported.

For the standard list class `rdf:List`, several RDF syntaxes provide ways to encode lists in a compact way. For this reason, as well as for interoperability reasons, the `ac:AudioCollection` can point to a `rdf:List` representation using the `ac:collectionAsList` property. The ontology constraints the usage of `rdf:first` and `rdf:rest` on the class `ac:AudioCollectionNode` (a member of an `ac:AudioCollection`) so that they "behave well" (e.g., they are functional) and are compatible with the formalisation of `ac:AudioCollection`. Instances of `ac:AudioCollection` can thus be represented either by using our formalism or by using standard RDF lists. They are formally equivalent and hence theoretically interoperable. In practice the transformation from one form to the other requires OWL-DL reasoning and it would not be always feasible or desirable. In that case it makes sense to chose one of the two options and run an ad hoc conversion if needed.

7.5 Evolution of the Ontology

Building an ontology that would encompass the whole audio domain (and all other domains connected with it) in all its complexity would be a very significant task that is beyond the scope of this work. The Audio Commons ontology is, for this reason, an implementation driven ontology evaluated and evolved in use. This means that the ontology will be growing depending on the demand for new services in the Audio Commons ecosystem (F.1). On the technical level, the last version of the ontology will always be accessible at the AC ontology URI, while past versions will accessible using an URI scheme including the version id (F.3). For reasons of backward compatibility, all the defined concepts will remain in the ontology and keep their current meaning. In case at some point the ontology maintainers decide that a concept is "not to be used any more", it will be annotated as *deprecated* (F.2).

8 Evaluation

We carried out an assessment of the ontology by using formal methods as well as checking its fitness for our domain and purposes.

8.1 Metrics and Formal Validation

The AC ontology defines in total 21 classes, 18 object properties (of which 5 functional), and 5 data properties (all of them functional). No individuals are defined (E.3). Every class and property defined has a textual description (`rdfs:comment`) and a label (`rdfs:label`), both are in English (E.7). For every property, domain and range are defined, except for three where only the range is defined, as they can be applied to individuals of a variety of types (`ac:homepage`, `ac:image`, and `ac:audioCategory`). For each entity defined in the ontology, the IRI is dereferenceable and leads via content negotiation either to an OWL syntax or a webpage documenting it (E.11).

The Audio Commons ontology has been checked for correctness, logical consistency, and alignment with established ontology design guidelines (G.1). The correctness of the ontology and its serializations has been checked first by loading it in the widely used ontology editor Protégé [10] and second through the VOWL copy[14] of the online validation service originally developed by the University of Manchester [5]. The logical consistency has been checked by running two reasoners, HermiT (version 1.3.8.413) [15] and FaCT++ (version 1.6.5) [17]. No inconsistencies have been found.

To validate the ontology with respect to existing good practices, we used the OntOlogy Pitfall Scanner! (OOPS!) online service [12]. This service, based on the existing relevant literature, checks for common pitfalls in ontology design. No pitfalls have been detected in the Audio Commons ontology.

8.2 Evaluation

The AC ontology is evaluated (G.2) by checking that it can (1) be used to reply to the competency questions described in Sect. 5.2, (2) fit in the current Audio Commons ecosystem, and (3) bring added value to it.

Answering Competency Questions. The questions can be formalised as queries from the data sources (the audio content repositories), for example using SPARQL (the standard query language for RDF). For simplicity and conciseness here the formalisation is described at a higher level, using just bits of SPARQL syntax for the graph patterns. Represented in the vocabulary defined by the AC ontology, all the competency questions consist in queries that get a set of individuals of type `ac:AudioClip`, say values of `?audioClip` where the triple `?audioClip rdf:type ac:AudioClip` exists. The set returned is determined by some filters that are applied on all the individuals available from the data source. Most filters can be represented as belonging to a certain category of a classification (mood, genre, instrumentation, ...), hence can be formalised as the existence of a triple `?audioClip ac:audioCategory <category1>`. Some filters (sample rate, bits per sample, ...) require to assess a numeric value. To require that the sample rate is certainly higher than 40 KHz, both triples `?audioClip ac:publicationOf ?signal` and

[14] http://visualdataweb.de/validator/validate.

`?signal ac:sampleRate ?sampleRate` have to exist and must satisfy `?sampleRate > 40E3`.

Fitting in the Audio Commons Ecosystem. The main application of the AC ontology is to provide a common way to represent multiple data models and APIs in the context of the AC ecosystem. As described in Sect. 2.2, a common API has already been defined in the context of the ecosystem. This is the AC API. It integrates multiple repositories by calling their specific APIs and is currently consumed by multiple client applications. The main endpoint of the API is the search endpoint[15] that offers search functionality on audio content that may be in any of the integrated services. Listing 1 is a sample response.

```
{ "Jamendo": {"num_results": 0, "results": [] },
  "Freesound": {
    "num_results": 1,
    "results": [
      { "id": "Freesound:385285", "name": "01 - Cars across_city.wav",
        "description": "Recorded in Moscow. Cars across on street.",
        "duration": 127.321, "author": "semenov_nick",
        "channels": 2, "bitdepth": 32, "samplerate": 48000.0,
        "bitrate": 3001, "filesize": 48915012, "format": "wav",
        "collection_url": "https://freesound.org/apiv2/packs/21409/",
        "license_deed_url": "http://creativecommons.org/publicdomain/zero/1.0/",
        "preview_url": "https://freesound.org/data/previews/385/385285_3596390-hq.ogg",
        "tags": ["city", "cars"] }
    ]
  },
  "Europeana": {"num_results": 0, "results": [] } }
```

Listing 1: An example JSON response from a search on the AC API

```
[ { "schema:actionStatus": "CompletedActionStatus", "schema:target": "services:Jamendo",
    "schema:result": {"membersCount": 0, "collectionAsList": [] } },
  { "schema:actionStatus": "CompletedActionStatus", "schema:target": "services:Freesound",
    "schema:result": {
      "membersCount": 1,
      "collectionAsList": [
        { "@id": "audioClips:Freesound/385285", "dc:title": "01 - Cars across_city.wav",
          "dc:description": "Recorded in Moscow. Cars across on street.",
          "duration": 127.321, "audioCategory": ["fstags:city", "fstags:cars"],
          "isPublishedBy": "users:Freesound/semenov_nick",
          "encodes": { "bitsPerSample": 31, "channels": 2, "sampleRate": 48000.0 },
          "availableAs": {
            "@id": "https://freesound.org/data/previews/170/170992_142024-hq.ogg",
            "@type": "AudioFile", "ebu:hasEncodingFormat": "ebuformat:wav",
            "ebu:bitRate": 3001, "ebu:fileSize": 48915012
          },
          "isMemberContentOf": "packs:Freesound/21409/",
          "cc:license": "http://creativecommons.org/publicdomain/zero/1.0/" }
      ]
    } },
  { "schema:actionStatus": "CompletedActionStatus", "schema:target": "services:Europeana",
    "schema:result": {"membersCount": 0, "collectionAsList": [] } } ]
```

Listing 2: JSON-LD response from a search on the AC API

[15] https://m.audiocommons.org/api/v1/search/text/.

Listing 2 shows the output of the next version of the AC mediator. The same content is represented in RDF using AC ontology concepts and serialized as JSON-LD [16] (G.5). It can be seen that the new JSON-LD format is still close to the original JSON format. The need to map the properties to the ontology forces a slightly more structured JSON representation however that could also facilitate API documentation and other uses even without considering RDF interpretation. The associated JSON-LD context, not shown, maps the prefixes with the corresponding namespaces and set the AC ontology namespace as default. Detailed technical discussion of the mapping can be found in the wiki pages of the AC mediator, the software component exposing the AC API[16].

Bringing Added Value. While the general usage context is broader, it can be shown that there is already added value in just using the return format of the AC API as described above. A "semantic client" will not consume the new format as pure JSON; rather, it will use a JSON-LD processor to interpret the result set as an RDF graph. Moreover, responses to different searches may be merged in a richer RDF graph. The graph model and the unique identification of entities that are potentially repeated across results (e.g., audio categories, authors, media formats) enable organising or ordering the result set(s) in multiple ways according to the needs (e.g., group results by author). Furthermore, this information may be enriched by adding linked information from other sources (e.g., a music instruments taxonomy) or even creating new annotations as a local RDF graph. These functionalities, which are quite straightforward using semantic web technologies and the AC ontology, would need to be explicitly programmed if the "old" JSON output was used.

9 Conclusions

The Audio Commons ontology has been designed to model the audio content production and publishing domain. Its aim is to facilitate integration and serendipitous reuse of audio, through an ecosystem centred on this model and composed of multiple repositories and agents. The AC ontology is related to existing relevant ontologies and models. The evaluation shows that it is consistent, follows good practices, and is functional to the ecosystem. We are planning a test with users, based on client applications that make use of the ontology. Furthermore, as the ontology is disseminated and the ecosystem expands, more feedback is expected in the near future. These inputs will allow to evolve the ontology based on potentially unexpected use cases and conduct a more in-depth evaluation.

Acknowledgement. This work was supported by the European Commission H2020 research and innovation grant AudioCommons under grant agreement number 688382.

[16] https://github.com/AudioCommons/ac-mediator/wiki/JSON-LD-mapping.

References

1. Fazekas, G., Raimond, Y., Jakobson, K., Sandler, M.: An overview of Semantic Web activities in the OMRAS2 Project. J. New Music Res. Spec. Issue Music Inf. OMRAS2 Project **39**(4), 295–311 (2011)
2. Fernández-López, M., Gómez-Pérez, A., Juristo, N.: METHONTOLOGY: from ontological art towards ontological engineering. In: Proceedings of the Ontological Engineering AAAI-97 Spring Symposium Series. American Asociation for Artificial Intelligence (1997)
3. Font, F., et al.: Audio commons: Bringing creative commons audio content to the creative industries. In: Audio Engineering Society Conference: 61st International Conference: Audio for Games, February 2016. http://www.aes.org/e-lib/browse.cfm?elib=18093
4. Gemmeke, J.F., et al.: Audio Set: an ontology and human-labeled dataset for audio events. In: Proceedings of the IEEE ICASSP 2017 (2017)
5. Horridge, M., Bechhofer, S.: The OWL API: a Java API for working with OWL 2 ontologies. In: Proceedings of the OWLED 2009, vol. 529, pp. 49–58. CEUR-WS.org (2009)
6. Matentzoglu, N., Malone, J., Mungall, C., Stevens, R.: Miro: guidelines for minimum information for the reporting of an ontology. J. Biomed. Semant. **9**(1), 6 (2018). https://doi.org/10.1186/s13326-017-0172-7
7. McBride, B.: Functional Requirements for Bibliographic Records, Final Report. UBCIM Publications, New Series 19 (1998)
8. Miles, A., Bechhofer, S.: SKOS Simple Knowledge Organization System - Reference. https://www.w3.org/TR/skos-reference/
9. Motik, B., et al.: OWL 2 Web Ontology Language: Structural Specification and Functional-Style Syntax, 2nd edn. W3C REC 11 December 2012 (2012)
10. Musen, M.A., the ProtégéTeam: The Protégé project: a look back and a look forward. AI Matt. **1**(4), 4–12 (2015)
11. Nakatani, T., Okuno, H.G.: Sound ontology for computational auditory scence analysis. In: AAAI/IAAI, pp. 1004–1010 (1998)
12. Poveda-Villalón, M., Gómez-Pérez, A., Suárez-Figueroa, M.C.: Oops!(ontology pitfall scanner!): an on-line tool for ontology evaluation. Int. J. Semant. Web Inform. Syst. (IJSWIS) **10**(2), 7–34 (2014)
13. Raimond, Y., Abdallah, S.: The event ontology. Technical report (2007). http://motools.sourceforge.net/event (2007)
14. Raimond, Y., Abdallah, S.A., Sandler, M.B., Giasson, F.: The music ontology. In: ISMIR, vol. 422, Vienna, Austria (2007)
15. Shearer, R., Motik, B., Horrocks, I.: HermiT: A Highly-Efficient OWL Reasoner. In: Proceedings of the OWLED 2008, vol. 432, p. 91 (2008)
16. Sporny, M., Longley, D., Kellogg, G., Lanthaler, M., Lindström, N.: JSON-LD 1.0: a JSON-based Serialization for Linked Data. W3C Recommendation 16 January 2014 (2014)
17. Tsarkov, D., Horrocks, I.: Fact++ description logic reasoner: System description. Automated reasoning, pp. 292–297 (2006)

Wiki-MID: A Very Large Multi-domain Interests Dataset of Twitter Users with Mappings to Wikipedia

Giorgia Di Tommaso[1], Stefano Faralli[2], Giovanni Stilo[1(✉)], and Paola Velardi[1]

[1] Department of Computer Science, University La Sapienza of Rome, Rome, Italy
{ditommaso,stilo,velardi}@di.uniroma1.it
[2] Unitelma-Sapienza, Rome, Italy
stefano.faralli@unitelmasapienza.it

Abstract. This paper presents Wiki-MID, a LOD compliant multi-domain interests dataset to train and test Recommender Systems, and the methodology to create the dataset from Twitter messages in English and Italian. Our English dataset includes an average of 90 multi-domain preferences per user on music, books, movies, celebrities, sport, politics and much more, for about half million users traced during six months in 2017. Preferences are either extracted from messages of users who use Spotify, Goodreads and other similar content sharing platforms, or induced from their "topical" friends, i.e., followees representing an interest rather than a social relation between peers. In addition, preferred items are matched with Wikipedia articles describing them. This unique feature of our dataset provides a mean to categorize preferred items, exploiting available semantic resources linked to Wikipedia such as the Wikipedia Category Graph, DBpedia, BabelNet and others.

Keywords: Semantic recommenders · Twitter · Wikipedia · Users' interest

1 Introduction

Recommender systems are widely integrated in online services to provide suggestions and personalize the on-line store for each customer. Recommenders identify preferred items for individual users based on their past behaviors or on other similar users. Popular examples are Amazon [1] and Youtube [2]. Other sites that incorporate recommendation engines include Facebook, Netflix, Goodreads, Pandora and many others.

Despite the vast amount of proposed algorithms, the evaluation of recommender systems is very difficult [3]. In particular, if the system is not operational and no real users are available, the quality of recommendations must be evaluated on existing datasets, whose number is limited and what is more, they are

Permanent URL: https://doi.org/10.6084/m9.figshare.6231326.

D. Vrandečić et al. (Eds.): ISWC 2018, LNCS 11137, pp. 36–52, 2018.
https://doi.org/10.1007/978-3-030-00668-6_3

focused on specific domains (i.e., music, movies, etc.). Since different algorithms may be better or worse depending on the specific purpose of the recommender, the availability of multi-domain datasets could be greatly beneficial. Unfortunately, real-life cross-domain datasets are quite scarce, mostly gathered by "big players" such as Amazon and eBay, and they not available to the research community[1].

In this paper we present a methodology for extracting from Twitter a large dataset of user preferences - that we call Wiki-MID - in multiple domains and in two languages, Italian and English. To reliably extract preferences from users' messages, we exploit popular services such as Spotify, Goodreads and others. Furthermore, we infer many other preferences from users' friendship lists, identifying those followees representing an interest rather than a peer friendship relation. In this way we learn, for any user, several interests concerning books, movies, music, actors, politics, sport, etc. The other unique feature of our dataset, in addition to multiple languages and domains, is that preferred items are matched with corresponding Wikipedia pages, thus providing the possibility to generalize users' interests exploiting available semantic resources linked to Wikipedia, such as the Wikipedia Category Graph, BabelNet, DBpedia, and others.

The paper is organized as follows: Sect. 2 summarizes previous research on creating datasets for recommender systems, Sects. 3, 4 and 5 present the methodology to create Wiki-MID, Sect. 6 is dedicated to dataset statistics and evaluation, and Sect. 7 describes the released resource, which has been designed on top of the Semantically-Interlinked Online Communities (SIOC) core ontology. Finally, in Sect. 8 we provide a summary of distinctive features of our resource and some directions for future work.

2 Related Work

Recommender systems are based on one of three basic approaches [4]: *collaborative filtering* [5] generates recommendations collecting preferences of many users, *content-based filtering* [6] suggests items similar to those already chosen by the users, and *knowledge-based* recommendation [7] identifies a semantic correlation between user's preferences and existing items. Hybrid approaches are also widely adopted, e.g., [8]. All approaches share the need of sufficiently large datasets to learn preferences and to evaluate the system, a problem that is one of the main obstacles to a wider diffusion of recommenders [9], since only a small number of researchers can access real users data, due to privacy issues.

To overcome the lack of datasets, challenges as RecSys have been launched[2], and dedicated web sites have been created (e.g., SNAP[3] or Kaggle[4]), where

[1] https://recsys.acm.org/wp-content/uploads/2014/10/recsys2014-tutorial-cross_domain.pdf.
[2] https://recsys.acm.org/.
[3] http://snap.stanford.edu/data.
[4] https://www.kaggle.com/datasets/?sortBy=hottest&group=all.

researchers can upload their datasets and make them available to the community. However it is still difficult to find appropriate data for novel types of recommenders, as the majority is focused on a single topic, like music [10, 11], food [12, 13], travel [14, 15] and more [16]. Furthermore, while a small number of large datasets are available, such as Movielens [17], Million song dataset [18] and Netflix Prize Dataset [19], many others are quite small and based on very focused experiments.

Concerning the source of data for extracting preferences, social networks are often used (mainly Twitter, Facebook, Google+, LinkedIn or a combination of sources, such as in [20]), since their content is freely available with more or less severe restrictions. The interested reader can refer to [21] for a detailed survey of methods adopted in literature to collect social data for the purpose of inferring and enhancing users' interests profiles. Preferences are induced from users' profiles (e.g., [22]), authoritative (topical) friendship relations [23], followee biographies [24], and messages ([25], [26], [27] and many others).

Data extraction from Twitter messages is a popular strategy, however, it is also computationally expensive and error-prone, since it requires natural language processing techniques to analyze the text. To overcome this difficulty, a number of studies exploited platforms (e.g., Youtube, Spotify) that integrate among their services the ability to post the user's personal content on the most popular social network sites, such as movies that users are watching. Sharing this information is done in a simple and predefined way. Depending on the social network chosen, the content, for example a Youtube video, will be shared with a pre-formatted message formed by the video name, a link, a self-generated text and, if provided, a numerical rating (e.g. "How It's Made: Bread" https://youtu.be/3UjUWfwWAC4 via YouTube). The message can also be enriched and personalized by the user. In [25] these types of messages are extracted from Twitter, to detect music interests. The dataset is based on 100,000,000 tweets with the #nowplaying main tag. Tweets are extracted via Twitter APIs over 3-years and next, MusicBrainz and Spotify are used to add more details. Other studies extract data about music [27] or sport [28] events. However, all the datasets generated in this way concern only one domain of interest.

To the best of our knowledge, the only really multi-domain dataset is presented in [29], where pre-structured tweets about three domains - movies, books and video-clips - are extracted respectively from IMDb (Internet Movies Database), Youtube and Goodreads. With respect to this work, we collect a much wider number of interests, since in addition to pre-formatted messages based on a number of available services, we reliably extract many additional types of interests exploiting users' followees lists. Furthermore, as shown in Sect. 7, we collected many interest *types* for each user, while the dataset released in [29] includes only 7 users with at least 3 types of interests.

3 Workflow

This section summarizes the data sources and workflow to create the WikiMID multi-domain resource. We extract preferences (with unary ratings) from a user's messages and from his/her friendship list, identifying those followees who represent an interest rather than a peer friendship relationship. The process is in three steps:

1. *Extracting interests from users' textual communications.* Using textual features extracted from users' communications, profiles or lists seems a natural way for modeling their interests. However, this information source has several drawbacks when applied to large data streams, such as the set of Twitter users. First, it is computationally very demanding to process millions of daily tweets in real time; secondly, the extraction process is error prone, given the highly ungrammatical nature of micro-blogs. To reliably extract preferences from Twitter users' messages, in line with other works surveyed in Sect. 2, we use a number of available services, described hereafter, that allow to share activities and preferences in different domains - movies, books etc. - using pre-formatted expressions (e.g., for Spotify: #NowPlaying) followed by the url of a web site, from which we can extract information without errors. The drawback is that a relatively small number of users access these services and in addition, preferences are extracted only in few domains.
2. *Extracting interests from users' friendship lists.* In [30] the authors argue that users' interests can also be implicitly represented by the authoritative (*topical*) friends they are linked to. This information is available in users' profiles and does not require additional textual processing. Furthermore, interests inferred from topical friends are less volatile since, as shown in [31], "common" users tend to be rather stable in their relationships. Topical friends are therefore both relatively stable and readily accessible indicators of a user's interest. Another advantage is that average Twitter users have hundreds of followees, many of which, rather than genuine friends, are indicators of a variety of interests in different domains, such as entertainment, sport, art and culture, politics, etc.
3. *Mapping interests onto Wikipedia pages.* The final step is to associate each interest, either extracted from messages or inferred from friendship relations, with a corresponding Wikipedia page, e.g.:
@nytimes ⇒ WIKI:EN:The_New_York_Times
(in this example, @nytimes is a Twitter account extracted from a user's friendship list). Although not all interests can be mapped on Wikipedia, our experiments show that this is possible in a large number of cases, since Wikipedia articles are created almost in real-time in correspondence with virtually any popular entity, either book, or song, actor, event, etc.

We applied this workflow to two Twitter streams in two languages, English and Italian, as explained in the next Sections.

4 Extraction of Users' Interests

4.1 Extracting Interests from Messages

Everyday a huge number of people uses on-line platforms (e.g. Yelp, Foursquare, Spotify, etc.) that allow to share activities and preferences on different domains on a social network in a standard way. Among the most popular services accessed by Twitter users, we selected those providing pre-formatted messages:

- **Spotify:** Spotify is a music service offering on-demand streaming of music, both desktop and mobile. Users can also create playlists, share and edit them in collaboration with other users. In addition to accessing the Spotify web site, users can retrieve additional information such as the record label, song releases, date of release etc. Since 2014, Spotify is widely used in America, Europe and Australia. Spotify is among the services allowing to generate self-generated content shares in Twitter. An example of these tweets is: "#Now-Playing The Sound Of Silence by Disturbed https://t.co/d8Sib5EDVf". The standard form of these tweets is:
  `#NowPlaying <title> by <artist > <URL>`
  By filtering the tweets stream and using Twitter APIs for hashtag detection, we generated a stream of all the users who listened music using Spotify.
- **Goodreads and aNobii:** Similarly to Spotify, a number of platforms allows to share opinions and reviews on books. In these platforms, users can share both titles and ratings. Similarly to Spotify, generated tweets have a predefined structure and point to an URL. In the book domain, we use Goodreads (10 million users and 300 million books in the database) and for Italian, the more popular aNobii service.
- **IMDb and TVShowTime:** In the domain of movies, currently there are no dominant services. Popular platforms in this area are Flixter, themoviedb.org and iCheckMovies. However, many of these platforms use the IMDb database, owned by Amazon, which handles information about movies, actors, directors, TV shows, and video games. We also use the TvShowTime service for Italian users.

In order to extract users' preferences from these services, we first collect in a Twitter stream TS all messages including a hashtag related to one of the above mentioned services (#NowPlaying, #IMDb ..). Then, we extract from TS the music, movie and book preferences for the set of users U who accessed these services. Unlike [29], we avoid parsing tweets using specific regular expressions, since users are free to insert additional text in the pre-formatted message. Rather, in line with [32], we exploit an element that most pre-formatted tweets have: the URL, e.g., #NowPlaying High by James Blunt https://t.co/7EiepE2Bvz.

Accordingly, we collect all tweets containing the selected hashtags and discard those which do not include an URL. The reason for extracting the information from the URL (which is computationally more demanding) rather than from the tweet itself is twofold: (i) Tweets can be ambiguous or malformed, and furthermore, users can insert additional text in the pre-formatted message, e.g., "*#NowPlaying Marty*. This guy is amazing.♫ http://t.co/jwxvLiNenW". Scraping the html page at the URL address ensures that we extract data *without errors*, even for complex items such as book and movie titles; (ii) The URL includes additional information (e.g., not only the title of a song, but also the singer and the record label), which provide us a context to reliably match the extracted entity (song, book, movie) with a Wikipedia article, as detailed in

Sect. 5.1. Since the URL in tweets is a short URL, we first extend the original URL so that all URLs belonging to a given platform can be identified (for example, https://t.co/oShYDc6DeL → http://spoti.fi/2cTPn0U). Next, we access the web site and scrape its content. For each platform we obtain the following data:

- **Music:** <Title, Author (eg. singer, band)>
- **Books:** <Title, Author>
- **Movie:** <Title, Year of production, Type (eg. movie, tv series)>

4.2 Extracting Interests from Users' "topical" Friends

In addition to preferences extracted from users' messages, we also induce interests from their *topical friends*, a notion that we first introduced in [23]. We denote as topical friends those Twitter accounts in a user's followees list representing popular entities (celebrities, products, locations, events ...). For example, if a user follows @David_Lynch, this means that he/she likes his movies, rather than being a genuine friend of the director. There are several clues to identify topical friends in a friendship list: first, topical relations are mostly *not reciprocated*, second, popular users have a high in-degree. However, these two clues alone do not allow to distinguish e.g., bloggers or very social users from truly popular entities. To learn a classification model to distinguish between topical and peer friends, we first collected a network of Verified Twitter Accounts. Verified accounts[5] are authentic accounts of public interest. We started from a set of seed verified contemporary accounts in 2016, and we then crawled the network following only verified friends, until no more verified accounts could be found. This left us with a network of 107,018 accounts of verified contemporary users (V), representing a "training set" to identify authoritative users' profiles. To learn a model of authoritativeness, we used the set V and a random balanced set of $\neg V$ users. For each account in V and $\neg V$, we extracted three structural features (in degree, out degree and their ratio) and one binary textual feature (presence in the user's account profile of role words such as *singer, artist, musicians, writer..*). Then, we used 80% of these accounts to train a SVM classifier with Laplacian kernel and the remaining 20% for testing with cross-validation, obtaining a total accuracy of 0.88 (true positive rate 0.95 and true negative rate 0.82).

Next, from the set U of users in our Twitter datasets (separately for the English and Italian streams), we collected the set F of Twitter accounts such that, for any $f \in F$ there is at least one $u \in U$ such that u follows f. The previously learned classifier was used to select a subset $F_t \subseteq F$ of *authoritative* users representing "candidate" topical friends.

Finally, an additional filtering step is applied to identify "true" topical friends in F_t, i.e., genuine users' *interests*, which consists in determining which members of the set F_t have a matching Wikipedia page. This step is described in Sect. 5. The intuition is that, if one such match exists, the entity to which the Twitter

[5] https://developer.twitter.com/en/docs/api-reference-index.

account belongs is indeed "topical"[6]. Although this filtering step may affect the recall of the method, it provides high accuracy, as demonstrated in Sect. 6.

5 Mapping Interests to Wikipages

The last step of our methodology consists in mapping the collected users' interests to Wikipages. This step has both the advantage of improving the precision of detected users' interests, and providing a mean to categorize them. We use different mapping methodologies for interests extracted from messages and those induced from users' friendship lists.

5.1 Mapping Movies, Songs and Books

Mapping interests extracted from users' messages to Wikipedia pages is a very reliable process, given the additional contextual information extracted from the URL (see previous Sect. 4.1). Wikipedia mapping is obtained by a cascade of weighted boolean query on a Lucene Index, as in the example below, used to search the Wikipage of an item: $<TITLE \in WikiTitle>^{w_1} \wedge <AUTHOR \in WikiGloss>^{w_2} \wedge ((<WORDS \in WikiTitle>^{w_3} \vee <AUTHOR \in WikiTitle>^{w_4} \vee \neg(<WORDS \in WikiTitle>^{w_3} \vee <AUTHOR \in WikiTitle>^{w_4} \vee <WORDS \in WikiText>^{w_5})) \vee <WORDS \in WikiText>^{w_5})$ $<WORDS>$ for music = {"song"} $<WORDS>$ for books = {"books", "novel", "saga" ...} $<WORDS>$ for movie = {"film", "series", "TV series", "episode" ...}
where w_i is a weight assigned to a query. When the page doesn't exist or is not available, we search the page of the item's author, using similar queries.

5.2 Mapping Topical Friends

Matching interests extracted from a user's friendship list with corresponding Wikipedia pages is far more complex, because of homonymy, polysemy and ambiguity. Furthermore, the information included in a user's Twitter profile is very sketchy and in some case misleading, therefore it may not provide sufficient context to detect a similarity with the correspondent Wikipedia article. For example, Bill Gate's description field[7] in his Twitter profile is: "*Sharing things I'm learning through my foundation work and other interests...*" which has little in common with his Wikipedia page: "*William Henry Gates III (born October 28, 1955) is an American business magnate, investor, author, philanthropist, humanitarian and co-founder of the Microsoft Corporation along with Paul Allen.*"

We note that other studies have considered this task. For example, the authors in [33]) use an heuristics based on the overlap coefficient of last 20 topical followees' tweets and the Wikipedia article summary, which is rather data

[6] We do not directly attempt a match of all $f \in F$ with Wikipedia, since it is very computationally demanding and has a reduced precision.

[7] As retrieved on January 2018.

demanding. In [34] the authors use a methodology which is similar to the one we firstly presented in [23], based on a comparison between Twitter description fields and the content of a Wikipage. As previously noted (see the Bill Gates example), this might not be sufficient in many cases. In the present work, to reliably assign a Wikipedia page to a large fragment of users in the set F_t of $U's$ authoritative friends, we use an *ensemble* of methods, with adjudication by majority voting. The methodology is described in what follows.

1. Task Description and Data - Given a set $F_t = \{f_1, f_2, ..., f_n\}$ of candidate "topical" Twitter profiles and a set of Wikipages $W = \{w_1, w_2, ..., w_m\}$ we define a mapping function $M : F_t \Rightarrow W \cup \{\lambda\}$ where the value of the function M for a given Twitter profile f_i is a Wikipage w_j, which is the corresponding Wikipage of the entity having the twitter profile f_i or λ, where λ means "no match".

We define an ensemble of three mappers exploiting the information included in Twitter profiles and in DBpedia entities associated to Wikipedia.

Profiles of Twitter users provide, among the others, the following information:

- **profile address:** e.g., https://twitter.com/katyperry;
- **user_ID:** a numeric value to uniquely identify a user (not visible on the rendered web page);
- **screen_name** a string that can be used to refer to a user when posting a message (e.g. @katyperry);
- **name** the extensive name of the owner of the profile (e.g. "Katy Perry");
- **url**: the link to a profile-relevant homepage (e.g. "katyperry.com"). Only a fragment of profiles have an URL to a homepage;
- **description:** a short description to describe the user and welcome profile visitors.

Furthermore, from each wikipage w_j (e.g., Fig. 1, upper right, shows the Wikipedia page of the singer "Katy Perry") it is possible - thanks to DBpedia - to collect additional information, here is a small subset:

- **title:** the title of the page (e.g. "Katy Perry");
- **content:** the textual content of the page;
- **homepage:** a property (collected and included on DBpedia from infoboxes) which (*when present*) links to a web page (homepage) related to the main entity described in w_j (e.g., "katyperry.com");
- **links extracted from the homepage:** are those links included on the source html of the above mentioned homepage, e.g., in the html of the webpage at katyperry.com we find: https://facebook.com/katyperry, https://twitter.com/katyperry, ...

2. Mapping Methods - We rely on an ensemble of three different methodologies (M_1, M_2 and M_3) of association between the set F_t of Twitter profiles and Wikipages. The first is based on text mining and structural properties of

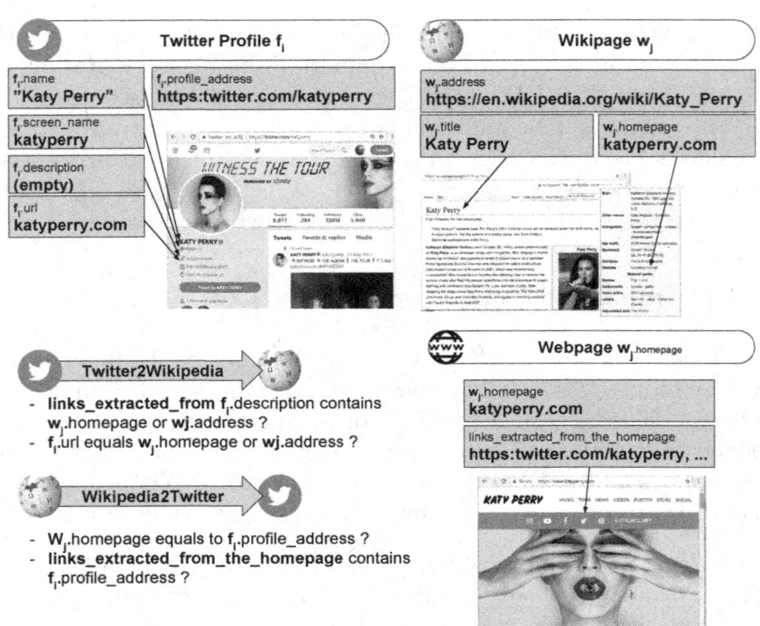

Fig. 1. Example of Twitter2Wikipedia and Wikipedia2Twitter mapping

the social network, the other two are based on finding direct correspondences between the field *url* in a Twitter profile and the property *homepage* in a DBpedia entity.

1. M_1 - **Context Based mapping:** We use the methodology that we first presented in [35], summarized in what follows:
   (a) Selection of candidate senses: For any f_i in F_t, find a (possibly empty) list of *candidate wikipages*, using BabelNet [36] synonym sets (in BabelNet, each "BabelSynset" points to a unique Wikipedia entry). For example, *@kathyperry* has candidates *Katy_Perry* and *Katy_Perry_discography*, but there are cases with dozens of candidates (e.g., https://en.wikipedia.org/wiki/John_Williams_(disambiguation));
   (b) BoW Disambiguation: Compute the bag-of-words (BoW) similarity between the user description in f_i's Twitter account and each candidate wikipage. The BoW representation for each wikipage is obtained from its associated BabelNet relations (relations are described in [37]);
   (c) Structural Similarity: If no Wikipages can be found with a sufficient level of similarity (as for the previous example of Bill Gates description field), select from f_i's friendship list those friends already mapped to a wikipage -if any- and compute the similarity between those wikipages and candidate wikipages. For example, to correctly map the Twitter account of Bill Gates to Wikipedia, profile information of the following Twitter users in his friendship

list are used: Paul Allen, Melinda Gates, TechCrunch, Microsoft Foundation, and more. Note that Paul Allen is explicitly mentioned in the first sentence of Bill Gate's wikipage.

2. M_2 - **Twitter2Wikipedia:** as sketched in Fig. 1, we first collect a set of URL from a given profile f_i including: the link (if any) in the field *url* and all the links extracted from the profile *description* filed (In the example of Fig. 1, since the profile description is empty, we collect only the link katyperry.com). Second, we search a Wikipage w_j (if any) for which one of the links collected for f_i (in our example we collected the link katyperry.com) matches with the link provided in the *homepage* property, (in our example the Wikipage with title "Katy Perry" has a property *homepage* whose value matches exactly the link katyperrry.com), or directly with the address of the page itself (e.g. https://en.wikipedia.org/wiki/Katy_Perry). Note that this mapping method is error prone: for example, from the Twitter profile of Paul Gilmour we extract the following url: skysports.com matching with the homepage property of the following wikipage: https://en.wikipedia.org/wiki/Sky_Sports. Although related, this is not Paul Gilmour's page https://en.wikipedia.org/wiki/Paul_Gilmour.
3. M_3 - **Wikipedia2Twitter:** M_3 is symmetric to M_2. As shown in Fig. 1, we map a given f_i to a Wikipage w_j if the *homepage* property, or one of the links extracted from the source html of the homepage in w_j, matches the Twitter *profile address*. Like for Twitter2Wikipedia, this mapping method is error prone.

For each of the above three approaches we add three additional mapping functions ESM_1, ESM_2, ESM_3 where each mapping function is defined as:

$$ESM_k(t_i) = \begin{cases} w_j, & \text{if } M_k(t_i) = w_j \text{ and } t_i.name = w_j.title \\ \lambda, & \text{otherwise} \end{cases} \tag{1}$$

In other words, ESM_k "reinforces" the result of M_k if the *name* field in the Twitter profile perfectly matches with the *title* of the Wikipedia page. Note that this is often not the case, as for *@realDonaldTrump*.

3. Ensemble Voting - For a given Twitter profile f_i the ensemble voting mechanism selects the Wikipage w_j for which there is maximum agreement among the 6 mapping functions (M_1, M_2, M_3, ESM_1, ESM_2, ESM_3), and there are at least 2 M_j, M_k in agreement ($j \neq k$). The threshold 2 has been empirically selected to obtain the best compromise between number of mapped interests and precision, as detailed in Sect. 6.

6 Wiki-MID Statistics and Evaluation

The outlined process has been applied to two streams of Twitter data, in English and Italian, extracted during 6 months (April-September 2017) using

Twitter APIs. We collected the maximum allowed Twitter traffic of English users mentioning service-related hashtags (e.g., #NowPlaying for Spotify), and the full stream of messages in Italian, since they do not exceed the maximum. As a final result, we obtained for a large number of users a variety of interests along with their corresponding Wikipedia pages. An excerpt of a Twitter user's interests is shown in Table 1. In the example, we selected two interests from each of the four sources from which they have been induced: IMDb (movies), Goodreads (books), Spotify (music) and the user's topical friends. Although a detailed analytics of interest categories is deferred to further studies, the example shows the common trend that a user's interests, either extracted from his/her messages or from topical friends, are strongly related, and in same case identical. For example, the user in Table 1 frequently accesses the IMDb and Spotify services, and he/she is also a follower of the IMDb and Spotify Twitter accounts. Furthermore, his/her interest in the band The Magnetic Field emerges from both source types.

Table 1. Excerpt of a Twitter user's interests

USER ID:787930***		
Source	Interest	Wikipage
IMDb	Eyes Wide Open - 2009 - movie	WIKI:EN:Eyes_Wide_Open_(2009_film)
	Okja - 2017 - movie	WIKI:EN:Okja
Goodreads	The Beautifull Cassandra - Jane Austen	WIKI:EN:Jane_Austen
	The Beach - Alex Garland	WIKI:EN:The_Beach_(novel)
Spotify	I Don't Know What I Can Save You From - Kings of Convenience	WIKI:EN:Kings_of_Convenience!
	Nothing Matters When We're Dancing - **The Magnetic Fields**	WIKI:EN:The_Magnetic _Fields
Topical friends	**@IMDb**	WIKI:EN:IMDb
	@UNICEF_uk	WIKI:EN:UNICEF_UK
	@TheMagFields	WIKI:EN:The_Magnetic _Fields
	@BarackObama	WIKI:EN:Barack_Obama
	@Spotify	WIKI:EN:Spotify

Overall, we followed 444,744 English-speaking and 25,135 italian-speaking users (the set U) who accessed at least one of the services mentioned in Sect. 4.1. Tables 2 and 3 show general statistics of interests extracted from users' messages respectively, for English and Italian speaking user. In the English dataset we crawled more than 20M tweets from these users, of which, about 2.7M could be associated to the URL of a corresponding book, movie or song. On average, we collected 6 interests per user. What is more, several users have interests in at least two of the three domains. Figure 2 compares the Venn diagram of interest types in our dataset (left) with that reported in [29] (right), to demonstrate the superior coverage of our dataset, even when considering only preferences extracted from users' messages. The last line of Tables 2 and 3 (precision) shows that the methodology to extract and map preferences from messages is very reliable. We evaluated the precision (two judges with adjudication) on a randomly selected balanced sample of 1200 songs, books, and movies in English, obtaining a precision of 96% with a k-Fleiss Inter Annotator Agreement (IAA) of 1[8]. For the Italian dataset, we evaluated 750 songs, books, and movies, obtaining a precision of 98%, and a k-Fleiss of 0.97.

Table 2. 6-months (April–September 2017) statistics on **message-based** interests extracted from English-speaking users

message-based interests ($\|U\|$ = 444,744 English speaking users)				
Platform:	Music	Books	Movie	Total
	Spotify	Goodreads	IMDb	All
#crawled tweets (tweets with selected hashtags)	19,941,046	693,975	97,772	20,732,793
#cleaned tweets (tweets fro which an URL was extracted)	2,519,166	139,882	88,355	2,747,403
# of unique interests with a mapping to a Wikipage	253,311	20,710	8,282	282,303
average #interests per user	6	8	6	6
average #users per interest	7	3	7	6
precision of Wikipedia mapping (on 3 samples of 400 items each)	94%	96%	97%	**96%**

Table 3. 6-months (April–September 2017) statistics on **message-based** interests extracted from Italian-speaking users

message-based interests ($\|U\|$ = 25,135 Italian speaking users)				
Platform	Music	Books	Movie	Total
	Spotify	ANobii	IMDb TVShowTime	All
#crawled tweets (tweets with selected hashtags)	273,256	12,198	2,229	287,683
#cleaned tweets (tweets for which an URL was extracted)	70,330	12,193	2,119	84,642
# of unique interests with a mapping to a Wikipage	9,926	4,690	279	14,895
average #interests per user	3	9	7	6
average #users per interest	5	2	5	4
precision of Wikipedia mapping (on 3 samples of 250 items each)	96%	98%	100%	**98%**

The number and variety of extracted preferences is mostly determined by the interests induced from users' topical friends, as shown in Table 4 (Table 5 for the Italian dataset). The average number of interests induced for each user is as high as 82, and the distribution is shown in Fig. 3, left (English stream), and right (Italian stream). Figure 3 (left) shows, e.g., that there are 100,000 users in U with $\geq$100 interests induced from their topical friends. As far as the

[8] The evaluation is rather straightforward, as readers may verify inspecting the released dataset and mappings.

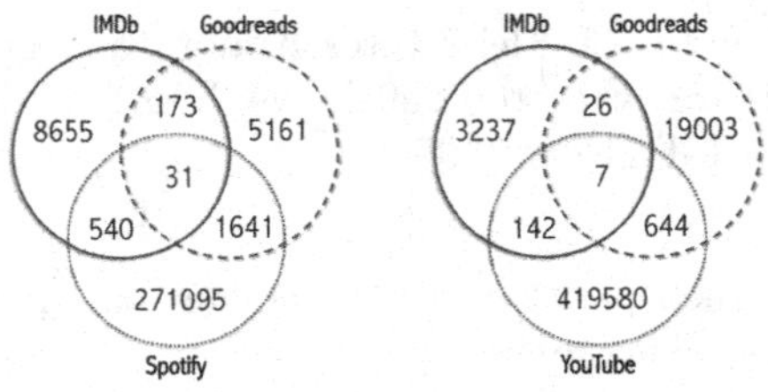

Fig. 2. Venn Diagram of message-based interest *types* for our English dataset (left) and the dataset in Dooms et al. (right)

Table 4. 6-months (April-September 2017) statistics on interests induced from **topical friends** of English-speaking users

Interests induced from topical friends ($\|U\|$ = 444,744 English speaking users)	
# of topical friends F'_t with indegree ≥ **40** in U	**409,743**
# of unique interests with a mapping to a Wikipage	58,789
average #interests per user	82
precision of Wikipedia mapping (tested on a sample of 1,250 items in F'_t)	90%

topical interests mapping performance is concerned, in [35] we estimated that inducing interests from topical friends and subsequent mapping to Wikipedia with mapping method $M1$ has an accuracy of 84%. Since our aim in this work is to generate a highly accurate dataset, first, we used an ensemble of methods, as detailed in Sect. 5.2, and furthermore, we considered only the subset F'_t in F_t with indegree (with respect to our population U) higher than 40. In fact, we noted that less popular topical friends may still include bloggers or Twitter users for which, despite some popularity, a Wikipage does not exist. In these cases, our methodology may suggest false positives. When applying the indegree filter, the precision - manually evaluated with adjudication on 1250 accounts randomly chosen in this restricted population F'_t - is as high as 90%, as shown in the last line of Tables 4 and 5. The k-Fleiss IAA are 0.95 and 0.92, respectively. We remark that we are not concerned here with measuring the recall, since the objective is to release a dataset with *high precision and high coverage*, in terms of number of interests per user, over the considered populations. To this end, the indegree threshold 40 was selected upon repeated experiments to obtain the best

Table 5. 6-months (April–September 2017) statistics on interests induced from **topical friends** of Italian-speaking users

Interests induced from topical friends ($\|U\|$ = 25,135 Italian speaking users)	
# of topical friends F'_t with indegree ≥ **42** in U	**29,075**
# of unique interests with a mapping to a Wikipage	4,580
average #interests per user	41.96
precision of Wikipedia mapping (tested on a sample of 1,250 items in F'_t)	90%

trade-off between the distribution of interests in the population U and precision of Wikipedia mapping.

Concerning coverage, when merging the two sources of information, our English dataset includes an average of 90 interests per user for about 450k users, and a total of $282,303 + 58,789 = 341,092$ unique interests in a large variety of domains. As a comparison, even when considering single domains, the largest available datasets[9], like MovieLens and Bookcrossing, do not exceed 150,000 users and 250,000 items, with a much lower density in terms of interests per user -although these resources provide ranked preferences rather than unary, as in WikiMED. Even the popular Million Songs Dataset Challenge [18] consists of a larger set of users (1.2 million users) but a comparable number of unique interests in a single domain (380,000 songs). To the best of our knowledge, *this is the largest freely available multi-domain interest dataset reported in literature, and furthermore, we provide the unique feature of a reliable mapping to Wikipedia.*

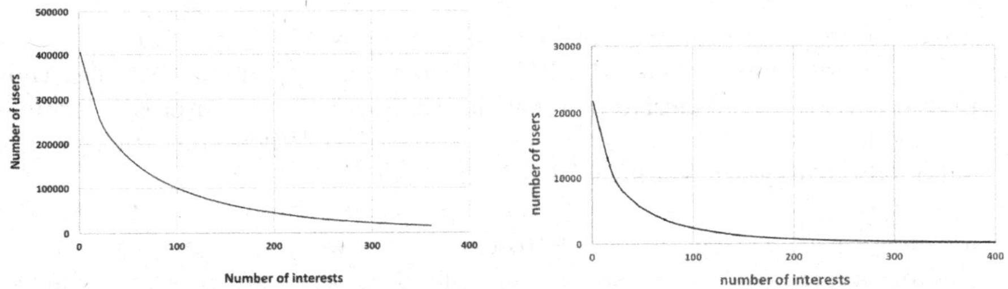

Fig. 3. Distribution of interests induced from users' topical friends: English dataset (left) and Italian dataset (right).

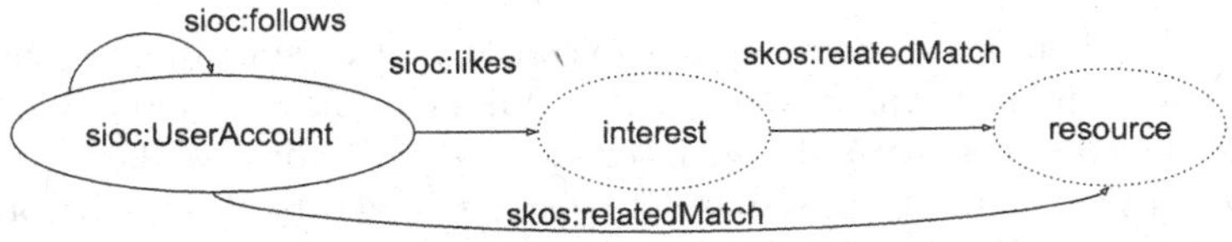

Fig. 4. The data model adopted for the design of our resource.

7 The Wiki-MID Resource

Our resource is designed on top of the Semantically-Interlinked Online Communities (SIOC) core ontology.[10] The SIOC ontology favors the inclusion of data

[9] https://www.kdnuggets.com/2016/02/nine-datasets-investigating-recommender-systems.html.

[10] http://rdfs.org/sioc/spec/sioc.html.

mined from social networks communities into the Linked Open Data (LOD) cloud. As shown in Fig. 4 we represent Twitter users as instances of the SIOC *UserAccount* class. Topical users and message based user interests are then associated, through the usage of the Simple Knowledge Organization System Namespace Document (SKOS)[11] predicate *relatedMatch*, to a corresponding Wikipedia page as a result of our automated mapping methodology. We release at http://wikimid.tweets.di.uniroma1.it/wikimid/ both the dataset and the related software under Creative Commons Attribution-Non Commercial-Share Alike 4.0 License.

8 Concluding Remarks

In this paper we presented Wiki-MID, a LOD-compliant resource that captures Twitter users' interests in multiple domains. With respect to other available datasets for Recommender Systems, our resource has several unique features:

(1) users' interests are induced from their messages and authoritative ("topical") friends, and associated with corresponding Wikipedia articles, thus providing a mean to derive a semantic categorization of interests through the exploitation of available resources linked to Wikipedia, such as the Wikipedia Category Graph, DBPedia, BabelNet, and others;

(2) for every user, we are hence able to extract in two languages (English and Italian) a variety of interests in multiple categories, such as art, science, entertainment, politics, sport and more;

(3) the dimension of the dataset is comparable with the largest single-domain interest datasets in literature, and the average number of multi-domain interests per user is far more large than other multi-domain datasets.

Further note, as shown in Sect. 6, that extracting interests from messages and topical friends, and subsequent mapping to Wikipedia, is a very reliable process (4% error rate for message-induced interests and 10% for friendship-induced). In addition, the availability of semantic resources linked to Wikipedia offers the possibility to identify for each user the "dominant" interest *categories*, on which recommenders could rely when suggesting new items. We leave to future research the exploitation of these features.

Acknowledgments. This work has been supported by the IBM Faculty Award #2305895190 and by the MIUR under grant "Dipartimenti di eccellenza 2018–2022" of the Department of Computer Science of Sapienza University.

[11] http://www.w3.org/2004/02/skos/core.html.

References

1. Linden, G., Smith, B., York, J.: Amazon.com recommendations: item-to-item collaborative filtering. IEEE Internet Comput. **7**(1), 76–80 (2003)
2. Davidson, J., Liebald, B., Liu, J., et al.: The youtube video recommendation system. In: Proceedings of the 4th RecSys, pp. 293–296. ACM (2010)
3. Fouss, F., Saerens, M.: Evaluating performance of recommender systems: an experimental comparison. In: International Conference on WI-IAT 2008, vol. 1, pp. 735–738. IEEE (2008)
4. Felfernig, A., Jeran, M., Ninaus, G., Reinfrank, F., Reiterer, S., Stettinger, M.: Basic approaches in recommendation systems. In: Robillard, M.P., Maalej, W., Walker, R.J., Zimmermann, T. (eds.) Recommendation Systems in Software Engineering, pp. 15–37. Springer, Heidelberg (2014). https://doi.org/10.1007/978-3-642-45135-5_2
5. Schafer, J.B., Frankowski, D., Herlocker, J., Sen, S.: Collaborative filtering recommender systems. In: Brusilovsky, P., Kobsa, A., Nejdl, W. (eds.) The Adaptive Web. LNCS, vol. 4321, pp. 291–324. Springer, Heidelberg (2007). https://doi.org/10.1007/978-3-540-72079-9_9
6. Pazzani, M.J., Billsus, D.: Content-based recommendation systems. In: Brusilovsky, P., Kobsa, A., Nejdl, W. (eds.) The Adaptive Web. LNCS, vol. 4321, pp. 325–341. Springer, Heidelberg (2007). https://doi.org/10.1007/978-3-540-72079-9_10
7. Trewin, S.: Knowledge-based recommender systems. Encycl. Libr. Inf. Sci. **69**(Suppl. 32), 180 (2000)
8. Burke, R.: Hybrid recommender systems: survey and experiments. User Model. User Adapt. Interact. **12**(4), 331–370 (2002)
9. Gunawardana, A., Shani, G.: A survey of accuracy evaluation metrics of recommendation tasks. JMLR **10**, 2935–2962 (2009)
10. Dror, G., Koenigstein, N., Koren, Y., Weimer, M.: The yahoo! music dataset and KDD-cup'11. In: Proceedings of KDD Cup 2011, pp. 3–18 (2012)
11. Shepitsen, A., Gemmell, J., Mobasher, B., Burke, R.: Personalized recommendation in social tagging systems using hierarchical clustering. In: RecSys 2008. ACM (2008)
12. Kamishima, T., Akaho, S.: Nantonac collaborative filtering: a model-based approach. In: Proceedings of the 4th RecSys, pp. 273–276. ACM (2010)
13. Sawant, S., Pai, G.: Yelp food recommendation system (2013)
14. Wang, H., Lu, Y., Zhai, C.: Latent aspect rating analysis on review text data: a rating regression approach. In: Proceedings of the 16th ACM SIGKDD, pp. 783–792 (2010)
15. Mavalankar, A.A., et al.: Hotel recommendation system. Internal Report (2017)
16. Çano, E., Morisio, M.: Characterization of public datasets for recommender systems. In: IEEE 1st International Forum on RTSI, pp. 249–257. IEEE (2015)
17. Harper, F.M., Konstan, J.A.: The movielens datasets: history and context. In: TiiS 2016 (2016)
18. McFee, B., Bertin-Mahieux, T., Ellis, D.P., Lanckriet, G.R.: The million song dataset challenge. In: Proceedings of the 21st WWW, pp. 909–916. ACM (2012)
19. Bennett, J., Lanning, S., et al.: The netflix prize. In: Proceedings of KDD, New York (2007)
20. Yan, M., Sang, J., Xu, C.: Mining cross-network association for youtube video promotion. In: Proceedings of the 22nd ACM MM, pp. 557–566. ACM (2014)

21. Piao, G., Breslin, J.G.: Inferring user interests in microblogging social networks: a survey. arXiv:1712.07691v3 (2017)
22. Chaabane, A., Acs, G., Kaafar, M.A., et al.: You are what you like! information leakage through users' interests. In: Proceedings of the 19th NDSS Symposium (2012)
23. Faralli, S., Stilo, G., Velardi, P.: Large scale homophily analysis in Twitter using a twixonomy. In: Proceedings of 24th IJCAI, Buenos Aires, 25–31 July 2015, pp. 2334–2340 (2015)
24. Piao, G., Breslin, J.G.: Inferring user interests for passive users on Twitter by leveraging followee biographies. In: Jose, J.M. (ed.) ECIR 2017. LNCS, vol. 10193, pp. 122–133. Springer, Cham (2017). https://doi.org/10.1007/978-3-319-56608-5_10
25. Pichl, M., Zangerle, E., Specht, G.: #Nowplaying on #Spotify: leveraging spotify information on Twitter for artist recommendations. Current Trends in Web Engineering. LNCS, vol. 9396, pp. 163–174. Springer, Cham (2015). https://doi.org/10.1007/978-3-319-24800-4_14
26. Kapanipathi, P., Jain, P., Venkataramani, C., Sheth, A.: User interests identification on Twitter using a hierarchical knowledge base. In: Presutti, V., d'Amato, C., Gandon, F., d'Aquin, M., Staab, S., Tordai, A. (eds.) ESWC 2014. LNCS, vol. 8465, pp. 99–113. Springer, Cham (2014). https://doi.org/10.1007/978-3-319-07443-6_8
27. Schinas, E., et al.: Eventsense: capturing the pulse of large-scale events by mining social media streams. In: Proceedings of the 17th PCI, pp. 17–24. ACM (2013)
28. Nichols, J., Mahmud, J., Drews, C.: Summarizing sporting events using Twitter. In: Proceedings of the 2012 International Conference on Intelligent User Interfaces, pp. 189–198. ACM (2012)
29. Dooms, S., De Pessemier, T., Martens, L.: Mining cross-domain rating datasets from structured data on Twitter. In: Proceedings of the 23rd WWW, pp. 621–624. ACM (2014)
30. Barbieri, N., Bonchi, F., Manco, G.: Who to follow and why: link prediction with explanations. In: Proceedings of the 20th ACM SIGKDD, pp. 1266–1275. ACM (2014)
31. Myers, S.A., Leskovec, J.: The bursty dynamics of the Twitter information network. In: Proceedings of the 23rd WWW, pp. 913–924. ACM (2014)
32. Pichl, M., Zangerle, E., Specht, G.: Combining spotify and Twitter data for generating a recent and public dataset for music recommendation. In: Grundlagen von Datenbanken, pp. 35–40 (2014)
33. Besel, C., Schlötterer, J., Granitzer, M.: Inferring semantic interest profiles from Twitter followees: does twitter know better than your friends? In: SAC 2016 (2016)
34. Nechaev, Y., Corcoglioniti, F., Giuliano, C.: SocialLink: linking DBpedia entities to corresponding Twitter accounts. In: d'Amato, C., et al. (eds.) ISWC 2017. LNCS, vol. 10588, pp. 165–174. Springer, Cham (2017). https://doi.org/10.1007/978-3-319-68204-4_17
35. Faralli, S., Stilo, G., Velardi, P.: Automatic acquisition of a taxonomy of microblogs users' interests. J. Web Semant. **45**, 23–40 (2017)
36. Navigli, R., Ponzetto, S.P.: BabelNet: the automatic construction, evaluation and application of a wide-coverage multilingual semantic network. AI **193**, 217–250 (2012)
37. Delli Bovi, L., Telesca, L., Navigli, R.: Large-scale information extraction from textual definitions through deep syntactic and semantic analysis. TACL **3**, 529–543 (2015)

HeLiS: An Ontology for Supporting Healthy Lifestyles

Mauro Dragoni(✉), Tania Bailoni, Rosa Maimone, and Claudio Eccher

FBK-IRST, Trento, Italy
{dragoni,tbailoni,rmaimone,cleccher}@fbk.eu

Abstract. The use of knowledge resources in the digital health domain is a trending activity significantly grown in the last decade. In this paper, we present HeLiS: an ontology aiming to provide in tandem a representation of both the food and physical activity domains and the definition of concepts enabling the monitoring of users' actions and of their unhealthy behaviors. We describe the construction process, the plan for its maintenance, and how this ontology has been used into a real-world system with a focus on "Key to Health": a project for promoting healthy lifestyles on workplaces.

1 Introduction

Chronic diseases, such as heart disease, cancer, and diabetes, are responsible for approximately 70% of deaths among Europe and U.S. each year and they account for about 75% of the health spending[1,2]. Such chronic diseases can be largely preventable by eating healthy, exercising regularly, avoiding (tobacco) smoking, and receiving preventive services. Prevention at every stage of life would help people stay healthy, avoid or delay the onset of diseases, keep diseases they already have from becoming worse or debilitating; it would also help people lead productive lives and, at the end, reduce the costs of public health.

People can start their own prevention process by simply monitoring their lifestyles, in terms of dietary habits and physical activities they do. In order to support this, structured resources able to combine all information and to support the integration of monitoring facilities have to be developed.

In this paper, we address this challenge from the knowledge perspective by presenting HeLiS[3], an ontology aiming to provide an integrated representation of foods, physical activities, good practices, user preferences and habits in order to support the promotion of healthy lifestyles. The relevance of the HeLiS ontology with respect to the state of the art pivots around the integrated model representing (i) a fine-grained description of foods at a level that is not present in the state of the art; (ii) physical activities at the metabolic level enabling the definition of relationships with food entities; and, (iii) user profiles described through

[1] http://www.who.int/nmh/publications/ncd_report_full_en.pdf.
[2] https://www.cdc.gov/media/releases/2014/p0501-preventable-deaths.html.
[3] http://w3id.org/helis.

D. Vrandečić et al. (Eds.): ISWC 2018, LNCS 11137, pp. 53–69, 2018.
https://doi.org/10.1007/978-3-030-00668-6_4

their physical status and possible allergies or pathologies. Moreover, the HeLiS ontology provides a flexible support to rules modeling that can be used for the reasoning on data provided by users. Besides the conceptual model per se, the HeLiS ontology represents a valuable resource for the healthcare domain thanks to the knowledge included into the provided resource.

Section 2 provides a brief overview of the main ontologies concerning the domains of food and physical activities. Then, in Sect. 3 we describe the HeLiS ontology by illustrating the methodology we followed and the main entities of the conceptual model. In Sect. 4, we show how to get and reuse the ontology together with examples of future projects that will integrate the ontology, while Sect. 5 presents an instantiation example of the HeLiS ontology into a real-world application. Section 6 discusses the sustainability and maintenance aspects, and, finally, Sect. 7 concludes the paper.

2 Related Work

We provide in this Section a brief summary of the most relevant work on ontologies describing both the food and the physical activity domains.

In [1] the authors describe food intake patterns identified by applying new food categories, in particular: (i) nutrient composition and energy density, (ii) current scientific evidence of health benefits, and (iii) culinary use of each food. In [2], a process is presented for a rapid prototyping of a food ontology oriented to the nutritional and health care domain that is used for sharing existing knowledge. However, unfortunately, this resource is no longer available.

The contribution presented in [3] discusses the design and development of a food-oriented ontology-driven system (FOODS), used for food or menu planning in a restaurant, clinic/hospital, or at home. FOODS comprises (i) a food ontology, (ii) an expert system using such an ontology and some knowledge about cooking methods and prices, and (iii) a user interface suitable for users with different levels of expertise. Its aim is to support the management of treatment plans for patients affected by type 1 or type 2 diabetes. Instead, the work presented in [4] focuses on the integration of different domain ontologies, like food, health, and nutrition, in order to help personalized information systems to retrieve food and health recommendations based on the user's health conditions and food preferences. Recently, the work presented in [5] describes an ontology modeling the protected names of brands, from the raw materials to the production process.

A set of ontologies have been proposed that collect information about packaged food. Examples are Open Food Facts[4] and Food Product Ontology [6]. However, their focus on categorizing and describing packaged food led to low coverage of concepts describing food compositions.

Recently, the FoodOn ontology[5] has been released. This ontology represents foods from a different perspective with respect to the HeLiS ontology. Instead of

[4] Open Food Facts. Available online: http://world.openfoodfacts.org/who-we-are.

[5] http://foodontology.github.io/foodon/.

focusing on food composition, they aim to realize a food description system that registers food manufacturers. Indeed, the FoodOn ontology includes, for each product, its origin, the physical attributes, processing, packaging, dietary uses and geographical origin.

Finally, in [7] the design steps are described, the working mechanism, and the case of use of the Ontology-Driven Mobile Safe Food Consumption System (FoodWiki) using semantic matching. This resource aims to address problems similar to the HeLiS ontology. However, no information about physical activities and their correlation with food categories are included in the ontology nor the possibility of modeling in a flexible way rules users should follow and the possible associated violations.

Concerning physical activity, we report two ontologies, both available through the BioPortal[6] website.

The first one is the SMASH (Semantic Mining of Activity, Social, and Health data)[7] ontology. The goal of the SMASH ontology is to describe concepts correlating physical activities and social networks. The system developed upon this ontology aims to sustain weight loss with continued intervention with frequent social contacts. The coverage of the SMASH ontology is very limited. Indeed, only 18 activities are defined.

The second one is the Ontology of Physical Exercises (OPE)[8]. Here, physical activities are modeled from the functional perspective. Thus, exercises are described in terms of movements, how the different musculoskeletal parts of the human body are engaged, and which are the expected health outcomes. Also in this case, the coverage of the physical activity domain is limited because only general categories of activities, like *AerobicExercise* or *IsotonicExercise*, are defined.

With respect to the state of the art, the main contribution of the HeLiS ontology is twofold. First, the coverage is definitely wider in both the food and the physical activity domains. Second, the HeLiS ontology defines concepts enabling the monitoring of users' actions and the representation of their unhealthy behaviors.

3 The HeLiS Ontology

The development of the HeLiS ontology followed the need of providing a knowledge artifact able not only to provide a representation of domains concerning healthier lifestyles, but, also, to support further activities like, for example, remote medical monitoring. The proposed ontology has been modeled with a focus on the connection between diet and physical behaviors with people health.

The process for building the HeLiS ontology followed the METHONTOLOGY [8] methodology. This approach is composed by seven stages: Specification,

[6] http://bioportal.bioontology.org/.
[7] http://bioportal.bioontology.org/ontologies/SMASHPHYSICAL.
[8] https://bioportal.bioontology.org/ontologies/OPE.

Knowledge Acquisition, Conceptualization, Integration, Implementation, Evaluation, and Documentation. The overall process involved three knowledge engineers and six domain experts from the Trentino Healthcare Department. More precisely, two knowledge engineers and four domain experts participated to the ontology modeling stages (hereafter, the modeling team). While, the remaining knowledge engineer and two domain experts were in charge of evaluating the ontology (hereafter, the evaluators).

The choice of METHONTOLOGY was driven by the necessity of adopting a lifecycle split in well-defined steps. The development of the HeLiS ontology requires the involvement of the experts in-situ. Thus, the adoption of a methodology having a clear definition of the tasks to perform was preferred. Other methodologies, like DILIGENT [9] and NeOn [10], were considered before starting the construction of the HeLiS ontology. However, the characteristics of such methodologies, like the emphasis on the decentralized engineering, did not fit our scenario well.

Specification. The purpose of the HeLiS ontology is two-fold. On the one hand, we want to provide a detailed and integrated model of the food and physical activity domain. On the other hand, we want to support a synergistic analysis of user data leading to different exploitation possibilities. Examples range from a simple report concerning dietary information to a complex analysis of users' behaviors according to a set of rules provided by domain experts.

The HeLiS ontology is modeled with a *high* granularity level. Concerning food representation, we modeled food composition till micro-nutrients level. While concerning physical activities, we classify them by their categories or by their effort levels. The latter allows to report precise information about the calories spent in one minute for each kilogram of body weight. Thanks to this granularity level, we favor the integration of the HeLiS ontology into several solutions going from simple mobile applications to expert systems.

Knowledge Acquisition. The acquisition of the knowledge necessary for building the HeLiS ontology has been done in two steps: (i) the analysis of unstructured resources containing information of interest about the food and physical activity domains; and, (ii) discussions with domain experts for deciding how to model classes, individuals and properties exploited to support the monitoring activity.

Information concerning both the food and physical activity domains has been modeled by starting from the following unstructured resources: (i) the archives of the Italian Minister of Agriculture[9] and of the Italian Epidemiological department[10] to collect information about the composition of basic foods and nutrients; (ii) the Turconi's atlas [11] to acquire information about recipes and to map recipe's ingredients with basic food's instances; and, (iii) the Compendium of Physical Activities[11] to create the taxonomy of physical activities and model all

[9] http://nut.entecra.it/.
[10] http://www.bda-ieo.it/.
[11] https://sites.google.com/site/compendiumofphysicalactivities/home.

information concerning the associated effort. In this step, we drafted, with the support of the domain experts, the main properties that have to be associated with entities.

The second step consisted in defining the proper entities enabling the reasoning on the data provided by users. Here, we defined two main entities (the *MonitoringRule* and the *Violation* concepts, described in Sect. 3.1), that (i) support the definition of rules used for monitoring users' behavior, and, (ii) allow the representation of violations associated with these rules. Violations instances can then be used as input for other services or applications. Properties associated with these concepts and their possible values have been designed according with the guidelines provided by the experts.

Conceptualization. The conceptualization of the HeLiS ontology was split into two steps. The first one was covered by the knowledge acquisition stage, where most of the terminology is collected and directly modeled into the ontology. Examples are the food and physical activity categories and the name of nutrients. While the second step consisted in deciding how to represent, as classes or as individuals, the information we collected from unstructured resources. Then, we modeled the properties used for supporting all the requirements.

During this stage we relied on several ontology design patterns (ODP) [12]. However, in some cases we renamed some properties upon the request of domain experts. In particular, we exploit the logical patterns *Tree* and *N-Ary Relation*, the alignment pattern *Class Equivalence*, and the content patterns *Parameter*, *Time Interval*, *Action*, *Classification*.

Integration. The integration of the HeLiS ontology has two objectives: (i) to align it with a foundational ontology, and (ii) to link it with the Linked Open Data (LOD) cloud. The first objective was satisfied by aligning the main concepts of the HeLiS ontology with ones defined within the DOLCE [13] top-level ontology. While, the second objective was satisfied by aligning our ontology with AGROVOC[12]. Recently AGROVOC has been included within the LOD cloud. This way, it may work as a bridge between the latter and the HeLiS ontology.

Implementation. The HeLiS ontology is represented by using the RDF/XML language in order to provide a formal representation enabling the check of inconsistencies, the visualization of ontology structure, and the ease of maintenance. The editing of the ontology is demanded to the MoKi tool [14], while the exposure of the ontology is granted by the services available from the HeLiS ontology website.

Evaluation. The evaluation procedure was conducted by one knowledge engineer and two domain experts that did not participate to the modeling process. To evaluate our ontology we adopted the metrics described in [15–19]: *Accuracy*, *Adaptability*, *Clarity*, *Completeness*, *Computational Efficiency*, *Conciseness*, *Consistency/Coherence*, and *Organizational fitness*.

[12] http://aims.fao.org/vest-registry/vocabularies/agrovoc.

The overall *Accuracy* of the ontology has been judged as good. The knowledge of the domain experts was in-line with the complexity of the use axioms. Indeed, within the HeLiS ontology there are not very complex axioms. Then, by considering the representation of the real world, the evaluators agreed on the correctness of the ontology in describing the domain.

Concerning the *Adaptability* of the ontology, the evaluators focused on the possible extension aspects. They verified that the ontology can be extended and specialized monotonically. Here, the question has to be addressed from two perspectives. Firstly, concerning the extension of the ontology from the content perspectives (i.e., adding new foods, recipes, activities, dietary profiles, etc.), the result was positive because any extension of the ontology did not require to remove any axiom. Secondly, concerning the representation of users' profiles, the update of the ontology was not monotonic because if a user is associated to a new profile the old association is removed. Anyway, the ontology does not react negatively to these changes because its consistency is preserved.

About the *Clarity* of the ontology, the evaluators agreed with the strategy decided by the modeling team about using concept labels communicating the intended meaning of each concept and the use of definitions and descriptions of the main concepts of the ontology, especially for the root concepts of each branch. Moreover, each definition has been well documented within the ontology in order to make the meaning of each concept understandable by who uses the ontology.

The experts agreed about the *Completeness* of the HeLiS ontology. However, they distinguished among the TBox and the ABox. Indeed, concerning the TBox, the evaluators agreed about the completeness of the ontology and the lexical representations of the concepts. In particular, they verified that all the represented nutrients appropriately cover the health domain and that all the information needed for the realization of tools supporting a healthy lifestyle were modeled within the ontology. While, regarding the ABox, the evaluators highlighted the necessity of including individuals concerning commercial products. This observation is interesting, especially, if we consider the possibility of developing end-users applications. Indeed, the presence of commercial products will improve the overall user engagement.

In order to verify the *Computational efficiency* of HeLiS, we observed how the ontology behaved within the scenario described in Sect. 5. Indeed, the HeLiS ontology itself does not contain axioms representing a criticism for reasoners. On the contrary, the final aim of the ontology is to be used for analyzing data provided by users. In Sect. 5 we show an example about how the ontology is used and we provide statistics regarding the amount of time needed for completing the reasoning activity with respect to the dimension of elaborated data.

The evaluators judged the HeLiS ontology *Concise* because all the axioms included are relevant with regards to the healthy lifestyle domain and there are no redundancies.

The HeLiS ontology has been judged, also, *Consistent* and *Coherent*. Consistent because no contradictions were found by the evaluators. Coherent because the evaluators observed little bias between the documentation containing the informal description of the concepts and their formalization.

Finally, concerning the *Organizational fitness*, the HeLiS ontology has been deployed within the organization as a web service in order to make it easily accessible by the community and potential stakeholders. Moreover, as described in Sect. 6, the ontology has been deployed also, within external architectures. A focus group has been organized with both the modeling team and the evaluators for discussing about the adopted methodology, that was judged appropriate by considering the necessity of working in-situ all together and of synchronizing the commitments of all the people involved.

Documentation. The documentation of the HeLiS ontology has been done from two perspectives. First, during the whole modeling process, a document has been prepared by the people involved in the construction process. This activity was necessary because, as we will mention in Sect. 6, the development of the HeLiS ontology and its sustainability are granted by a public funding program. Thus, all performed steps were documented and archived within the funding dossier. Second, in order to ease the readiness of the ontology for users, we provided a different documentation file generated by using the LODE[13] system and available on the ontology website.

3.1 Inside the HeLiS Ontology

The ontology contains six root concepts: *Activity*, *Food*, *MonitoringEntity*, *Nutrient*, *TemporalEvent*, and *UserEvent*. Beside these, we also defined the *User* concept that does not play the role of superclass of any concept but that is fundamental for associating specific events with the people did them.

Figure 1 shows a general overview of the ontology with the main concepts.

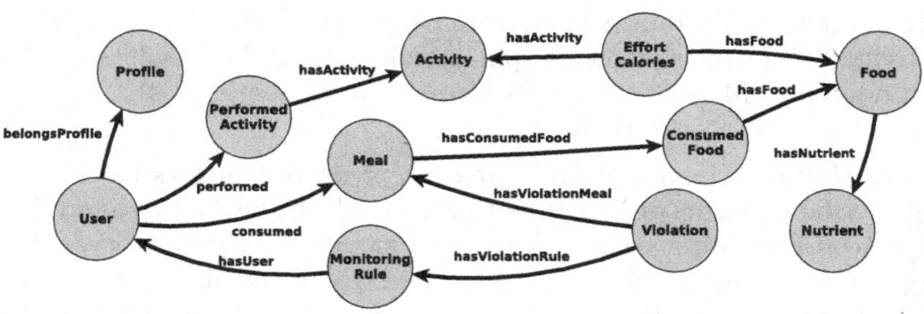

Fig. 1. Overview of the HeLiS ontology.

Below, we detail the entities associated with each root concept by providing the semantic meaning of the most important entities.

Food and Nutrient. The *Food* root concept subsumes two macro-groups of entities descending from *BasicFood* and *Recipe* concepts. Instances of the

[13] http://www.essepuntato.it/lode.

BasicFood concept describe foods for which micro-information concerning nutrients (carbohydrates, lipids, proteins, minerals, and vitamins) is available, while instances of the *Recipe* concept describe the composition of complex dishes (such as *Lasagna*) by expressing them as a list of instances of the *RecipeFood* concepts. This concept reifies the relationships between each *Recipe* individual, the list of *BasicFood* it contains and the amount of each *BasicFood*. Besides this dual classification, instances of both *BasicFood* and *Recipe* concepts are more fine-grained categorized. While, concerning the number of individuals, currently, the HeLiS ontology contains 986 individuals of type *BasicFood* and 4408 individuals of type *Recipe*.

Entities subsumed by the *BasicFood* concepts are the range of the *hasMonitoredEntityType* object property linking each individual of the *MonitoringRule* concept (described below) with a specific category of food. We can also notice that all instances of *BasicFood* and *Recipe* concepts are within the domain of the *hasPositiveProperty* and *hasNegativeProperty* data properties having as range a string value.

Beside the food-related concepts, the classification of nutrients is also defined. The *Nutrient* concept subsumes 81 different type of nutrients properly categorized. Nutrients are instantiated with through individuals describing a specific amount of a nutrient. Then each *BasicFood* is linked to all the necessary nutrients' individuals through the *hasNutrient* object property.

Activity. The second groups of entities relates to physical activities. The *PhysicalActivity* concepts subsumes 21 subclasses representing likewise physical activity categories and a total of 856 individuals each one referring to a different kind of activity. For each activity, we defined datatype properties providing the amount of calories consumed in one minute for each kilogram of weight and the MET (Metabolic Equivalent of Task) value expressing the energy cost of the activity. MET values allow to split activities in *LightActivity* (MET < 3), *ModerateActivity* (MET $[3, 6]$), and *VigorousActivity* (MET > 6).

TemporalEvent. The *TemporalEvent* concept defines entities used for representing specific moments or delimited timespan which the data to analyze refers to. These concepts are used in two ways. First, when users provide data concerning the food consumption, these data have to be associated with a specific temporal event that, in the case of food consumption, is the *Meal* concept. In turn, the *Meal* concept subsumes other concepts defining specific kind of meals (i.e. *Breakfast*, *Snack*, *Lunch*, and *Dinner*). Second, the other descendant of the *TemporalEvent* concept is the *Timespan* one. Instances of the children of *Timespan* are used for driving the data selection and reasoning operations to a specific portion of data. Indeed, as shown later in this Section and also in Sect. 5, the definition of a rule must contain the link a *TemporalEvent* instance.

UserEvent. This concept subsumes the conceptualization of information that a user can provide, i.e. food consumption and performed activities, and also links them with the possible violation that can be generated after their analysis. Concerning the representation of users' activities and personalized information, we modeled the *ConsumedFood* and the *PerformedActivity* concepts. Both concepts are used as reification of the fact that a user consumed a specific quantity of a food or performed an activity for a specific amount of time. In the first case, every *Meal* is associated with a list of *ConsumedFood* through the *hasConsumedFood* object property. While, in the second case, instances of the *PerformedActivity* concept associate a user with the amount of time he/she spent in performing a specific activity. Here, we did not use a concept for grouping a list of activities a user is routinely doing (including sleeping), thus a further concept for grouping activities is useless. Then, we included also the possibility of providing user-specific information representing the energy consumption equivalence, e.g. how much a specific user has to run for burning 100 grams of pasta. This information is represented by instantiating the *EffortCaloriesEquivalent* concept (EffortCalories in Fig. 1). Then, instances of the *User* concept are used as object for the *hasViolationUser* object property defined on the *Violation* concept (described below).

For the sake of clarity, we provide below two examples presenting how these concepts are used. Figure 2 shows are the *User* and the *ConsumedFood* concepts are linked. While Fig. 3 shows this link exists between the *User* and the *PerformedActivity* concepts.

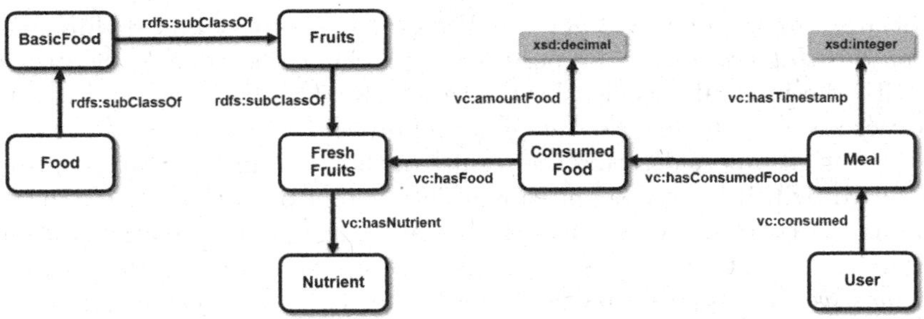

Fig. 2. Example describing the representation of a user eating some fresh fruits.

MonitoringEntity. Concepts subsumed by the *MonitoringEntity* one are responsible for modeling the knowledge enabling the monitoring of users' behaviors. Here, we can appreciate five concepts: *MonitoringRule*, *Violation*, *Profile*, *Goal*, and *Interval*. The *MonitoringRule* concept provides a structured representation of the parameters inserted by the domain experts for defining how users should behave. First of all, it is necessary to determine the entities affected by the monitoring rule, and the time period to be considered during the rule evaluation process. This information is provided through two object properties and one annotation property: *monitoredEntity* (e.g. *Corn* or *Walk*) and *monitoredEntityType* (e.g. *Food* or *Activity*) object properties; while the time period is provided

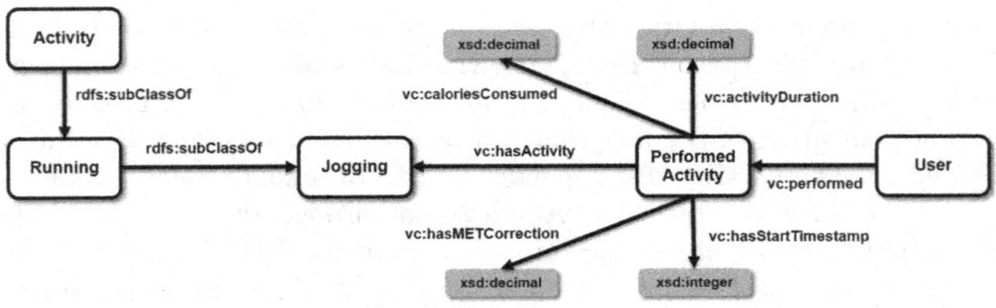

Fig. 3. Example describing the representation of a user performing a running session.

through the *timing* annotation property that may contain the URI of the *Timespan* concept. *MonitoringRule* instances are the directives that can be exploited by a reasoner for analyzing user data. The content of the *command* datatype property specifies how the reasoner has to behave when it analyzes data of type *monitoredEntity*. The *command* is accompanied by the *hasOperator* datatype property that specifies the kind of comparison that the reasoner has to make with respect to the value/s specified through the *hasMonitoredValue* datatype property or the *hasMonitoredInterval* object property. In the first case, a classic comparison is performed between provided data and the values contained in the monitoring rule, while, the second case indicates to the reasoner that the value specified into the rule is not a fixed value, but an interval. If the second case occurs, the reasoner will get the *ValueInterval* object linked through the *hasMonitoredInterval* object property and will check the value of the provided data with the interval specified by the *lowerBound* and *upperBound* datatype properties associated with the *ValueInterval* object.

Violation instances describe the results of the reasoning activities and they can be used by third party applications. The content of each *Violation* instance is computed according to the user data that triggered the violation. Information materialized at runtime is contained in the *hasViolationHistory* and *hasViolationLevel* datatype properties. The former contains the number of times the *MonitoringRule* associated with the generated violation has been already triggered by the user. The latter represents the severity of the violation. Indeed, the knowledge base contains a set of pre-modeled intervals representing different levels of violations, expressed in terms of percentage with respect to the monitored value defined within the rule. When a rule is violated, the reasoner queries the violation intervals for knowing the level of the generated violation. Finally, the *hasTimestamp* datatype property contains the timestamp in which the violation instance is generated.

Each *MonitoringRule* is linked to at least one *Profile* concept. A *Profile* represents a set of rules which a *User* should follow for being compliant with an alimentary guideline. An example of *Profile* is the Mediterranean Diet that contains around 220 dietary rules. While, in case of diseases, a *Profile* may contain a set of rules specifically thought for managing that disease.

The *Goal* concepts represents a specific objective, in the context of a *Profile*, that a user is expected to achieve within a given timestamp. In practice, a *Goal* is composed by a subset of the *MonitoringRule* instances linked to a *Profile*. However, with respect to a *Profile*, that is only linked to a *User* through the *belongsProfile* object property, the definition of a *Goal* includes more information. First of all, a *Goal* is associated with a *Profile* through the *appliesTo* object property. Then, a *Goal* is associated with one or more *MonitoringRule* through the *hasMonitoringRule* object property. The temporal validity of a *Goal* is expressed by the *timing* object property that specifies for how much time a *User* has to respect the violations' limits associated with the *Goal*. The range of this object property is the *TemporalEvent* concept. Finally, we mentioned above that each *Goal* specifies a limit of violations that a *User* must not reach. Thus, a *Goal* is associated with comparison operator (*hasOperator* datatype property) and with a value to monitor (*hasMonitoredValue*). As example, by assuming to have the rule "MR1", the timing of "28 days", the comparison operator "less" and the monitored value 2, it means that a *User* associated with this goal does not have to violate the rule "MR1" more than two times within a period of 28 days.

Finally, the *Interval* concepts subsumes concepts used for describing interval of values. Beside the *ValueInterval* concept introduced above, the current version of HeLiS defines the *ViolationInterval* one. Instances of this concept allows to transform the percentage representing the difference between expected and observed values into discrete levels representing how much a *MonitoringRule* has been violated.

4 Availability and Reusability

The HeLiS ontology is licensed under the Creative Commons Attribution-NonCommercial-ShareAlike 4.0[14] and it can be downloaded from the PerKApp project website at the link reported at the beginning of the paper. The rational behind the CC BY-NC-SA 4.0 is that the Trentino Healthcare Department, that funds the project in which the HeLiS ontology has been developed, was not in favor of releasing this ontology for business purposes. Hence, they force the adoption of this type of license for releasing the ontology. The HeLiS ontology can be downloaded in two different modalities: (i) the conceptual model only, where the user can download a light version of the ontology that does not contain any individual, or (ii) the full package, where the ontology is populated with all the individuals we have already modeled. The HeLiS ontology is constantly updated due to the project activities using the ontology as core component.

The ontology is available also as web service. Detailed instructions are provided on the ontology website. Briefly, the service exposes a set of *informative* methods enabling the access to a JSON representation of the individuals included into the ontology.

[14] https://creativecommons.org/licenses/by-nc-sa/4.0/.

The reusability aspect of the HeLiS ontology can be seen from three perspectives. First, the HeLiS ontology contains structured supervised knowledge about the food and physical activity domains. Provided information has a high granularity and the consistency of the modeled data have been validated by domain experts. Hence, the ontology represents a valuable artifact for the digital health domain. Second, the ontology model represents a relevant resource of medical knowledge. In particular, the ontology contains a set of rules, modeling good practices, related to what a person should eat and which physical activities a person should do for maintaining a good health, and the relationships between food's allergies or intolerances and specific food categories. This kind of knowledge is presented, for the first time, in a structured way and it can be reused in several third-party applications for different purposes. Third, the HeLiS ontology enables the construction of health-based (but not limited to this domain) applications exploiting the whole content of the ontology as well as the sole conceptual model.

5 HeLiS Ontology in Action

The topic area of the HeLiS ontology is timely for researchers and developers working on digital health applications (including mobile) who are now exploring the use of Semantic Web technology in the form of knowledge graphs and rules to analyze the nutrition intake over time, activities, and their association with health risks and symptoms related to chronic diseases.

The HeLiS ontology can be integrated within applications going beyond the mere access to the resource for informative purposes. Indeed, the *Monitoring* branch of the ontology can be populated to properly respond to the needs of the solutions integrating the HeLiS ontology. An instantiation example is the PerKApp platform.

The PerKApp platform[15] develops a personalized healthy lifestyle recommendation system addressing the challenge of monitoring people behaviors with the aim of promoting possible behavioral changes. Here, the HeLiS ontology is used to support the reasoning activities on the data provided by users. The collection and modeling of domain knowledge is performed integrating specific user facilities into the MoKi tool [14], this tool has been used to support the work of domain experts that are responsible for modeling the rules used to validate user data. The inference engine is powered by the RDFPro tool [20], a tool that allows us to consider the aforementioned aspects by supporting out-of-the-box OWL 2 RL and the fixed point evaluation of `INSERT... WHERE...` SPARQL-like entailment rules that leverage the full expressivity of SPARQL (e.g., `GROUP BY` aggregation, negation via `FILTER NOT EXISTS`, derivation of RDF nodes via `BIND`).

Summing-Up Example. Let us consider the following scenario. After a colloquium with her physicians, Michelle has to follow these two rules: (i) the total

[15] http://perkapp.fbk.eu/.

amount of calories for each meal has to be lower than 1000 and (ii) the number of portions of the *SweetBeveragesAndJuices* food category should be restricted to maximum one per day. At first, the doctor created a user associated to Michelle and inserted profile's details. Then, he defined the two rules that the PerKApp system has to validate every day.

```
:Hyperglycemia a :Profile.

:Michelle a :User; :hasUserId "493853"^^xsd:integer; :belongsProfile :Hyperglycemia.

:MR1 a :MonitoringRule;
    :appliesTo :Hyperglycemia; :timing :Meal; :monitoredEntity :Food;
    :command "hasCalories"; :hasOperator "lower"; :hasMonitoredValue "1000"^^xsd:integer;
    :hasRuleId "1"^^xsd:integer; :hasPriority "2"^^xsd:integer.

:MR2 a :MonitoringRule;
    :appliesTo :Hyperglycemia; :timing :Day; :monitoredEntity :SweetBeveragesAndJuices;
    :command "portion"; :hasOperator "greater"; :hasMonitoredValue "1"^^xsd:integer;
    :hasRuleId "2"^^xsd:integer; :hasPriority "1"^^xsd:integer.
```

For the first two days, Michelle inserted the data about her food intake correctly as shown below (for brevity, we reported only the meals, or snacks, that trigger persuasive actions):

```
:Michelle :consumed :Breakfast-493853-1, :Snack-493853-2, :Snack-493853-3, :Dinner-493853-4.
:Michelle :consumed :Breakfast-493853-5, :Meal-493853-6, :Snack-493853-7, :Dinner-493853-8.

:Breakfast-493853-1 a :Breakfast;
    :hasTimestamp "2016-12-14T07:19:00Z";
    :hasConsumedFood [ :hasFood :AlmondMilk; :amountFood "250"^^xsd:integer ],
                     [ :hasFood :RiceFlakes; :amountFood "100"^^xsd:integer ].

:Snack-493853-3 a :Snack;
    :hasTimestamp "2016-12-14T11:34:00Z";
    :hasConsumedFood [ :hasFood :CannedOrangeSoda; :amountFood "300"^^xsd:integer ],
                     [ :hasFood :Apple; :amountFood "150"^^xsd:integer ].

:Breakfast-493853-5 a :Breakfast;
    :hasTimestamp "2016-12-15T07:23:00Z";
    :hasConsumedFood [ :hasFood :FruitJuices; :amountFood "200"^^xsd:integer ],
                     [ :hasFood :RiceFlakes; :amountFood "100"^^xsd:integer ].

:Snack-493853-7 a :Snack;
    :hasTimestamp "2016-12-15T15:52:00Z";
    :hasConsumedFood [ :hasFood :FruitJuices; :amountFood "200"^^xsd:integer ],
                     [ :hasFood :Sandwich; :amountFood "150"^^xsd:integer ].

:Dinner-493853-8 a :Dinner;
    :hasTimestamp "2016-12-15T19:45:00Z";
    :hasConsumedFood [ :hasFood :CocaCola; :amountFood "330"^^xsd:integer ],
                     [ :hasFood :Pizza; :amountFood "450"^^xsd:integer ].
```

Based on the provided data, on the ontology currently available in the PerKApp platform, and on the RDF encoding of the monitoring rules, the PerKApp reasoner determines that the amount of kilo-calories consumed during each meal, except for *Dinner-493853-8*, satisfies the rule *MR1*. However, the reasoner determines that the rule *MR2* has been violated in both days. This event triggers into the knowledge base the assertion of the following *Violation* individuals:

```
:violation-493853-1-20161214 a :Violation;
    :hasViolationRule :MR2; :hasViolationUser :Michelle; :hasViolationMeal :Breakfast-493853-1;
    :hasViolationMeal :Snack-493853-3; :hasViolationQuantity 2;
    :hasViolationExpectedQuantity 1; :hasViolationLevel 1;
    :hasTimestamp "2016-12-14T22:30:00Z"; :hasPriority 1; ...

:violation-493853-1-20161214 a :Violation;
    :hasViolationRule :MR2; :hasViolationUser :Michelle; :hasViolationMeal :Breakfast-493853-5;
    :hasViolationMeal :Snack-493853-7; :hasViolationMeal :Dinner-493853-8;
    :hasViolationQuantity 3; :hasViolationExpectedQuantity 1;
    :hasViolationLevel 2; :hasTimestamp "2016-12-15T22:30:00Z"; :hasPriority 1; ...

:violation-493853-8-20161215 a :Violation;
    :hasViolationRule :MR1; :hasViolationUser :Michelle; :hasViolationMeal :Dinner-493853-8;
    :hasViolationQuantity 1356; :hasViolationExpectedQuantity 1000; :hasViolationLevel 2;
    :hasTimestamp "2016-12-15T19:45:00Z"; :hasPriority 2; ...
```

Among the violations shown above, one is selected by the PerKApp platform as the most appropriate one to generate the feedback, then sent to Michelle. In this particular case, Michelle is advised that she is consuming too much sweet beverages and juices. This kind of feedback aims at reminding Michelle that she has to follow the doctor's suggestions in order to avoid the possible insurgence of diabetes, due to the excessive consumption of sugar.

Violation individuals are generated through reasoner integrated into RDF-pro. Reasoning activities on the HeLiS ontology have the goal of verifying that a user's lifestyle is consistent with the monitoring rules defined by domain experts, detecting and possibly materializing violations in the knowledge base, upon which further actions may be taken. Reasoning can be implemented via the fixed point, forward chaining evaluation of IF-THEN *entailment rules* (cf. *monitoring rules*, which are RDF individuals in the PerKApp ontology) that implement the semantics of OWL 2 RL (to account for TBox declarations in the ontology) and of monitoring rules by matching non-conforming patterns in RDF data and asserting corresponding *Violation* individuals.

Below, we show one of the SPARQL queries we implemented for the PerKApp project. In particular, we report the query used to detect if the portions of a specific food category consumed by a user exceeded the daily quota or not.

```
:check_contains_food_portion_less_day a rr:Rule, rr:NonFixpointRule;
  rr:phase  10;
  rr:insert """ ?v :hasViolationRule ?rule; :hasViolationGoal ?goal; :hasViolationUser ?user;
                   :hasViolationQuantity ?quantity; :hasViolationConstraint ?operator;
                   :hasViolationEntityType ?et; :hasViolationLevel ?level;
                   :hasViolationStartTime ?minTimestamp; :hasViolationEndTime ?timestamp;
                   :hasTimestamp ?timestamp. """;
  rr:where  """ {
                 SELECT ?rule ?goal ?user ?et ?mv (MAX(?mealTs) AS ?timestamp)
                        (MIN(?mealTs) AS ?minTimestamp) (COUNT(DISTINCT ?cf) AS ?quantity)
                 WHERE {
                   ?rule a :MonitoringRule; :timing ?timing; :command "portion";
                           :monitoredEntityType :Food; :hasMonitoredValue ?mv;
                           :monitoredEntity ?class.
                   FILTER EXISTS {?rule :hasOperator "less"}
                   FILTER EXISTS {?rule :timing :Day}
                   {SELECT DISTINCT ?rule ?goal ?user WHERE {
                     {?rule :appliesTo ?user} UNION
                     {?rule :appliesTo ?goal. ?goal ^:belongsProfile ?user.}}}
                   ?meal :hasUser|^:consumed ?user; :hasTimestamp ?mealTs.
                   {?timing rdfs:subClassOf :Timespan} UNION {?meal a ?timing}
```

```
22.         BIND (:mintEntityType(?class) AS ?et)
23.                 ?cf   ^:hasConsumedFood ?meal; :hasFood ?food; :amountFood ?amount .
24.                 FILTER(?amount > 0.0)
25.                 ?food a ?class. }
26.               GROUP BY ?rule ?goal ?user ?et ?mv }
27.             ?user :hasUserId ?userId.
28.             ?rule :hasOperator ?operator; :hasMonitoredValue ?value; :hasRuleId ?ruleId.
29.             FILTER (?operator = "less" && ?quantity >= ?value)
30.             BIND (:computeViolationLevel(?mv, ?quantity) AS ?level)
31.             BIND (:mintViolation(?ruleId, ?userId, ?timestamp) AS ?v) """.
```

Rows from 3 to 7 contain the definition of the *Violation* individual. From row 12 to row 16 the reasoner selects which rules to validate, in this case the ones focusing on the consumption of the number of *portions* of foods (instead of quantities) that should not exceed (*less* operator) the daily (timing *:Day*) limit. Rows from 17 to 19 allows to verify if the current user is associated to a profile linked to the selected *MonitoringRule*. Rows from 23 to 25 check if the food consumed by the current user is of the same type of the food monitored by the rule. Rows 28 and 29 perform trivial checks about the coherence between the rule's operator and the computed quantity. Finally, at rows 30 and 31 we included two functions we implemented to compute the violation level based on the difference between the detected number of portions and the expected one, and for generating the *Violation* identifier.

6 Resource Sustainability and Maintenance

As mentioned in the previous section, the presented ontology is the result of a collaborative work between several experts. While, on the one hand this collaboration led to the development of an effective and useful ontology, on the other hand the sustainability and the maintenance of the produced artifact represent a criticality.

Concerning the sustainability, this ontology has been developed in the context of the *Key to Health* project[16]. The goal of this project is to "combine efforts of employers, employees, and society to improve the mental and physical health and well-being of people at work"[17] and it aims at preventing the onset of chronic diseases related to an incorrect lifestyle through organizational interventions directed to workers. Actions might concern the promotion of correct diet, physical activity, social and individual well-being, as well as the discouragement of bad habits, such as smoking and alcohol consumption. This project, recently started within FBK, aims to promote healthy behaviors on workplaces and it is part of the "Trentino Salute 4.0" framework promoted by the Trentino's local government. One of the goals of this framework is to promote the integration of artificial intelligence solutions into digital health platforms with the long term goal of improving the life quality of citizens. The presented ontology is part of the core technologies used in this framework. The overall sustainability plan for

[16] https://sites.google.com/fbk.eu/keytohealth.

[17] Luxembourg Declaration on Workplace health promotion in the European Union. 1997.

the continuous update and expansion of the HeLiS ontology is granted by this framework and by the projects mentioned in Sect. 4.

The maintenance aspect is managed by the infrastructure available within FBK from both the hardware and software perspectives. In particular, we enable the remote collaboration between experts thanks to the use of the MoKi [14] tool (details about the tool are out of the scope of this paper). Here, it is important only to remark that this tool implements the support for the collaborative editing of ontologies by providing different views based on the kind of experts (domain expert, language expert, ontology engineer, etc.) that has to carry out changes to the ontology.

The canonical citation for the HeLiS ontology is "Dragoni M., Bailoni T., Maimone R., and Eccher C. HeLiS: An Ontology For Supporting Healthy Lifestyles (2018) http://w3id.org/helis".[18]

7 Conclusions and Future Work

In this paper, we presented the HeLiS ontology: a knowledge artifact for the digital health domain specifically developed to support healthy lifestyles. The knowledge modeled within the HeLiS ontology combines information extracted from unstructured resources with the ones collected from domain experts coming from the medical domain. We described the process we followed to build the ontology and which information we included. Then, we presented how the ontology can be reused and we briefly introduced the projects and use cases that will adopt the HeLiS ontology.

Future work will focus on the integration of our model with classification schemata describing the food and physical activity domains from different perspectives, like LanguaL[19] and OPE (see Footnote 8). Moreover, our intent is to populate the HeLiS ontology with data describing the nutritional information of commercial products. Finally, we aim to integrate knowledge from the mindfulness domain in order to support the correlation between the mental and the physical health status of people.

References

1. Grafenauer, S., Tapsell, L., Beck, E.: Beyond nutrients: classification of foods to identify dietary patterns for weight management. In: 16th International congress of Dietetics (2012)
2. Cantais, J., Dominguez, D., Gigante, V., Laera, L., Tamma, V.: An example of food ontology for diabetes control. In: Proceedings of the International Semantic Web Conference 2005 Workshop on Ontology Patterns for the Semantic Web (2005)
3. Snae, C., Bruckner, M.: Foods: a food-oriented ontology-driven system, pp. 168–176 (2008)

[18] DOI of the ontology file will be provided in case of acceptance in order to include possible refinements suggested by Reviewers.

[19] http://langual.org/.

4. Helmy, T., Al-Nazer, A., Al-Bukhitan, S., Iqbal, A.: Health, food and user's profile ontologies for personalized information retrieval. In: Shakshuki, E.M. (ed.) Proceedings of the ANT 2015, London, pp. 1071–1076, 2–5 June 2015
5. Peroni, S., Lodi, G., Asprino, L., Gangemi, A., Presutti, V.: FOOD: food in open data. In: Groth, P., et al. (eds.) ISWC 2016. LNCS, vol. 9982, pp. 168–176. Springer, Cham (2016). https://doi.org/10.1007/978-3-319-46547-0_18
6. Kolchin, M., Zamula, D.: Food product ontology: initial implementation of a vocabulary for describing food products. In: Proceedings of the 14th Conference of Open Innovations Association FRUCT. FRUCT 2014, Helsinki, FRUCT Oy, pp. 191–196 (2013)
7. Çelik Ertuğrul, D.: FoodWiki: a mobile app examines side effects of food additives via semantic web. J. Med. Syst. **40**(2), 1–15 (2016)
8. Fernández-López, M., Gómez-Pérez, A., Juristo, N.: Methontology: from ontological art towards ontological engineering. In: Proceedings of AAAI-Spring Symposium on Ontological Engineering of AAAI (1997)
9. Pinto, H.S., Staab, S., Tempich, C.: DILIGENT: towards a fine-grained methodology for distributed, loosely-controlled and evolving engineering of ontologies. In: de Mántaras, R.L., Saitta, L. (eds.) Proceedings of the 16th Eureopean Conference on Artificial Intelligence, ECAI 2004, Including Prestigious Applicants of Intelligent Systems, PAIS 2004, Valencia, 22–27 August 2004, pp. 393–397. IOS Press (2004)
10. Suárez-Figueroa, M.C.: NeOn methodology for building ontology networks: specification, scheduling and reuse. Ph.D. thesis, Technical University of Madrid (2012)
11. Turconi, G., Roggi, C.: Atlante fotografico alimentare. Uno strumento per le indagini nutrizionali
12. Hitzler, P., Gangemi, A., Janowicz, K., Krisnadhi, A., Presutti, V. (eds.): Ontology Engineering with Ontology Design Patterns - Foundations and Applications. Studies on the Semantic Web, vol. 25. IOS Press (2016)
13. Gangemi, A., Guarino, N., Masolo, C., Oltramari, A., Schneider, L.: Sweetening ontologies with DOLCE. In: Gómez-Pérez, A., Benjamins, V.R. (eds.) EKAW 2002. LNCS (LNAI), vol. 2473, pp. 166–181. Springer, Heidelberg (2002). https://doi.org/10.1007/3-540-45810-7_18
14. Dragoni, M., Bosca, A., Casu, M., Rexha, A.: Modeling, managing, exposing, and linking ontologies with a wiki-based tool. In: LREC, pp. 1668–1675 (2014)
15. Obrst, L., Ceusters, W., Mani, I., Ray, S., Smith, B.: The Evaluation of Ontologies, pp. 139–158. Springer, Boston (2007). https://doi.org/10.1007/978-0-387-48438-9_8
16. Gangemi, A., Catenacci, C., Ciaramita, M., Lehmann, J.: Modelling ontology evaluation and validation. In: Sure, Y., Domingue, J. (eds.) ESWC 2006. LNCS, vol. 4011, pp. 140–154. Springer, Heidelberg (2006). https://doi.org/10.1007/11762256_13
17. Gómez-Pérez, A.: Ontology evaluation. In: Staab, S., Studer, R. (eds.) Handbook on Ontologies. International Handbooks on Information Systems, pp. 251–274. Springer, Heidelberg (2004). https://doi.org/10.1007/978-3-540-24750-0_13
18. Gruber, T.R.: Toward principles for the design of ontologies used for knowledge sharing? Int. J. Hum. Comput. Stud. **43**(5–6), 907–928 (1995)
19. Grüninger, M., Fox, M.: Methodology for the design and evaluation of ontologies. In: Workshop on Basic Ontological Issues in Knowledge Sharing, IJCAI 1995, 13 April 1995
20. Corcoglioniti, F., Rospocher, M., Mostarda, M., Amadori, M.: Processing billions of RDF triples on a single machine using streaming and sorting. In: ACM SAC, pp. 368–375 (2015)

SABINE: A Multi-purpose Dataset of Semantically-Annotated Social Content

Silvana Castano[1], Alfio Ferrara[1], Enrico Gallinucci[2], Matteo Golfarelli[2(✉)], Stefano Montanelli[1], Lorenzo Mosca[3], Stefano Rizzi[2], and Cristian Vaccari[4,5]

[1] DI, University of Milan, Milan, Italy
{silvana.castano,alfio.ferrara,stefano.montanelli}@unimi.it
[2] DISI, University of Bologna, Bologna, Italy
{enrico.gallinucci,matteo.golfarelli,stefano.rizzi}@unibo.it
[3] DFCS, University of Roma Tre, Rome, Italy
lorenzo.mosca@sns.it
[4] Royal Holloway University of London, Egham, UK
cristian.vaccari@rhul.ac.uk
[5] University of Bologna, Bologna, Italy

Abstract. Social Business Intelligence (SBI) is the discipline that combines corporate data with social content to let decision makers analyze the trends perceived from the environment. SBI poses research challenges in several areas, such as IR, data mining, and NLP; unfortunately, SBI research is often restrained by the lack of publicly-available, real-world data for experimenting approaches, and by the difficulties in determining a ground truth. To fill this gap we present SABINE, a modular dataset in the domain of European politics. SABINE includes 6 millions bilingual clips crawled from 50 000 web sources, each associated with metadata and sentiment scores; an ontology with 400 topics, their occurrences in the clips, and their mapping to DBpedia; two multidimensional cubes for analyzing and aggregating sentiment and semantic occurrences. We also propose a set of research challenges that can be addressed using SABINE; remarkably, the presence of an expert-validated ground truth ensures the possibility of testing approaches to the whole SBI process as well as to each single task.

Keywords: Dataset · Social technologies · Sentiment analysis
Text analysis

1 Introduction

During the last decade, an enormous amount of *user-generated content* (UGC) related to people's tastes, opinions, and actions has been made available due

Supported by the Italian Ministry of Education "Future in Research 2012" initiative for the project "Building Inclusive Societies and a Global Europe Online: Political Information and Participation on Social Media in Comparative Perspective" (www.webpoleu.net).

D. Vrandečić et al. (Eds.): ISWC 2018, LNCS 11137, pp. 70–85, 2018.
https://doi.org/10.1007/978-3-030-00668-6_5

to the omnipresent diffusion of social networks and portable devices. This huge wealth of information has raised an intense interest from decision makers because it can give them a timely perception of the market mood and help them explain the phenomena of business and society. *Social Business Intelligence* (SBI) is the discipline that aims at combining corporate data with UGC to let decision makers analyze and improve their business based on the trends and moods perceived from the environment [9].

In the context of SBI, the most widely used category of UGC is the one coming in the form of textual *clips*. Clips can either be messages posted on social media or articles taken from on-line newspapers and magazines or even customer comments collected on the corporate CRM. Digging information useful for decision makers out of textual UGC requires first crawling the web to extract the clips related to a *subject area* (e.g., politics), then enriching them in order to let as much information as possible emerge from the raw text. Enrichment activities may simply identify the structured parts of a clip, such as its author, or even use NLP techniques to interpret each sentence, find the *topics* it mentions, and if possible assign a *sentiment* (i.e., positive, negative, or neutral) to it [12]. For instance, the tweet "UKIP's Essex county councillors stage protest against flying of EU flag", in the subject area of EU politics, mentions topics "UKIP" and "protest" and has positive sentiment. Figure 1 sketches the overall SBI process.

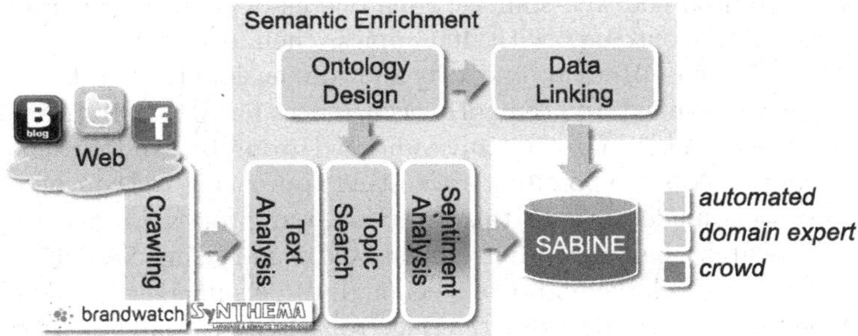

Fig. 1. The functional architecture of the SBI process which created SABINE

From a scientific point of view, SBI stands at the crossroads of several areas of Computer Science such as Database Systems, Information Retrieval, Data Mining, Natural Language Processing, and Human-Computer Interaction. Though the ongoing research in these single fields has made available a bunch of results and enabling technologies for SBI, an overall view of the related problems and solutions is still missing. Besides, the peculiarities of SBI systems open new research problems in all the previous areas. On the other hand, research developments in SBI are often restrained by the lack of publicly-available, real-world data for experimenting approaches, and by the inherent difficulties in determining a ground truth for assessing the effectiveness of an approach.

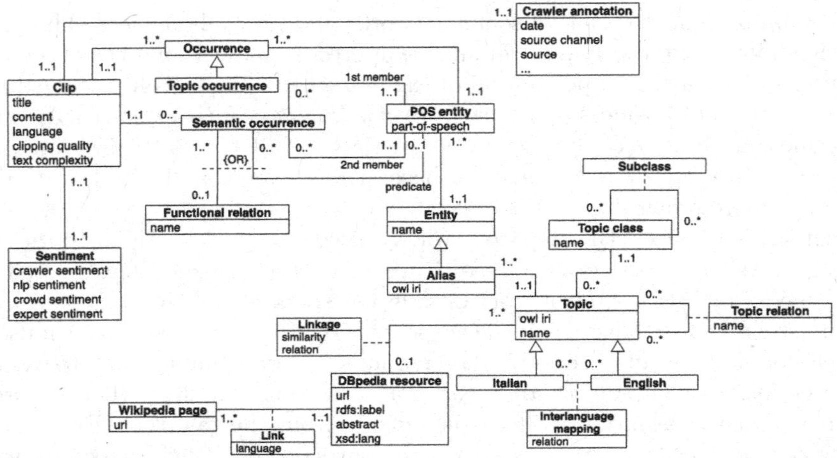

Fig. 2. UML model of SABINE

To fill this gap, in this paper we present SABINE (SociAl Business INtelligence datasEt), a modular benchmark in the domain of European politics with specific reference to the 2014 European elections. SABINE includes: 6 millions bilingual clips crawled from 50 000 web sources, each one associated with metadata and sentiment scores; an ontology with 400 topics, their occurrences in the clips, and their mapping to DBpedia; and two multidimensional cubes for analyzing and aggregating sentiment and semantic occurrences for SBI analytics purposes. Remarkably, the presence of a manually-validated ground truth for each phase of the SBI process ensures the possibility of testing approaches to the whole process as well as to each single task. In this direction, our proposal is complemented by a set of research challenges that can be addressed using SABINE; the task selection we propose is large and diverse enough to be sufficiently representative of a wide range of research tasks, ranging from content analysis to the more comprehensive SBI analytics.

The paper outline is as follows. In Sect. 3 we describe the benchmark content. In Sect. 4 we discuss the techniques adopted for building SABINE. In Sect. 5 we propose a set of SBI-related research tasks for SABINE. Finally, in Sect. 6 we draw the conclusions.

2 Related Literature

As remarked throughout the paper, the research challenges supported by SABINE span several research areas. Table 1 presents a (non-comprehensive) list of datasets available in such areas, emphasizing the novelty of SABINE as a multi-purpose dataset.

A first set of datasets comes from the information retrieval area. Probably the most popular testbed and dataset series in information retrieval was

Table 1. Functional comparison of datasets

Benchmark	Sentiment analysis	Topic search	Document classif.	Cross-language analysis	Topic discovery	Data linking	Multidim. modeling	SBI analytics
Reuters [2]	✗	✗	✓	✓	✓	✗	✗	✗
20 Newsgroups	✗	✗	✓	✗	✗	✗	✗	✗
TREC	✓	✓	✓	✓	✓	✗	✗	✗
CORA	✗	✓	✓	✗	✓	(✓)	✗	✗
CustomerReview [11]	✓	✓	✗	✗	✗	✗	✗	✗
MovieReview [17]	✓	✗	✗	✗	✗	✗	✗	✗
KDD Cup	✗	✗	✓	✓	✗	✓	✗	✗
OAEI [1]	✗	✗	✗	✓	✗	✓	✗	✗
SemEval [14]	✓	✗	✗	✓	✗	✗	✗	✗
SABINE	✓	✓	✓	✓	✓	✓	✓	✓

promoted by the Text Retrieval Conference (TREC, http://trec.nist.gov/data.html) since 1992. The datasets therein contained have been used for a variety of tasks, spanning from sentiment analysis to topic search and discovery. Similar datasets include the Reuters [2] collection (about 112,000 documents in five different languages associated with one or more thematic categories), 20 Newsgroup (http://qwone.com/~jason/20Newsgroups/), which contains 20,000 newsgroup documents, partitioned across 20 different categories, including politics and religion, and CORA (https://people.cs.umass.edu/~mccallum/data.html), a collection of different datasets supporting topic search and discovery, document classification, and information extraction.

In the fields of sentiment analysis and opinion mining, the publication of novel algorithms and techniques is often coupled with the release of datasets. CustomerReview [11] provides 4000 product reviews from Amazon and C—net.com, manually labeled as to whether an opinion is expressed and comprising the rating of the reviewer. These reviews are also grouped according to the mentioned products—whose occurrence is detected by using data mining and natural language processing techniques. Similarly, MovieReview [17] provides four datasets based on movie reviews, comprising a total of 7,000 documents and 20,000 sentences. Annotations to these reviews include the sentiment polarity (positive or negative) obtained from the original website, the rating of the user (e.g., 2.5 out of 5), and a subjectivity status (i.e., "subjective" or "objective" if it is a plot summary or a review).

Besides these datasets, we mention also international competitions that additionally provide a set of evaluation metrics and a methodology for comparing the systems that participate in the contest. KDD Cup (http://www.kdd.org/kdd-cup) is the Data Mining and Knowledge Discovery competition organized by the ACM Special Interest Group on Knowledge Discovery and Data Mining. KDD Cup is mainly oriented towards data mining and prediction, but some editions provided data for tasks related to document classification and data linking. The Ontology Alignment Evaluation Initiative (OAEI http://oaei.ontologymatching.org/) is the main reference for ontology matching. It is specifically devoted to semistructured data linking (in particular to ontology alignment and instance matching) and supports several tasks spanning from cross-language ontology matching to instance matching. In 2016, SABINE has been used as the reference dataset for the task of inter-linguistic mapping and data linking within the OAEI Instance Matching Track [1]. Finally, the International Workshop on Semantic Evaluation (SemEval) proposes different tasks to evaluate computational semantic analysis systems, ranging from cross-lingual similarity to humor and truth detection. In particular, the competition on sentiment analysis proposed in [14] was based on a set of about 30,000 tweets, whose polarity had been manually assigned by a consensus of users via the crowdsourcing platforms Amazon's Mechanical Turk and CrowdFlower.

SABINE Contribution. With respect to these datasets, the main contribution of SABINE is the coverage of a wide range of different but related tasks and the homogeneity of the evaluation environment. The growing interest for data-intensive information systems and the coexistence of unstructured, semistructured, and structured data (not only on the web but also in the enterprise context) motivates the need for an integrated evaluation environment, where applications for search, linking, classification, analysis, and multidimensional modeling of data may be tested over the same data following a homogeneous methodological approach. This is not feasible with any of the datasets of Table 1, in that each of them provides different sets of data, different ground truth collections, and/or different evaluation methodologies. Conversely, all the research tasks supported by SABINE are built from the same collection of data by following the methodology discussed in Sects. 4 and 5 to the end of providing a comprehensive and homogeneous dataset.

3 The Content of SABINE

SABINE has been built as one of the results of the WebPolEU project (webpoleu.altervista.org), whose goal was to investigate the connection between politics and social media. SBI was used in the project as an enabling technology for analyzing the UGC generated in Italian and English during a timespan ranging from March, 2014 to May, 2014 (the 2014 European Parliament Election was held on May 22–25, 2014). By analyzing digital literacy and online political participation, the research evaluated the inclusiveness, representativeness, and quality of the online political discussion. The UML model of the SABINE content (except for the multidimensional part, whose content is described by Fig. 4) is shown in Fig. 2, while its quantitative features are summarized in Table 2 (see [7] for a more detailed profiling of the clips). The main content components of SABINE can be described as follows.

3.1 Topics and Mappings

SABINE provides about 400 relevant topics organized in a topic ontology built by domain experts (a team of five socio-political researchers). The topic ontology

Table 2. SABINE figures

Figure	ENG	ITA	*Figure*	ENG	ITA	*Figure*	ENG	ITA
# Web Sources	23K	25K	# Entities	2868K	1242K	# Entity Occurrences	511M	218M
# Topics	409	434	# Clips	3275K	2394K	# Topic Occurrences	23M	14M
# Topics Aliases	709	798	Avg Chars per Clip	2026	1677	# Semantic Occurrences	48M	35M

(modeled by classes Topic, Topic class, Subclass, and Topic relation in Fig. 2) represents the set of concepts and relationships that, on the domain experts' judgement, are relevant to the subject area; its role in the SBI process is twofold: to act as a starting point for designing effective crawling queries on the one hand, and to support analyses based on relevant concepts (e.g., how often the public debt policy is mentioned) and on their aggregations (e.g., how often the sector of economics and its policies are discussed) on the other. The class diagram for the topic ontology of the socio-political subject area of SABINE benchmark is shown in Fig. 3; for instance, topics "public debt" and "austerity " are instances of topic class Policy and are related to topic "economic policy" (Sector). To enable more accurate analyses, a large set of topic aliases (class Alias in Fig. 2) has been identified and is available for topics (e.g., "tory" is an alias for "conservative").

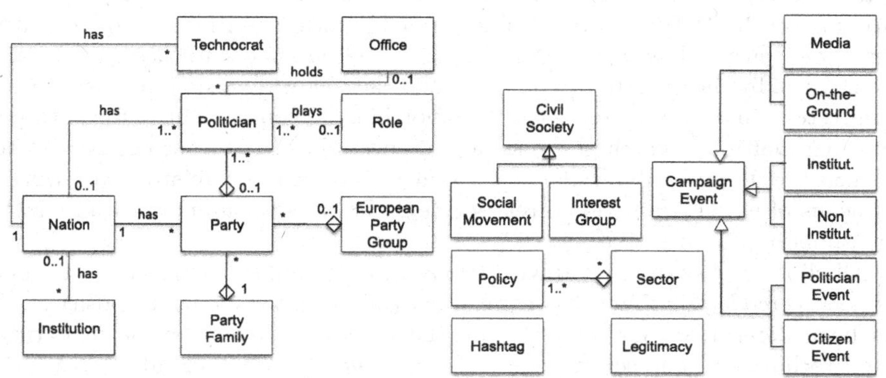

Fig. 3. The topic ontology represented as a UML class diagram

Inter-Language mappings (class Interlanguage mapping) between corresponding topics in the two benchmark languages have been manually created by the domain experts. In most cases these mappings simply express an exact translation (e.g., "immigration" is mapped onto "immigrazione" with semantics owl:sameAs), whereas they are based on weaker semantic relationships when a concept is differently expressed in the two languages (e.g., "immigration" is mapped onto "migrante irregolare" —which means illegal migrant— with semantics sabine:related). A mapping has been found only in 86% of cases since, according to the experts' judgement, some topics are specific of either UK or

Italy (e.g., "Scottish National Party" and "Quirinale"). Furthermore, topics have been linked to their corresponding DBpedia resources (classes DBpedia resource, Wikipedia page, Link, and Linkage). Linkage has been carried out automatically as described in Sect. 4.6 and then validated by domain experts.

3.2 Clips and Annotations

The benchmark provides a large corpus (around 6 millions) of raw clips (class Clip) extracted by the Brandwatch crawler from a broad set (almost 50 000) of web sources including social networks, blogs, and web sites. The most frequent clip sources are Facebook (53.8% of the clips) and Twitter (27.5% of the clips). The corpus is bilingual and *comparable*, i.e., it includes text in two languages (English and Italian) regarding similar topics [6,16]. Each clip is associated with a set of metadata (class Crawler annotation); 40 attributes overall are provided, partly returned by the crawler (e.g., title, date, source MozRank, author information, and geo-localization) and partly manually annotated by the domain experts (e.g., source type).

Clips are enriched with other relevant information resulting from clip text analysis. In particular, each clip is associated with all its occurring entities and their parts-of-speech or POSs (classes POS entity, Entity, and Occurrence). An *entity* is a concept emerging from text analysis, which is not necessarily a topic; parts-of-speech (POSs) are the roles taken by entities within a clip sentence (e.g., *noun*, *verb*, *preposition*). Among the set of entity occurrences, a relevant role is taken by the occurrences of topics and their aliases (class Topic occurrence). Finally, text analysis also led to the detection of more complex linguistic patterns involving multiple entities in the same sentence (classes Semantic occurrence and Functional relation). In particular, each semantic occurrence relates two entities by means of either a functional relation (e.g., *agent* or *qualifier*) or a predicate corresponding to an entity.

All clips are also annotated with two sentiment values (class Sentiment). The first one (*crawler sentiment*) is categorical (i.e., negative, neutral, positive); it has been determined for each clip by the Brandwatch crawler through rule-based techniques. The second one (*NLP sentiment*) is numerical; it has been determined by the SyN semantic engine for each clip sentence, then averaged for each clip (see Sect. 4.5). Finally, a subset of 2400 clips have also been labeled with a *crowd sentiment*, and half of these 2400 clips have further been labeled by domain experts (*expert sentiment*). This subset of manually-labelled clips has been created using a stratified sampling strategy based on the type of clip source (e.g., social network and blog) and on the clip sentiment.

Example 1. Here is an example of a SABINE clip: "Another compassionate conservative. Making fun of a parkinson's victim. Michael J Fox has more courage than you will ever hope to have". Some metadata for this clip are source="facebook.com", channel_type="facebook", source_type="Social network/Social media", country="US", and fb_role="audience". The only occurring topic is "conservative"; among the occurring entities we mention "compassionate", "victim", and "courage" (with POSs *adjective*, *noun*, and *noun* respectively). Text analysis led to find different semantic occurrences of entities with their POSs, for instance the one between "compassionate" and "conservative" (with POS *adjective* and *noun*, respectively, and functional relation *qualifier*) and the one between "have" and "courage" (with POS *verb* and *noun*, respectively, and functional relation *object*). The expert sentiment for this clip is -1 (i.e., negative), while both the crawler and the NLP sentiment are positive (because neither of the approaches was able to detect the irony in the sentence "Another compassionate conservative"). Another example of clip is "US President Barack Obama criticized Russia [...]", which shows a semantic occurrence between entities "Barack Obama" (POS *proper noun*) and "Russia" (POS *proper noun*) involving entity "criticize" as a predicate.

3.3 Multidimensional Cubes

These are ROLAP cubes providing an easy-to-query representation of the clip content and of the outcome of the clip enrichment process. The first cube, Sentiment, is centered on clips, and it represents the set of topics appearing in each clip as well as the sentiment values computed for that clip. The second cube, Semantic Occurrence, is centered on the semantic occurrence of POS entities within clips and explicitly models couples of entities in the same sentence together with an optional predicate. The conceptual schemas of the Sentiment cube is depicted in Fig. 4 using the DFM notation [10], where cube measures are listed inside the box, dimensions are circles directly attached to the box, and hierarchies are

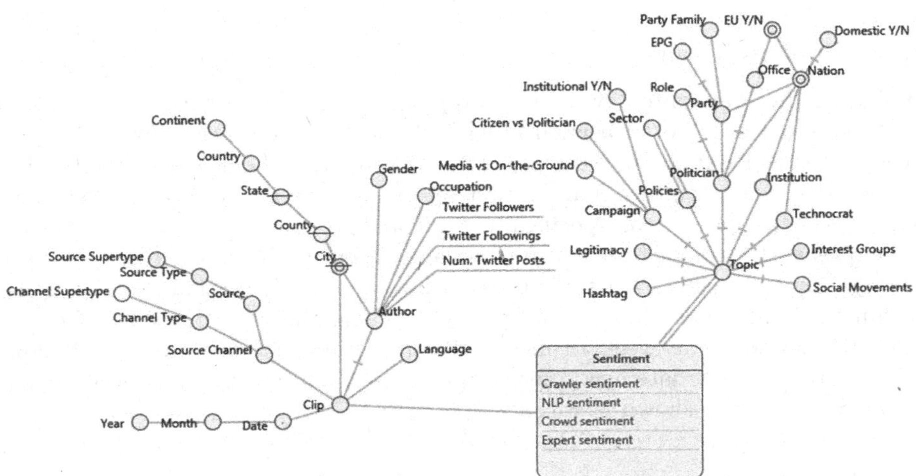

Fig. 4. A DFM representation of the Sentiment cube (for drawing simplicity, some levels of the topic hierarchy are hidden)

shown as DAGs of dimension levels. In particular, the hierarchy built on dimension Clip includes the crawler annotations, while the one built on Topic (called *topic hierarchy* from now on) derives from the topic ontology of Fig. 3 and enables topic-based aggregations of clips in the OLAP front-end. For instance, a roll-up from Politician to Party and Party Family allows to obtain the opinions about a wing as an average of the opinions about all the politicians belonging to the parties of that wing.

4 SABINE Construction Techniques

To develop SABINE we followed the methodology described in [8], which has been conceived on the one hand to support and speed up the initial design of an SBI process, on the other hand to maximize the effectiveness of the experts' analyses by continuously optimizing and refining all the design tasks. Quick tuning iterations are probably the most distinctive feature of this type of projects, and they are necessary to cope with the high fickleness of the topics covered in social conversations. The initial step consists of a *Macro Analysis* aimed at defining the project border, the main subject areas and topics. This information is the starting point for the *Ontology Design* and *Source Selection* steps, aimed at creating a topic ontology and at defining the list of web sites and social channels to be monitored, respectively. Topics and web sources are the input for *Crawling Design*, which is aimed at creating the keyword queries to be submitted to the crawling engine. To extract semantics from raw clips, a *Semantic Enrichment* phase is then triggered. At this stage, it is possible to test and tune the overall process.

In the following subsections we give further details on the techniques adopted for each task of the SBI process, using Fig. 1 as a reference.

4.1 Ontology Design

Designing the topic ontology of the European politics subject area was mainly a methodological issue. Consistently with the methodology we followed [8], we initially carried out a *macro-analysis* to identify the themes relevant to the subject area (e.g., "culture") and a first set of topics (e.g., "school"). Then, the *ontology design* phase was specifically dedicated to collecting, for each theme, a comprehensive set of topics and to arrange them within an ontology (using Protégé) by expressing inter-topic relationships (e.g., to state that "school" falls within the context of "educational policy"). Along the whole project lifetime, the topic ontology was weekly tuned and refined (in collaboration with domain experts) to accommodate new topics, topic classes, and relationships; the model of the final result is shown in Fig. 3.

4.2 Crawling

The two main design activities related to crawling are *source selection* and *crawling query design*. For crawling we adopted the Brandwatch service to ensure a satisfactory coverage of web sources along the project duration (three months). Brandwatch adopts a template-based engine, that is, it extracts only the informative UGC by detecting and discarding advertisements and banners (a process called *clipping*); it also drops duplicate clips using content aggregators. As to source selection, the Brandwatch source base has been extended with more than 100 additional domain-specific web sites suggested by our domain experts.

Keyword-based queries in Brandwatch rely on a set of 23 operators (ranging from Boolean ones to proximity ones) that allow to express filters on both textual features (e.g., the maximum distance between two words) and metadata (e.g., the author's country and the web site name). To enhance the quality of the crawling result, we weekly run a review of the set of queries initially created plus a *content relevance analysis* aimed at discarding off-topic clips.

4.3 Text Analysis

For text analysis we used the *SyN-Semantic Center* commercial engine. SyN was used for splitting clips in sentences and for extracting the single entities, their part-of-speech, and their semantic occurrences. Moreover, several techniques have been adopted to ensure homonyms and homographs identification: (i) keyword-based crawling queries have been designed to properly identify topics by explicitly excluding homonyms (e.g., to avoid occurrences of "Osborne" referring to the musician "Ozzie" rather than to the politician "George"); (ii) SyN-Semantic Center includes a module for homonyms and homographs handling based on ontological disambiguation. In detail, the linguistic and semantic text analysis made by SyN is based on morpho-syntactic, semantic, semantic role, and statistical criteria. At the heart of the lexical system lies McCord's theory of slot grammars [13]. The system analyzes each sentence, cycling through all its possible constructions and trying to assign the context-appropriate meaning to each word by establishing its context. Each slot structure can be partially or fully instantiated and it can be filled with representations from one or more statements to incrementally build the meaning of a statement. The core of the system is the SyN ontology, developed through twenty years of experiences and projects.

4.4 Topic Search

With this term we refer to the task of indexing all the occurrences of a topic (or one of its aliases) within a clip. We relied on two different techniques for searching topic occurrences in SABINE. The first one is a simple text matching technique that retrieves the exact occurrences of topics and aliases, implemented in-house as a Java algorithm. The second one is based on the results of the text analysis made by SyN, which extracts all the occurring entities in a clip; in this case, topic occurrences are obtained by linking topics and aliases to the corresponding entities.

Clearly, both techniques have pros and cons. By avoiding all kinds of text analysis, the first technique typically trades a better performance in terms of speed with a lower accuracy of the results. In particular, the results tend to include the occurrences where a topic (or an alias) is used in the clip with different semantics from the one originally meant in the topic ontology. This problem arises when topics (or aliases) in the ontology are too generic. The second technique presents the opposite challenge: by carrying out an in-depth comprehension of the clip semantics, the entities produced by SyN tend to be very specific, possibly leading to the pulverization of the same concept into a wide set of entities. Therefore, this problem arises when topics (or aliases) in the ontology are intentionally generic.

The adoption of both techniques enabled us to double-check the results and to track down the causes of conflicting results. Eventually, mismatches were manually solved in most cases, yielding a 91% agreement between the two techniques (over 14 millions occurrences).

4.5 Sentiment Analysis

Sentiment analysis is probably the hardest task in SBI; for this reason we included in SABINE both system-based and human-based sentiment scores. While system-based scores can be used as a baseline for testing other automatic techniques, human-based scores represent the ground truth.

Crawler Sentiment. This score, computed by Brandwatch, tags each clip of SABINE. The sentiment analysis component of Brandwatch is based on mining rules specifically developed for each language supported.

NLP Sentiment. SyN includes its own sentiment analysis component [15] whose score takes into account the negative or positive polarization of words and concepts, as well as the syntactical tree of the sentence being analyzed. Each clip of SABINE is tagged with this score as well.

Expert Sentiment. This score was defined for a sample of 1200 clips (600 English + 600 Italian) by asking our domain experts to manually tag them. The clips are equally divided by media type and NLP sentiment (as computed by SyN). Besides defining the clip sentiment as either negative, neutral, or positive, the domain experts were also asked to rate, for each clip, its *clipping quality* (i.e., the amount of non-relevant text present in the clip due to an inadequate template used by the crawler when clipping), which could impact on the difficulty of assigning the right sentiment, and its intrinsic *text complexity* (i.e., the effort of a human expert in assigning the sentiment due to irony, incorrect syntax, abbreviations, etc.).

Crowd Sentiment. This score was given to a sample of about 2400 clips (1200 + 1200, including the clips tagged by experts) through a crowdsourcing process. To this end, we selected a crowd of around 900 workers within a class of bachelor-degree students in the field of humanities and political science at the University of Milano (average worker age is 21). Crowdsourcing activities were performed during one month and each worker tagged 46 clips on average. As a support we employed our Argo system (island.ricerca.di.unimi.it/projects/argo/, in Italian), which provides crowdsourcing functionalities based on *multi-worker task assignment* and *consensus evaluation* techniques [5].

4.6 Data Linking

The goal of data linking is to link ontology topics to the Linked Data Cloud. In SABINE, this has been done by coupling automated techniques with manual validation and revision by domain experts. As a first step, topic aliases were used to retrieve a set of candidate DBpedia resources for each topic t through the DBpedia Lookup Service. The degree of similarity between t and the retrieved candidates (if any) was evaluated through the HMatch matching algorithm [3]. HMatch takes into account both the linguistic information available for t (i.e., its aliases) and the ontological information (i.e., the topic class of t). Then, topic t was linked to the DBpedia resource e, among the candidates, yielding the highest degree of similarity. The link between t and e is formally defined as a 4-tuple of the form $\mathcal{L}_{t,e} = \langle t, e, \sigma_{t,e}, \rho_{t,e} \rangle$, where $\sigma_{t,e}$ is a real number in the range $[0, 1]$ representing the degree of similarity between t and e, and $\rho_{t,e}$ represents the semantics of the relation holding between t and e. Each resulting link $\mathcal{L}_{t,e}$ was submitted to domain experts to specify the most suitable semantics $\rho_{t,e}$, choosing among (i) owl:sameAs (t and e have exactly the same meaning); (ii) sabine:narrower/sabine:broader (the meaning of t is more specific/generic than the one of e); (iii) sabine:related (there is a positive association between the meanings of t and e[1]. If none of the previous options was deemed suitable by the domain experts, the link was marked as *incorrect*. The links with semantics different from owl:sameAs and all the incorrect links were submitted to a second validation round, where domain experts manually found additional DBpedia resources to be associated with the corresponding topic with owl:sameAs semantics. This procedure is crucial to ensure the quality of the resulting links. However, due to the effort required to the domain experts, this process has been feasible only for the 400 topics, but not for the entities because SABINE provides over 4 million entities, which would have resulted in an overwhelming manual activity for the experts. Table 3 shows some statistics about validation and refinement of English topics.

Example 2. As an example, we propose the link ⟨"school", dbpedia:State_school, 0.75, owl:sameAs⟩ between topic "school" and DBpedia resource dbpedia:State_school. In the first round of validation, experts confirmed that "school"

[1] In SABINE, topics are not modeled as SKOS concepts because they include concrete entities (specific parties, institutions, people) that are better represented as OWL Named Individuals in order to keep the distinction between concepts and instances in the ontology (whereas, according to the W3C, skos:Concepts should only represent abstract entities or conceptual knowledge). Besides, not even the DBpedia resources that our topics are related to are instances of skos:Concept. Thus, we could not reuse the skos:broader and skos:narrower properties because (although their meaning is similar to the one we intend) they are used only to describe relationships between skos:Concepts.

Table 3. Results of the domain expert validation and revision for the English topics

Relation semantics	# links	Avg. similarity
owl:sameAs	252	0.812
sabine:narrower	7	0.721
sabine:broader	39	0.756
sabine:related	17	0.781
incorrect	62	0.22
sub-total	377	
expert-provided resource	135	0.433
total	512	

Most of the automatically retrieved links have been considered correct with owl:sameAs semantics (67%); 17% of the remaining links have been evaluated as correct but with semantics different from owl:sameAs. In particular, the automatic linking procedure tends to provide specific (rather than generic) DBpedia resources for topics. The automatic approach was incorrect in 16% of cases. We note also a positive correlation between the average degree of similarity associated with links and the positive evaluation provided by experts. This is important for associating a reliable degree of similarity to the links in the final dataset. Finally, for 135 topics, domain experts provided a DBpedia resource as an owl:sameAs counterpart for the topic.

can be linked to dbpedia:State_school, but with semantics sabine:broader (since school is broader than dbpedia:State_school). The resulting link was ⟨"school", dbpedia:State_school, 0.75, sabine: broader⟩. Since the semantics is different from owl:sameAs, the link was submitted to the second validation round, where we asked experts to manually find a DBpedia resource which actually has an owl:sameAs relation with "school". Experts found the DBpedia resource dbpedia:School, which leads to the addition of a second link for topic "school". The links resulting from the two validation rounds are then

$$\langle \text{"school"}, \text{dbpedia:State\_school}, 0.75, \text{sabine:broader} \rangle$$
$$\langle \text{"school"}, \text{dbpedia:School}, 1.0, \text{owl:sameAs} \rangle$$

5 Research Tasks

In this section we describe the main research tasks that are supported by the data provided in SABINE. To enable partial, ad-hoc downloads for each task, we subdivided SABINE into separate components shown as packages in Fig. 5. A package models a component of the dataset that can be downloaded separately;

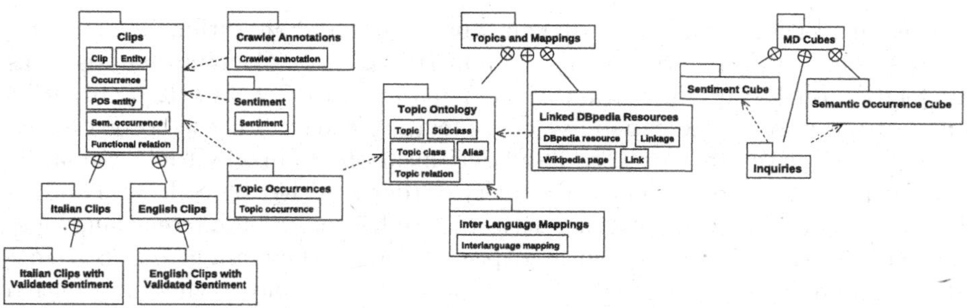

Fig. 5. Components of SABINE and their composition and dependency relationships

Table 4. Task overview

Task	Input packages	Ground truth packages	Ground truth
Content analysis tasks			
Sentiment analysis	Clips	Sentiment	Sentiment manually defined by crowd and experts
Topic search	Clips; Topic Ontology	Topic Occurrences	Topic occurrences obtained by automatic techniques and validated by experts
Document classification	Clips	Crawler Annotations	Classifications induced by the crawler annotations
Semantic analysis tasks			
Cross-language analysis	Clips; Topic ontology	Inter-Language Mappings	Mappings manually defined by experts
Topic discovery	Clips	Topic Ontology; Topic Occurrences	Topic ontology manually defined by experts and topic occurrences obtained by automatic techniques and validated by experts
Data linking	Topic Ontology	Linked DBPedia Resources	Links obtained by an automatic procedure and validated by experts
SBI analytics tasks			
Multidimensional modeling	Clips; Topic Occurrences; Topic Ontology; Sentiment; Crawler Annotations	MD Cubes	Cube schemas and instances
SBI analytics	Clips	Inquiries	Datasets resulting from the inquiries and validated by experts

one package depends on another one when an object of the former references an object of the latter. For each task, in Table 4 we summarize the SABINE component(s) to be taken in input and the ground truth we provide together with the packages where it is contained.

The main idea of SABINE research tasks is that different combinations of subsets of the main dataset (i.e., packages) can be used either as a training or as a testing set for a variety of different approaches and algorithms in the areas of *content analysis*, *semantic analysis*, and *SBI analytics*. Content analysis tasks are focused on the interpretation of the clips provided by SABINE in order to automatically assess the capability of associating them with the correct sentiment, the effectiveness of retrieval and classification of clips by topic. Semantic analysis is mainly focused on the discovery of topics in the clip collection and on the discovery of semantic relations between SABINE topics in different languages (i.e., English and Italian) and with DBpedia. Finally, the main goal of SBI analytics is to enable complex analyses of social content by integrating information obtained from a semantic enrichment process that includes techniques coming from different research fields. For this reason, the tasks in this subsection play a special role in SABINE because, unlike those described above which are related to each single phase of enrichment, they concern the capabilities of integrating the results.

6 Conclusion

In this paper we have presented SABINE, a dataset of semantically annotated social content in the domain of European politics. SABINE aims to constitute a publicly-available, real-world dataset for experimenting and comparing the most commonly performed SBI tasks, crossing the various involved research fields ranging from Database Systems, Information Retrieval, Data Mining, up to Natural Language Processing and Human-Computer Interaction. SABINE has been designed and properly packaged for modular download to enable the evaluation of a wide variety of research tasks, either separately or in combination, ranging from those more focused on content analysis, to those related to semantic analysis up to more comprehensive SBI analytics. The SABINE components related to data linking and cross-language analysis have been used for the tasks of inter-linguistic mapping and data linking within the OAEI Instance Matching Track [1]. A main technical advance of SABINE is the availability of multiple, complementary, and validated enrichments of the social content (i.e., textual clips) in form of metadata, annotations, sentiment scores, and DBpedia mappings. The availability of a user-validated ground truth, either by domain experts or by crowdsourcing or both, for each enrichment phase represents a further technical advance of SABINE. In our future work, we plan to undertake a new crowdsourcing project to manually annotate an increasingly larger amount of clips.

SABINE [4] is available for download at http://purl.org/sabine under the CC BY-NC 4.0 license; packages are made available as compressed archive files containing JSON files (the Clips package), OWL files (the Topics and Mappings package and all its sub-packages), and CSV files (all other packages).

References

1. Achichi, M., et al.: Results of the ontology alignment evaluation initiative 2016. In: CEUR Workshop Proceedings, vol. 1766, pp. 73–129. RWTH (2016)
2. Amini, M., Usunier, N., Goutte, C.: Learning from multiple partially observed views-an application to multilingual text categorization. In: Advances in Neural Information Processing Systems, pp. 28–36 (2009)
3. Castano, S., Ferrara, A., Montanelli, S.: Matching ontologies in open networked systems: techniques and applications. In: Spaccapietra, S., Atzeni, P., Chu, W.W., Catarci, T., Sycara, K.P. (eds.) Journal on Data Semantics V. LNCS, vol. 3870, pp. 25–63. Springer, Heidelberg (2006). https://doi.org/10.1007/11617808_2
4. Castano, S., et al.: SABINE: a multi-purpose dataset of semantically-annotated social content. In: Vrandečić, et al. (eds.) ISWC 2018, Part II. LNCS, vol. 11137, pp. 70–85 (2018). http://purl.org/sabine
5. Castano, S., Ferrara, A., Genta, L., Montanelli, S.: Combining crowd consensus and user trustworthiness for managing collective tasks. Futur. Gener. Comput. Syst. **54**, 378–388 (2016)
6. Chu, C., Nakazawa, T., Kurohashi, S.: Iterative bilingual lexicon extraction from comparable corpora with topical and contextual knowledge. In: Gelbukh, A. (ed.) CICLing 2014. LNCS, vol. 8404, pp. 296–309. Springer, Heidelberg (2014). https://doi.org/10.1007/978-3-642-54903-8_25
7. Francia, M., Gallinucci, E., Golfarelli, M., Rizzi, S.: Social business intelligence in action. In: Nurcan, S., Soffer, P., Bajec, M., Eder, J. (eds.) CAiSE 2016. LNCS, vol. 9694, pp. 33–48. Springer, Cham (2016). https://doi.org/10.1007/978-3-319-39696-5_3
8. Francia, M., Golfarelli, M., Rizzi, S.: A methodology for social BI. In: Proceedings of IDEAS, pp. 207–216 (2014)
9. Gallinucci, E., Golfarelli, M., Rizzi, S.: Advanced topic modeling for social business intelligence. Inf. Syst. **53**, 87–106 (2015)
10. Golfarelli, M., Rizzi, S.: Data Warehouse Design: Modern Principles and Methodologies. McGraw-Hill, New York City (2009)
11. Hu, M., Liu, B.: Mining and summarizing customer reviews. In: Proceedings of the SIGKDD, Seattle, USA, pp. 168–177 (2004)
12. Liu, B., Zhang, L.: A Survey of Opinion Mining and Sentiment Analysis. In: Aggarwal, C., Zhai, C. (eds.) Mining Text Data, pp. 415–463. Springer, Boston (2012). https://doi.org/10.1007/978-1-4614-3223-4_13
13. McCord, M.C.: Slot grammar. In: Studer, R. (ed.) Natural Language and Logic. LNCS, vol. 459, pp. 118–145. Springer, Heidelberg (1990). https://doi.org/10.1007/3-540-53082-7_20
14. Nakov, P., Ritter, A., Rosenthal, S., Sebastiani, F., Stoyanov, V.: SemEval-2016 task 4: sentiment analysis in twitter. In: Proceedings of SemEval@NAACL-HLT, San Diego, USA, pp. 1–18 (2016)
15. Neri, F., Aliprandi, C., Capeci, F., Cuadros, M., By, T.: Sentiment analysis on social media. In: Proceedings of ASONAM, pp. 919–926 (2012)
16. Otero, P.G.: Learning bilingual lexicons from comparable English and Spanish corpora. In: MT Summit xI, pp. 191–198 (2007)
17. Pang, B., Lee, L.: Seeing stars: exploiting class relationships for sentiment categorization with respect to rating scales. In: Proceedings of ACL, Michigan, USA, pp. 115–124 (2005)

Querying Large Knowledge Graphs over Triple Pattern Fragments: An Empirical Study

Lars Heling(✉), Maribel Acosta, Maria Maleshkova, and York Sure-Vetter

Institute AIFB, Karlsruhe Institute of Technology (KIT), Karlsruhe, Germany
{lars.heling,maribel.acosta,maria.maleshkova,york.sure-vetter}@kit.edu

Abstract. Triple Pattern Fragments (TPFs) are a novel interface for accessing data in knowledge graphs on the web. So far, work on performance evaluation and optimization has focused mainly on SPARQL query execution over TPF servers. However, in order to devise querying techniques that efficiently access large knowledge graphs via TPFs, we need to identify and understand the variables that influence the performance of TPF servers on a fine-grained level. In this work, we assess the performance of TPFs by measuring the response time for different requests and analyze how the requests' properties, as well as the TPF server configuration, may impact the performance. For this purpose, we developed the *Triple Pattern Fragment Profiler* to determine the performance of TPF server. The resource is openly available at https://doi.org/10.5281/zenodo.1211621. To this end, we conduct an empirical study over four large knowledge graphs in different server environments and configurations. As part of our analysis, we provide an extensive evaluation of the results and focus on the impact of the variables: triple pattern type, answer cardinality, page size, backend and the environment type on the response time. The results suggest that all variables impact on the measured response time and allow for deriving suggestions for TPF server configurations and query optimization.

1 Introduction

Accompanied by the proliferation of knowledge graphs on the web as Linked Data [11], storage, and management solutions are constantly being newly developed or improved in order to support the necessity for accessing knowledge graphs online. A variety of interfaces to access and query RDF knowledge graphs have been proposed, including SPARQL endpoints and Triple Pattern Fragments (TPFs) [14]. Conceptually, the main difference among these interfaces is the expressivity of the requests they are able to handle: endpoints support the execution of SPARQL queries while TPFs are able to evaluate triple patterns. Furthermore, TPFs allow for querying RDF knowledge graphs which may be stored in different sources or backends. The evaluation of a triple pattern

D. Vrandečić et al. (Eds.): ISWC 2018, LNCS 11137, pp. 86–102, 2018.
https://doi.org/10.1007/978-3-030-00668-6_6

against a TPF server produces a sequence of RDF triples that match the given triple pattern; this is called a fragment. In addition, fragments may be partitioned into several fragment pages which contain a fixed maximum number of triples defined as the page size. The page size is configured by the data provider. To retrieve an entire fragment, clients must iterate (or paginate) over the TPF pages.

In order to devise efficient querying techniques over knowledge graphs on the web, it is necessary to understand the factors that impact the performance of the different interfaces, in particular, their response time. The factors that impact the response time of SPARQL endpoints have been extensively studied [2,6,8,10]. Therefore, in this work, we tackle the problem of identifying the variables that impact on the response time of TPFs when querying large RDF knowledge graphs. Our work aims to contribute to a better understanding of the costs related to retrieving data by querying RDF knowledge graphs over TPFs. An array of variables impacting the performance of TPFs have been the subject of previous studies [4,5,13,14], however these evaluations mostly focus on a higher level of query evaluation. Therefore, we provide a fine-grained study on the performance of TPFs and analyze further variables potentially impacting the performance. Concretely, we focus on the following research questions:

RQ1 How does the type of triple pattern impact the response time?
RQ2 What are the effects of the answer cardinality of triple patterns on response time?
RQ3 What is the impact of using different page sizes on performance?
RQ4 How does the response time differ when comparing different backends?
RQ5 What are the effects on response time when TPFs serve as an interface to several knowledge graphs simultaneously?
RQ6 What differences in performance can be observed between querying TPFs in controlled and real-world environments?

We investigate these research questions by first devising a profiler that generates triple patterns from sampling knowledge graphs over TPFs. Our profiler then executes the generated triple patterns and records the performance of the TPF servers. The outcome of the profiler allows for a fine-grained analysis of the factors that may impact on TPF performance. Subsequently, we conduct an empirical study which evaluates the costs of querying four well-known knowledge graphs over TPFs. In summary, we make the following contributions:

- A methodology for generating triple patterns from sampling knowledge graphs over TPFs;
- A fine-grained and extensive evaluation to analyze the impact of several independent variables on the performance of TPFs;
- Finally, we support the reproducibility of our results by providing the raw data as well as the *Triple Pattern Fragment Profiler*[1].

[1] https://github.com/Lars-H/tpf_profiler.

The herein presented evaluation setup was specially designed in order to ensure reuse and open access for the community, as well as reproducibility of the experimental results. The used settings can be replicated, in order to enable the verification of our analysis but also to provide the basis for further experiments and further work by other researchers in the field. The results of our study potentially allow for deriving a relationship between the properties of a requested triple pattern and the corresponding response time. In a subsequent step, the information derived from our analysis may be applied for constructing an empirical cost estimation model to be used in (federated) query engines and help to improve the configuration when setting up TPF servers.

The remainder of this paper is structured as follows. Section 2 presents related work and in Sect. 3 we present our methodology including the sample generation and the TPF profiler. The setup of our study is detailed in Sect. 4. The results of our empirical study are presented and discussed in Sect. 5. Finally, in Sect. 6 we provide our conclusions and an outlook to future work.

2 Related Work

Montoya et al. [8] identified the independent and dependent (or observed) variables that impact on the performance of querying federations of SPARQL endpoints. Following a similar classification, the independent variables that may impact on the performance of TPFs can be grouped into four dimensions: Query, Knowledge Graph (KG), Triple Pattern Fragment Configuration, and Platform. Table 1 summarizes the independent variables studied in the literature and in our work. In this work, we focus on analyzing the dependent variable response time, i. e., the elapsed time between the client contacting the server and the first response arriving. In the following, we position our work with respect to experimental studies that have analyzed the impact of independent variables on the response time (or cost) of querying KGs via SPARQL endpoints or TPF servers.

Query. This dimension includes variables with regard to the structure of the query. Most of the works have studied the impact of the query shape with different types of joins (e. g., subject-subject, subject-object, etc.) on the performance of endpoints [2,8] and TPFs [1,13,14], as well as the effect of specifying SPARQL query operators with different complexity [2,6,8] on the response time of endpoints. Nonetheless, little attention has been paid to studying the impact of individual triple patterns and their instantiations [2,14] or answer size [1] on server performance. In this work, we conduct a fine-grained study of independent variables in the query dimension at the level of triple patterns to better understand the behavior of TPFs when evaluating different types of requests.

Knowledge Graph (KG). The variables in this dimension characterize the KG or the data, including the number of statements in the KG, the distribution of nodes and relationships, and partitioning or replication of the data. Analogous to other studies [2,6,8], in our work we study the response time of servers when querying real-world KGs with different sizes and data distributions.

Table 1. Comparison of empirical studies that analyze the impact of independent variables on the response time of SPARQL endpoints or TPFs.

Independent variables	Endpoints				Triple Pattern Fragments					
	[8]	[6]	[2]	[10]	[13]	[1]	[14]	[4]	[5]	Our Work
Query										
Query shape	✓	✗	✓	✗	✓	✓	✓	✗	✗	✗
Query complexity	✓	✓	✓	✗	✗	✗	✗	✗	✓	✗
Triple pattern type	✗	✗	✓	✗	✗	✗	✗	✗	✗	✓
Number of constants	✗	✗	✓	✗	✗	✗	✓	✗	✗	✓
Answer cardinality	✗	✗	✗	✗	✗	✓	✗	✗	✗	✓
Knowledge Graph (KG)										
KG size	✓	✓	✓	✗	✗	✗	✓	✓	✗	✓
Data frequency distribution	✓	✗	✗	✗	✗	✓	✓	✗	✗	✓
KG partitioning/replication	✓	✗	✗	✓	✗	✗	✗	✗	✗	✗
Triple Pattern Fragment Configuration										
Backend type	–	–	–	–	✗	✗	✗	✗	✗	✓
Page size	–	–	–	–	✗	✗	✗	✗	✗	✓
Pagination	–	–	–	–	✗	✗	✗	✗	✗	✓
Relation KGs/TPF instance	–	–	–	–	✗	✗	✗	✗	✗	✓
Platform										
Server workload	✗	✗	✓	✗	✓	✗	✓	✓	✓	✓
Network delays	✓	✗	✓	✗	✗	✓	✗	✗	✗	✓
Caching	✗	✗	✓	✗	✗	✗	✗	✓	✓	✗
Data serialization	✗	✗	✗	✗	✗	✗	✓	✗	✗	✗
Hardware configuration	✗	✓	✗	✗	✗	✗	✗	✗	✗	✗

Triple Pattern Fragment Configuration. This dimension focuses on the variables that are particular to the TPFs. Based on the current definition and implementations of TPFs [13,14] we identify four variables in this dimension: backend type, page size, pagination, relation KGs/TPF instance. Current implementations of TPF servers support different backends including HDT files [3], SPARQL endpoints, and RDF documents.[2] Another important feature of TPFs is that they partition the result of evaluating a triple pattern into fragment pages; each page contains a fixed maximum number of triples. In this work, we will focus on studying the behavior of TPF servers with configurations, suitable for querying large knowledge graph. Another variable in this dimension is pagination, i. e., the cost of iterating the fragment pages to completely evaluate a triple pattern. Lastly, the KGs/TPF variable captures the correspondence of the KGs accessed via an instance of a TPF server. From all the independent

[2] https://github.com/LinkedDataFragments/Server.js.

variables listed in the TPF configuration dimension, only the work by Hartig and Buil-Aranda [5] investigated the impact of varying the page size on cache hits but not on response time. In contrast, our work studies the variables pagination and the relation KGs/TPF by studying the cost of dereferencing different pages of a given fragment and quantifying the cost of querying TPFs when KGs are accessed individually or simultaneously through a TPF instance, respectively.

Platform. The platform dimension as defined by Montoya et al. [8] comprises the variables that describe the computing infrastructure. In this dimension, the most studied variables are: server workload [2,4,5,13,14], network delays [1,2,8], and caching [2,4,5]. In our work, we study the impact of server workload and network delays when performing requests in two different environments: a controlled environment with one client and no network delays (best-case scenario), and a real-world environment when querying public TPF servers.

3 Our Approach

We devise an approach to measure the response time of Triple Pattern Fragment (TPF) servers under different conditions. Our approach is independent of the underlying knowledge graph (KG) and TPF server configuration and can be applied to conduct empirical studies on any KG accessible via TPF servers. The main components of our approach are the **Triple Pattern Sampling Component** and the **Evaluation Component**. In the following, we describe the sampling component and then present how the samples are used to capture the performance of TPFs.

The goal of the proposed Triple Pattern Sampling Component is generating random triple patterns with different characteristics to evaluate the performance of TPF servers under different conditions. The input of the sampling method is a KG G composed of a set of RDF triples accessible via a TPF server, and a sample size m. The output of this method is a sample S, with S corresponding to a set of triple patterns such that the evaluation of a triple pattern $tp \in S$ over G produces answers, i.e., $[[tp]]_G \neq \emptyset$. The core idea of our sampling method is to select a set of RDF triples in a given KG and derive triple patterns by replacing constants (RDF terms) with existential variables. Assuming R the set of RDF terms – IRIs, literals, and blank nodes – and V the universe of variables [9], the triple patterns in the generated sample are of the following 2^3 types: $\{\langle r1, r2, r3\rangle$, $\langle v1, v2, v3\rangle$, $\langle v1, v2, r3\rangle$, $\langle v1, r2, v3\rangle$, $\langle r1, v2, v3\rangle$, $\langle r1, v2, r3\rangle$, $\langle r1, r2, v3\rangle$, $\langle v1, r2, r3\rangle\}$ with $r1, r2, r3 \in R$ and $v1, v2, v3 \in V$.

For a given KG G accessible via a TPF server, the sampling component produces a set of RDF triples $G^* \subseteq G$, such that $|G^*| = m$. In order to build G^*, the proposed component performs random sampling over the KG such that G^* contains RDF triples that capture different characteristics of G in terms of data distributions. Furthermore, the sampling component exploits the features of TPFs to construct G^* from G. For instance, when evaluating a triple pattern

tp against G over a TPF server, the server provides the set[3] G_{tp} of RDF triples that match tp as well as metadata that includes an approximation to the total number of answers of tp. Furthermore, TPFs partition G_{tp} into subsets called pages [14]. Let $\mathscr{P}$ be the set of pages of G_{tp}, each page $P_i \in \mathscr{P}$ contains a fixed maximum number of triples, i.e., $|P_i| \leq p_{max}$. To retrieve the complete answer G_{tp} of triple pattern tp, it is necessary to iterate over all the pages in $\mathscr{P}$, where $|\mathscr{P}| = \left\lceil \frac{|G_{tp}|}{p_{max}} \right\rceil$. To generate G^*, the sampling method proceeds as follows:

1. The sampling component evaluates the triple pattern tp with variables in subject, predicate and object position (i. e. $tp = \langle v1, v2, v3 \rangle$) against the KG G via a TPF server. The result of this evaluation is a set of pages $\mathscr{P}$ with RDF triples G_{tp} that match tp.
2. From the set of pages $\mathscr{P}$, the sampling method randomly selects m pages following a random uniform distribution, i.e., all the pages in $\mathscr{P}$ have the same probability of being selected.
3. For each page $P_i \in \mathscr{P}$, the sampling method randomly selects an RDF triple $\langle s, p, o \rangle \in P_i$ following a random uniform distribution and adds it to G^*.
4. From the RDF triples in G^* the triple pattern sample S is generated by replacing the RDF terms with variables: $S = \bigcup_{\langle s,p,o \rangle \in G^*} \{v1, s\} \times \{v2, p\} \times \{v3, o\}$, with $v1, v2, v3 \in V$.

It is important to note that during the sampling of RDF triples, two different RDF triples may lead to the generation of the same triple pattern. For instance, consider the RDF triples $\langle s1, p, o1 \rangle$ and $\langle s2, p, o2 \rangle$ with $s1, s2, p, o1, o2 \in R$, $s1 \neq s2$, and $o1 \neq o2$. In this case, both triples produce the common triple pattern $\langle v1, p, v3 \rangle$ with $v1, v3 \in V$. Nonetheless, according to step 4 of the sampling method, $\langle v1, p, v3 \rangle$ occurs once in S. The reason for restricting S to unique triple patterns is to avoid unwanted caching effects when measuring the performance of TPF servers, which may happen when requesting the same triple pattern several times sequentially. As every triple pattern tp with $|[[tp]]_G| \in [1, |G|]$, $\forall tp \in S$ is unique in the sample, it follows that $m \leq |S| \leq 2^3 m$. This means that the input parameter m of the triple pattern sampling component corresponds to a lower bound on the size of S. Furthermore, we also want to examine the response time for triple patterns which produce no results for the KG (i. e., $[[tp]]_G = \emptyset$). Thus, we extend the sampling component to add m randomly generated patterns to the sample S with $[[tp]]_G = \emptyset$. Patterns that produce empty result sets are generated by randomly selecting triple patterns from the sample and randomly replacing constants with URIs not contained in the KG.

The aforementioned sampling method is implemented in our Triple Pattern Fragment Profiler (see Footnote 1) to conduct the empirical studies. The integration of the sampling methodology in the TPF Profiler to examine the performance of TPF servers is shown in Fig. 1. The required input is the sample size m as well as the URL of the TPF server to access the KG G. Based on this

[3] Although the formal definition specifies that G_{tp} is a sequence of RDF triples [14], for the sake of simplicity, we define G_{tp} as a set of RDF triples.

input, the triple pattern sample set S is generated. Thereafter, in the Evaluation Component, each triple pattern in the sample is requested sequentially at the TPF server and the response time for each request is recorded. A detailed presentation of the implementation and the setup of the experimental studies is given in the following section.

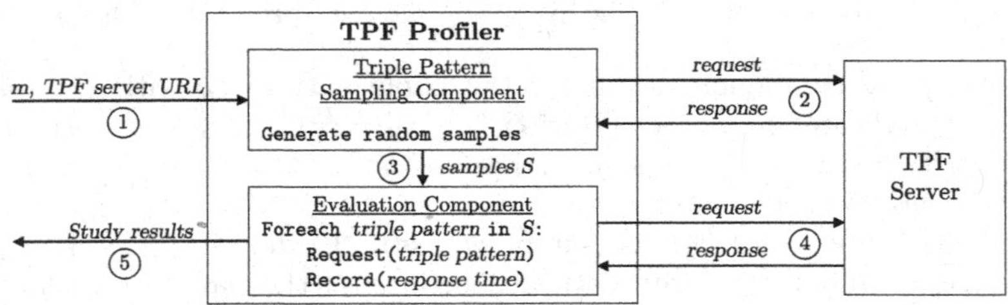

Fig. 1. Overview of the Triple Pattern Fragment Profiler. The numbers indicate the execution sequence.

4 Experimental Settings

We provide a detailed description of the settings used to assess the performance of TPF server in different conditions using the presented approach. This includes environment and implementation, backend and page size, the selection of the knowledge graphs, sample size determination, and reported metrics. To ensure repeatability of our study, the TPF Profiler, as well as the HDT files containing the KGs, are available at https://doi.org/10.5281/zenodo.1211621 under the BSD 3-Clause license. In addition, we provide the raw data and the analysis tools to ensure reproducibility of our experimental results (see Footnote 9) under CC BY 4.0.

Environment and Implementation. We conducted our study in two types of environments: a real-world environment by accessing autonomous TPF servers, and a controlled environment using a dedicated server. As a result, we compare the querying costs in two environments which differ in networking conditions, server workload, and possibly hardware capabilities. In the real-world environment, we accessed the TPF servers available at the official portal of Linked Data Fragments[4]. For the controlled environment, we deployed the TPF server v2.2.3 [12] on a Debian GNU/Linux 8.6 64-bit machine with CPU AMD Opteron 6204 3.3 GHz (4 physical cores) and 32 GB RAM. The TPF profiler is implemented in `Python 2.7.9` and executed on the same server instance to avoid network latencies accessing the TPF server in the controlled environment.

Backend and Page Size. We consider different backends as well as page sizes for each backend to provide a detailed insight into how the configuration of the

[4] http://linkeddatafragments.org.

TPF server impact on its performance. To our knowledge, the backend used in the publicly available servers is HDT files and thus is the only backend type in the real-world environment. In the controlled environment, both HDT files as well as SPARQL endpoints are used as backends since they are both suitable for querying large KGs. The SPARQL endpoints are set up using Virtuoso Open Source Edition Version 7.2.4[5]. Since the settings of the real-world server may not be changed, the page size p_{max} is set to 100 answers per page. In the controlled environment, four different page size settings are investigated: $p_{max} = \{100, 500, 1000, 10000\}$. The configuration files for both the TPF server and the Virtuoso SPARQL endpoint are provided in our repository (see Footnote 1).

Table 2. Characterization of the knowledge graphs studied in the evaluation. The namespace `ldf` corresponds to http://data.linkeddatafragments.org/.

Knowledge graph	# Triples	# Subjects	# Predicates	# Objects	Server IRI
DBLP	88,150,324	5,125,936	27	36,413,780	`ldf:dblp`
DBpedia	377,367,913	30,458,591	57,465	145,396,686	`ldf:dbpedia2014`
GeoNames	123,020,821	8,345,450	26	42,728,317	`ldf:geonames`
Wiktionary	64,358,375	10,163,240	27	21,554,657	`ldf:wiktionary`

Knowledge Graphs (KGs). Having both a controlled and a real-world environment requires KGs which are both publicly accessible via TPF servers and are available for download to be hosted locally in the controlled environment. Therefore, we selected four well-known KGs available at the official portal of Linked Data Fragments (see Footnote 4) from different knowledge domains: publications (DBLP), geography (GeoNames), linguistics (Wiktionary), and cross-domain (DBpedia). The selected KGs differ in their size (i. e., the number of RDF triples) as well as the number of distinct subjects, predicates and objects (cf. Table 2). As the basis for the controlled environment, we use the identical HDT files as in the real-world environment to ensure comparability. The N-Triples files for Virtuoso are generated from these HDT files using the `hdt2rdf` tool[6] and the characterization in Table 2 is derived from the HDT files using the `hdtInfo` tool (see Footnote 6).

Generated Samples. For our study, we generated samples of triple patterns to be executed over the selected KGs following the sampling method described in Sect. 3. A key aspect of the evaluation settings for sampling is determining an appropriate sample size, i.e., the parameter m. In the following, we describe how we determined m empirically such that m fulfills the conditions: (i) m is large enough to cover a variety of data to represent the general characteristics of the KG, and (ii) m is small enough to efficiently assess the performance of the servers in a feasible fashion. A basic approach to set m is a sample size relative to the size

[5] https://virtuoso.openlinksw.com.
[6] https://github.com/rdfhdt/hdt-cpp.

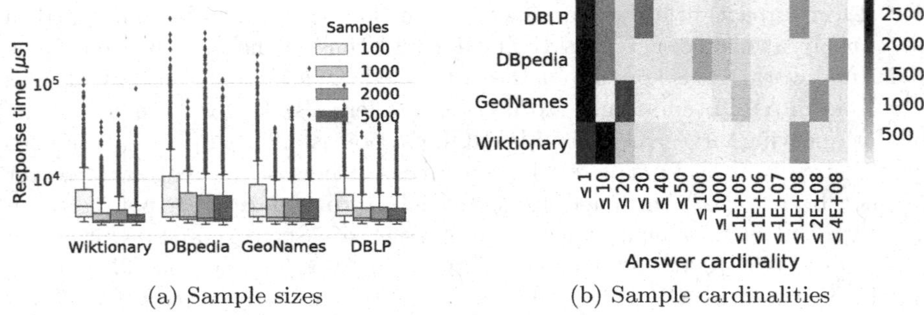

(a) Sample sizes (b) Sample cardinalities

Fig. 2. Sample properties. On the left, the overall response times with respect to the sample size m are shown. On the right, the distribution of the answer cardinalities for a sample of size $m = 1000$ is shown.

of the KG (e. g., 1% of all triples). However, the major drawback of this approach is that for large datasets, a large number of requests needs to be performed on the TPF to acquire the sample. For instance, a sample size with 1% of DBpedia with ~400M RDF triples would require approximately 4M requests. Firstly, this is not feasible with respect to the overall run time of the study and, secondly, a larger sample size does not necessarily entail a more representative sample.[7] To verify this, we measured the performance of TPFs in a controlled environment (HDT backend) when varying the sample size $m = \{100, 1000, 2000, 5000\}$ on the studied KGs. The results reported in Fig. 2a reveal that there is no substantial difference in the response times while increasing m (for $m \geq 1000$) in all KGs. Furthermore, we inspected the answer cardinality distribution produced by the triple patterns in the sample obtained with 1000 RDF triples. Figure 2b indicates that the sample generated for each studied KG contains triple patterns with a wide range of answer cardinalities. Therefore, in this study, we set $m = 1000$.

Metrics. The metric to assess the cost of querying TPF server is the response time. In this work, response time t is defined as the elapsed between sending a request and the arrival of its response. Therefore, the time for retrieving the first fragment page of a requested triple pattern is measured. To measure t in our implementation, we use the Python library `requests`[8] v2.18.4. in which the elapsed time between the request and the response merely considers the time until the parsing of the headers is completed and therefore, is not affected by the size of the response's content. We report the measurements of response time in microseconds (μs).

We conducted an extensive experimental study to answer the research questions stated in Sect. 1. At the core of the study is the identification of the independent variables and their effect on TPF performance. Table 3 summarizes the five

[7] Since the likelihood of samples having RDF terms in common increases for larger samples resulting in the same triple pattern derived from the sample.

[8] http://docs.python-requests.org/en/master.

Table 3. Overview of the independent variables analyzed per knowledge graph in the evaluation of the experimental study. R and V correspond to RDF terms and variables, respectively. G denotes a knowledge graph and tp a triple pattern.

Independent variable	# Levels	Levels
Environment	2	{controlled, real-world}
Backend	2	{HDT, Virtuoso}
Page size	4	$p_{max} = \{100, 500, 1000, 10000\}$
Triple pattern type	2^3	$(R \cup V) \times (R \cup V) \times (R \cup V)$
Knowledge graph	4	{DBLP, DBpedia, Geonames, Wiktionary}
Answer cardinality	$\|G\|$	$\|[[tp]]_G\| \in [0, \|G\|]$

variables analyzed and their levels with respect to their impact on the response time in the experimental evaluation. The results of our experiments investigating the different levels of these variables yield ≈590000 measurements.

5 Empirical Results

In this section, we present and analyze the results of our experimental studies. Essential to the evaluation of the experimental studies are the methods for analyzing the results. The goal of the analysis is determining the impact of the independent variables on the dependent variable, i. e., the response time. For this purpose, we propose the use of several statistical methods. The method selection primarily depends on the type of the independent variable and dependent variable. For the analysis in our study, the dependent variable response time is continuous. The independent variable, however, is discrete for the answer cardinality and page number and categorical for all others. Therefore, we apply a correlation analysis for the answer cardinality and paginating. For all other independent variables, we report on the significance of the difference between the categories. In our case, the observed variable response time is not normally distributed[9] and thus, we use the non-parametric Kruskal-Wallis test [7] to test our hypothesis. For the sake of brevity, we provide a rather graphical evaluation in this paper. Nevertheless, the complete statistical analysis as well all visualizations of the following evaluation are provided online (see Footnote 9).

Triple Pattern Type. First, we address the research question, whether the triple pattern type has an impact on the response time. In this evaluation, we merely consider the page size $p_{max} = 100$ to allow for the comparability with the real-world environment. In Fig. 3 the results are visualized as a boxplot separated by triple pattern type including `empty` for triple patterns with an empty answer set. The results are listed for the different KGs in both the controlled (with HDT and Virtuoso backend) and real-world environment. Additionally, the mean

[9] https://doi.org/10.5281/zenodo.1211621.

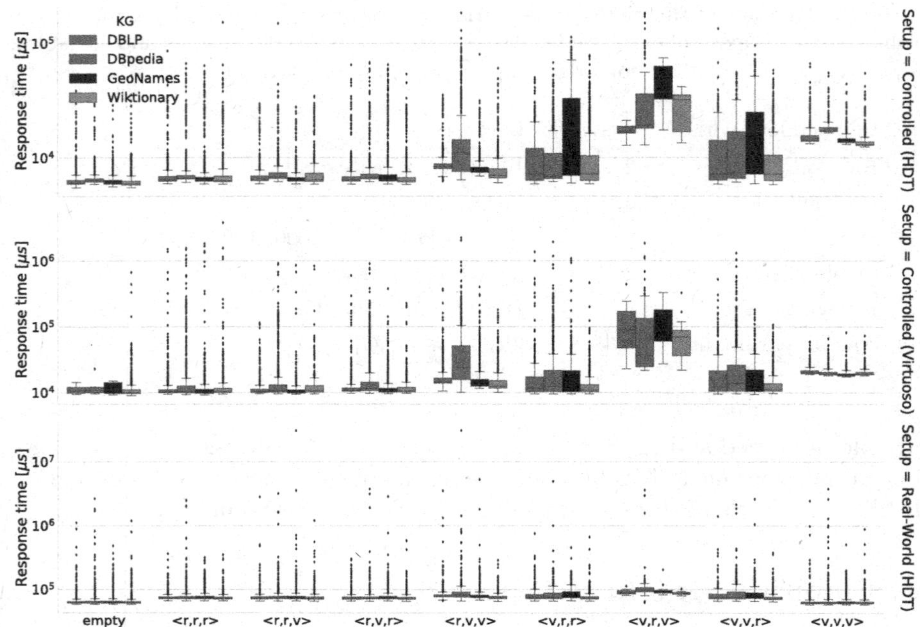

Fig. 3. Boxplot of the response time for the different pattern types.

number of answers per pattern type is listed in Table 4 for the different KGs. The results reveal a difference in response time for the different pattern types. Conducting a Kruskal-Wallis test, we find that the difference between the groups (i. e. pattern types) is statistically significant at a level $\alpha = 0.05$ in both the controlled and real-world environment. In more detail, the response times for some pattern types differ more prominently from the other pattern types. For instance, the triple pattern type $\langle v, r, v\rangle$ shows the highest (median) response times for all KGs in all environments. Note that $\langle v, r, v\rangle$ denotes the triple pattern composed of variable, constant, variable (in that order), but the variables are not necessarily the same. Furthermore, the $\langle v, v, v\rangle$ pattern type yields the second highest response time in the controlled environment (for both HDT and Virtuoso backend), except for DBpedia in which case the $\langle r, v, v\rangle$ yields a higher response time for the Virtuoso backend. Intriguingly, this is not true in the real-world environment, in which the $\langle v, v, v\rangle$ pattern type has one of the lowest response times. This might be due to the fact, that the results for the pattern are cached in the real-world environment as it is requested more frequently. Comparing the backends in the controlled environment, it can be observed that the pattern types have a similar impact on the response. However, the average response time with the Virtuoso backend (21.7 ms) is more than twice as high than for the HDT backend (10.2 ms).

Moreover, compared to the other KGs, the response times for GeoNames are notably higher for the pattern types $\langle v, r, r\rangle$ and $\langle v, v, r\rangle$. Taking the mean number of answers for these pattern types into consideration, we observe that these pattern types also yield the most answers on average. The previous observation may lead to the assumption that merely the higher answer cardinalities are the reason for higher response times.

Table 4. Mean answer cardinality of the triple patterns in the sample listed for the different triple pattern types and the different KGs.

	$\langle r, r, r\rangle$	$\langle r, r, v\rangle$	$\langle r, v, r\rangle$	$\langle r, v, v\rangle$	$\langle v, r, r\rangle$	$\langle v, r, v\rangle$	$\langle v, v, r\rangle$	$\langle v, v, v\rangle$
DBLP	1.00E + 00	2.37E + 00	1.19E + 01	2.23E + 01	3.92E + 03	4.18E + 06	3.94E + 03	8.82E + 07
DBpedia	1.00E + 00	4.85E + 00	2.82E + 01	5.55E + 01	2.09E + 03	9.58E + 05	3.50E + 03	3.77E + 08
GeoNames	1.00E + 00	1.06E + 00	8.11E + 00	1.50E + 01	3.36E + 04	5.13E + 06	3.58E + 04	1.23E + 08
Wiktionary	1.00E + 00	5.83E + 00	6.95E + 00	1.32E + 01	4.57E + 04	3.50E + 06	1.18E + 05	6.44E + 07

Answer Cardinality. The response times with respect to the answer cardinality are shown in Fig. 4. The results reveal a similar trend for both the controlled and the real-world environment. There is an increase in the response time up to ≈100 answers and thereafter the response times appear to be rather steady. As the page size in this visualization is $p_{max} = 100$ answers per page, the results suggest that the difference may be related to the page size. To quantify this relation, we report on the correlation coefficient ρ, which allows for measuring the strength and direction of the linear correlation between two variables. Table 5 lists the correlation coefficients for answer cardinality and response time for samples when: (i) the answers fit in one fragment page ($\rho_{\leq p_{max}}$), (ii) pagination is required to dereference the fragment ($\rho_{>p_{max}}$), and (iii) the overall correlation coefficient (ρ). The correlation is reported for the different backends and page sizes in the controlled environment as well as the correlation for the HDT backend and page size $p_{max} = 100$ in the real-world environment.

In Table 5, a weak positive correlation ($\rho \in [0.5, 0.75]$) is highlighted with a light color and a stronger positive correlation ($\rho \in [0.75, 1]$) with a dark color. The results for Virtuoso reveal more often and stronger positive correlations between the answer cardinality and the response time. This indicates that HDT is more efficient in querying the patterns regardless of the answer cardinality. This is also visible in Fig. 4, where there are more outliers with respects to the overall trend for HDT backend in the controlled environment. In contrast, the Virtuoso backend appears to have an increasing lower bound of the response time for higher answer cardinalities. Moreover, it can be observed that the page size also influences the correlation in the controlled environment since the correlation is only present for page size $p_{max} > 100$. In the real-world environment, no correlation can be observed as most correlation indices are close to zero. This suggests that exogenous factors, e.g. network delays and server load, affect the response time, such that the differences induced by the answer cardinality vanish.

Page Size. We measure the performance of TPFs for different page size settings. We report on throughput, i.e., the number of answers produced per time unit.

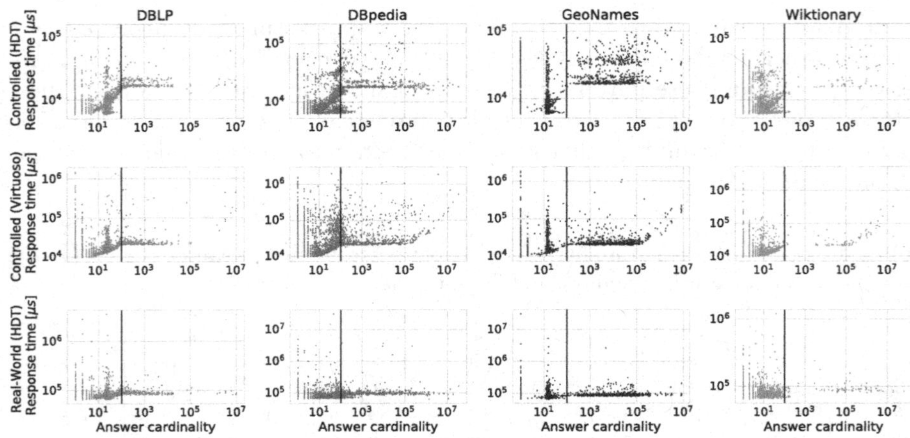

Fig. 4. Response time in the controlled environment and real-world environment with respect to the number of answers for all samples and each KG for all samples except the $\langle v, v, v\rangle$ pattern type. The black line indicates 100 results (page size).

Table 5. Correlation of answer cardinality and response time in the controlled and real-world environment for varying page size and backend. Correlation coefficient for samples whose answers fit in one fragment page ($\rho_{\leq p_{max}}$), when pagination is required to dereference the fragment ($\rho_{>p_{max}}$), and the overall correlation (ρ).

Environment		Controlled								Real-World
Page Size p_{max}		100		500		1000		10000		100
Backend		HDT	Virt.	HDT	Virt.	HDT	Virt.	HDT	Virt.	HDT
DBLP	$\rho_{\leq p_{max}}$	0.266	0.078	0.141	0.842	0.858	0.913	0.979	0.734	0.039
	$\rho_{>p_{max}}$	0.100	0.441	0.262	0.885	0.229	0.896	0.049	0.727	-0.059
	ρ	0.100	0.166	0.051	0.522	0.269	0.420	0.475	0.451	0.004
DBpedia	$\rho_{\leq p_{max}}$	0.297	0.143	0.301	0.709	0.301	0.918	0.916	0.679	0.020
	$\rho_{>p_{max}}$	-0.014	0.185	0.039	0.735	0.039	0.576	-0.000	0.171	0.000
	ρ	0.036	0.129	0.101	0.525	0.101	0.381	0.130	0.219	0.002
Geo-Names	$\rho_{\leq p_{max}}$	-0.032	0.041	0.130	0.777	0.878	0.963	0.983	0.997	-0.006
	$\rho_{>p_{max}}$	0.169	0.292	0.158	0.890	0.318	0.701	0.354	0.328	0.004
	ρ	0.179	0.127	0.136	0.465	0.238	0.295	0.276	0.253	0.001
Wikt-ionary	$\rho_{\leq p_{max}}$	0.047	0.013	0.207	0.388	0.239	0.302	0.667	0.986	0.071
	$\rho_{>p_{max}}$	-0.027	0.207	-0.019	0.204	-0.063	0.211	-0.263	0.025	-0.077
	ρ	0.059	0.029	0.061	0.281	0.124	0.267	0.193	0.237	0.009

We compute the throughput per fragment page (denoted Θ) when evaluating each triple pattern $tp \in S$ over a KG G as follows:

$$\Theta(tp) := \frac{\min\{|[[tp]]_G|, p_{max}\}}{t(tp)} \; [answers/\mu s], \tag{1}$$

where $|[[tp]]_G|$ is the answer cardinality, p_{max} the page size and $t(tp)$ the response time. Figure 5 shows the mean Θ for all KGs, different backends and page size in the controlled environment. The relative changes in throughput are listed in Table 6. For the HDT backend, the results reveal an increase in throughput for bigger page sizes in most cases. The biggest increase is achieved by setting the TPF page size from 100 to 500 answers per page, in all KGs. When increasing the page size further from 1000 to 10000, the throughput even slightly decreases. Similar results are observable for the Virtuoso backend. In contrast to HDT, however, the throughput is not affected as strongly: it merely improves half as much when increasing the page size from 100 to 500 answers per page. Thus, the improvement is less significant than it is for the HDT backend.

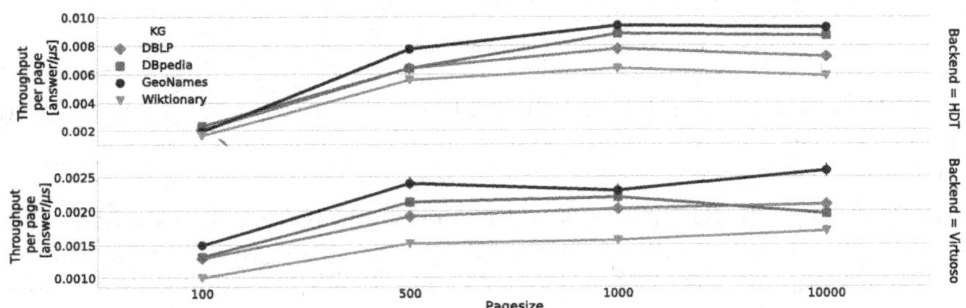

Fig. 5. Mean TPF throughput (Θ) for all KGs, different backends and page sizes.

Table 6. Relative changes in throughput Θ for increasing the TPF page size.

	DBLP		DBpedia		GeoNames		Wiktionary	
	HDT	Virtuoso	HDT	Virtuoso	HDT	Virtuoso	HDT	Virtuoso
From 100 to 500	0.677	0.329	0.636	0.380	0.751	0.376	0.710	0.345
From 500 to 1000	0.174	0.052	0.274	0.031	0.175	−0.047	0.124	0.029
From 1000 to 10000	−0.072	0.033	−0.013	−0.119	−0.011	0.115	−0.085	0.083

Paginating. As the preceding evaluation shows, the page size has an impact on the throughput of TPFs. Moreover, the magnitude of the impact varies for the backend types. The throughput merely considers retrieving at most the first p_{max} answers. However, for $|[[tp]]_G| > p_{max}$ paginating is required to obtain all answers. To examine how the response time varies when paginating, we dereference the first 10000 pages for the $\langle v, v, v\rangle$ pattern. The results are presented in Fig. 6. It can be observed that for the HDT backend, paginating yields a rather constant response time, while a steady increase can be observed for the Virtuoso backend. These observations are supported by the correlation coefficient ρ, which clearly indicates a strong positive correlation between page number and response time for the Virtuoso backend with $\rho = 0.920$ and no correlation for the HDT backend with $\rho = -0.036$. The two previous evaluations suggest that the page size configuration for TPF server needs to consider the backend. For instance, for the Virtuoso backend increasing the page size might be suitable as it increases throughput and paginating is increasingly costly. As the response time for paginating is rather constant for the HDT backend, bigger page sizes may allow for exploiting the increased throughput and further reduce the necessity of paginating. However, the adjustments need to take the answer cardinality distribution of the KG into consideration as well.

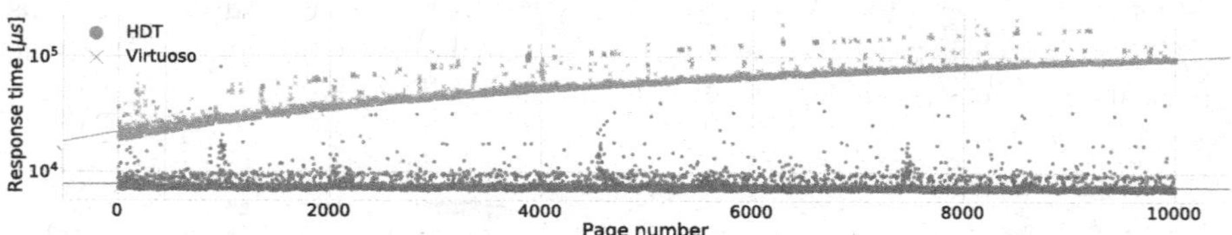

Fig. 6. TPF response time for paginating the first 10000 pages for the triple pattern $\langle v, v, v \rangle$ for both HDT and Virtuoso backend in the controlled environment.

KG/TPF Instance Relation. Finally, we examine the results when making all KGs simultaneously available in the TPF server for each backend. The mean response time for all KGs (Multiple KG) and one KG (Single KG) available at a time are presented in Fig. 7. The results show an increase in the TPF response time when making multiple KGs simultaneously available regardless of the backend. Intriguingly, the increase is higher for the smaller KGs and lower for larger KGs for the HDT backend and the opposite holds for the Virtuoso backend. These observations may be due to the index created by the TPF server when making several KGs available simultaneously.

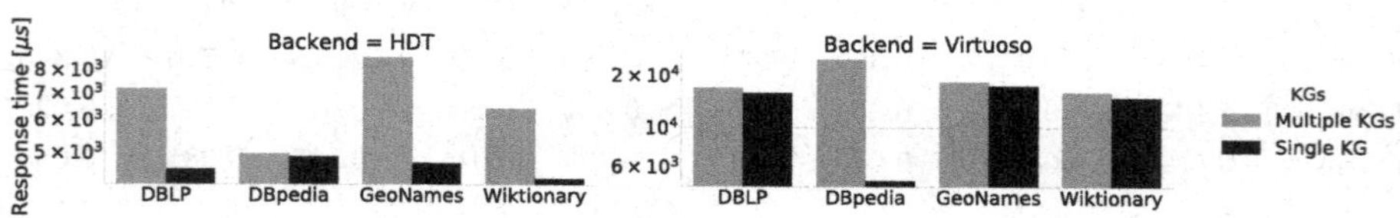

Fig. 7. Relation KGs/TPF instance. Mean response time for both backends with all KGs available and single KGs available at a time.

6 Conclusions and Future Work

We have proposed an approach to assess the cost of querying knowledge graphs (KGs) over Triple Pattern Fragments (TPFs). The presented TPF Profiler includes a sampling component able to generate triple patterns from KGs and an evaluation component to capture the response time for the sampled triple patterns. The results allow for conducting fine-grained analyses identifying factors that impact on server performance. We conducted an empirical study using the proposed approach to evaluate the TPF servers in controlled and real-world environments using diverse well-known KGs and different TPF configurations. Thereafter, we conducted a fine-grained analysis studying the impact of triple pattern type, answer cardinality, backend, page size, environment type, and KGs per TPF server on the response time. To conclude our findings, we answer the research questions stated in Sect. 1.

Answer to RQ1. Our empirical study confirms that the type of triple pattern has a significant impact on the response time of TPF servers regardless of the backend.

Answer to RQ2. Empirical results reveal different behaviors of TPFs depending on the answer cardinality of the triple patterns. For triple patterns that produce answers fitting in one page, answer cardinality is rather positively correlated with response time.

Answer to RQ3. The results of the experimental studies indicate an improved throughput for increasing page sizes. However, the throughput decreases for large page sizes again and the relative improvement is higher for the HDT backend.

Answer to RQ4. Our experimental study reveals significant differences in the response times between different backend types. Overall, the HDT backend outperforms the Virtuoso backend and allows for querying triple patterns more efficiently.

Answer to RQ5. Empirical results suggest that in real-world environments the impact of the analyzed variables on response time is reduced as exogenous factors increasingly affect the response times of TPF servers. In addition, querying autonomous real-world servers can be orders of magnitude more costly.

Answer to RQ6. Our experimental study reveals that accessing multiple KGs through a single TPF server negatively impacts on server performance.

At first sight, the absolute measured impact (in seconds) of some factors on the response time might not appear very high. However, the results in our study report on the response time for retrieving the first page of a fragment. Thus, the response times for paginating to retrieve all answers of a request needs to be considered as well. Moreover, the evaluation of SPARQL queries over TPFs typically requires submitting a large number of requests to the TPF server, thus, leveraging the observations may improve the overall query execution time.

Our future work will focus on integrating more variables into our analysis, studying different variable levels (e. g., different SPARQL endpoint implementations) and gathering additional data for KGs with different characteristics.

Acknowledgments. The authors thank Ruben Verborgh for providing feedback and the KG dumps and Javier Fernández for the fruitful discussions about HDT. This work was carried out with the support of the German Research Foundation (DFG) within the project "Sozial-Raumwissenschaftliche Forschungsdateninfrastruktur (SoRa)".

References

1. Acosta, M., Vidal, M.-E.: Networks of linked data eddies: an adaptive web query processing engine for RDF data. In: Arenas, M., et al. (eds.) ISWC 2015. LNCS, vol. 9366, pp. 111–127. Springer, Cham (2015). https://doi.org/10.1007/978-3-319-25007-6_7
2. Buil-Aranda, C., Hogan, A., Umbrich, J., Vandenbussche, P.-Y.: SPARQL web-querying infrastructure: ready for action? In: Alani, H., et al. (eds.) ISWC 2013. LNCS, vol. 8219, pp. 277–293. Springer, Heidelberg (2013). https://doi.org/10.1007/978-3-642-41338-4_18
3. Fernández, J.D., Martínez-Prieto, M.A., Gutierrez, C.: Compact representation of large RDF data sets for publishing and exchange. In: Patel-Schneider, P.F., et al. (eds.) ISWC 2010. LNCS, vol. 6496, pp. 193–208. Springer, Heidelberg (2010). https://doi.org/10.1007/978-3-642-17746-0_13
4. Folz, P., Skaf-Molli, H., Molli, P.: CyCLaDEs: a decentralized cache for triple pattern fragments. In: Sack, H., Blomqvist, E., d'Aquin, M., Ghidini, C., Ponzetto, S.P., Lange, C. (eds.) ESWC 2016. LNCS, vol. 9678, pp. 455–469. Springer, Cham (2016). https://doi.org/10.1007/978-3-319-34129-3_28
5. Hartig, O., Buil-Aranda, C.: Bindings-restricted triple pattern fragments. In: Debruyne, C. (ed.) OTM 2016. LNCS, vol. 10033, pp. 762–779. Springer, Cham (2016). https://doi.org/10.1007/978-3-319-48472-3_48
6. Kjernsmo, K., Tyssedal, J.S.: Introducing statistical design of experiments to SPARQL endpoint evaluation. In: Alani, H., et al. (eds.) ISWC 2013. LNCS, vol. 8219, pp. 360–375. Springer, Heidelberg (2013). https://doi.org/10.1007/978-3-642-41338-4_23
7. Kruskal, W.H., Wallis, W.A.: Use of ranks in one-criterion variance analysis. J. Am. Stat. Assoc. **47**(260), 583–621 (1952)
8. Montoya, G., Vidal, M.-E., Corcho, O., Ruckhaus, E., Buil-Aranda, C.: Benchmarking federated SPARQL query engines: are existing testbeds enough? In: Cudré-Mauroux, P. (ed.) ISWC 2012. LNCS, vol. 7650, pp. 313–324. Springer, Heidelberg (2012). https://doi.org/10.1007/978-3-642-35173-0_21
9. Pérez, J., Arenas, M., Gutierrez, C.: Semantics and complexity of SPARQL. In: Cruz, I., et al. (eds.) ISWC 2006. LNCS, vol. 4273, pp. 30–43. Springer, Heidelberg (2006). https://doi.org/10.1007/11926078_3
10. Rakhmawati, N.A., Karnstedt, M., Hausenblas, M., Decker, S.: On metrics for measuring fragmentation of federation over SPARQL endpoints. In: WEBIST, pp. 119–126 (2014)
11. Schmachtenberg, M., Bizer, C., Paulheim, H.: Adoption of the linked data best practices in different topical domains. In: Mika, P. (ed.) ISWC 2014. LNCS, vol. 8796, pp. 245–260. Springer, Cham (2014). https://doi.org/10.1007/978-3-319-11964-9_16
12. Verborgh, R.: Linkeddatafragments/server.js: v2.2.2, May 2017. https://doi.org/10.5281/zenodo.570148
13. Verborgh, R., et al.: Querying datasets on the web with high availability. In: Mika, P., et al. (eds.) ISWC 2014. LNCS, vol. 8796, pp. 180–196. Springer, Cham (2014). https://doi.org/10.1007/978-3-319-11964-9_12
14. Verborgh, R., et al.: Triple pattern fragments: a low-cost knowledge graph interface for the web. Web Semant. Sci. Serv. Agents World Wide Web **37**, 184–206 (2016)

Drammar: A Comprehensive Ontological Resource on Drama

Vincenzo Lombardo[1(✉)], Rossana Damiano[1(✉)], and Antonio Pizzo[2]

[1] CIRMA and Dipartimento di Informatica, Università di Torino, Torino, Italy
rossana.damiano@unito.it

[2] CIRMA and Dipartimento di Studi Umanistici, Università di Torino, Torino, Italy

Abstract. This paper reports about the release of a comprehensive ontological resource on drama, called Drammar. Drama is pervasive across cultures and is realized through disparate media items. Drammar has been designed with the goals to describe and encode the core dramatic qualities and to serve as a knowledge base underlying a number of applications. The impact of the resource is displayed through its direct application in a few tasks and its extension to serve in novel projects in the digital humanities.

Keywords: Drama · Wiki · Digital humanities

1 Introduction

A drama is a story conveyed through characters who perform live actions: for example, theatrical plays (e.g., Shakespeare's *Hamlet* in screenplay, performance, movie formats, respectively), TV series (HBO's *Sopranos*[1]), reality shows (CBS's *Survivor*[2]), and some videogames (Ubisoft's *Assassin's Creed*[3]). Drama is pervasive across cultures and ages [26] as well as across media, the latter named dramatic media in [9]. A single drama can assume several forms, fulfilling a number of its core conditions. For example, the abstraction of the oral tale *Cinderella* has, e.g., Perrault's [34] and Disney's [1] versions.

This paper presents an ontology for describing the domain of drama, called Drammar. The encoding of the major concepts and relations of the drama domain must address a vast field of research where scholars have addressed several topical notions, such as genre or writing style. Drammar, in particular, addresses the so–called *dramatic qualities*, that is those elements that are necessary for the existence of a drama, shared by a number of analyses of drama scholarship, e.g. [14,18,38]. Such element, namely story units, characters or agents, actions, intentions or plans, goals, conflicts, values at stake, and emotions are partially taken into account in a number of annotation projects, where media

[1] http://www.hbo.com/the-sopranos, visited on 11 June 2018.

[2] http://www.cbs.com/shows/survivor/, visited on 11 June 2018.

[3] https://www.ubisoft.com/en-US/game/assassins-creed/, visited on 11 June 2018.

D. Vrandečić et al. (Eds.): ISWC 2018, LNCS 11137, pp. 103–118, 2018.
https://doi.org/10.1007/978-3-030-00668-6_7

chunks (e.g., text paragraphs, video segments, etc.) are annotated for the sake of studying the relationships between the linguistic expressions and the drama content.

Drammar[4] is the first comprehensive ontology of the dramatic qualities; it makes the knowledge about drama available as a vocabulary for the linked interchange of drama encodings and readily usable by automatic reasoners.

2 Related Work on Drama Domain Encoding

In the last decade, the emerging technologies for media indexing and retrieval have prompted a number of initiatives that leverage structured representations of the dramatic content. Elson has introduced a template-based non-standardized representation language for describing the content of narrative texts, with the goal of creating a corpus of annotated stories, called DramaBank [8]; more recently, minimal annotation schemata have been targeted at grasping the regularities of written and oral narratives at the discourse level, by relying on quantitative approaches, which can overcome the difficulties of recruiting annotators [36]. All these initiatives, however, introduce representations that are task-oriented, i.e. they tend to focus on the realization of narratives though a specific medium (written tales), and lack the capability to represent the universal elements of dramatic narration that go behind the expressive characteristics of each medium, with no attempts for standardization and data linking.

In parallel with these trends in text annotation, the general media annotation has evolved towards the use of ontologies to describe the contents, given the languages and resources made available by the Semantic Web [5]. Ontologies and vocabularies have appeared that support the representation of the media content according to a shared semantics, available across the Web according to the paradigm of Linked Data [15]. In particular, semantic resources such as VERL (the Video Event Representation Language, described in [11]), provide tools for the structured description of events that can be applied also to the description of incidents in stories. A media–independent model of story is provided by the OntoMedia ontology, exploited in the Contextus Project[5] (see [16] and [19]) to annotate the narrative content of media objects which range from written literature to comics and TV fiction. This project encompasses some concepts that are relevant for the description of drama, such the notion of character; however, being mainly targeted at the comparison of story events and timelines across media in crossmedia contexts, it lacks the capability of representing the core notions of drama. In the field of cultural heritage dissemination, the StorySpace ontology is an ontology that supports museum curators in linking the content of artworks through stories, with the ultimate goal of enabling the generation of user tailored content retrieval (see also [29]). However, the representation of story provided by StorySpace is functional to the creation of story repertories

[4] http://purl.org/drammar.

[5] Registered at http://lov.okfn.org/dataset/lov/vocabs/stories, visited on 11 June 2018.

for curating activities; as such, it is not committed to a comprehensive account of the narratological theory, and lacks some crucial elements for drama ontology, such as the notion of character.

3 Drama Domain and Drammar Ontology

Drama is media independent, including Sofocles' texts, reality shows, and even some videogames within the same domain. Our approach avoids references to style and artistic qualities by aiming at representing the elements shared by different, cross-media manifestations of drama, the so–called intangible elements of such a cultural heritage form [24]. Bazin speaks of "dramatic elements" as"interchangeable between one art and another" [4]. The model of drama assumed here lays at the basis of the system intended to produce a dramatic manifestation, "an action played live by characters" [41]. Any drama, beyond the form it takes, produces in the audience the perception of something, intuitively called 'story', directly enacted by characters. Drama, differently from literature, must show some characters in their actions and such actions should not be reduced to the mere description of a movement, but to a manifestation of some intention, as discussed by Styan [39]. So, actions, organized into bounded story segments, stem (more or less straightforwardly) from characters' internal motivations and, at the same time, provide information on the characters themselves and their goals. Above all, stand conflicts: Styan opens his essay by showing the difference between an ordinary conversation and a dramatic dialogue.

3.1 The Dramatic Qualities

The dramatic qualities we are taking into account have been distilled after a thorough analysis of the drama literature and discussions between drama scholars and ontology engineers in a wiki[6]. So, although the cultural object known as "drama" includes many features that we have neglected in the representation (such as genre, topic, writing style, and even *Weltanschauung*), Drammar includes those elements that are deemed as necessary in the literature, and that can be grouped under the following four categories: *action*, *agent*, *conflict*, and *segmentation*. The description of these elements has provided the requirements for the design of the ontology; the vast literature on drama that has been reviewed is surveyed and discussed in the wiki.

Action. The word action signifies an intentional, purposive, conscious and subjectively meaningful activity. It is done by an agent and it is the expression of a will, thus involving a goal, an intentionality. It is a key concept of dramatic theory: the etymological roots of the words "drama" and"dramaturgy" themselves stem from the ancient Greek verb dràò, which means "to do","to act", intended as performing an action following a human deliberation.

[6] https://www.di.unito.it/wikidrammar, visited on 11 June 2018.

From the literature, we know that action is a foundational element of drama, responding to a logic of cause and effect and motivated by a character's goal. Actions are part of a character behavior that has some degree of unity and wholeness, the enactment of a character's deliberation, defining the character function in the plot.

Agent. The term "agent" is preferred to "character" because of the focus on the structural elements of the dramatic action rather than the psychological, moral, social, or political entity that comes out of the narrative as a cognitive product of the audience. From our modeling point of view, the notion of agent does not take into account the historical differences between the hero in the Greek Tragedy, the romantic protagonist, or the modern character, but we focus on its agentive qualities.

From the literature, we know that an agent has intentions and mental states, so to motivate his/her behavior, does actions, as initiators or as reactions to others' actions. An agent interacts with the environment and the other agents, and displays emotions. An agent is the medium of representation in drama.

Conflict. Conflict is the fundamental principle of dramatic theory, ubiquitous in the history of dramatic theory and critics, the expression of a tension, achieved through the opposition of characters. Conflict is traditionally indicated as the force that motivates the character's changes. Nevertheless, it reached its modern meaning only during the growth of the new "serious genre" (late 18th century), when it took on more specific and definite traits.

From the literature, we have that conflict is an opposition between agents that arises from the presence of differences in agents' goals and/or (moral, ethical, political) values or between an agent and some situation occurring. A conflict is represented by an obstacle and provokes an emotional response in the agent.

Segmentation. Since its origins, dramatic theory has considered drama as a unitary whole, but consisting of different parts. Consequently, it should be possible to segment the dramatic works in parts and analyze how these are organized in order to create the wholeness of the work. Although the literature has adopted different naming rules for the segmentation (beats, scene, sequence, acts, episode), we know that the parts of the drama are organized hierarchically; each part, at each level, has the form of the whole drama (fractal recursion). So, we resort to a more generic term to name such parts and we call them *units*. In our model of drama, the units are the containers of the agents' actions and involve reciprocal relationships, both with other units located at the same structural level, and with other higher or lower structural levels.

A short sample from Shakespeare's *Hamlet*, the so called "nunnery" scene, can clarify these elements. In this scene, situated in the Third Act, Ophelia is sent to Hamlet by Polonius (her father) and Claudius (Hamlet's uncle, the king) to confirm the assumption that Hamlet's madness is caused by his rejected love.

According to the two conspirators, Ophelia should induce him to talk about his inner feelings. At the same time, Hamlet tries to convince Ophelia that the court is corrupted and that she should go to a nunnery. In the climax of the scene, Hamlet puts Ophelia to a test to prove her honesty: guessing (correctly) that the two conspirators are hidden behind the curtain, he asks the girl to reveal where her father Polonius is. She decides to lie, by replying that he is at home. Hamlet realizes from the answer that also Ophelia is corrupted and consequently becomes very angry, realizing that there is no hope to redeem the court (and Ophelia too). The climactic incident in the scene consists of a question-answer pair:

- Hamlet: "Where is your father?"
- Ophelia: "At home, my Lord!"

This is a (very relevant) story unit: boundaries are decided through the detection of a specific goal pursuit, distinct from the goals pursued in the previous unit. Here Hamlet, one of the two characters in the unit, is pursuing the goal of proving Ophelia's honesty. Honesty is a value for Hamlet, and Ophelia's behavior is putting at stake such a value. So, he decides to pursue the goal of proving Ophelia honesty through a plan in which he asks a question he knows the answer of, i.e. the current location of her father Polonius (Hamlet is correctly convinced that Polonius is in the same room, behind a curtain), and Ophelia lies, by answering with a false location, i.e. Polonius' home. So, we can list the following elements for this unit (descriptions are provided informally, see next section for formal representations):

```
Action
- Dialogue between Hamlet and Ophelia (Question/Answer)
Agents
- Hamlet
  - Value at stake: Honesty
  - Goal: Prove Ophelia honesty
  - Plan or Intention: Asking Ophelia a rhetorical question
  - Plan accomplishment: failure
  - Emotions: Distress, Reproach, Anger
- Ophelia
   - Value at stake: Father's authority
   - Goal: Respect father's authority
   - Plan or Intention: Lying about presence of Polonius in the room
   - Plan accomplishment: success
   - Emotions: Disappointment, Joy, Shame
Conflict
- Hamlet who searches for honesty VS. Ophelia who lies
Segmentation
- Unit: Hamlet tests Ophelia for honesty
- Scene:
  - Ophelia tries to prove Hamlet madness is caused by rejected love
  - Hamlet tries to save Ophelia from corruption in Elsinor court
```

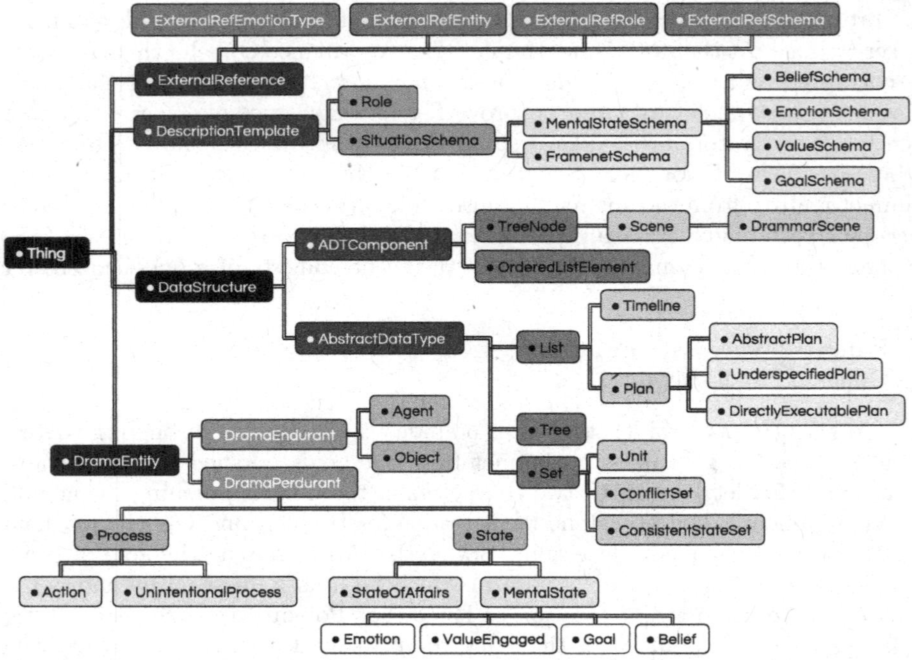

Fig. 1. Taxonomy of ontology Drammar. Colors distinguish sections of the taxonomy (four major sections); box colors desaturate while going to specific classes. (Color figure online)

3.2 The Drammar Ontology

In this section, we introduce the ontology, from the classes of the ontology organized in a taxonomy to the transversal relations over them. The resource has been growing through a number of projects that have dealt with the annotation of dramatic media [21,23], the rule-based calculation of characters' emotions [20], the characterization of drama as a form of intangible cultural heritage [24]. The resource described here is the result of a continuous stabilization due to these several projects.

The design of the taxonomy (Fig. 1) follows the well-known principle according to which a class specifies into subclasses depending on the value of a specific trait, or property. As an example, consider the class `Belief`: the concept of `Belief` is defined as the sum of the traits accumulated top-down along the taxonomy: *temporally extended entity*, for being a type of DramaPerdurant, *stative*, for being a State, *related to some agent's mind*, for being an Belief.

The top-level of Drammar contains four classes:

- `DramaEntity` is the class of the dramatic entities, i.e the entities that are peculiar to drama;

- `DataStructure` is the class that organises the elements of the ontology into common structures (namely, list, sets and trees);
- `DescriptionTemplate` contains the patterns for the representation of instantiated drama in terms of roles;
- `ExternalReference` is the class that bridges Drammar to commonsense concepts situated in external resources.

DramaEntity. `DramaEntity` groups all the peculiar elements that belong to the drama domain. It is divided into two subclasses, each describing specific drama elements: `DramaPerdurant` represents the temporally extended entities of a drama (subclass, `rdfs:subClassOf` of the class `perdurant` in DOLCE-Lite[7], as described in [12]), further subdivided into the `Process` class and the `State` class. The `Process` class (subclass of class `process` in DOLCE-Lite), represents what occurs in a drama, and subsumes `Action`, i.e. the intentional processes, and `UnintentionalProcess`; intentional processes (namely, actions) are also the basic elements of agents' plans. States are interleaved with timelines (sequences of processes grouped into units) to form the dynamics of drama. As part of plans, states form their preconditions and effects. The `State` class (subclass of `state` in DOLCE-LITE) further divides according to the entity type to which the state is attributed. The entity of attribution can be agent or world, thus yielding two subclasses: `MentalState` and `StateOfAffairs`. The subclasses of the `MentalState` class, then, acknowledges the rational vs. irrational distinction. Inspired by [10], mental states are the core of the description of the intentional behavior of agents and they belong to one of the following classes: `Belief`: the agent's subjective view of the world; `Emotion`: the emotions felt by the agent; `ValueEngaged`: the values of an agent, which are engaged (put at stake or in balance, respectively) by the unfolding of the plot; `Goal`: the objectives that motivate the actions of the agents.

`DramaEndurant` represents the time independent entities that participate into drama perdurants (corresponds to, that is `rdfs:subClassOf` the class `endurant` in DOLCE-Lite [12]); such entities are agents (class `Agent`) and objects (class `Object`), kept distinct from each other by the feature of intentionality: agents intentionally perform actions, while objects are simply involved in the actions (often called "props" in drama production).

DataStructure. Class `DataStructure` encodes the structures that provide an organization to the elements of drama. This class includes abstract data types (subsumed by the `AbstractDataType` class), i.e. ordered lists (`List` class), unordered sets (`Set` class), and hierarchical trees (`Tree` class), and their components (subsumed by the class of abstract data type components, `ADTComponent` class), such as list elements (`OrderedListElement`), set members (`SetMember`), and tree nodes (`TreeNode`). `List` is inspired by a well-known ontology, the Ordered List Ontology[8], originally developed as part of the Music Ontology [37],

[7] http://www.loa.istc.cnr.it/ontologies/DOLCE-Lite.owl.
[8] http://purl.org/ontology/olo/core#.

being music an intrinsically sequential medium; its implementation in Drammar makes some simplifications (e.g., indexes are not included), while defining a number of structures as subclasses of the `List` class. Drammar List is a subclass of the Ordered List class in OLO (`drammar:List rdfs:subClassOf olo:OrderedList`); then, plans (class `Plan`) are lists of simpler plans (or actions), timelines (class `Timeline`) are lists of units. Precedence relations are encoded for the list elements, and the first and last element of a list can be marked. Plans are further subdivided into abstract plans (class `AbstractPlan`), which represent long term intentions consisting of simpler plans, and directly executable plans (class `DirectlyExecutablePlan`), which represent short term intentions consisting of immediate actions to execute. `Set` includes structure types that gather elements of the same type (instances of the `SetMember` class, a type of data type component, or `ADTComponent` class), but where ordering is irrelevant. This is the case of units (`Unit` class), i.e., sets of processes which compose the timelines, and of state sets, which provide the precondition and effects of timelines and plans. State sets can be internally consistent (`ConsistentStateSet`), or can include conflicting elements (`ConflictSet`). The first type provides the preconditions and effects of timeline, which are typically internally consistent (in drama, as in the real world), while the latter serves the function of modeling the conflicts which may arise from the intentions (i.e., plans in Drammar) of different characters. `Tree` represents tree-like structures. In drama, tree-like structures are needed to represent the notion of scene: a scene, of larger or smaller granularity, can subsume other scenes, and can be subsumed by larger scenes. A tree contains instances of the `TreeNode` class, a type of `ADTComponent`. A `Scene` is a type of TreeNode; a `DrammarScene` is a scene defined on conflicts.

ExternalReference. Class `ExternalReference` bridges the representation of drama onto the commonsense and linguistic concepts stored in external resources. External vocabularies, such as SUMO (Standard Upper Merged Ontology, [32]) or FrameNet [3], are not directly re-used in the ontology, The `ExternalReference` class is characterised by data properties whose values point to the IRIs (or identifiers, where an IRI is not available, as in the case of WordNet[9]) of the concepts in the external resources. These properties, all subsumed by the `quale` datatype property, are intended by design to point at specific vocabularies: *quale_YAGOSUMO_concept* for the YAGOSUMO ontology; *quale_MWNSense* for MultiWordNet senses (the multilingual alignment of WordNet [35]); *quale_framenetFrame* for FrameNet. By doing so, the representation of every manifestation of drama, will be unambiguously linked to a vocabulary term on the web (or to some inner identification system of a publicly available resource when the resource itself is not available as Linked Data).

[9] https://wordnet.princeton.edu.

The ExternalReference class is divided into subclasses: ExternalRefEntity maps a perdurant (process or state) or an endurant (agent or object) onto its description: for instance, the mapping target may be the identifier of a lexical entry for describing a process (e.g., "kill" in WordNet), or the IRI of an ontology class for describing an object (e.g., concept "Weapon" in SUMO). ExternalRefSchema maps a process or state onto a verbal frame that describes it according to a role structure, with the ExternalRefRole to map the single roles onto their description in the frame (for example, the frame for "Killing" and the "Killer" role, respectively, in FrameNet). ExternalRefEmotionType maps the emotions of the characters (e.g., Fear) onto a reference model of emotions, namely Ortony, Clore, and Collins' model, known as OCC [31]. Here is a representation sample, namely the representation of attribute *prince* for individual Agent *Hamlet* in the instantiated drama ontology (the complete encoding example is published on the web site):

```
1 ###  http://www.cadmos.cirma.unito.it/drammar.owl#ExtRef_prince
2  drammar:ExtRef_prince rdf:type owl:NamedIndividual ,
3  drammar:ExternalRefEntity ;
4  drammar:originalTerm "principe"^^xsd:string ;
5  drammar:quale_MWNSense "{n#07498573} prince"^^xsd:string ;
6  drammar:quale_YAGOSUMO_concept "yago:YagoLegalActor"^^xsd:string .
```

The individual ExtRef_prince works as a pointer to the synset n#07498573 in MultiWordNet (line 5), which corresponds to the meaning intended by the annotator with the word "principe" (Italian for "prince", stored in the originalTerm data property, line 4), and, by broader match, to the concept of LegalActor in YAGO (line 6), which is retrieved from the endpoint SPARQL of the ontology.

DescriptionTemplate. Class DescriptionTemplate contains the patterns for encoding the role-based schemata. It has the purpose of mapping a situation (as intended in [13]), be it a process or state, onto its linguistic description. Its subclasses, namely Role and SituationSchema, provide the primitives to realize a role schema for describing the situation. The SituationSchema class represents the description of a situation in terms of the roles involved in it (see the Situation Description ontology pattern [13]). This class is related to the Role class through specific properties. Class SituationSchema divides into subclasses for representing specific schema types: FramenetSchema, for mapping the description of entities onto the linguistic reality encoded in lexical-semantic resources, e.g., FrameNet [3]. MentalStateSchema, for mapping the description of a mental state onto specific schemata for the different types of mental states, each of which is committed to a specific model. The MentalStateSchema further specifies into BeliefSchema, EmotionSchema, GoalSchema, ValueEngagedSchema. For example, the ValueSchema relates an agent's value engaged in a given timeline or plan (ValueEngaged) with some reference value system (which may be shared by agents).

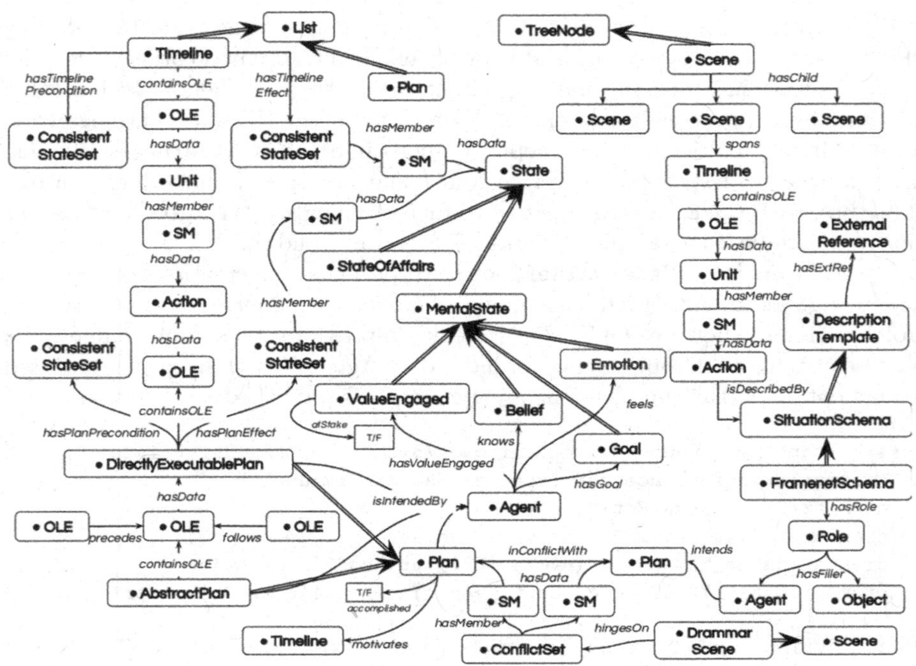

Fig. 2. Overview of Drammar. Double lines represent subsumption (fragments of the taxonomy above); solid lines represent object properties; OLE stands for class OrderedListElement; SM stands for class SetMember.

3.3 Design and Engineering of Drammar

The axioms of the Drammar ontology encode each drama element through a pattern of classes and properties, resorting to Artificial Intelligence theories and models. Figure 2 provides an overview of the ontology properties: on the left side, the timeline of incidents grouped into units (upper part, left), connected with the agents' intentions (or plans, lower part, left) through the concept of action (middle part, left); on the right side, the hierarchical scene structure (upper part, right), connected to the patterns for describing actions (lower part, right), which assign roles to agents; the middle of the figure describes the agent, with its conflicts (lower part, middle), and mental states (middle). The *Timeline* is the closest element to the drama document (a literary text or an audiovisual medium), a succession of the incidents (or *Actions*) that happen in the drama. Incidents are assembled into discrete structures, called *Units*. Each succession of incidents forms a sub-timeline of the whole timeline of the drama. This level is formalized through the Situation Calculus paradigm [27]: with sub-timelines that function as operators advancing the story world from one state to another (states aggregated in *ConsistentStateSets*), that work as preconditions and effects of some sub-timeline of incidents. The actions result from a deliberation process of the

Agent, which centers upon the notion of the character's intention in achieving (or trying to achieve) a *Goal*. The intention, or the commitment of the character, is represented by a *Plan*, which consists of the actions that are to be carried out in order to achieve some goal; plans are organized hierarchically, with high-level behaviors (*AbstractPlans*) formulated as lists of lower-level plans, or subplans, until the *DirectlyExecutablePlans*, which directly contain actions. Goals originate from the values of the characters that are put at stake and need to be restored (*ValueEngaged*), given the *Beliefs* (i.e. the knowledge) of the agents. This level is formalized through the rational agent paradigm, or BDI (Belief, Desire, Intention) paradigm [6] (which is also applied in the computational storytelling community [30,33]. So, an agent is characterized by goals, beliefs, values engaged, emotions, and plans; values can be *atStake* (true) or in balance (*atStake* false); plans can be in conflict with other plans, possibly of other agents; a conflict set aggregates all the plans, agents, and goals that determine a dramatic scene (*DrammarScene*), through the game of alternate accomplishments. A plan *motivates* the existence of a (sub)timeline, has preconditions and effects, which are consistent sets of states, and can be *accomplished* or not. Finally, scenes, defined by the author or perceived by the audience, to appropriately segment the timeline, are recursively composed of daughter scenes. A scene *spans* a timeline, that is a sequence of units. Some scenes are *DrammarScenes*, meaning that they are motivated by some conflict over the characters' intentions.

The development of Drammar can be described through the NeOn ontology engineering methodology [40], thanks to its flexibility and its focus on the relation of the ontology with non-ontological resources, such as linguistic and other semantic resources. Geared to a networked world, NeOn consists of a set of scenarios for the development of ontologies in a collaborative way. Briefly, the NeOn methodology maps a set of *activities* onto 9 *scenarios* for building and maintaining ontologies and ontology networks. In line with the spirit of the NeOn, only some specific scenarios and activities specifically apply to the design and development of Drammar. In particular, Scenario 1 (*From Specification to Implementation*), Scenario 2 (*Reusing and re-engineering non-ontological resources*), Scenario 5 (*Reusing and Merging Ontological Resources*), Scenario 7 (*Reusing ontology design patterns*) and Scenario 9 (*Localizing Ontological Resources*) were relevant for Drammar. For example, in accordance with the prescribed activities for the scenario 2 *Reusing and re-engineering non-ontological resources* (*Searching non-ontological resources*, *Assessing the candidate non-ontological resources* and *Selecting the most appropriate ontological resources*), the available resources were surveyed and selected by taking into account the requirements for drama description: in particular, given the focus on the representation of incidents in the annotation of drama, linguistic resources for the representation of processes and events were privileged with respect to less structured lexical-syntactic resources. Based on the survey, we selected a set of non-ontological semantic resources, such as lexical–semantic resources, to cope with linguistic counterpart of the elements of drama, and we developed an interface with these resources: WordNet and MultiWordNet [28,35], for the lexical description of incidents (actions and events); FrameNet, for their description in terms of their argument structure [3]; VerbNet [17], for the verbal forms not indexed by FrameNet.

The concepts and relations of the ontology Drammar are written in OWL DL; the extension in [20] was encoded into OWL2 RL (Rule Language). Ontology Drammar is available at the url http://purl.org/drammar, under the license Attribution 4.0 International (CC BY 4.0). This license allows anyone to share the original ontology but prohibits to distribute modifications of the ontology. Though this may limit the reusability of this resource, it is important to notice that Drammar has a theoretical background that addresses a specific scholarship; this is why we decided to disallow free changes in the distribution. Of course, people can build a novel ontology and depart from some core aspects of Drammar. The canonical citation for Drammar is "Vincenzo Lombardo, Rossana Damiano, Antonio Pizzo. Drammar: a comprehensive ontology of drama (2018). http://purl.org/drammar". A LODE documentation[10] is online and reachable from the resource url[11]. Drammar was also submitted for inclusion to LOV registry.

4 Impact of Drammar

Drammar is the first extensive ontology on drama and storytelling that covers whatever is intended as a dramatic quality, as demonstrated by the thorough analysis produced by the wiki (See Footnote 6). The wiki, through the analysis of a vast literature on drama, claims a number of statements. These statements have been translated into proto-axioms (i.e., axioms expressed in a controlled natural language) and then into formal axioms. These were conceived through a collaboration between two ontology engineers and one drama scholar (namely, the authors); then, on the one side, postdoc scholars from the humanities have validated the resource through the encoding of Stanislavsky's Action Analysis, useful in perspective for supporting actor rehearsals and drama staging [2], and the creation of a sample of metadata annotations for teaching purposes [22]; on the other side, researchers from computer science have applied the metadata annotation for the devise of SWRL rules for the computation of characters' emotions [20], the realization of printed charts of the characters' intentions aligned with the timeline of incidents [25] the characterization of the status of an annotated document [21], the preservation of drama as a form of intangible cultural heritage [24]. Further extensions of the resource, as well as the correction of errors and inconsistencies, can be addressed by starting from the update of the wiki, which is easily accessible to drama scholars, with limited competence on formal languages.

Resources of this kind are claimed to be of great importance for the researchers in the digital humanities: as discussed by Varela [42], semantic web technologies and ontologies in particular are suitable to represent disagreement in performance interpretations. Leveraging on the Richard Schechner's web diagram for the representation of the theory of performance, Varela claims that the

[10] See http://www.essepuntato.it/lode, visited on 11 June 2018.

[11] http://purl.org/drammar/lode.

semantic web is better suited to represent the knowledge of theatre and performance; in particular its fractal nature addresses the atmosphere of "sophisticated disagreement" that characterizes performance research [42, p. 136]. The notion of theatre and drama does not manifest in an item or an event sufficiently unified and standardized to be represented via a conventional database. In Varela's terms "the advantage of an ontology is that several aspects of these disagreements could be captured in a structured, systematic way" [42, p. 139]. Semantic web technologies, and ontologies in particular, are suitable to represent performance interpretations through the possibility of sharing the terminology through several approaches and the possibility for instances to belong to multiple classes scattered through several ontologies, though maintaining the original meaning cross-culturally. The ontological approach may also overtake a probabilistic-quantitative methods, as claimed by [7]: "they [the quantitative methods] fail to catch the intrinsic intentional and semantic nature of many literary phenomena [that can] be attained and made explicit and computable using a mixed human-machine approach, like that required by ontology modeling" [7, p. 30]. In other terms, the mixed human-machine approach described by Ciotti, accounts for the need of a human interpretation of the cultural object represented, and therefore pairs with the "sophisticated disagreement" described by Varela. The computational ontology is claimed to be the right method to get a representation that describes the domain of drama (and its intention-based actional nature) and that includes the human variations without disrupting the consistence of the model. Finally, there is a vast interest in the media and storytelling community for structured and semi-structured data sources. For storytelling (a larger category than drama), the effort has been in the creation of semi-structured resources that are available as specialized web sites, such as TV tropes[12] and fan fiction[13]. These sources benefit from the publication of the Drammar ontology because of the realization of a common ground for the definition of tropes, on top of the core dramatic elements and the capability to link several sources.

About maintenance and sustainability, we have proceeded through two initiatives. First, the latest release of the resource has appointed the CIRMA institution of the University of Turin[14] as publisher and responsible for the maintenance of the resource through its Scientific Committee, semesterly called to debate about improvements and updates due to local annotation projects. Second, we have launched the POP-ODE initiative [22], for the collection of a large corpus of encoded works through crowdsourcing. The POP-ODE toolkit consists of a number of tools and workflow (POP-ulating Ontology Drammar Encodings)[15] It includes a friendly interface and a visualizer to ease the work of annotators from the area of humanities.

---

[12] http://www.tvtropes.org.

[13] https://www.fanfiction.net.

[14] http://www.cirma.unito.it.

[15] Downloadable at the url http://www.cirma.unito.it/drammar/popode/popode_folder.zip.

A few well-known models have been reused: for example, the Ordered List Ontology (see above), in the case of the sequential structures (e.g., for timeline and plans). However, for other well-known cases, we decided not to proceed. Drammar is a specific core ontology for drama; probably, in the encoding of a specific drama (in the annotation task) other well-known models (such as, e.g., FOAF, for describing people, or PROV-O, for provenance information) can be relevant (e.g., in the drama "Hamlet", king Claudius is the uncle of Hamlet and assassin of his father); however, we have focused on the structural components of drama and their relations; so, for the annotation task, all the models that refer to commonsense knowledge should be included.

The Drammar ontology is also the base for a cataloguing and access portal project carried out by a consortium of companies from the movie industry as well as in the ICT sector. Ongoing project Smart-DH[16] (Smart Digital Heritage) aims at building an archive of Italian movies owned by the Augustus Color company, segmented for scenes and tagged according to an annotation schema built on top of the Drammar ontology.

5 Conclusion

This paper has described the ontological resource on drama called Drammar. Drammar has been realized through a collaboration of computer scientists and drama scholar through a wiki platform, for the exchanging of definitional ideas and the encoding of axioms. The ontology has been applied to a few tasks, namely reasoning about characters' emotions, graphical display of characters' intentions, encoding of action analysis for rehearsals. The creation and maintenance of software tools for easing the annotation tasks prompt a crowdsourcing initiative for the gathering of a tagged dataset of drama, for research and teaching purposes in the digital humanities.

References

1. Cinderella. animated movie produced by Walt Disney, distributed by RKO Radio Pictures Release (1950)
2. Albert, G., Pizzo, A., Lombardo, V., Damiano, R., Terzulli, C.: Bringing authoritative models to computational drama (encoding knebel's action analysis). In: Nack, F., Gordon, A.S. (eds.) ICIDS 2016. LNCS, vol. 10045, pp. 285–297. Springer, Cham (2016). https://doi.org/10.1007/978-3-319-48279-8_25
3. Baker, C., Fillmore, C., Lowe, J.: The berkeley framenet project. In: Proceedings of the 36th Annual Meeting of the Association for Computational Linguistics and 17th International Conference on Computational Linguistics, vol. 1. pp. 86–90. Association for Computational Linguistics (1998)
4. Bazin, A., Gray, H.: What is Cinéma?: Essays Selected and Translated by Hugh Gray. University of California Press, Berkeley (1967)

[16] http://illogic.xyz/2018/05/10/smart-dh/.

5. Berners-Lee, T., Hendler, J., Lassila, O.: The semantic web. Sci. Am. **284**(5), 34–43 (2001)
6. Bratman, M.E.: Intention, Plans, and Practical Reason. Harvard University Press, Cambridge (1987)
7. Ciotti, F.: Toward a formal ontology for narrative. Matlit Revista do Programa de Doutoramento em Materialidades da Literatura **4**(1), 29–44 (2016)
8. Elson, D.K.: Dramabank: annotating agency in narrative discourse. In: Proceedings of the Eighth International Conference on Language Resources and Evaluation (LREC 2012), Istanbul, Turkey (2012)
9. Esslin, M.: The Field of Drama. Methuen, London (1988)
10. Ferrario, R., Oltramari, A.: Towards a computational ontology of mind. In: Varzi, A.C., Vieu, L. (eds.) Proceedings of the International Conference FOIS 2004, Torino, Italy, pp. 287–297 (2004)
11. François, A.R., Nevatia, R., Hobbs, J., Bolles, R.C.: VERL: an ontology framework for representing and annotating video events. IEEE MultiMed. **5**, 76–86 (2005)
12. Gangemi, A., Guarino, N., Masolo, C., Oltramari, A., Schneider, L.: Sweetening ontologies with DOLCE. In: Gómez-Pérez, A., Benjamins, V.R. (eds.) EKAW 2002. LNCS (LNAI), vol. 2473, pp. 166–181. Springer, Heidelberg (2002). https://doi.org/10.1007/3-540-45810-7_18
13. Gangemi, A., Presutti, V.: Ontology design patterns. In: Staab, S., Studer, R. (eds.) Handbook on Ontologies. IHIS, pp. 221–243. Springer, Heidelberg (2009). https://doi.org/10.1007/978-3-540-92673-3_10
14. Hatcher, J.: The Art and Craft of Playwriting. Story Press, Cincinnati (1996)
15. Heath, T., Bizer, C.: Linked data: Evolving the web into a global data space. Synthesis Lectures on the Semantic Web: Theory and Technology, pp. 1–136 (2011)
16. Jewell, M., et al.: OntoMedia: an ontology for the representation of heterogeneous media. In: Proceedings of SIGIR workshop on Mutlimedia Information Retrieval. ACM SIGIR (2005)
17. Kipper, K., Korhonen, A., Ryant, N., Palmer, M.: Extending VerbNet with novel verb classes. In: Proceedings of LREC, vol. 2006, p. 1 (2006)
18. Lavandier, Y.: La dramaturgie. Le clown et l'enfant, Cergy (1994). http://www.clown-enfant.com/leclown/dramaturgie/
19. Lawrence, K.F.: Crowdsourcing linked data from shakespeare to Dr who. In: Web Science International Conference (2011)
20. Lombardo, V., Battaglino, C., Pizzo, A., Damiano, R., Lieto, A.: Coupling conceptual modeling and rules for the annotation of dramatic media. Semant. Web J. **6**(5), 503–534 (2015)
21. Lombardo, V., Damiano, R., Pizzo, A., Terzulli, C.: The intangible nature of drama documents: an FRBR view. In: Proceedings of the 2017 ACM Symposium on Document Engineering, pp. 173–182. ACM (2017)
22. Lombardo, V., et al.: Annotation of metadata for dramatic texts: the POP-ODE initiative. In: NL4AI@AI*IA. CEUR Workshop Proceedings, vol. 1983, pp. 30–42. CEUR-WS.org (2017)
23. Lombardo, V., Pizzo, A.: Multimedia tool suite for the visualization of drama heritage metadata. Multimed. Tools Appl. **75**(7), 3901–3932 (2014)
24. Lombardo, V., Pizzo, A., Damiano, R.: Safeguarding and accessing drama as intangible cultural heritage. ACM J. Comput. Cult. Herit. **9**(1), 1–26 (2016)
25. Lombardo, V., Pizzo, A., Damiano, R., Terzulli, C., Albert, G.: Interactive chart of story characters' intentions. In: Nack, F., Gordon, A.S. (eds.) ICIDS 2016. LNCS, vol. 10045, pp. 415–418. Springer, Cham (2016). https://doi.org/10.1007/978-3-319-48279-8_39

26. Mamet, D.: Three Uses of the Knife: On the Nature and Purpose of Drama. The Columbia Lectures on American Culture. Columbia University Press (1998). http://books.google.it/books?id=YZRODCVYACsC
27. McCarthy, J.C.: Mental situation calculus. In: Proceedings of the 1986 Conference on Theoretical Aspects of Reasoning About Knowledge. pp. 307–307. TARK 1986, Morgan Kaufmann Publishers Inc., San Francisco (1986). http://dl.acm.org/citation.cfm?id=1029786.1029815
28. Miller, G.: WordNet: a lexical database for english. Commun. ACM **38**(11), 39–41 (1995)
29. Mulholland, P., Collins, T.: Using digital narratives to support the collaborative learning and exploration of cultural heritage. In: 2002 Proceedings of the 13th International Workshop on Database and Expert Systems Applications, pp. 527–531. IEEE (2002)
30. Norling, E., Sonenberg, L.: Creating Interactive Characters with BDI Agents. In: Proceedings of the Australian Workshop on Interactive Entertainment IE 2004 (2004)
31. Ortony, A., Clore, G., Collins, A.: The Cognitive Structure of Emotions. Cambrigde University Press, Cambrigde (1988)
32. Pease, A., Niles, I., Li, J.: The suggested upper merged ontology: a large ontology for the semantic web and its applications. In: Working Notes of the AAAI-2002 Workshop on Ontologies and the Semantic Web, Edmonton, Canada, vol. 28 (2002)
33. Peinado, F., Cavazza, M., Pizzi, D.: Revisiting character-based affective storytelling under a narrative BDI framework. In: Spierling, U., Szilas, N. (eds.) ICIDS 2008. LNCS, vol. 5334, pp. 83–88. Springer, Heidelberg (2008). https://doi.org/10.1007/978-3-540-89454-4_13
34. Perrault, C.: Contes, chap. Cendrillon, pp. 274–79. Flammarion, Paris (1989)
35. Pianta, E., Bentivogli, L., Girardi, C.: MultiWordNet: developing an aligned multilingual database. In: Proceedings of the First International Conference on Global WordNet, January 2002. http://multiwordnet.fbk.eu/paper/MWN-India-published.pdf
36. Rahimtoroghi, E., Corcoran, T., Swanson, R., Walker, M.A., Sagae, K., Gordon, A.: Minimal narrative annotation schemes and their applications. In: Seventh Intelligent Narrative Technologies Workshop. AAAI Publications (2014)
37. Raimond, Y., Abdallah, S.A., Sandler, M.B., Giasson, F.: The music ontology. In: ISMIR, pp. 417–422. Citeseer (2007)
38. Spencer, S.: The Playwright's Guidebook: An Insightful Primer on the Art of Dramatic Writing. Faber & Faber (2002). http://books.google.it/books?id=nDrHmckSqi4C
39. Styan, J.L.: The Elements of Drama. University Press, Cambridge (1963)
40. Suárez-Figueroa, M.C., Gómez-Pérez, A., Fernández-López, M.: The NeOn Methodology for ontology engineering. In: Suárez-Figueroa, M.C., Gómez-Pérez, A., Motta, E., Gangemi, A. (eds.) Ontology Engineering in a Networked World, pp. 9–34. Springer, Heidelberg (2012). https://doi.org/10.1007/978-3-642-24794-1_2
41. Szondi, P.: Theory of the moderna drama parts i-ii. Boundary 2 **11**(3), 191–230 (1983)
42. Varela, M.E.: Interoperable performance research promises and perils of the semantic web. Drama Rev. **60**(3), 136–147 (2016)

The SPAR Ontologies

Silvio Peroni[1(✉)] and David Shotton[2]

[1] Department of Classical Philology and Italian Studies, University of Bologna, Bologna, Italy
silvio.peroni@unibo.it
[2] Oxford e-Research Centre, University of Oxford, Oxford, UK
david.shotton@oerc.ox.ac.uk

Abstract. Over the past eight years, we have been involved in the development of a set of complementary and orthogonal ontologies that can be used for the description of the main areas of the scholarly publishing domain, known as the SPAR (Semantic Publishing and Referencing) Ontologies. In this paper, we introduce this suite of ontologies, discuss the basic principles we have followed for their development, and describe their uptake and usage within the academic, institutional and publishing communities.

Keywords: OWL ontologies · SPAR · Scholarly communication
Scholarly publishing · Semantic publishing

1 Introduction

The last decade has seen a new evolutionary step in scholarly publishing that has drastically changed the way of publishing and sharing research information. This is the use of Web and Semantic Web technologies for making published entities such as articles and scientific data machine-readable and easier to discover, browse and interact. Researchers in the field refer to it using the name *Semantic Publishing* [28]. This movement has actively involved people from both academia and industry, including (a) publishers such as Springer Nature (https://www.springernature.com), Elsevier (https://www.elsevier.com) and F1000 (https://f1000.com), (b) institutions for the assessment of the quality of research such as the Italian National Agency for the Evaluation of Universities and Research Institutes (http://www.anvur.it), and (c) broad academic communities such as Force11 (https://www.force11.org) and Linked Research (https://linkedresearch.org), all united in the goal of changing and improving the current practices of research communication.

While the invention of print permitted the written recording of scientific discoveries, the advent of the Web has allowed researchers and publishers to increase drastically their visibility by means of new communication channels and electronic publications. Semantic Publishing additionally brings machines (software agents, intelligent interfaces, Semantic Web reasoners, etc.) into the game. The

D. Vrandečić et al. (Eds.): ISWC 2018, LNCS 11137, pp. 119–136, 2018.
https://doi.org/10.1007/978-3-030-00668-6_8

more scholarly data (e.g. metadata of scholarly publications, their associated research data, and the experimental and computational workflows employed in the research) are available in machine-readable forms, the more those involved in the whole scholarly communication domain – researchers, publishers, reviewers, readers – will benefit. Such benefits will involve increased visibility, findability of related research, and discoverability of previously unknown links between works that were not explicitly related by citations.

Providing scholarly data that can be easily parsed, processed and interpreted computationally is a matter of having expressive shared machine-processable descriptions on the Web. Semantic Web technologies such as RDF, RDFS, OWL, and SPARQL provide the main building blocks towards that goal. Such tools can be considered a formal, semantically-oriented interlingua for machines to express and query data. In 2010, when we started to work actively on this topic, what were missing were appropriately rich ontologies for enabling the accurate and reasonably expressive description of all aspects of the scholarly publishing domain.

Since then, our work in this area has been dedicated to the development of a set of complementary and orthogonal ontologies that can be used for the description of the main areas of this publishing domain, from the metadata of scholarly artefacts to the specification of the workflow processes that result in the publication of a scholarly bibliographic product. This work, described here, has resulted in the development of the *SPAR (Semantic Publishing and Referencing) Ontologies* (http://www.sparontologies.net), a suite of ontological modules enabling one to record the scholarly publishing domain using Semantic Web technologies. Previous publications describing one or more of the SPAR Ontologies are listed at http://www.sparontologies.net/publications.

The rest of the paper is organised as follows. In Sect. 2, we introduce other existing Semantic Web vocabularies and ontologies that have been used over the past decade or so for the description of (parts of) the publishing domain. In Sect. 3, we describe the design principles we have followed for the development of the SPAR Ontologies. In Sect. 4, we introduce each of the current SPAR Ontologies, clarifying their scope and highlighting their most significative aspects. In Sect. 5 we show how external parties, including researchers, publishers and institutions, have used the SPAR Ontologies for describing bibliographic documents. Finally, in Sect. 6, we conclude the paper by sketching out some future work.

2 Existing Models

Our work on the SPAR Ontologies was not the first effort to provide Semantic Web descriptions of the publishing domain. The *Dublin Core Metadata Terms* (DCTerms) [8] is among the first international standards to describe bibliographic information on the Web. Going further than DCTerms is the *Functional Requirements for Bibliographic Records* (FRBR) [19], a relatively recent specification made by the International Federation of Library Association and Institution, that models the concept of a bibliographic entity according to four different but closely-related point of views called *work* (the conceptual idea), *expression* (the content), *manifestation* (the format), and *item* (the tangible object). These models, actively used today with others of similar kind including the *Publishing Requirements for Industry Standard Metadata* (PRISM) [20], should be considered top-level vocabularies rather than something developed to characterise specific aspects of scholarly publishing. Thus all of them lack the concepts of journal article, book chapter, conference paper, reference list, citation, editor and similar entities that are useful for describing the scholarly publication world in detail. Furthermore, they were not developed with the RDF/OWL data model in mind, but rather as merely documental specifications, although Semantic Web implementations of them have been provided in recent years.

While past proposals exist for the adoption of semantic technologies for the scholarly publishing domain (e.g. ScholOnto [2]), the *AKT Reference Ontology* (AKTRO) should probably be listed as the first ontology specifically developed by means of Semantic Web technologies for describing this domain. Originally developed in OCML and then converted in OWL (http://lov.okfn.org/dataset/lov/vocabs/akt), it provides a set of classes and properties that allow the description of different kinds of publications and agents involved in the publishing process.

A first serious attempt towards providing an OWL-native scholarly-oriented publication ontology was the *Bibliographic Ontology* (BIBO) [7], which introduced the concepts of article, journal, conference proceedings, etc. While still in use today, it falls short of being sufficiently comprehensive, lacking for instance the concepts of dataset, blog post, citation function and publication workflow. Additionally, BIBO is a 'flat' ontology, so that the class bibo:AcademicArticle conflated the two concepts Research Paper (a conceptual Work) and Journal Article (an Expression of such a Work). Since then, several Semantic Web ontologies have been proposed for addressing the description of additional aspects of the publishing domain. Notable examples are the *Semantic Web Journal (SWJ) ontology* [18], the *Semantic Web Conference* (SWC) *ontology* (http://data.semanticweb.org/ns/swc/ontology), *Semantically Annotated LaTeX* (SALT) ontologies [13], the *Nature Ontologies* [15], the *SciGraph Ontologies* [16], and the *Conference Ontology* [23].

Along with these, several new OWL-based top-level models have been releases, that provide general vocabularies for referring to aspects of the scholarly and publishing domain. Among the most notable are the *Resource Description and Access* (RDA) OWL vocabularies [17] and *BIBFRAME* (http://bibframe.org/vocab.rdf), which can be considered as an implementation of FRBR [19]. BIBFRAME has been recently extended by *bibliotek-o* (https://bibliotek-o.org), developed as a supplement to the core BIBFRAME ontology.

Finally, several groups have started to propose larger and more complex Semantic Web ontologies for describing the publishing domain as a whole, instead of focussing on top-level concepts or a few specific aspects. Apart from the SPAR Ontologies, which are described in the following sections, some of the first attempts in this direction are those of the *Semantic Web Applications in*

Neuromedicine (SWAN) project which proposed the *SWAN biomedical discourse ontology* [4], the *VIVO Integrated Semantic Framework* (VIVO-ISF) ontology (http://vivoweb.org/ontology/core), and the *Semanticscience Integrated Ontology* (SIO) [10] – all of which align with some of the entities described in the SPAR Ontologies.

3 Development Guidelines

In order to provide a set of ontologies to describe scholarly bibliographic information that built on our previous experience of developing the prototype Citation Typing Ontology (CiTO) [29], we started in 2010 to work on a comprehensive and complementary set of orthogonal OWL 2 DL ontologies that describe all the aspects of the publishing domain, namely the SPAR (Semantic Publishing and Referencing) Ontologies. The development process we followed complied with the following development principles (DP) that were derived after discussions with researchers and publishers.

DP1 - Enabling Adoption: Addressing Actors' Requirements. The involvement of domain experts is good practice in any robust methodology for ontology design, since it allows the developer to gather information about the important aspects to model and also to employ the terminology used in real practice. Thus, a model developed for describing the publishing domain, which aims at being adopted broadly in the scholarly communications community, should strongly take into consideration all the actors involved in such domain (authors, editors, publishers, members of academic communities, etc.), their needs, and the vocabularies they use in their day-to-day work for referring to publishing things. Within the scholarly communication domain, this requirement is also currently being addressed by worldwide initiatives such as Metadata 2020 (see http://www.metadata2020.org/projects/definitions/).

DP2 - Enabling Reuse: Interoperable Ontological Modules. Organising comprehensive ontological descriptions of a field by the use of a set of smaller orthogonal ontology modules is recognised good practice in ontology design, since it improves the reusability of such modules in different contexts and across domains [6]. For the SPAR ontologies, we thus developed several complementary ontological blocks, each describing a particular aspect of the publishing domain, which could be combined easily to meet the actual description needs. In this way, we enable a possible adopter to select and use consistently only those blocks describing, for instance, citations and bibliographic metadata, without caring about other aspects such as authors affiliations, contributors' roles, and the structural and rhetorical organisation of the document.

DP3 - Enabling Application in Different Contexts: Minimal Logical Constraints. OWL allows one to define several constraints on ontological entities. Those vary from the specification that certain relations can be used only among specific kinds of entities (domain and range constraints, e.g. that the concept of authorship can involve only a person and the scholarly paper (s)he has authored, but cannot involve cats or tractors) to more complex definitions (e.g. all authors of an academic document should be affiliated with at least one recognised institution or organization, the name of which must be explicitly stated in the authored document). On the one hand, the use of such constraints allows an accurate formal verification of a dataset: it is possible to understand whether it is compliant with the ground model defined by the ontology or whether it presents some inconsistencies that violates the constraints defined. However, on the other hand, excessive use of such constraints would make the ontology unusable in practice, in particular in the context of Linked Open Data (LOD), thus resulting only in a pure theoretical exercise in ontology construction. As confirmation, it is worth noticing that some of the most influential and (re-)used ontologies in the context of LOD nowadays are light-weight [5], containing only a very limited number of strictly necessary constraints that allow a respectable level of formal verification of data, while enabling their easy learning and re-use in different contexts. Thus, logical constraints must be used with caution if the goal is to make the ontology adopted worldwide in a variety of contexts.

DP4 - Enabling Interoperability: Reusing Existing Vocabularies. The reuse of existing models and vocabularies that already have widespread uptake and usage allows an improvement in the interoperability of the ontology itself with other ones [14] (particularly in the context of LOD), while avoiding "re-inventing the wheel". In addition, the inclusion of concepts from existing, shared and well-known vocabularies and/or the adoption of widely used design patterns is good practice for the development of a new ontology, since they can provide an initial more general stub from which to the develop more specific descriptions [26], and can speed the adopters' learning curve for understanding the new model.

DP5 - Enabling Human Understandability: Supporting People with Examples and Tools. Alongside the development of the formal ontology itself, one should work also to provide mechanisms for understanding and accessing the ontology by people who may not necessarily be experts in Knowledge Representation. Two main artefacts can be used in this direction. First, the creation of several clear examples that show how one can use the ontology for modelling particular scenarios. Second, the development of tools, particularly graphical and presentational tools, that assist people to understand the ontology with the minimum of effort and without having specific technical background of the ontology language used for its development.

Following these principles, we have developed three tools that have been systematically used since 2010 during the development of the SPAR Ontologies, and that have more recently also been adopted by several other projects. These are:

- the *Simplified Agile Methodology for Ontology Development* (SAMOD, https://github.com/essepuntato/samod) [24], an agile methodology for the development of ontologies by means of small steps in an iterative workflow that focuses on creating well-developed and well-documented models starting from exemplar domain descriptions;

- the *Live OWL Documentation Environment* (LODE, http://www.essepuntato.it/lode) [25], a service that automatically extracts classes, properties, individuals, annotations, general axioms and namespace declarations from OWL ontologies, and renders them in a human-readable HTML page designed for browsing and easy navigation by means of embedded links;
- the *Graphical Framework for OWL Ontologies* (Graffoo, http://www.essepuntato.it/graffoo) [11], an open source tool that can be used to present the classes, properties and restrictions within OWL ontologies, or sub-sections of them, in clear and easy-to-understand diagrams. Examples of Graffoo representations are given below.

4 Introducing the SPAR Ontologies

Our ontology development following these principles, which developed from a preliminary project in this area [29], resulted in the release, at the end of 2010, of eight complementary and interoperable core ontology modules under the umbrella name of the *SPAR (Semantic Publishing and Referencing) Ontologies* (http://www.sparontologies.net). Over the following years, we have extended these with additional modules to address other aspects of the publishing domain.

These SPAR Ontologies form a suite of orthogonal, non-overlapping and complementary OWL 2 DL ontology modules (all made available with a Creative Commons Attribution License 4.0) for the creation of comprehensive machine-readable RDF metadata covering every aspect of semantic publishing and referencing: document descriptions, bibliographic resource identifiers, types of citations and their related contexts, bibliographic references, document parts and status, agents' roles and contributions, bibliometric data and workflow processes. All these ontologies have been developed by means of SAMOD [24], and are accompanied by (a) a short descriptive page on the SPAR website, (b) Graffoo diagrams [11], (c) examples of usage in Turtle (available both on the SPAR website and in Figshare), (d) publication information (if any), and (e) the use of LODE to create HTML documentation of the ontologies [21]. Most of their terms are already listed in LOV (Linked Open Vocabularies) with an appropriate category tag (http://lov.okfn.org/dataset/lov/vocabs?tag=SPAR).

The SPAR Ontologies follows the FAIR principles for data publication [31], and reuse existing standards developed for describing bibliographic resources, such as FRBR [19] and PRISM [20]. They also reuse and import several existing models, among which are DCTerms [8], SKOS [22], FOAF [1], the Collections Ontology [3], PROV-O [21], and several Ontology Design Patterns [26]. An bird-eye view of all the SPAR Ontologies and their associated external ontologies and design patterns is shown in Fig. 1.

While we have already published various articles describing some of the SPAR Ontologies, we avoid mentioning them explicitly here, since they are all available on the SPAR Ontologies website at http://www.sparontologies.net/publications. This present paper is the first to present a bird-eye view of the whole suite of SPAR Ontologies, and will be used from now on as the canonical way for citing the SPAR Ontologies.

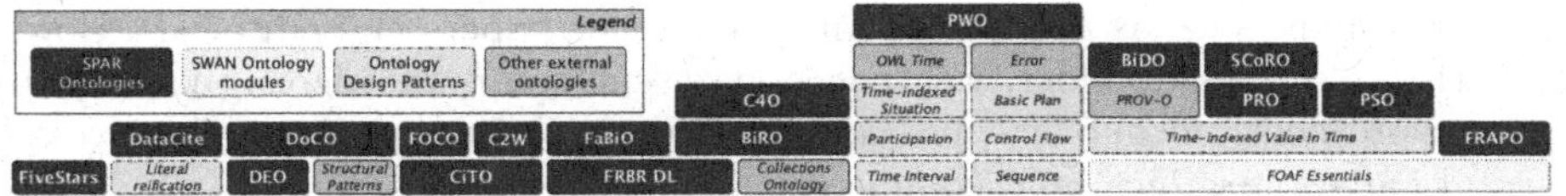

Fig. 1. The SPAR Ontologies and their relations with other models.

4.1 Ontologies for Describing Bibliographic Resources and Their Parts

Five ontological modules have been developed to permit the accurate description of bibliographic resources, their identifiers, and their internal components (paragraphs, sections, results, methods, etc.). These are:

- FaBiO (the FRBR-aligned Bibliographic Ontology, http://purl.org/spar/fabio),
- FRBR-DL (the Essential FRBR in OWL2 DL Ontology, http://purl.org/spar/frbr),
- DoCO (the Document Components Ontology, http://purl.org/spar/doco),
- DEO (the Discourse Elements Ontology; http://purl.org/spar/deo), and
- the DataCite Ontology (http://purl.org/spar/datacite).

Exemplar terms from these ontologies are shown in Fig. 2.

FaBiO is an ontology for recording and publishing descriptions of entities that are published or potentially publishable. FaBiO entities are primarily textual publications such as books, magazines, newspapers and journals, and items

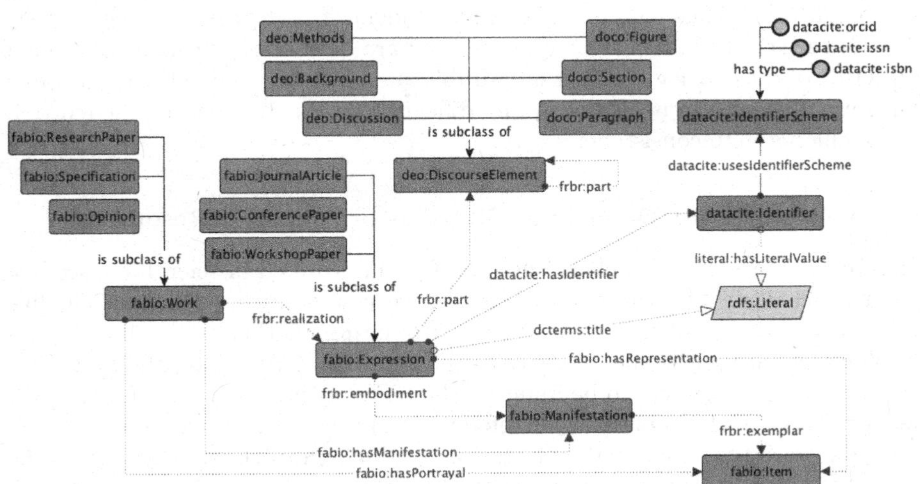

Fig. 2. A Graffoo diagram illustrating some of the ontological entities defined in FaBiO, FRBR-DL, DoCO, DEO and the DataCite Ontology.

of their content such as articles, poems, conference papers and editorials. However, they also include blogs, web pages, datasets, computer algorithms, experimental protocols, formal specifications and vocabularies, legal records, governmental papers, technical and commercial reports and similar publications, and also anthologies, catalogues and similar collections. FaBiO is based on FRBR, implemented via FRBR-DL (*the Essential FRBR in OWL2 DL Ontology*), which offers an OWL 2 DL view of the Davis and Newman FRBR RDF vocabulary (http://vocab.org/frbr/core).

DoCO is an ontology that provides a general-purpose structured vocabulary of document elements. DoCO has been designed as a general unifying ontological framework for describing different aspects related to the content of scientific and other scholarly texts. Its primary goal has been to improve the interoperability and shareability of academic documents (and related services) when multiple formats are actually used for their storage. It is based on a theory of structural pattern for descriptive documents (*Structural Patterns,* shown in Fig. 1) [9] and, by means of DEO (*the Discourse Elements Ontology*), also provides entities for describing the pure rhetorical characterisation of document components – e.g. Introduction, Background, Evaluation, Materials, Methods and Conclusion.

Finally, the DataCite Ontology, which we originally developed for describing in RDF the metadata properties of the DataCite Metadata Schema version 3.1 (https://semanticpublishing.wordpress.com/2016/02/08/mapping-datacite-3-1/), provides a flexible mechanism to define identifiers (DOI, ISSN, ORCID, etc.) for bibliographic resources (e.g., papers and datasets) and related entities (e.g., authors) as first-class data entities (by means of the *Literal Reification* design pattern shown in Fig. 1 [12]), instead of by using simple links between the owner of an identifier and the identifier string. This approach allows us better flexibility to extend the coverage of the ontology by adding additional individuals defining new kinds of identifier schemes (members of the class `datacite:IdentifierScheme`) as they are developed, without modifying the TBox of the ontology. This is a desirable aim, since once an ontology is stable it is best to minimize its structural modifications, so as to avoid the generation of possible usage inconsistencies.

4.2 Ontologies for Describing Citations of Scholarly Resources

The word "citation" is widely misused and misinterpreted in scientific discourse and the scholarly literature. For instance, we commonly use"citation" in five different ways to identify either (a) the act of citing another work, (b) a bibliographic reference put in the reference list at the end of a paper, (c) the particular in-text reference pointers to be found in the body of the citing work (e.g., "[3]") denoting a particular bibliographic reference, (d) the citation link between citing and cited work that is created by inclusion of a bibliographic reference, or (e) the published work itself that is the object of that citation. In order to avoid such ambiguities, we have made an effort to provide a clear vocabulary for describing all these different aspects.

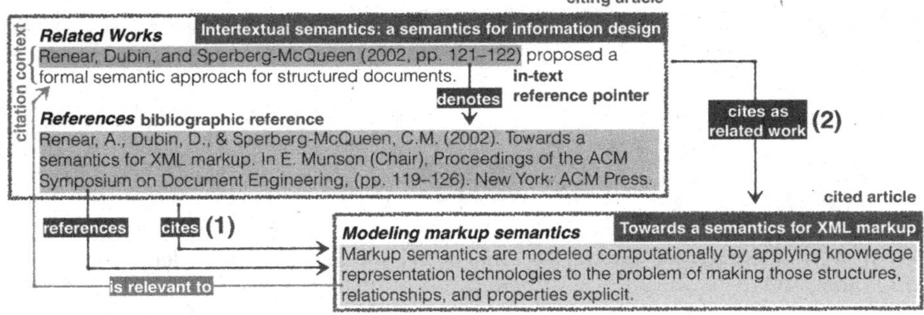

Fig. 3. The main components that allow a clear description of a citation and of all the related information.

In Fig. 3, we summarise visually the various components involved in the citation process, providing a clear nomenclature of distinct terms. In this context, a citation is a conceptual directional link from a citing entity to a cited entity, created by a human performative act of making a citation. It is worth mentioning that the citations instantiated by the inclusion of a bibliographic reference in the reference list (see (1) in Fig. 3) are "plain links" between the citing and cited entity, while those ones generated by using an in-text reference pointer (i.e. a local reference to a particular bibliographic reference in the reference list, see (2) in Fig. 3) have textual contexts that may reflect citation functions, i.e. the reason or reasons that an author cites the cited article at different places in the text of the citing article [30].

The SPAR Ontologies make available three primary ontological modules to allow one the accurate description of all the aforementioned entities. These are:

- CiTO (the *Citation Typing Ontology*, http://purl.org/spar/cito),
- BiRO (the *Bibliographic Reference Ontology*, http://purl.org/spar/biro), and
- C4O (the *Citation Counting and Context Characterisation Ontology*, http://purl.org/spar/c4o).

Exemplar terms from these ontologies are shown in Fig. 4.

CiTO makes it possible for authors (or others) to identify a citation link and to describe the citation intent (e.g., `cito:extends`, `cito:usesMethodIn`, `cito:supports`) when someone cites a particular publication, thus permitting the motives of an author when referring to another document to be captured. CiTO also allows one to create metadata describing citations that are distinct from metadata describing the citing or cited works themselves, thus enabling citations to be treated as first-class data entities (by means of the class `cito:Citation`, see https://opencitations.wordpress.com/2018/02/19/citations-as-first-class-data-entities-introduction/) with their own properties (e.g. `cito:hasCitationTimeSpan`).

Additionally, two further supplementary ontologies have been made available so as to classify all the CiTO properties according to their factual and

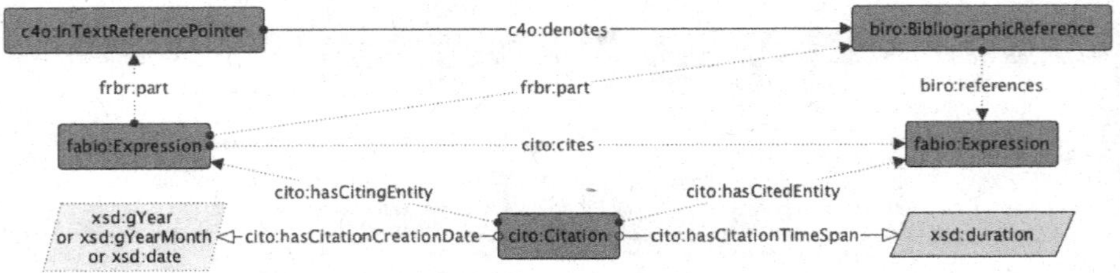

Fig. 4. A Graffoo diagram illustrating some of the ontological entities defined in CiTO, BiRO, and C4O.

positive/neutral/negative rhetorical functions (FOCO, the *Functions of Citations Ontology*) and to map them with appropriate Wordnet synsets (C2W, the *CiTO to Wordnet Ontology*).

BiRO allows the description of reference lists and bibliographic references themselves. The *Collections Ontology* [3] is additionally employed to permit the counting and ordering of bibliographic references in a reference list.

Besides defining references in machine-readable form, it is also useful to describe how these references are used within the citing paper. C4O has been developed to describe these aspects, e.g. the in-text reference pointers within the citing paper, the links denoted by in-text reference pointers to the bibliographic references, the total number of in-text reference pointers within the citing paper that denote the same bibliographic reference, how much the referenced article is globally cited (according to particular bibliographic citation service, e.g., Google Scholar, on a particular date), and the textual contexts involved in a citation – i.e., the textual phrase, sentence or paragraph in the citing article containing a particular in-text reference pointer, and the structural or rhetorical section of the citing article where this is found.

4.3 Ontologies for Describing the Publishing Workflow

Five other ontological modules, the first three of which are strongly based on Ontology Design Patterns [26] (shown in Fig. 1), have been developed to permit the description of contextual aspects of a publication, namely agents' roles, document statuses, steps in the publishing workflow, contributors' roles, and related academic administrative information. These are:

- PRO (the *Publishing Roles Ontology*, http://purl.org/spar/pro),
- PSO (the *Publishing Status Ontology*, http://purl.org/spar/pso),
- PWO (the *Publishing Workflow Ontology*, http://purl.org/spar/pwo),
- SCoRO (the *Scholarly Contributions and Roles Ontology*, http://purl.org/spar/scoro), and
- FRAPO (the *Funding, Research Administration and Projects Ontology*, http://purl.org/cerif/frapo).

Exemplars of use of terms from these ontologies is shown in Fig. 5.

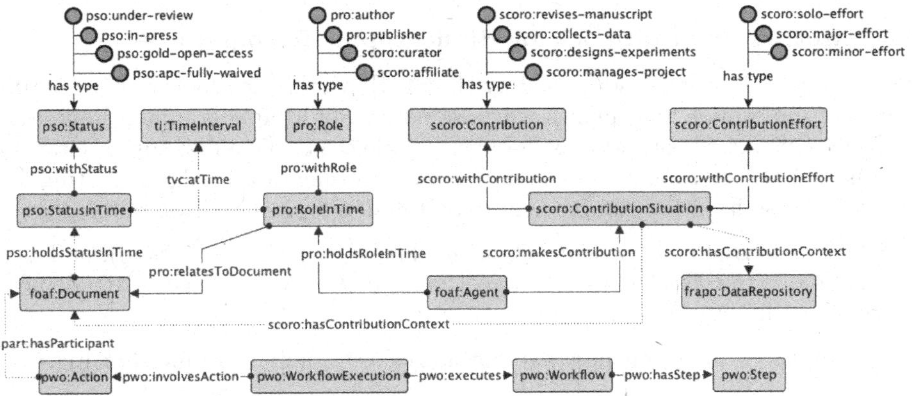

Fig. 5. A Graffoo diagram illustrating some of the ontological entities defined in PRO, PSO, PWO, SCoRO and FRAPO.

PRO is an ontology that permits the characterisation of the roles of agents – people (e.g. editors, publishers), corporate bodies and computational agents – in the publication process. Furthermore, it allows one to specify the role an agent has in relation to a particular bibliographic entity (as author, editor, reviewer, etc.) or to a specific institution (as publisher, librarian, etc.), over the specified time period during which each role is held.

This ontology is complemented by SCoRO, which extends PRO by allowing the description of the contributions and additional roles of scholars, and the organizations of which they are members, with respect to projects, research investigations, publications, and other academic activities and outcomes.

PSO characterises the publication status of scholarly-related entities (e.g. documents) at each of the different stages in the publishing process (e.g., draft, submitted, under review, rejected for publication, accepted for publication, version of record, peer reviewed, open access). Using PSO, it is possible to describe the status of a document and how it changes over time.

PWO enables a description of the logical steps in a workflow, such as the process of publication of a document. Each step may involve one or more events or actions that take place at a particular phase of the workflow and in a certain order (e.g., authors are writing the article, the article is under review, a reviewer suggests to revise the article, the article is modified and accepted for publication, the article is finally published).

Complementing these ontologies, FRAPO is an OWL 2 DL ontology inspired by CERIF (https://www.eurocris.org/cerif/main-features-cerif) for describing academic administrative information, particularly as it relates to grant funding and research projects. It can be used for the characterization of grant applications, funding bodies, research projects, project partners, etc. It can also be used to describe other types of projects, for example building projects and educational projects.

4.4 Metrics and Statistics for Bibliographic Resources

The final aspects that is addressed by the SPAR Ontologies concern the specification of qualitative and quantitative evaluations of a bibliographic resource or an agent (impact factor, h-index, e-index, article citation counts, author citation counts, conference ranking, etc.) and for the encoding of a Five Stars rating for articles. These have been implemented by the development of two ontologies:

- BiDO (the *Bibliometric Data Ontology*, http://purl.org/spar/bido), and
- FiveStars (the *Five Stars of Online Journal Articles Ontology*, http://purl.org/spar/fivestars).

Exemplar terms from these two ontologies are briefly introduced in Fig. 6.

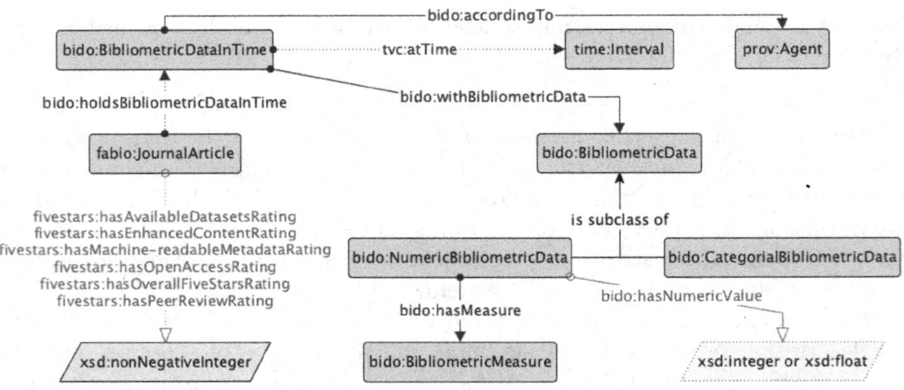

Fig. 6. A Graffoo diagram illustrating some of the ontological entities defined in BiDO and FiveStars.

BiDO is a modular OWL 2 ontology that allows the description of bibliometric data (either numerical data or categorial data) concerning people, articles, journals, and other scholarly-related entities. The core module of this ontology allows one to specify that these bibliometric data relate to specific times or time periods (using the same Ontology Design Patterns that have been adopted for use with the ontologies described in Sect. 4.3) and involve certain agents and events (by reusing PROV-O [21], shown in Fig. 1).

Finally, FiveStars is a simple ontology that is intended for use by publishers and others wishing to encode the *Five Stars ratings for published articles*, as proposed in [27], so they can accompany other machine-readable metadata for the article. The ontology includes twelve data properties, six for specifying the value of each factor (shown in Fig. 6) and six for specifying a comment related to each assigned value.

5 Community Uptake and Sustainability

The SPAR Ontologies have already been adopted by various communities and in several projects for describing data related to the publishing domain. Of those known to us, these include the US Global Change Information System, the Working Group on Document Standards of the High Level Committee Management of the United Nations, various W3C Working Groups, Springer Nature, OpenCitations, and Wikidata – a complete list is available at http://www.sparontologies.net/uptake. On the same page are also listed the 677 articles (according to Google Scholar, as of 20 March 2018) that link to one or more of the SPAR Ontologies or that cite one of our publications describing them (listed at http://www.sparontologies.net/publications).

The SPAR Ontologies website has been accessed more than 1 million times since its launch in December 2015, by users from several countries (identified by the IP address of the request, after excluding all the user agents that contained one of the following tokens: "crawler", "spider", "bot", "yahoo! Slurp", "bubing"), as shown in Fig. 7 – with the US being responsible for 78% of the accesses. It is worth mentioning that the pages related to the ontologies themselves (http://www.sparontologies.net/ontologies and related subpages) have together gained a very high percentage of the overall accesses (i.e. 88%), showing that the main reason people access the SPAR Ontologies website is to explore and use the ontologies.

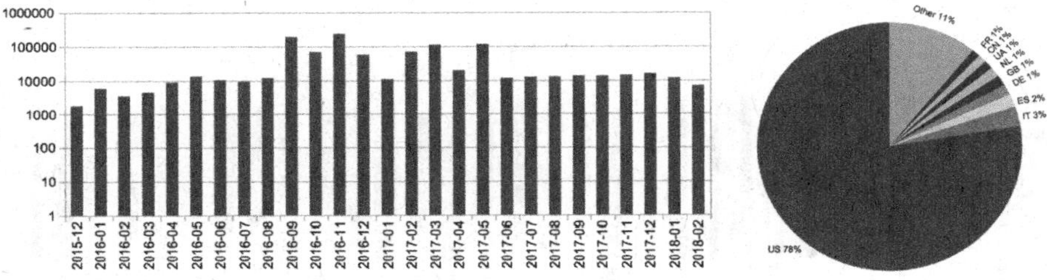

Fig. 7. The page accesses that the SPAR Ontologies website has received since its launch, grouped by month and by country (excluding accesses by automated web crawlers and bots).

We have also analysed the statistics related to the SPAR Ontologies Twitter account (https://twitter.com/sparontologies). As shown in Fig. 8, there was a clear increase in the account engagements from February 2017, possibly due to the beginning of our active publishing activity on Twitter starting from July 2016 – while in the previous months this Twitter account was available but was not intensively used for sharing purposes.

In Fig. 9, we show the statistics concerning the SPAR-related resources (i.e. the examples of usage in Turtle and the DataCite mapping document, all available at http://www.sparontologies.net/examples) that were made available on

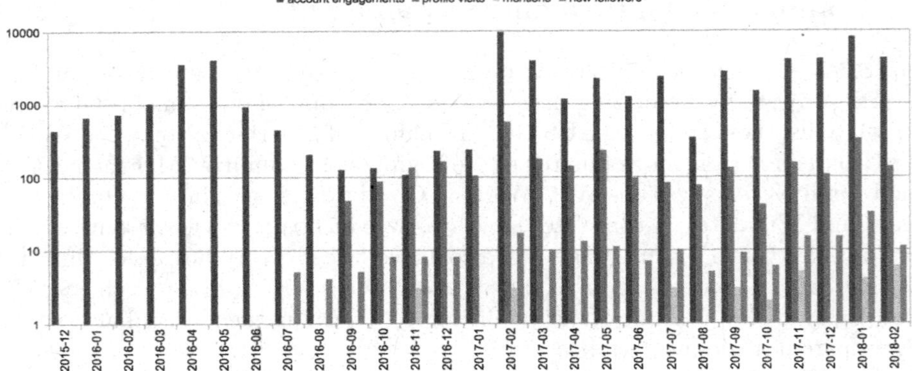

Fig. 8. The number of interactions that the SPAR Ontologies Twitter account has had over the past years.

Figshare some months before the SPAR Ontologies website launch in December 2015. These Figshare documents have obtained 9,575 views and 1,463 downloads overall. Similar to the previous statistics for the Twitter account, the Figshare graph shows a clear increase in the downloads from January 2017. In contrast, the number of views of the Figshare resources reached rough stability immediately after the launch of the SPAR website.

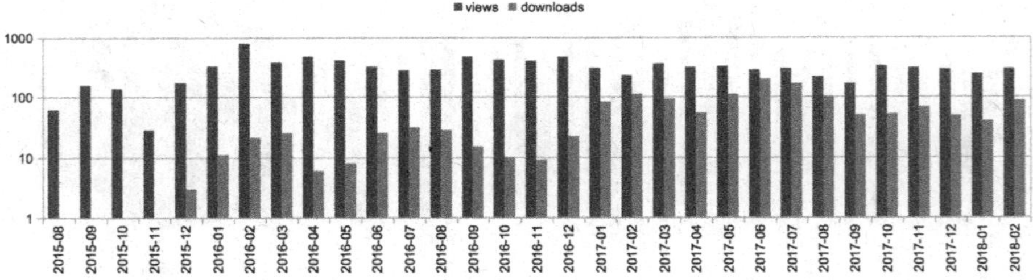

Fig. 9. The number of times SPAR-related resources (examples and the DataCite mapping document) released on Figshare have been visited and downloaded in the past months.

In order to guarantee better support for the SPAR community, all the SPAR Ontologies resources have recently been migrated from Sourceforge to GitHub (https://github.com/sparontologies), which includes several repositories, one for each ontology included in SPAR. This makes it easier to post and handle issues, and to gather new feedback from the community by using the GitHub issue tracker available in each repository.

While the IRI of all the current (and future) ontologies are specified by means of http://purl.org/spar (and http://purl.org/cerif) as base URLs, we have also activated a new https://w3id.org/spar domain as an alternative route for accessing them. While the use of PURL is necessary since it was originally used for defining all the ontology IRIs, the w3id.org service has been recently adopted additionally in order to guarantee a more fine-grained configuration of the content negotiation mechanism, since the approach it uses allows more complex operations via .htaccess (see https://github.com/perma-id/w3id.org/blob/master/spar/.htaccess) that PURL is able to offer.

We have also recently released a set of new contribution guidelines (https://sparontologies.github.io) that will permit members of the community to propose new ontologies to be added to the SPAR suite. In particular, we have defined a clear workflow – made possible using the new GitHub SPAR repository – for accepting such external contributions in a more structured way than was possible in the past.

We are very pleased to report that we are in the process of including the FAIR* Reviews ontology (https://w3id.org/fr/def/core) within the SPAR Ontologies, whose creators are our first adopters of these new SPAR contribution guidelines.

Finally, the sustainability of the SPAR Ontologies is being addressed in two ways. First, as described above, we have put in place contributor guidelines that permit others to join the SPAR community, and we respond swiftly to requests for modifications to the ontologies. While these are mostly stable and maintained without a specific revision schedule, individual ontologies, particularly FaBiO and CiTO are occasionally extended to meet specific requirements. Second, we are actively engaged in negotiations with two major scholarly institutions regarding the long-term future of OpenCitations (http://opencitations.net) as a scholarly infrastructure organization involving the wider community. This will directly improve the sustainability not only of the SPAR Ontologies but also of the other services being developed by OpenCitations.

6 Conclusions

In this paper, we have introduced our ongoing effort (over the past eight years) in ontology development for the scholarly publishing domain, by describing the SPAR (Semantic Publishing and Referencing) Ontologies, a suite of ontological modules recording different aspects of the scholarly publishing domain by means of Semantic Web technologies. We have briefly presented all the ontologies currently included within the SPAR suite (as of 31 March 2018), have listed their main development principles, and have described tools we have created to support their understanding and reuse. Finally, we have also given some Web site usage and social media statistics showing community interactions with the SPAR Ontologies, and have described their uptake within academic articles and development projects.

In the future, we plan to work actively on two specific aspects. First, we want to extend community involvement with the SPAR Ontologies, by suggesting to creators of additional specialist ontologies for the scholarly publishing domain that they consider making their ontological artefacts available as part of the SPAR Ontologies, which our new guidelines should facilitate. Additionally, we want to develop tools that authors and publishers can use for the (semi-) automatic production of SPAR-based metadata, so as to facilitate the Web publication of machine-readable descriptions of academic publications, structured according to a complete, well-developed and trusted data model for the scholarly publishing domain.

Acknowledgements. We would like to thank the many people that have contributed to the development of different aspects of the SPAR Ontologies, in particular those listed at http://www.sparontologies.net/about#collabs, and the creators of the FAIR* Review ontology, Idafen Santana-Pérez and María Poveda-Villalón. While the majority of the SPAR development work was undertaken without direct external financial support, we gratefully acknowledge the original Jisc funding to David Shotton that triggered these developments, and recent financial support provided to us by the Alfred P. Sloan Foundation for the OpenCitations Enhancement Project (grant number G-2017-9800), that has indirectly facilitated this work.

References

1. Brickley, D., Miller, L.: FOAF Vocabulary Specification 0.99. Namespace Document 14 January 2014 - Paddington Edition (2014). http://xmlns.com/foaf/spec/
2. Buckingham Shum, S., Motta, E., Domingue, J.: ScholOnto: an ontology-based digital library server for research documents and discourse. Int. J. Digit. Libr. **3**(3), 237–248 (2000). https://doi.org/10.1007/s007990000034
3. Ciccarese, P., Peroni, S.: The Collections Ontology: creating and handling collections in OWL 2 DL frameworks. Semant. Web **5**(6), 515–529 (2014). https://doi.org/10.3233/SW-130121
4. Ciccarese, P., et al.: The SWAN biomedical discourse ontology. J. Biomed. Inform. **41**(5), 739–751 (2008). https://doi.org/10.1016/j.jbi.2008.04.010
5. Corcho, O., Poveda-Villalón, M., Gómez-Pérez, A.: Ontology engineering in the era of linked data. Bull. Assoc. Inf. Sci. Technol. **41**(4), 13–17 (2015). https://doi.org/10.1002/bult.2015.1720410407
6. d'Aquin, M.: Modularizing ontologies. In: Suárez-Figueroa, M.C., Gómez-Pérez, A., Motta, E., Gangemi, A. (eds.) Ontology Engineering in a Networked World, pp. 213–233. Springer, Heidelberg (2012). https://doi.org/10.1007/978-3-642-24794-1_10
7. D'Arcus, B., Giasson, F.: Bibliographic Ontology Specification. Specification Document - 4 November 2009. http://bibliontology.com
8. DCMI Usage Board: DCMI Metadata Terms. Dublin Core Metadata Initiative (2012). http://dublincore.org/documents/dcmi-terms/
9. Di Iorio, A., Peroni, S., Poggi, F., Vitali, F.: Dealing with structural patterns of XML documents. J. Am. Soc. Inf. Sci. Technol. **65**(9), 1884–1900 (2014). https://doi.org/10.1002/asi.23088
10. Dumontier, M., Baker, C.J.O., Baran, J.: The Semanticscience Integrated Ontology (SIO) for biomedical research and knowledge discovery. J. Biomed. Semant. **2014**(5), 14 (2014). https://doi.org/10.1186/2041-1480-5-14

11. Falco, R., Gangemi, A., Peroni, S., Shotton, D., Vitali, F.: Modelling OWL ontologies with graffoo. In: Presutti, V., Blomqvist, E., Troncy, R., Sack, H., Papadakis, I., Tordai, A. (eds.) ESWC 2014. LNCS, vol. 8798, pp. 320–325. Springer, Cham (2014). https://doi.org/10.1007/978-3-319-11955-7_42
12. Gangemi, A., Peroni, S., Vitali, F.: Literal reification. In: Proceedings of WOP 2010, pp. 65–66 (2010). http://ceur-ws.org/Vol-671/pat04.pdf
13. Groza, T., Handschuh, S., Möller, K., Decker, S.: SALT - semantically annotated LaTeX for scientific publications. In: Franconi, E., Kifer, M., May, W. (eds.) ESWC 2007. LNCS, vol. 4519, pp. 518–532. Springer, Heidelberg (2007). https://doi.org/10.1007/978-3-540-72667-8_37
14. Hammar, K., Lin, F., Tarasov, V.: Information reuse and interoperability with ontology patterns and linked data. In: Abramowicz, W., Tolksdorf, R., Węcel, K. (eds.) BIS 2010. LNBIP, vol. 57, pp. 168–179. Springer, Heidelberg (2010). https://doi.org/10.1007/978-3-642-15402-7_23
15. Hammond, T., Pasin, M.: The nature.com ontologies portal. In: Proceedings of LISC 2015 (2015). http://ceur-ws.org/Vol-1572/paper2.pdf
16. Hammond, T., Pasin, M., Theodoridis, E.: Data integration and disintegration: Managing Springer Nature SciGraph with SHACL and OWL. In: Proceedings of the Posters, Demos and Industry Tracks of ISWC 2017 (2017). http://ceur-ws.org/Vol-1963/paper493.pdf
17. Hillmann, D., Coyle, K., Phipps, J., Dunsire, G.: RDA vocabularies: process, outcome, use. D-Lib Mag. **16**(1/2) (2010). https://doi.org/10.1045/january2010-hillmann
18. Hu, Y., Janowicz, K., McKenzie, G., Sengupta, K., Hitzler, P.: A linked-data-driven and semantically-enabled journal portal for scientometrics. In: Alani, H., et al. (eds.) ISWC 2013. LNCS, vol. 8219, pp. 114–129. Springer, Heidelberg (2013). https://doi.org/10.1007/978-3-642-41338-4_8
19. IFLA Study Group on the Functional Requirements for Bibliographic Records: Functional Requirements for Bibliographic Records - Final Report. International Federation of Library Associations and Institutions (2009). https://www.ifla.org/files/assets/cataloguing/frbr/frbr_2008.pdf
20. International Digital Enterprise Alliance: PRISM Introduction. PRISM - Publishing Requirements for Industry Standard Metadata, version 2.1 (2009). http://www.prismstandard.org/specifications/2.1/PRISM_introduction_2.1.pdf
21. Lebo, T., Sahoo, S., McGuinness, D.: PROV-O: The PROV Ontology. W3C Recommendation, 30 April 2013. https://www.w3.org/TR/prov-o/
22. Miles, A., Bechhofer, S.: SKOS Simple Knowledge Organization System - Reference. W3C Recommendation, 18 August 2009. https://www.w3.org/TR/skos-reference/
23. Nuzzolese, A.G., Gentile, A.L., Presutti, V., Gangemi, A.: Conference linked data: The ScholarlyData Project. In: Proceedings of ISWC 2016, pp. 150–158 (2016). https://doi.org/10.1007/978-3-319-46547-0_16
24. Peroni, S.: A simplified agile methodology for ontology development. In: Dragoni, M., Poveda-Villalón, M., Jimenez-Ruiz, E. (eds.) OWLED/ORE -2016. LNCS, vol. 10161, pp. 55–69. Springer, Cham (2017). https://doi.org/10.1007/978-3-319-54627-8_5
25. Peroni, S., Shotton, D., Vitali, F.: The live OWL documentation environment: a tool for the automatic generation of ontology documentation. In: ten Teije, A., et al. (eds.) EKAW 2012. LNCS, vol. 7603, pp. 398–412. Springer, Heidelberg (2012). https://doi.org/10.1007/978-3-642-33876-2_35

26. Presutti, V., Gangemi, A.: Content ontology design patterns as practical building blocks for web ontologies. In: Li, Q., Spaccapietra, S., Yu, E., Olivé, A. (eds.) ER 2008. LNCS, vol. 5231, pp. 128–141. Springer, Heidelberg (2008). https://doi.org/10.1007/978-3-540-87877-3_11
27. Shotton, D.: The five stars of online journal articles - a framework for article evaluation. D-Lib Mag. **18**(January/February issue) (2012). https://doi.org/10.1045/january2012-shotton
28. Shotton, D.: Semantic publishing: the coming revolution in scientific journal publishing. Learn. Publ. **22**(2), 85–94 (2009). https://doi.org/10.1087/2009202
29. Shotton, D.: CiTO, the Citation Typing Ontology. J. Biomed. Semant. **1**(Suppl 1), S6 (2010) https://doi.org/10.1186/2041-1480-1-S1-S6
30. Teufel, S., Siddharthan, A., Tidhar, D.: Automatic classification of citation function. In: Proceedings of EMNLP 2006, pp. 103–110 (2006). https://doi.org/10.3115/1610075.1610091
31. Wilkinson, M.D., Dumontier, M., Aalbersberg, I.J., Appleton, G., et al.: The FAIR Guiding Principles for scientific data management and stewardship. Sci. Data **3** (2016). https://doi.org/10.1038/sdata.2016.18

Browsing Linked Data Catalogs with LODAtlas

Emmanuel Pietriga[1(✉)], Hande Gözükan[1], Caroline Appert[1], Marie Destandau[1], Šejla Čebirić[2], François Goasdoué[2,3], and Ioana Manolescu[2]

[1] Univ. Paris-Sud, CNRS, Inria, Université Paris-Saclay, Orsay, France
{emmanuel.pietriga,hande.gozukan,marie.destandau}@inria.fr, appert@lri.fr
[2] Inria and École Polytechnique, Palaiseau, France
sejla.c@gmail.com, ioana.manolescu@inria.fr
[3] Univ. Rennes 1, Rennes, France
fg@irisa.fr

Abstract. The Web of Data is growing fast, as exemplified by the evolution of the Linked Open Data (LOD) cloud over the last ten years. One of the consequences of this growth is that it is becoming increasingly difficult for application developers and end-users to find the datasets that would be relevant to them. Semantic Web search engines, open data catalogs, datasets and frameworks such as LODStats and LOD Laundromat, are all useful but only give partial, even if complementary, views on what datasets are available on the Web. We introduce LODAtlas, a portal that enables users to find datasets of interest. Users can make different types of queries about both the datasets' metadata and contents, aggregated from multiple sources. They can then quickly evaluate the matching datasets' relevance, thanks to LODAtlas' summary visualizations of their general metadata, connections and contents.

Keywords: Linked Data Catalogs · Dataset Search · Visualization

1 Introduction

Open data catalogs, Semantic Web search engines and related services play an essential role in the development of the Web of Data. They enable a wide range of users to identify datasets relevant to their purposes, effectively supporting *"modern semantic approaches [that] leverage vastly distributed, heterogeneous data collection with needs-based, lightweight data integration"* [9]. Data publishers can find relevant datasets to link to, thus adding value to their data and enriching the overall ecosystem. Software developers can look for stable datasets to rely upon in their application. Ontology designers can identify and reuse existing

Resource Type: Software Framework, Research prototype/service
Permanent URL: http://purl.org/lodatlas
https://gitlab.inria.fr/epietrig/LODAtlas

D. Vrandečić et al. (Eds.): ISWC 2018, LNCS 11137, pp. 137–153, 2018.
https://doi.org/10.1007/978-3-030-00668-6_9

concepts from other vocabularies. Data analysts, data journalists and other end-user profiles can find the various datasets, ideally already linked, that will help them answer their questions. The Semantic Web community itself also makes use of these services for research purposes, this new *Web of Data* and its dynamics being interesting phenomena to study on their own [26].

Over the last fifteen years, we have seen a variety of resources emerge, some of which have played a foundational role, addressing obvious needs of the community: search engines such as Swoogle [18], SWSE [24], Sindice.com [33]; open data catalogs with some level of support for the specifics of linked data, such as CKAN[1]-based portals datahub.io, data.gov and europeandataportal.eu; services such as LODStats [20] and the LOD Laundromat [5].

Along with the proper means to describe linked datasets using VoID [2], this entire ecosystem should enable users from all of the above profiles to easily find the datasets that are of interest to them. But unfortunately, reality is somewhat different. According to Vandenbussche *et al.* [35], only 13.7% of the registered 562 public SPARQL endpoints have VoID descriptions.[2] Some services have been discontinued. Others are still available but no longer updated. Yet other services are evolving, but dropping support for the specifics of linked data in the process [3], as their focus is elsewhere.

The need for linked data catalogs has been asserted again very recently by the LOD community, following datahub.io's evolution (see `public-lod@w3.org` discussion thread [3]). The discussion also emphasizes the opportunity to move to a framework that would itself be more reliant on linked data technologies for the management and serving of the metadata describing available datasets. While that would certainly be highly relevant and useful, we would be missing an opportunity by focusing only on technical aspects, leaving aside the more human-centric dimension of dataset search. Indeed, one issue with the services aforementioned is that while they are quite useful, each of them, taken individually, only provides incomplete information. Users consequently have to gather information from multiple such services in order to find the datasets they need.

The LODAtlas project has been initiated to explore an alternative user interface, aimed at making it easier for a broad range of users to find datasets of interest. LODAtlas aggregates data about datasets from multiple sources. It then lets users explore the resulting linked data catalog in various ways, using keyword search and faceted navigation. Selection criteria can freely mix constraints on the datasets' metadata (*e.g.*, description, last modification date), the links that exist between them, and their schema-level [22] content, favoring visual representations of the result-sets using coordinated multiple views [36].

2 Background and Motivation

The visualization of linked data has been an active field of research for many years, with the development of so-called linked data browsers (*e.g.*, [8]) and

[1] Comprehensive Knowledge Archive Network.

[2] Numbers updated on 2018-03-29 from http://sparqles.ai.wu.ac.at/discoverability.

visualization tools, as well as supporting vocabularies [31] – see Dadzie *et al.*'s surveys [15,16]. Such user interfaces enable users to navigate on the Web of Data, displaying, in one form or another, the actual RDF statements contained in the datasets. Here, we are more interested in interfaces that enable users to identify sources servings datasets relevant to their purposes, that can then be browsed using one of the above tools.

Early Semantic Web keyword-based search engines, such as Swoogle [18] and Falcons [14], were already enabling users to identify data sources and vocabularies, even if indirectly: based on keywords input by the user, they would return vocabularies or *"documents"* containing instance data matching the search criteria. Those would be displayed to users as more-or-less flat lists of links to external resources (ontologies, RDF documents), or their content would be exposed as raw triples. Sindice.com [33] played a somewhat different role: given a certain RDF resource URI as input, the API would provide the client application (*e.g.*, a linked data browser) with links to additional data sources containing statements involving that resource URI as subject or object. The following generation of search engines, including SWSE [24] and Watson [17], provided significant improvements such as, *e.g.*, displaying the information contained in the retrieved statements in a much more human-friendly manner (SWSE); and providing useful metadata about the source (Watson). The general concept remained essentially the same, however.

A range of recent systems can assist users in the identification of datasets that suit their needs. As it is difficult to gain a clear understanding of the content of a dataset by looking at the raw triples, recent work has focused on providing visual summaries of the content of a given dataset. Given a SPARQL endpoint, LODEx [7] automatically generates a schema-centric, node-link diagram visualization of the content behind this endpoint. LODSight [19] and ExpLOD [27] follow conceptually similar approaches, representing similar information as node-link diagrams. The former provides more concise, but possibly less accurate summaries than LODEx as it might suggest possible relations that are not actually present in the data. The latter, ExpLOD, provides additional information about the interlinking between datasets. Loupe [30] also enables users to inspect the content of datasets. Rather than node-link diagrams, Loupe generates interactive summary tables based on explicit schema-level definitions and an analysis of how schema elements are actually used to describe instance data.

Aether [29] gives a complementary view on SPARQL endpoints, automatically generating a set of VoID-derived statistical charts (bar charts, pie charts) about namespace, class and property usage, also enabling the visual comparisons of two endpoints. LODStats [20] also provides statistical metadata about RDF datasets, at a wider scale, and makes those metadata themselves available as a linked dataset using the LDSO vocabulary, which extends VoID.

Other useful datasets and services include LODatio [22,28], a powerful data source search engine. Aimed at a more technical audience, it takes as input a raw SPARQL query that captures which types of resources and properties the user is interested in finding, and returns a ranked list of matching data sources.

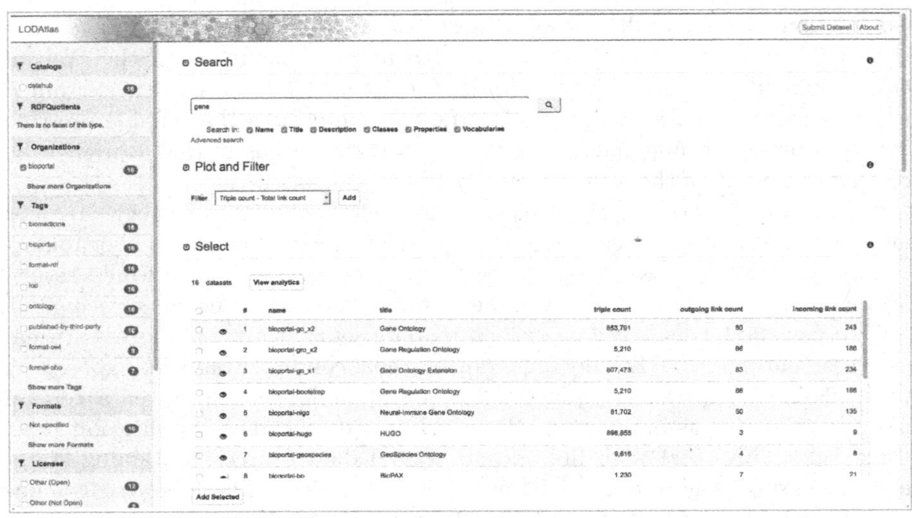

Fig. 1. Searching for datasets containing gene in their title, published by the bioportal.

LODatio also suggests alternative queries based on the one input to narrow or widen the result list. Of interest primarily to dataset creators and ontology engineers, the LOV portal [34] is a very valuable, curated source of information aimed at facilitating the reuse of vocabularies, that provides data about the interconnections between vocabularies and version history.

Finally, while it primarily serves other purposes, LOD laundromat [6], and more precisely the LOD Wardrobe [5], lets people browse through a list of *"cleaned"* versions of a significant proportion of the LOD datasets available publicly on the Web. The Wardrobe offers some query capabilities, statistical charts and can show raw data fragments.

LODAtlas does not aim at replacing the above services and datasets, but rather at integrating a coherent subset of them into a single Web-based UI to facilitate the search for linked datasets. As described in the next section, LODAtlas takes the perspective of a user *shopping for datasets* by expressing her various needs (catalog metadata, schema-level constraints, interlinks) using different means (keyword search, URI search, faceted navigation) and assessing candidate datasets through visual summaries of their properties and contents.

3 Browsing the LOD cloud with LODAtlas

LODAtlas lets users browse the datasets found in one or more catalogs. In the following we take, as a running example, dataset descriptions exported from the CKAN-based `datahub.io` portal *before* it evolved to the new version,[3] as this

[3] These descriptions are still available from old.datahub.io.

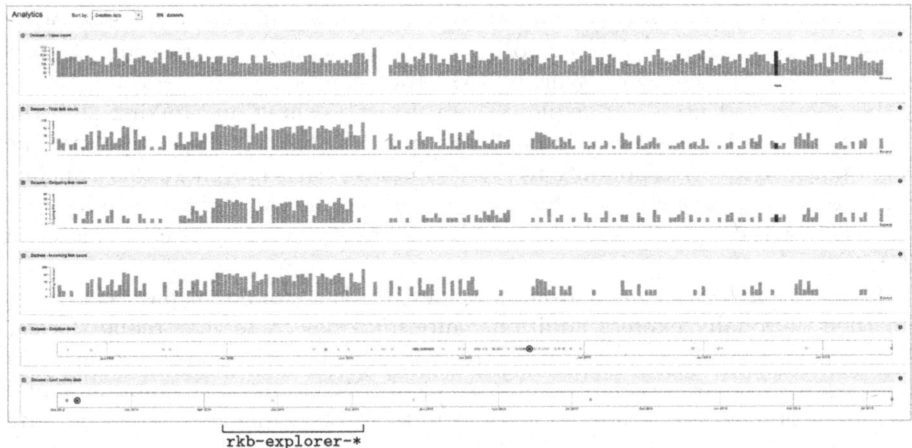

Fig. 2. Looking at all datasets from Linking Open Data Cloud, sorted by creation date. Hovering dataset near in one chart highlights it (black) in all charts (brushing). **`rkb-explorer`** datasets are discussed in Sect. 3.3.

older version remains for now one of the most important sources of information about linked open datasets. As discussed in Sect. 5, multiple data catalogs can be added to the same instance of LODAtlas, in which case the provenance of the dataset description (which catalog it was imported from) becomes an additional possible search criterion.

3.1 Overview

LODAtlas provides users with two means to browse datasets: using keyword/URI search, and using faceted navigation. Both can be used in conjunction, to iteratively refine the result list. Figure 1 shows the results of a basic search for keyword gene in the datasets' name or title, published by the bioportal.

Users can search for keywords and URIs in any combination of: dataset name, title and description; vocabularies, classes and properties used. Results are ordered to first show exact matches, and then partial ones, if any. When searching for classes or properties, LODAtlas looks for the input value in the class or property URI, as well as in the corresponding `rdfs:label` from the vocabulary definition. Only datasets that actually feature at least one instance of the property or class will be considered exact matches. For example, when searching for `foaf:knows` in Properties, LODAtlas will return as exact matches only the datasets that feature at least one statement whose property URI is `foaf:knows`.

From an initial list of candidate datasets obtained with keyword/URI search and faceted navigation, users can further refine the results based on other dataset characteristics, that are more efficiently represented and specified using simple visualization widgets. First, users can display charts that summarize (Fig. 2):

the number of triples in each considered dataset, the number of links to other datasets (incoming, outgoing, or both), and timelines showing creation and last update dates. All charts are synchronized: they can be sorted according to any of the above, and users can explore them using brushing and linking [36]: the dataset hovered by the cursor immediately gets highlighted in all views (see the single black item corresponding to dataset near in each bar chart and timeline in Fig. 2). This set of simple interactive visualizations can further help identify datasets of interest, and can yield interesting observations, as discussed later in Sect. 3.3.

Based on insights gained from this view on the candidate datasets, users can then optionally express additional filtering rules to further refine the list (Fig. 3a). Such rules, specified interactively by drawing selection regions in scatterplots and timelines, declare combinations of restrictions on the minimum and maximum numbers of: triples, counts of links to other datasets, creation date and last update date. Once satisfied, the user can then select some or all of the remaining datasets in the list, and put them in what we call the *dataset cart*, which is conceptually similar to customers' cart on e-commerce platforms.

The dataset cart is separate from the previous list of search results, the rationale being that users may want to first populate their cart with some datasets based on a set of selection criteria, and then add or remove datasets incrementally, based on other criteria. While it would theoretically be possible to capture the final dataset list with a single elaborate query, from the user's perspective this would be quite tedious. Making it possible for users to explicitly store datasets of interest in a cart, temporarily forget about them and continue exploring freely, strongly favors the exploration of the catalog.

In our case, there is obviously no intention to sell the datasets in the cart. The latter should only be seen as a metaphor that will be familiar to many users. *"Checking out"* on LODAtlas only means exporting the cart as simple VoID descriptions of the chosen datasets, for later re-use in any context. Those VoID exports contain a limited set of statements, relying on `foaf:homepage`, as an inverse functional property, to automatically connect to other descriptive statements about the datasets, found elsewhere on the Web.

Before checking out (which remains optional), the contents of the cart can also be visualized in more detail, helping users get a better idea of how the chosen datasets are interlinked and how much data they hold individually. Figure 3b shows some of the available visualizations. From left to right: a bar chart showing the triple count for each dataset (when hovering a dataset, the other ones change color depending on whether they feature incoming links, outgoing links, both, or none); an adjacency matrix giving an overview of which datasets are connected to which ones; a radial network layout showing the same information in a more intuitive, but less scalable, manner.

3.2 Visual Summaries of a Dataset's Contents

The selection of a dataset is not only based on triple count, number of links to other datasets, and presence of some keywords. In their search for datasets, users

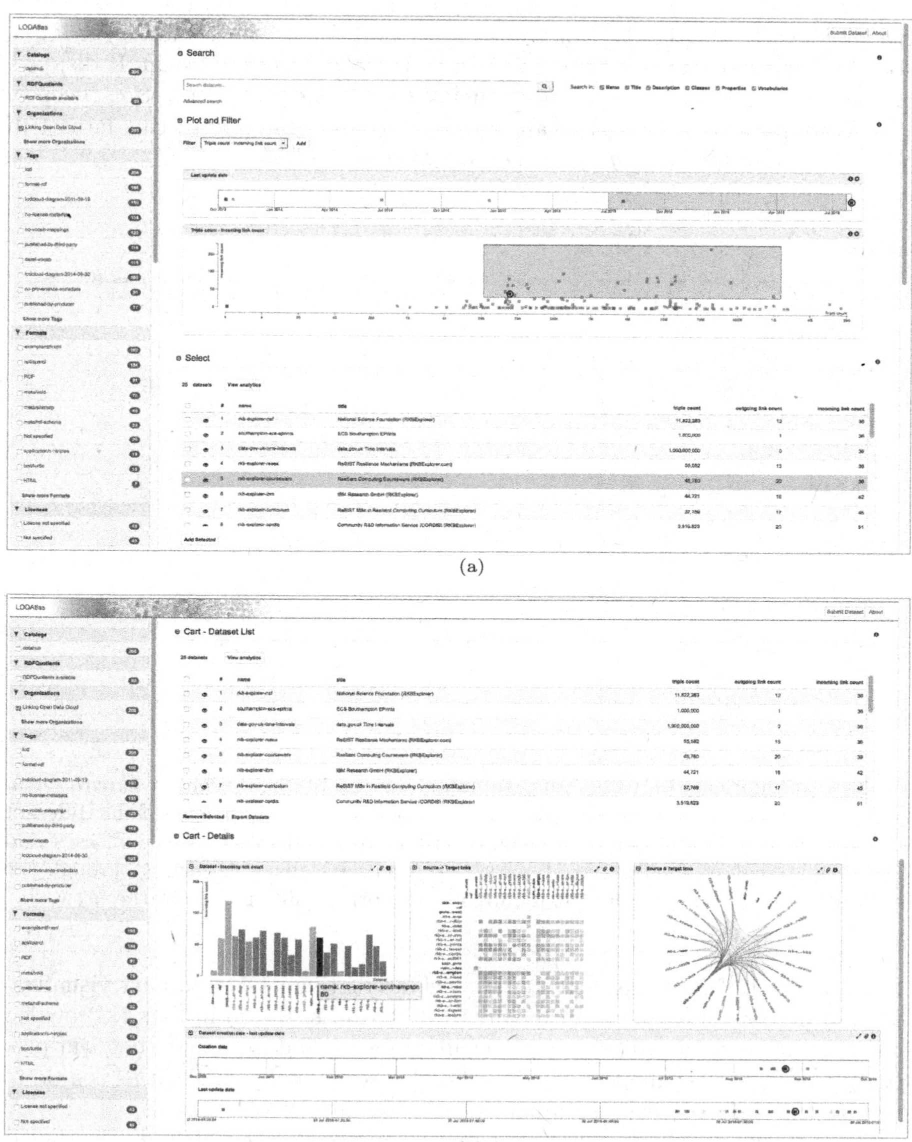

Fig. 3. (a) Filtering search results using visual, dynamic queries. (b) Putting the selected datasets in the user's cart and looking at their characteristics in more detail.

will often want to get more detailed information about what is in the dataset, as suggested by services such as LODSight [19] and LODatio [28].

Any dataset can be inspected in more detail by clicking on the eye-like icon associated with it (Fig. 1). This pops-up a new panel that features multiple tabs. The first one (not shown in the paper) is the dataset's ID card. It displays general metadata about the dataset, including its title and description, license, author and publisher, as well as all resource files associated with the dataset in the catalog description (*e.g.*, partial extracts, full dumps).

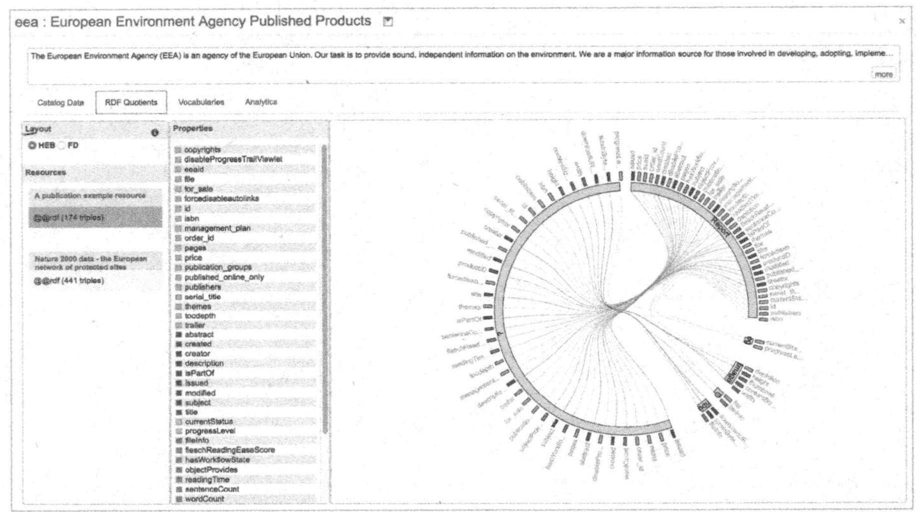

Fig. 4. RDFQuotients-derived visual summary of one of the *European Environment Agency*'s datasets. The summary shows how properties relate instances of the different classes (arcs sometimes represent instances that have multiple classes). Classes and properties are color-coded by vocabulary, based on namespace. Brushing through the sorted list of properties on the left highlights the corresponding edge in the network.

The next tab, RDFQuotients, features a novel interactive RDF summary visualization that has been designed specifically for LODAtlas, shown in Fig. 4. Provided that a dump, even a partial one, is available for a dataset, and that the processing workflow described in Fig. 7 completes successfully, LODAtlas is able to generate this type of visual summary of the contents of the dataset.

The visualization is directly based on a summarization of RDF graphs that is computed using the RDFQuotients framework [11,12]. RDFQuotients work on the standard semantics of an RDF graph G, which can be materialized as an RDF graph called its closure (a.k.a saturation), that comprises G's explicit triples, plus those derived from them and entailment rules from [23], *i.e.*, G's implicit triples. The framework defines a summary of G as a *quotient graph*, which is an RDF graph itself. In particular, it proposes four novel RDF node equivalence relations that allow quotient graphs (*i*) *summarizing both the structure and the semantics* of the original graphs and (*ii*) having more *compact summaries* than those relying on classical (non-RDF) node equivalence relations, *e.g.*, those based on backward and/or forward bisimulation.

Two of these equivalence relations, called *strong equivalence* and *weak equivalence*, only consider how nodes are connected to others using data properties, *i.e.*, different from the built-in RDF properties such as `rdf:type`, `rdfs:subClassOf`, *etc.* Two nodes are strongly equivalent whenever their incoming (resp. outgoing) data properties may cooccur on a single summary node, based on the input graph analysis; they are weakly equivalent whenever they have no incoming and outgoing edge, or their incoming or outgoing data properties may cooccur on a single summary node, or they are weakly equivalent to another node. These two equivalence relations are particularly useful for RDF graphs with untyped or poorly typed data. The two other equivalence relations, called *typed-strong equivalence* and *typed-weak equivalence*, consider only types for typed nodes and the aforementioned strong and weak equivalences for untyped nodes; typed nodes are equivalent whenever they have the same types.

The resulting quotient graphs are then transformed into JSON data structures more amenable to visualization with D3 [10]. They can be represented using a node-link diagram based on force-directed layout, or using a radial network layout based on hierarchical edge bundling [25]. The latter is less familiar and requires a bit of training to interpret, but usually scales better while conveying additional information. The hierarchy used as input for edge bundling is that of subsumption relationships between involved classes.

When multiple resource files are associated with a dump for a given dataset, LODAtlas tries to compute summaries for each such file individually. Each of them is listed in that tab, and users can select any one of them to get the corresponding visual summary. While in some cases the summaries will look very similar, there are also cases where the resource files associated with a single dataset dump contain complementary but very different subsets of the data. In such cases, having access to individual summaries seems more relevant than merging them all in a single, necessarily more complex one, since there was an attempt at modularizing the dataset in the first place.

The following tab, Vocabularies (not shown in this paper), lists all vocabularies actually used to describe RDF resources in the dataset, featuring direct links to the schemas or ontologies, as well as links to the corresponding entries in LOV (Linked Open Vocabulary [34]), when available. As discussed later, this tab may include more ontology-level information in the future, derived from Chen *et al.*'s minimal modules and best excerpts [13].

Finally, the Analytics tab (Fig. 5) features charts very similar to those in Fig. 2, but restricted to the datasets linked to the one being looked at in detail. In this context, the latter serves as a pivot, and all other datasets can be color-coded depending on the nature of their link to it, following the same convention as in the bar chart of Fig. 3b for incoming, outgoing, and two-way connections.

3.3 Examples of Use

This section illustrates some examples of use for LODAtlas:

- Performing advanced **searches that combine criteria about the datasets' metadata and their contents**. Conjunctions of constraints can be specified iteratively using different means, as illustrated in Fig. 3a. For instance, users could search for all datasets that (1) contain dbpedia in their description (by entering that string in the search field); (2) feature instances of class `foaf:Person` (by then selecting the corresponding value in facet Classes); (3) have been updated in the last three months (by adding the corresponding timeline plot and selecting the relevant time span); and finally, (4) feature at least 50,000 statements and more than 2 outgoing links to other datasets (by drawing a selection region in the corresponding scatterplot).
- **Monitoring datasets** recently added to the catalog or updated, that link to a particular dataset of interest. Figure 5 shows tab Analytics for dataset DBpedia. Using the first timeline, users can quickly find out which datasets have been recently added to the catalog, that feature links (incoming, outgoing, or both) to DBpedia. The second timeline gives similar information about when these datasets have been updated. Brushing in the timeline makes it possible to get a quick estimate about the size and interlinks of those datasets.
- **Spotting noteworthy events** in a selection of datasets. Going back to Fig. 2, sorting by creation date immediately reveals a time span that features

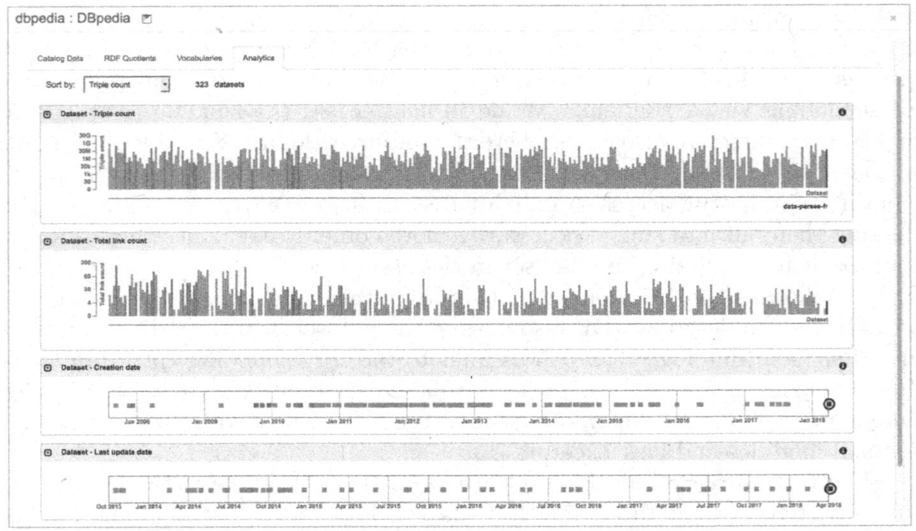

Fig. 5. According to CKAN data fetched from `datahub.io`, the last dataset added to the LOD cloud that links to DBpedia is data-persee-fr, a dataset about scientific publications: added March 21st, 2018 and last updated 10 days later, it features a larger-than-average number of triples compared to all datasets linking to DBpedia.

datasets with a significantly larger number of link counts. Brushing through the histograms indicates that this "surge" corresponds to the addition of RKB Explorer [21] entries in the catalog.

- **Comparing & contrasting the contents** of related datasets. The RDF-Quotients-based visual summaries show how instances of different classes are effectively described, and connected to, other instances, using which properties. Users can get a first impression about the suitability of different datasets for their purposes. These summaries can also help them understand how those datasets can work together to derive more data, or identify opportunities to link them when they are not already linked.

4 Implementation

LODAtlas is based on Java EE 7 Web Profile edition, and deployed on an Apache Tomcat 8 server. The following Javascript libraries play a key role on the front-end side: D3.js [10] for generating the SVG visualisations; Crossfilter.js for filtering the data presented in charts, which effectively enables the brushing and linking features described earlier; JQuery for AJAX calls to server-side REST endpoints; and Bootstrap for general page layout and icons.

Figure 6 gives an overview of LODAtlas' architecture. The backend is implemented in Java, adopting a layered architecture. An ElasticSearch server stores and indexes the data. The Web server's REST endpoint receives requests and forwards them to the ElasticSearch service, which processes the requests and returns results as Plain Old Java Objects (POJO). These are the converted to JSON and transmitted back to the client. The REST endpoint can also be queried directly by any external tool (http://lodatlas.lri.fr/api/).

The ElasticSearch index gets populated by an independent module called the LODAtlas Data Manager (DM for short). That module is a standalone Java application that creates an aggregated database using several APIs to harvest metadata from different catalogs, and to process dataset dumps when available.

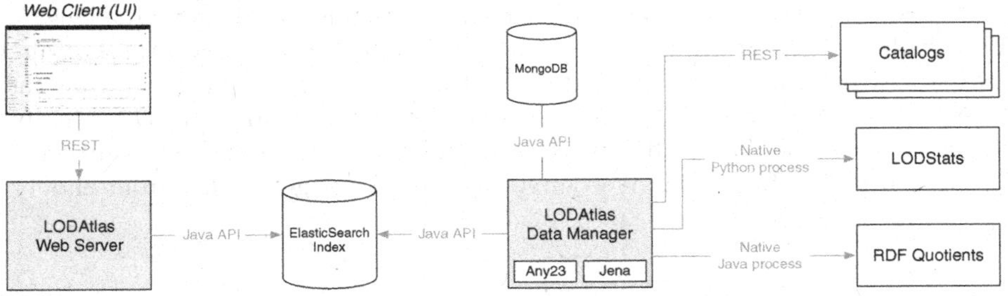

Fig. 6. LODAtlas - System architecture

The identification of relevant datasets in a catalog and fetching of the corresponding metadata is based on CKAN API v3.[4] Any23[5] and the Jena RIOT API[6] handle the conversion of dump files to N-Triples, providing support for a broad range of RDF serialization formats. LODStats [20] is used as an external service to extract classes, properties and vocabularies, and RDFQuotients [11] provide summaries of the RDF dumps.

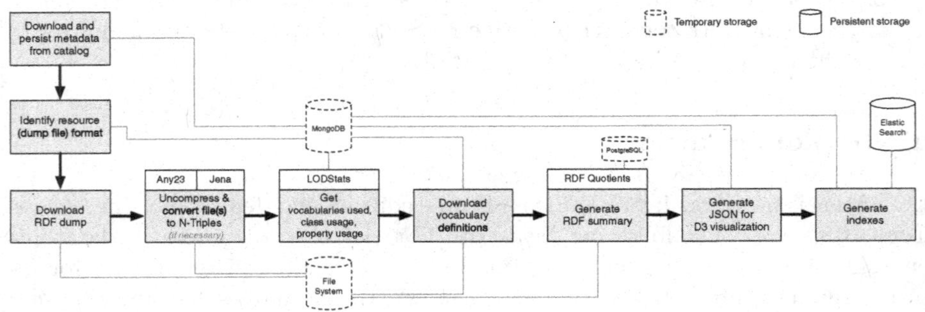

Fig. 7. LODAtlas - Dataset processing workflow.

Figure 7 illustrates the processing workflow of a dataset whose description has been found in a catalog and matches the requirements for being considered a *linked data* dataset (*e.g.*, on datahub.io, having `lod` as one of the declared tags). Once the JSON metadata has been downloaded from the catalog and temporarily stored in a MongoDB[7] instance, the DM checks for resource files associated with this dataset. Among these resource files, those that are using one of the supported RDF serializations are downloaded, uncompressed (if necessary), and converted to N-Triples. For each resource file, LODStats returns information about the vocabularies, classes and properties used. This information is also temporarily stored in MongoDB, and vocabulary definitions get concatenated in a single file for use by RDFQuotients to compute the summaries. RDFQuotients use their own local PostgreSQL database to make summary computations more efficient. The resulting RDF graph is transformed into a JSON data structure, that also gets stored in MongoDB. This data structure is optimized for generating the interactive summary visualization (Fig. 4) on the front-end using D3. Finally, the contents of the MongoDB instance get indexed in ElasticSearch, which will be queried by the LODAtlas Web server to generate pages for the front-end.

---

[4] http://docs.ckan.org/en/latest/api/.
[5] https://any23.apache.org.
[6] https://jena.apache.org/documentation/io/.
[7] https://www.mongodb.com.

5 Availability, Sustainability and Future Work

LODAtlas started as a research project initiated by team ILDA[8] at INRIA and LRI (Univ. Paris-Sud & CNRS), with contributions from INRIA team CEDAR.[9] The project began long before `datahub.io`'s recent, major overhaul, and subsequent loss of LOD entries in its catalog [3]. Our goal was to investigate alternative user interfaces for browsing linked data catalogs in order to facilitate the discovery of relevant datasets. As such, the project had no intention to replace `datahub.io` for the LOD community. The context has now changed, however: we were able to retrieve, process and store locally all LOD dataset metadata from http://old.datahub.io; LODAtlas' dataset processing workflow has been streamlined, and the service has gained maturity through an iterative design process of the user interface over several years; we now have access to more computing resources at INRIA for dataset processing.[10] In addition, the design of novel user interfaces for the Web of Data is a central topic of our research team, which means that we are committed to LODAtlas, not just as a service to be maintained, but as a research project aimed at evolving based on feedback from the community. As such, the main instance of LODAtlas at http://purl.org/lodatlas will be accepting new LOD-related dataset submissions. As is currently the case for LOV [34], we have opted for a lightweight curated model where each submission will be manually checked prior to inclusion by a LODAtlas team member, both for relevance and quality, before triggering the automatic processing of the new dataset. We may reconsider this choice if the service gains traction and the submission volume increases too much, in which case we would rather rely on a community effort.

Another element to consider is that LODAtlas is contributed to the community as much as a software framework as a research prototype/service. The code is hosted on GitLab at https://gitlab.inria.fr/epietrig/LODAtlas under the GNU General Public License (GPL) version 3.0, and is also made available as a Docker[11] bundle for deployment by anyone interested, for use with any CKAN-compatible catalog description. See the project's GitLab page for information about running the demo with `docker-compose`.

As summarized in Table 1, the main LODAtlas instance gathers descriptions from `datahub.io` and from `data.gov`. Catalog metadata can be processed for all relevant datasets, though some entries might be missing information depending on the completeness of the original description. LODStats and RDFQuotients

[8] http://ilda.saclay.inria.fr.

[9] https://team.inria.fr/cedar.

[10] The processing workflow can be run on any mid-range hardware configuration, but can also be parallelized. While we started the project with a single machine equipped with 2 CPUs and 16GB RAM, our current setup enables us to instantiate up to 4 virtual machines in parallel on the local cloud infrastructure, each machine having 8 CPUs, 1TB of disk space, 117GB RAM. While such computing power is not necessary for all datasets, the processing of some very large dumps – and more particularly the summarization – may require significant resources.

[11] https://www.docker.com.

Table 1. Catalogs featured in LODAtlas instance at http://purl.org/lodatlas

Catalog	Linked Datasets (entries created)	RDFQuotients (computed)	Resource files (total)	Resource files (success)
datahub.io	1,280[a]	417	72,303	23,996
data.gov	(↗)6,772[b]	(↗)6,011	(↗)6,365	(↗)6,099

[a] querying CKAN API for all datasets tagged with `lod`.
[b] querying CKAN API for all datasets featuring `rdf` in any metadata field.
(↗) 71% of `data.gov` datasets have been processed at the time of writing. The Web site will contain updated numbers, accounting for the whole 9,482 RDF-tagged datasets (as of 2018-06-14).

processing is more subject to failure (this does not impact the creation of the dataset's entry in LODAtlas, but means that some features will not be available, such as the visual summary). The processing of `datahub.io` is complete: we were able to compute RDFQuotients summaries for 33% of the datasets. The processing of `data.gov` is still ongoing at the time of writing. The current success rate for resource file processing yields RDFQuotients summaries for 89% of the datasets. Coverage thus varies significantly depending on the catalog. There can be many causes of failure: unavailability of any resource file, absence of resource file in one of the supported RDF serializations, failure to process a file for reasons such as, *e.g.*, syntax errors or size limitations (we are currently unable to process individual RDF dumps larger than 10GB).

Future work on LODAtlas will start by considering additional catalogs, such as https://www.europeandataportal.eu which, at the time of writing, is declaring 38,170 RDF datasets. We are also in the process of integrating a new version of RDFQuotients, which is providing cardinality information about the actual usage of classes and properties in resource files. This will enable us to: (1) extend search capabilities by adding criteria on the number of instances of a given class or property; and (2) enhance the summary visualizations, representing this cardinality information by adjusting the property edges' stroke width depending on the relative number of statements of each sort.

Another possibility we are considering is to show partial views on vocabulary definitions based on solutions such as Chen *et al.*'s minimal modules and best excerpts [13]. For a given dataset, relevant starting points (classes) could be identified in the instance data, that would serve as input to generate views on coherent subsets of vocabulary definitions, small enough to be meaningfully visualized and understood by users.

In the longer term, as interactive graph visualization is an active research topic in the team (see, *e.g.*, [4,32]), we are also contemplating the possibility to generate an advanced, interactive visualization similar in spirit to the Linking Open Data cloud diagram [1] using the dataset descriptions stored in LODAtlas. The prioritization of new features will depend on feedback from the community.

References

1. Abele, A., McCrae, J.P., Buitelaar, P., Jentzsch, A., Cyganiak, R.: Linking Open Data cloud diagram (2017). http://lod-cloud.net
2. Alexander, K., Cyganiak, R., Hausenblas, M., Zhao, J.: Describing Linked Datasets with the VoID Vocabulary. W3C Interest Group Note, March 2011. https://www.w3.org/TR/void/
3. public-lod@w3.org thread datahub.io, February 2018. https://lists.w3.org/Archives/Public/public-lod/2018Feb/0001.html
4. Bach, B., Pietriga, E., Fekete, J.D.: Graphdiaries: animated transitions and temporal navigation for dynamic networks. IEEE Trans. Vis. Comput. Graph. **20**(5), 740–754 (2014). https://doi.org/10.1109/TVCG.2013.254
5. Beek, W., Rietveld, L., Bazoobandi, H.R., Wielemaker, J., Schlobach, S.: LOD laundromat: a uniform way of publishing other people's dirty data. In: Mika, P., et al. (eds.) ISWC 2014. LNCS, vol. 8796, pp. 213–228. Springer, Cham (2014). https://doi.org/10.1007/978-3-319-11964-9_14
6. Beek, W., Rietveld, L., Schlobach, S., van Harmelen, F.: LOD Laundromat: why the semantic web needs centralization (Even If We Don'T Like It). IEEE Internet Comput. **20**(2), 78–81 (2016). https://doi.org/10.1109/MIC.2016.43
7. Benedetti, F., Bergamaschi, S., Po, L.: Visual querying lod sources with lodex. In: Proceedings of the International Conference on Knowledge Capture, K-CAP 2015, pp. 12:1–12:8. ACM (2015). https://doi.org/10.1145/2815833.2815849
8. Berners-Lee, T., et al.: Tabulator: exploring and analyzing linked data on the Semantic Web. In: Proceedings of the 3rd international Semantic Web User Interaction Workshop (2006)
9. Bernstein, A., Hendler, J., Noy, N.: A new look at the semantic web. Commun. ACM **59**(9), 35–37 (2016). https://doi.org/10.1145/2890489
10. Bostock, M., Ogievetsky, V., Heer, J.: D3: Data-driven documents. IEEE Trans. Vis. Comput. Graph. **17**(12), 2301–2309 (2011). https://doi.org/10.1109/TVCG.2011.185
11. Čebirić, Š., Goasdoué, F., Manolescu, I.: A framework for efficient representative summarization of RDF graphs. In: ISWC (Posters & Demonstrations) (2017). http://ceur-ws.org/Vol-1963/paper512.pdf
12. Čebirić, Š., Goasdoué, F., Manolescu, I.: Query-Oriented Summarization of RDF Graphs. Research Report RR-8920, INRIA (2017). https://hal.inria.fr/hal-01325900
13. Chen, J., Ludwig, M., Ma, Y., Walther, D.: Zooming in on ontologies: minimal modules and best excerpts. In: d'Amato, C., et al. (eds.) ISWC 2017. LNCS, vol. 10587, pp. 173–189. Springer, Cham (2017). https://doi.org/10.1007/978-3-319-68288-4_11
14. Cheng, G., Ge, W., Qu, Y.: Falcons: Searching and browsing entities on the semantic web. In: Proceedings of the International Conference on World Wide Web, pp. 1101–1102. ACM (2008). https://doi.org/10.1145/1367497.1367676
15. Dadzie, A.S., Pietriga, E.: Visualisation of linked data - reprise. Semant. Web J. **8**(1), 1–21 (2017). https://doi.org/10.3233/SW-160249
16. Dadzie, A.S., Rowe, M.: Approaches to visualising linked data: a survey. Semant. Web J. **2**(2), 89–124 (2011). https://doi.org/10.3233/SW-2011-0037
17. d'Aquin, M., Motta, E.: Watson, more than a semantic web search engine. Semant. Web J. **2**(1), 55–63 (2011). https://doi.org/10.3233/SW-2011-0031

18. Ding, L., et al.: A search and metadata engine for the semantic web. In: Proceedings of the International Conference on Information and Knowledge Management, CIKM 2004, pp. 652–659. ACM (2004). https://doi.org/10.1145/1031171.1031289
19. Dudáš, M., Svátek, V., Mynarz, J.: Dataset summary visualization with LODSight. In: Gandon, F., Guéret, C., Villata, S., Breslin, J., Faron-Zucker, C., Zimmermann, A. (eds.) ESWC 2015. LNCS, vol. 9341, pp. 36–40. Springer, Cham (2015). https://doi.org/10.1007/978-3-319-25639-9_7
20. Ermilov, I., Lehmann, J., Martin, M., Auer, S.: LODStats: the data web census dataset. In: Groth, P., et al. (eds.) ISWC 2016. LNCS, vol. 9982, pp. 38–46. Springer, Cham (2016). https://doi.org/10.1007/978-3-319-46547-0_5
21. Glaser, H., Millard, I.C., Jaffri, A.: RKBExplorer.com: a knowledge driven infrastructure for linked data providers. In: Bechhofer, S., Hauswirth, M., Hoffmann, J., Koubarakis, M. (eds.) ESWC 2008. LNCS, vol. 5021, pp. 797–801. Springer, Heidelberg (2008). https://doi.org/10.1007/978-3-540-68234-9_61
22. Gottron, T., Scherp, A., Krayer, B., Peters, A.: Lodatio: using a schema-level index to support users infinding relevant sources of linked data. In: Proceedings of the International Conference on Knowledge Capture, K-CAP 2013, pp. 105–108. ACM (2013). https://doi.org/10.1145/2479832.2479841
23. Hayes, P.J., Patel-Schneider, P.F.: RDF 1.1 Semantics (2014). https://www.w3.org/TR/rdf11-mt/
24. Hogan, A., Harth, A., Umbrich, J., Kinsella, S., Polleres, A., Decker, S.: Searching and browsing linked data with SWSE: the semantic web search engine. Web Semant. **9**(4), 365–401 (2011). https://doi.org/10.1016/j.websem.2011.06.004
25. Holten, D.: Hierarchical edge bundles: visualization of adjacency relations in hierarchical data. IEEE Trans. Vis. Comput. Graph. **12**(5), 741–748 (2006). https://doi.org/10.1109/TVCG.2006.147
26. Käfer, T., Abdelrahman, A., Umbrich, J., O'Byrne, P., Hogan, A.: Observing linked data dynamics. In: Cimiano, P., Corcho, O., Presutti, V., Hollink, L., Rudolph, S. (eds.) ESWC 2013. LNCS, vol. 7882, pp. 213–227. Springer, Heidelberg (2013). https://doi.org/10.1007/978-3-642-38288-8_15
27. Khatchadourian, S., Consens, M.: ExpLOD: summary-based exploration of interlinking and RDF usage in the linked open data cloud. In: Aroyo, L., et al. (eds.) ESWC 2010. LNCS, vol. 6089, pp. 272–287. Springer, Heidelberg (2010). https://doi.org/10.1007/978-3-642-13489-0_19
28. Leka, M., Schmidt, H., Blume, T., Vagliano, I., Scherp, A.: Searching for Sources of Data on the Web with LODatio+ (2018). http://lodatio.informatik.uni-kiel.de
29. Mäkelä, E.: Aether – generating and viewing extended VoID statistical descriptions of RDF datasets. In: Presutti, V., Blomqvist, E., Troncy, R., Sack, H., Papadakis, I., Tordai, A. (eds.) ESWC 2014. LNCS, vol. 8798, pp. 429–433. Springer, Cham (2014). https://doi.org/10.1007/978-3-319-11955-7_61
30. Mihindukulasooriya, N., Poveda Villalon, M., Garcia-Castro, R., Gomez-Perez, A.: Loupe - an online tool for inspecting datasets in the linked data cloud. In: Proceedings of the ISWC 2015 Posters & Demonstrations Track (2015). http://ceur-ws.org/Vol-1486/paper_113.pdf
31. Pietriga, E., Bizer, C., Karger, D., Lee, R.: Fresnel: a browser-independent presentation vocabulary for RDF. In: Cruz, I., et al. (eds.) ISWC 2006. LNCS, vol. 4273, pp. 158–171. Springer, Heidelberg (2006). https://doi.org/10.1007/11926078_12
32. Romat, H., Appert, C., Bach, B., Henry-Riche, N., Pietriga, E.: Animated edge textures in node-link diagrams: a design space and initial evaluation. In: Proceedings of the CHI Conference on Human Factors in Computing Systems. ACM (2018). https://doi.org/10.1145/3173574.3173761

33. Tummarello, G., Delbru, R., Oren, E.: Sindice.com: weaving the open linked data. In: Aberer, K., et al. (eds.) ASWC/ISWC -2007. LNCS, vol. 4825, pp. 552–565. Springer, Heidelberg (2007). https://doi.org/10.1007/978-3-540-76298-0_40
34. Vandenbussche, P.Y., Atemezing, G.A., Poveda, M., Vatant, B.: Linked open vocabularies (LOV): a gateway to reusable semantic vocabularies on the web. Semant. Web J. **8**(3), 437–452 (2016). https://doi.org/10.3233/SW-160213
35. Vandenbussche, P.Y., Umbrich, J., Matteis, L., Hogan, A., Buil-Aranda, C.: Sparqles: Monitoring public sparql endpoints. Semant. Web J. **8**(6), 1049–1065 (2017). https://doi.org/10.3233/SW-170254
36. Wang Baldonado, M.Q., Woodruff, A., Kuchinsky, A.: Guidelines for using multiple views in information visualization. In: Proceedings of the Working Conference on Advanced Visual Interfaces, AVI 2000, pp. 110–119. ACM (2000). https://doi.org/10.1145/345513.345271

A Framework to Build Games with a Purpose for Linked Data Refinement

Gloria Re Calegari, Andrea Fiano, and Irene Celino(✉)

Cefriel, Viale Sarca 226, 20126 Milano, Italy
{gloria.re,andrea.fiano,irene.celino}@cefriel.com

Abstract. With the rise of linked data and knowledge graphs, the need becomes compelling to find suitable solutions to increase the coverage and correctness of datasets, to add missing knowledge and to identify and remove errors. Several approaches – mostly relying on machine learning and NLP techniques – have been proposed to address this *refinement* goal; they usually need a *partial gold standard*, i.e. some "ground truth" to train automatic models. Gold standards are manually constructed, either by involving domain experts or by adopting crowdsourcing and human computation solutions.

In this paper, we present an open source software framework to build *Games with a Purpose for linked data refinement*, i.e. web applications to crowdsource partial ground truth, by motivating user participation through fun incentive. We detail the impact of this new resource by explaining the specific *data linking* "purposes" supported by the framework (creation, ranking and validation of links) and by defining the respective *crowdsourcing tasks* to achieve those goals.

To show this resource's versatility, we describe a set of *diverse applications* that we built on top of it; to demonstrate its reusability and extensibility potential, we provide references to detailed documentation, including an entire *tutorial* which in a few hours guides new adopters to customize and adapt the framework to a new use case.

1 Introduction

In the era of data-driven technologies, evaluating and increasing the quality of data is an important topic. In the Semantic Web community, the emergence of the so-called knowledge graphs has been welcomed as a success of the linked data movement but, in the meantime, has raised a number of research challenges about their management, from creation to verification and correction. The term knowledge graph refinement [16] has been used to indicate the process to increase the quality of knowledge graphs in terms of finding and removing errors or adding missing knowledge. The same refinement operation is a challenge also for any linked dataset of considerable size.

Addressing at scale the linked data refinement problem requires a trade-off between purely-manual and purely-automatic methods, with approaches that, on

D. Vrandečić et al. (Eds.): ISWC 2018, LNCS 11137, pp. 154–169, 2018.
https://doi.org/10.1007/978-3-030-00668-6_10

the one hand, adopt human computation [18] and crowdsourcing [14] to collect manually labeled training data and, on the other hand, employ statistical or machine learning to build models and apply the refinement operation on a larger scale. Indeed, active learning [20] and other recent research approaches in the artificial intelligence area have put back the "human-in-the-loop" by creating mixed human-machine approaches.

In this paper, we present an open source and reusable resource to build human computing applications in the form of games with a purpose [27] aimed to solve linked data refinement tasks. The presented resource consists of both a software framework and a crowdsourcing approach, that can be customized and extended to address different data linking issues.

The paper is organized as follows: Sect. 2 presents related work; Sect. 3 defines the refinement purpose addressed by our framework and Sect. 4 explains the crowdsourcing task; the software resource is presented in Sect. 5 and some applications built on it are illustrated in Sect. 6; since it is a recently released resource, to explain its potential customization and to simplify its adoption, we set up an entire tutorial, briefly introduced in Sect. 7; the code of the framework and the tutorial are available on GitHub and documented on Apiary (links throughout the paper); Sect. 8 concludes the paper.

2 Related Work

Data linking is rooted in the record linkage problem studied in the databases community since the 1960s [9]; for this reason, in the Semantic Web community, the term is often used to name the problem of finding equivalent resources on the Web of linked data [10]; in this meaning, data linking is the process to create links that connect subject- and object-resources from two different datasets through a property that indicate a correspondence or an equivalence (e.g. `owl:sameAs`).

We prefer to generalize the concept of *data linking* extending it to the task of creating links in the form of triples, without limitation to specific types of resources or predicates, nor necessarily referring to linking across two different datasets or knowledge graphs (data linking can happen also within a single dataset or knowledge graphs). In this sense, data linking can be interpreted as a solution to a *linked data refinement* issue, i.e. the process to create, update or correct links in a dataset. As defined in [16] with respect to knowledge graphs, with data linking we do not consider the case of constructing a dataset or graph from scratch, but rather we assume an existing input dataset which needs to be improved by adding missing knowledge or identifying and correcting mistakes.

The Semantic Web community has long investigated the methods to address the data linking problem, by identifying linked dataset quality assessment methodologies [30] and by proposing manual, semi-automatic or automatic tools to implement refinement operations [11,12]. The large majority of refinement approaches, especially on knowledge graphs in which scalable solutions are needed, are based on different statistical and machine learning techniques [8,17,23,24].

Machine learning methods, however, need a partial *gold standard* to train automated models; those training sets are usually created manually by experts: while this usually leads to higher quality trained models, it is also an expensive process, so those "ground truth" datasets are usually small. Involving humans at scale in an effective way is, on the other hand, the goal of crowdsourcing [14] and human computation [18]. Indeed, microtask workers have been employed as a mean to perform manual quality assessment of linked data [1,21].

Among the different human computation approaches, games with a purpose [27] have experienced a wide success, because of their ability to engage users through the incentive of fun. A GWAP is a gaming application that exploits players' actions in the game to solve some (hidden) tasks; users play the game for fun, but the "collateral effect" of their playing is that the comparison and aggregation of players' contributions are used to solve some problems, usually labelling, classification, ranking, clustering, etc. Also in the Semantic Web community, GWAPs have been adopted to solve a number of linked data management issues [3–5,7,13,22,25,28,29], from multimedia labelling to ontology alignment, from error detection to link ranking.

While the general guidelines and rules to build a GWAP have been described and formalized [15] (game mechanics like double player, answer aggregation like output agreement, task design, etc.), building a GWAP for linked data management still requires time and effort. To the best of our knowledge, source code was made available only for the labelling game Artigo [29].

3 Data Linking Purpose

As explained in Sect. 2, with *data linking* we refer to the general problem of creating links in the form of triples. In this section, we provide the basic definitions and we illustrate the cases that our framework supports as purpose of the games that can be built on it.

3.1 Basic Definitions

The following formal definitions will be used throughout the paper.

Resources. *$\mathcal{R}$ is the set of all resources (and literals), whenever possible also described by the respective types. More specifically: $\mathcal{R} = \mathcal{R}_s \cup \mathcal{R}_o$, where $\mathcal{R}_s$ is the set of resources that can take the role of subject in a triple and $\mathcal{R}_o$ is the set of resources that can take the role of object in a triple; as said above the two sets are not necessarily disjoint, i.e. it can happen that $\mathcal{R}_s \cap \mathcal{R}_o \neq \emptyset$.*

Predicates. *$\mathcal{P}$ is the set of all predicates, whenever possible also described by the respective domain and range.*

Links. *$\mathcal{L}$ is the set of all links; since links are triples created between resources and predicates it is: $\mathcal{L} \subset \mathcal{R}_s \times \mathcal{P} \times \mathcal{R}_o$; each link is defined as $l = (r_s, p, r_o) \in \mathcal{L}$ with $r_s \in \mathcal{R}_s, p \in \mathcal{P}, r_o \in \mathcal{R}_o$. $\mathcal{L}$ is usually smaller than the full Cartesian product* *of* $\mathcal{R}_s, \mathcal{P}, \mathcal{R}_o$*, because in each link* (r_s, p, r_o) *it must be true that* $r_s \in domain(p)$ *and* $r_o \in range(p)$.

Link Scores. σ *is the score of a link, i.e. a value indicating the confidence on the truth value of the link; usually* $\sigma \in [0, 1]$*; each link* $l \in \mathcal{L}$ *can have an associated score.*

One final note on subject resources: in the following sections, as well as in the framework implementation, we always assume that any subject entity can be shown to players in the game through some *visual representation*; if the entity is a multimedia element (image, video, audio resource) this requirement is automatically satisfied; in other cases, some additional information about the subject may be required: e.g., a place could be represented on a map, a person through his/her photo, a document with its textual content, etc.

3.2 Data Linking Cases

Given the previous definitions, we can split the general data linking problem in a set of more specific cases as follows.

Link Creation. *A link* l *is created: given* $\mathcal{R} = \mathcal{R}_s \cup \mathcal{R}_o$ *and* $\mathcal{P}$*, the link* $l = (r_s, p, r_o), r_s \in \mathcal{R}_s, p \in \mathcal{P}, r_o \in \mathcal{R}_o$ *is created and added to* $\mathcal{L}$.

All three components of the link to be created exist, i.e. they are already included in the sets $\mathcal{R}$ and $\mathcal{P}$. It is important to note that ***classification*** can be seen as a special case of link creation in which, given a resource $r_s \in \mathcal{R}_s$ to be classified and the predicate $p \in \mathcal{P}$ indicating the relation between the resource and a set of possible categories $\{cat_1, cat_2, \ldots, cat_n\} \subset \mathcal{R}_o$, the resource $r_o \in \{cat_1, cat_2, \ldots, cat_n\}$ is selected to create the link $l = (r_s, p, r_o)$.

For example, this is the case of music classification: given a list of resources of type "music tracks" in $\mathcal{R}_s$, the predicate `mo:genre` $\in \mathcal{P}$ and a set of musical styles in $\mathcal{R}_o$, the task is to assign the music style to each track by creating the link $(track, genre, style)$.

The case of link creation in which new resources and/or predicates are added to $\mathcal{R}$ and/or $\mathcal{P}$ (e.g. free-text labelling of images) is currently not supported by our framework, but it could be one of its possible extensions.

Link Ranking. *Given the set of links* $\mathcal{L}$*, a score* $\sigma \in [0, 1]$ *is assigned to each link* l*. The score represents the probability of the link to be recognized as true. Links can be ordered on the basis of their score* σ*, thus obtaining a ranking.*

In other words, we consider a Bernoulli trial in which the experiment consists in evaluating the "recognizability" of a link and the outcome of the experiment is "success" when the link is recognized and "failure" when the link is not recognized. Under the hypothesis that the probability of success is the same every time the experiment is conducted, the score σ of a link l is the estimation for the binomial proportion in the Bernoulli trial.

In the case of human computation, crowdsourcing or citizen science, each trial consists of a human user that evaluates the link and states that, in his/her opinion, the link is true (success) or false (failure); the human evaluators, if suitably selected, can be considered a random sample of the population of all humans; therefore, aggregating the results of the evaluations in the sample, we can estimate the truth value of a link for the entire population, by computing the probability of each link to be recognized as true. Then, ordering links on the basis of their score means having a metrics to compare different links on their respective "recognizability".

For example, this could be the case of ranking photos depicting a specific person (e.g. an actor, a singer, a politician): given a set of images of the person, human-based evaluation could be employed to identify the pictures in which the person is more recognizable or more clearly depicted.

Link Validation. *Given the set of links* $\mathcal{L}$*, a score* $\sigma \in [0, 1]$ *is assigned to each link* l*. The score represents the actual truth value of the link. A threshold* $t \in [0, 1]$ *is set so that all links with score* $\sigma \geq t$ *are considered true.*

The difference between link validation and the previous case of link ranking is twofold: first, in link validation each link is considered separately, while in link ranking the objective is to compare links; secondly, while in link ranking human judgment is used to estimate the subjective opinion of the human population, in the case of link validation the hypothesis is that, if a link is recognized as true by the population of humans (or by a sample of that population), this is a good estimation of the actual truth value of the link. The latter is also the reason for the introduction of the threshold t: while the truth value is binary (0 = false, 1 = true), human validation is more fuzzy, with "blurry" boundaries; the optimal value for the threshold is very domain- and application-dependent and it is usually empirically estimated.

An example of link validation would be assessing the correct music style in audio tracks: it is well-known that sometimes music genres overlap and identifying a music style could also be subjective (e.g. there is no strict definition of what "rock" is); employing humans in this validation would mean attributing the most shared evaluation of a music track's genre.

As mentioned before, in the last two cases, the human evaluation of a link can be considered a Bernoulli trial: each link l is assessed n times by n different users u; the link is recognized as true X times (with of course $X \leq n$); each user u_i can be more or less reliable and, in some cases, it is possible to estimate his/her reliability ρ_i. Therefore, the score of a link is $\sigma = f(n, X, \rho)$, i.e. it is a function of the number of trials n, the number of successes X and the reliability values of the involved users $\rho = \{\rho_1, \rho_2, \ldots, \rho_n\}$.

4 Crowdsourcing Tasks for Data Linking

Our framework allows to design and develop GWAP applications to solve data linking issue. In other words, the games built on top of our framework ask players to solve atomic data linking issues as basic tasks within the gameplay.

It is worth noting that building a GWAP does not automatically guarantee to collect enough players/played games to solve the data linking problem at hand; however, in our experience, if the task to be solved is properly embedded in a simple game mechanics and if the game is targeted to a specific community of interest, a GWAP is a valuable means to collect a "ground truth" dataset to train machine learning algorithms [19].

4.1 Game Basics

Each GWAP built with our framework is a simple casual game organized in rounds; each round is formed by several levels and each level requires the player to perform a single action, which corresponds to the creation, ranking or validation of a link. According to the definition of von Ahn [27], each GWAP is an *output-agreement double-player game*: users play in random pairs and the game score is based on the agreement between the answers provided by the players (i.e. if they agree, they get points). Our framework does not require users to play simultaneously, because it implements a common strategy in this kind of games, in which a user plays with a "recorded player", so the game scoring is obtained by matching answers provided at different times.

Our framework allows for both time-limited game rounds, in which players can answer to a variable number of tasks per round depending on their speed, and for level-limited rounds, in which players have a maximum number of tasks to address in each round; the choice of either option depends on considerations related to the specific task difficulty and to the game incentive mechanism.

The game adopts a *repeated linking* approach by asking different players to address the same data linking task; conversely, the same task is never given twice to the same player. The "true" solution of a data linking task, therefore, comes from the aggregation of the answers provided by all the users who "played" the task in any game round. The number of players required to solve a data linking task depends on the aggregation algorithm as explained in the following.

It is worth noting that the game scoring (i.e. points gained by players) is not directly related to the data linking scoring (i.e. the attribution of a score σ to a link l): the former is an engagement mechanism to retain players, the latter is the very purpose of the game.

4.2 Atomic Tasks and Truth Inference

As mentioned above, the atomic task in any GWAP built with our framework is an individual data linking task. For example, in the case of music classification, a player could be given an audio track (the resource r_s), the relation `mo:genre` (the predicate p) and some options for music genres (e.g., classical, pop, rock, electronic, representing the potential objects of the link l); the action for the player would be to choose the genre (the resource r_o, say 'rock') that better describes the audio track. By performing this action (the atomic task), the player is saying that he believes that the link $l = (r_s, p, r_o)$ is "true"; the game can therefore alter the truth score σ of the link l by incrementing it.

Inspired by record linkage literature [9], we assume each link score to start from 0 and to be incrementally increased at each user contribution.

The aggregation algorithm (also known in literature as *truth inference* algorithm) implemented in our framework computes the score σ as follows:

$$\sigma = \delta_+ \cdot \sum_i \rho_i$$

where δ_+ is the increment and ρ_i is the reliability of the ith player which provided a solution to the task.

In record linkage, where scores are assigned at each possible couple of records, the "matching" score is increased while the "non-matching" scores are decreased respectively. This is also supported by our framework, which allows to decrease the score of the links evaluated as "false" by the players. In the example above, if the music genres were mutually exclusive, the game could also alter the scores $\bar{\sigma}$ of the links $\bar{l} = (r_s, p, r_{\bar{o}})$ where $r_{\bar{o}} \neq r_o$ by decrementing them. The more general aggregation formula then becomes:

$$\sigma = \delta_+ \cdot \sum_i \rho_i - \delta_- \cdot \sum_j \rho_j$$

where δ_+ and δ_- are the increment and decrement quantities, and ρ_i and ρ_j the reliability of the players judging the link as true or false, respectively.

The reliability of the player introduced in the equations above represents the level of trustworthiness in the collected answers and is used to weight players' contributions in the truth inference process. Intuitively, the answers provided by "reliable" players should count more than those provided by users who simply give random answers.

As common practice [26], we adopt assessment tasks to quantify reliability: in each game, we compute player reliability as a function of the number of errors the player makes on those assessment tasks. The reliability value is 1 if the player makes no mistakes and decreases toward zero with increasing errors. As proposed in [4], by default our framework computes the player local reliability ρ as follows:

$$\rho = e^{-k \cdot m}$$

where m is the number of mistakes per game round and k a suitable constant.

When the score σ of a link overcomes a defined threshold t, the respective data linking task is considered solved and it is removed from the game, i.e. no other player will be asked to solve the same task. A data linking solved task is then used again in the game as assessment task to measure player reliability.

Interested readers can find examples of the input tasks, player contributions and aggregated output results of the Night Knights game (cf. Sect. 6.3) at https://github.com/STARS4ALL/Night-Knights-dataset.

4.3 Tuning Parameters

Clearly, the truth inference mechanism explained before requires setting a number of parameters. The most suitable values depend on a mix of domain-specific considerations and empirical evaluation. Hereafter, we give some basic guidance.

- **Reliability parameter** k: how much to penalize a player on the basis of his mistakes on the assessment tasks; this parameter should be tuned based on (1) how many assessment tasks are given in each round and (2) how much the reliability should be decreased on user's mistakes. For example, if each round contains q assessment tasks, k could be empirically set to $3/q$, so that, with q errors, reliability r decreases to $e^{-3} \approx 0.05$ (i.e. player's answers are weighted very low because the user mistook all assessment tasks).
- **Score increment** δ_+ **(and decrement** δ_-**)**: how much the score should be incremented/decremented at each player answer; this usually depends on the minimum number of distinct users desired for truth inference (as well as on the threshold, cf. below). For example, if we think that n players are needed, δ_+ could be set to $1/n$, so that exactly n players with full reliability are sufficient to reach score $\sigma = 1$.
- **Threshold** t: the value of the score σ that must be overcome to consider a link solved; this parameter is also very context-specific and could be set to 1 or to a value close to it. Of course, this threshold depends also on the setting of the above parameters.

5 The GWAP Enabler Framework

To give a better idea of our software framework, we give some details on its technical internals. The GWAP Enabler is released as open source with an Apache 2.0 license and is made available at https://github.com/STARS4ALL/gwap-enabler.

5.1 Structure of the Framework

Our GWAP Enabler is a template Web application formed by some basic software building blocks, which provide the basic functionalities to build a linked data refinement game; by customizing the "template", any specific GWAP application can be built. The three main components are the User Interface (UI), the Application Programming Interface (API) and the Data Base (DB).

The main table of the database is named `resource_has_topic` and contains all the links $l = (r_s, p, r_o) \in \mathcal{L}$. Each link has a score $\sigma \in [0, 1]$ which is updated every time it is played by a user. The subject r_s and object r_o resources of links are stored respectively in the tables named `resource` and `topic`. For example, referring to the music classification case illustrated in the previous section, the `resources` table should contain the audio tracks and the `topics` table should list all the possible music styles. These three tables together contain all the data linking problem information and they are initially filled accordingly to the purpose of the specific GWAP.

In addition to those tables, the database contains further information to customize the game according to the desired behaviour: the `configuration` table to change the truth inference parameters (cf. Sect. 4.3) and the `badge` table to change the badges given as reward to players during the gameplay. The remaining tables manage internal game data such as users, rounds, leaderboard

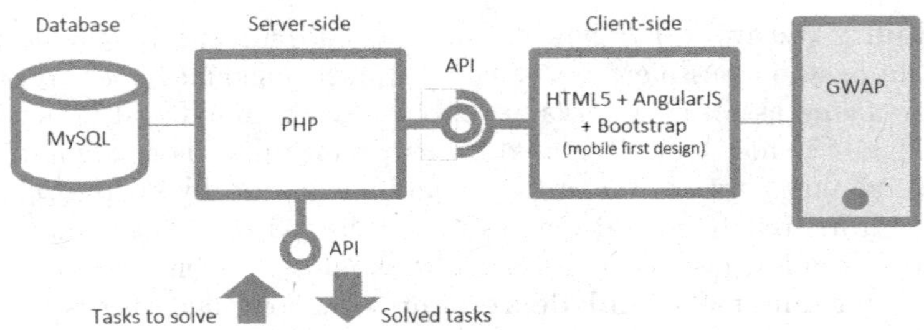

Fig. 1. Architecture of the GWAP enabler

or logs and are automatically filled during the gameplay; as such, they do not need to be modified or filled: they can of course be freely adapted at developers' will, being aware that altering those tables will require also modifying the code.

From a technical point of view, the GWAP Enabler is made up of an HTML5, AngularJS and Bootstrap frontend, a PHP backend to manage the logic of the application and a MySQL database, as shown in Fig. 1. The communication between frontend and backend happens through an internal Web-based API, whose main operations consist in retrieving the set of links to be shown to players in each game round, updating the link score according to the agreement and disagreement of players and updating the points of the players to build leaderboards and assign badges.

Another external Web-based API is also set up to feed the game with new data and to access the GWAP results aggregated by the truth inference algorithm; in particular the API methods allow submitting new tasks to be solved, retrieving solved tasks (= refined links) in both JSON and JSON-LD formats, getting the number of tasks to be completed and listing the main KPIs of the GWAP (e.g., throughput, ALP and expected contribution [27]) for the game evaluation. This API is exhaustively documented at https://gwapenablerapi.docs.apiary.io/.

Further details about the framework architecture can be found in the GitHub repository; the code of the enabler is in the "App" folder whereas the scripts to create the database are in the "Db" folder. Further details about the installation steps, the technical requirements, the database structure and some customization option are given in the wiki pages of the repository at https://github.com/STARS4ALL/gwap-enabler/wiki.

5.2 How to Build a GWAP by Using the Framework

To create a game instance out of the provided template, a developer should perform a series of operations and changes that affects the three building blocks constituting the framework.

First of all, the basic data linking case (cf. Sect. 3) and atomic crowdsourcing task (cf. Section 4) should be designed to address the specific use case. Then, the database has to be filled up with data by adding the core resources and links (`resource`, `topic` and `resource_has_topic` tables), the GWAP aggregation parameters (`configuration` table) and badges information (`badge` table). Please, note that a pre-processing phase is required to prepare the data and a careful analysis of the specific refinement purpose is an essential step to find and tune the proper parameters, thus this initial step could be long and complex depending on the context.

Once data are in the DB, the code can be run as is or it can be tailored to address specific requirements; for example, game mechanics could be altered, further data to describe resources or links can be added (e.g., maps/videos/sounds) or a different badge/point assignment logic can be defined. Finally, the UI should be customized with the desired look and feel and the specific game elements (points, badges, leaderboards, etc.) could be modified to give the game a specific flavour or mood.

6 Existing Applications Built on Top of the Framework

We used the enabler to build three GWAPs that address the three different classes of data linking. Indomilando game aims to rank a set of cultural heritage assets of Milano, based on their recognizability; Land Cover Validation is an example of link validation game in which users are involved in checking land use information produced by two different (and disagreeing) sources; Night Knights is a game for both link creation and validation in which a set of images has to be classified into a predefined set of categories.

6.1 GWAP for Link Ranking: Indomilando

Indomilando[1] [6] is a web-based Game With a Purpose aimed to rank a quite heterogeneous set of images, depicting the cultural heritage assets of Milan. In each round the game shows the name of an asset and four photos, in which one represents the asset and the other three are put as distractors. The user has to choose the right picture related to the asset and, as an incentive, he gains points for each correct answer. A photo is removed from the game when it is correctly chosen three times. All the given answers on the photo (selection or non-selection of the picture) are recorded and analysed ex-post to measure "how much" the picture represents the asset: the intuition is that, the more a photo is correctly identified by players, the more recognizable it is.

In Indomilando, we have a set of links l that connects each photo with the asset it refers to; the assets and the photos are the subjects r_s and objects r_o of the links to be ranked. By counting and suitably weighting the number of times the pictures has been recognized (or not), we calculate the scores σ of these links. Since they represent the probability of the links to be recognized as true, by ordering them we can rank the links, and thus the pictures of the cultural heritage assets of Milan, on the basis of their recognisability.

[1] Cf. http://smartculture-games.innovationengineering.eu/indomilando/.

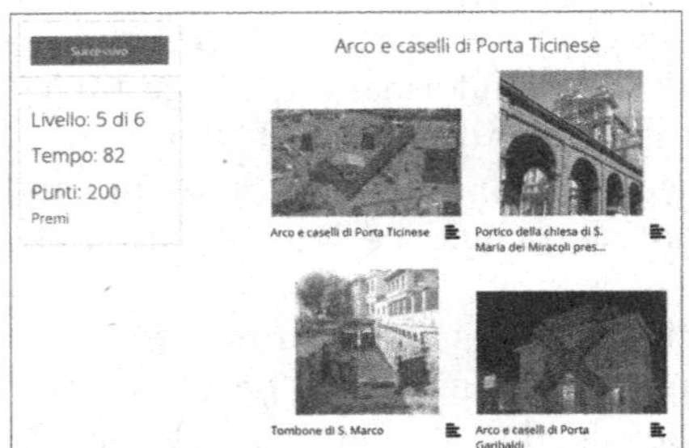

Fig. 2. Indomilando: gameplay (left) and asset visualisation on a map (right).

The output of this game can be employed for various goals: selecting the best pictures representing an asset, understanding if an asset would benefit from further photo collection or evaluating if an asset may require additional promotional campaign because it is less recognized.

From a gamification point of view, users gain points for each correct answer and can challenge other players in a global leaderboard. Another incentive we give to players is the possibility to view on a map the assets they played with and to display their historic and cultural description, as shown in Fig. 2: this is an additional learning reward that Indomilando players showed to appreciate. These incentives were very effective and the game had a great success: all the 2100 pictures we put in the game were played and ranked by 151 users, with a throughput of 125 photo ranked/hour.

6.2 GWAP for Link Validation: Land Cover Validation Game

The Land Cover Validation game[2] [2] is designed to engage Citizen Scientists in the validation of land cover data in the Como municipality area (in Lombardy, Italy). The player is requested to classify the areas in which two different land cover maps disagree: the DUSAF classification[3] made by Lombardy Region and GlobeLand30[4] provided by a Chinese agency. The validation is presented to the user as a classification task: given an aerial photo of a 30×30 square area (pixel), the player has to select the right category from a predefined list of land use types (i.e. residential, agricultural areas, etc.), as shown in Fig. 3. As regards the incentives and the entertaining environment, players gains points and badges if they agree with one of the existing classifications and they can challenge other players in the leaderboard.

From the data linking perspective, each pixel is the subject r_s of two links, one connecting the pixel with its land cover defined by DUSAF ($r_{o_{DUSAF}}$) and the other with the GlobeLand30 classification ($r_{o_{GL30}}$). Each time one of the two land cover options is selected, the score σ of the corresponding link is increased.

[2] Cf. http://landcover.como.polimi.it/landcover/.

[3] Cf. http://www.geoportale.regione.lombardia.it/en/home.

[4] Cf. http://www.globallandcover.com/GLC30Download/index.aspx.

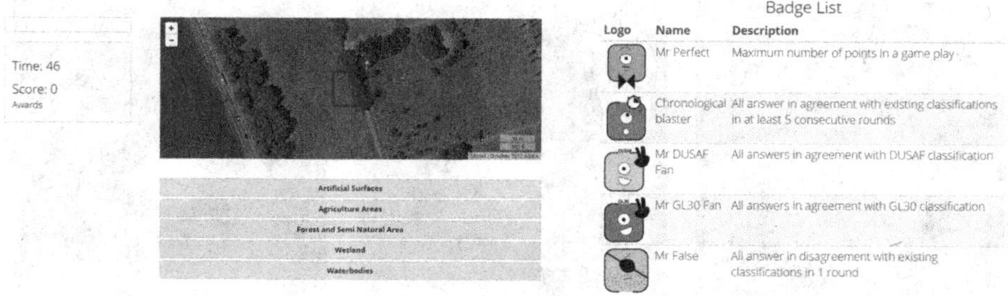

Fig. 3. Land cover validation game: pixel classification (left) and badges (right).

This score represents the link truth value and a threshold is set so that all links with a score higher than this value are considered true. When a link score exceeds the threshold, the corresponding pixel is removed from the game.

The game completely fulfilled its goal, since all the target 1600 aerial pixels were validated, thanks to 68 gamers that played more than 20 hours during the FOSS4G Europe Conference in 2015.

6.3 GWAP for Link Creation and Validation: Night Knights

Night Knights[5] [19] is a GWAP designed to classify images taken from the International Space Station (ISS) into a fixed set of six categories, developed within a project that aims to increase the awareness about the light pollution problem.

Each human participant plays the role of an astronaut that, coupled with another random player, has the mission of classifying as many images as possible playing one-minute rounds. As Fig. 4 shows, if players agree on the same classification they get points, which are collected to challenge other users in a global leaderboard.

The hidden goal of the game is to create new links between each image r_s and its correct category r_o, by cross-validating them using the contributions of multiple users, suitably aggregated. The link creation algorithm works as follow: starting from a set of links connecting each image with all the available categories, the score σ of a link is increased if a player chooses the corresponding category. Each image is offered to multiple players, whose contributions are weighted according to their reliability (measured with assessment tasks) and aggregated in the link score; once the score overcomes a specific threshold, the image is classified and removed from the game. By design, a minimum of four agreeing users are required to reach the classification threshold. More than 35,000 images have been classified since the launch of the game in February 2017 by 1,000+ players, with a throughput of 203 images classified/hour.

[5] Cf. http://www.nightknights.eu/.

(a) Classify an image (b) Agreement (c) Disagreement

Fig. 4. Night Knights: the game play

7 A Step-by-Step Tutorial to Reuse the Framework

In the previous section, we showed how the GWAP Enabler was successfully employed to implement games to create and validate links through an image classification process and to rank links based on their recognizability. Since it is a new resource, in this section, we introduce a tutorial that guides new adopters to customize and adapt the framework to a new use case. All the required changes to the GWAP Enabler sources are explained step-by-step in the GitHub repository at https://github.com/STARS4ALL/gwap-enabler-tutorial.

The goal of this tutorial is twofold; on the one hand, we provide developers with a guided example of an "instantiation" of our framework; on the other hand, we demonstrate how this GWAP Enabler could be used and adapted to build a crowdsourcing web application to enrich and refine OpenStreetMap data (and, consequently, its linked data version LinkedGeoData), in which users are motivated to participate through fun incentives.

More specifically, the GWAP application built in the tutorial is about classifying OpenStreetMap restaurants on the basis of their cuisine type. Data about restaurants are selected in a specific area (we give the example of the city of Milano, but developers can easily change it to their preferred location); those restaurants with an already specified type of cuisine are taken as "ground truth" (for the assessment tasks to compute player reliability), whereas all the remaining one are the subject resources, target of the classification process.

Players are randomly paired and are shown the restaurant name and position on a dynamic map; the game consists in finding an agreement on the cuisine type, selected in a set of predefined categories (the most widely used in

OpenStreetMap). As a result, players contributions are aggregated in the truth inference process that implements the link collection and validation.

To create such an application, some changes to the framework's core functionalities are required; while in the GWAP Enabler by default each resource is displayed to players in the form of an image, in this tutorial scenario we want to show developers how to display both a textual information (the restaurant's name) and an interactive map (the restaurant's position). Therefore, this requires (1) to correctly store the relevant information in the database, (2) to modify the API code to retrieve the additional data and (3) to modify the UI code to display name and map.

In the tutorial, we provide some basic instruction and we explain how to embed the map by using Leaflet[6], an open-source JavaScript library for mobile-friendly interactive maps. We do not detail the graphical aspect, letting developers define their desired look-and-feel to give the game a more personal flavour or mood. By going through the wiki instructions, developers can get their up and running GWAP in about half a day and they can gain enough knowledge about the framework to be able to reuse it for their own purpose, since the tutorial touches upon all the relevant modifications.

8 Discussion and Conclusion

In this paper, we presented an open source software framework to build Games With a Purpose embedding a crowdsourcing task for linked data refinement. The framework is aimed to help in the process of collecting manually-labelled gold standard data, which are needed as training set for automatic learning algorithms to implement refinement operations (link collection, ranking and validation) on large scale linked datasets and knowledge graphs. In other words, the presented framework helps to simplify the tedious and expensive human process of data collection, letting researchers focus on the subsequent steps of their scientific study and experimentation.

We introduced the data linking cases implemented by the framework to explain its level of generality and potential for reuse; we illustrated the crowdsourcing task and truth inference process to clarify its design and possible customizations. Then, we gave some technical details about the internals of the GWAP Enabler, designed and developed accordingly to the most common Web development best practices, and we demonstrated the framework versatility by describing the diverse applications we built on top of it. Finally, we presented a step-by-step tutorial as a more detailed documentation and as a means to ease the reuse of this new resource; following the tutorial, a developer is guided to build an entirely new GWAP in a few hours, saving significant coding effort.

We provided all the references to get access to the framework code (released under an Apache 2.0 license) and its online documentation which consists of data

[6] Cf. http://leafletjs.com/.

schemas, API specification and sample input/output data, technical requirements and installation instructions, guided instruction to customize the GWAP Enabler.

The framework could be further extended to cover other refinement cases like free text labelling (i.e., insertion of new literal objects) or data linking issues related to the choice of different predicates.

Acknowledgments. This work was partially supported by the STARS4ALL project (H2020-688135) co-funded by the European Commission.

References

1. Acosta, M., Zaveri, A., Simperl, E., Kontokostas, D., Auer, S., Lehmann, J.: Crowdsourcing linked data quality assessment. In: Alani, H., Kagal, L., Fokoue, A., Groth, P., Biemann, C., Parreira, J.X., Aroyo, L., Noy, N., Welty, C., Janowicz, K. (eds.) ISWC 2013. LNCS, vol. 8219, pp. 260–276. Springer, Heidelberg (2013). https://doi.org/10.1007/978-3-642-41338-4_17
2. Brovelli, M.A., Celino, I., Fiano, A., Molinari, M.E., Venkatachalam, V.: A crowdsourcing-based game for land cover validation. Appl. Geomat. **10**(1), 1–11 (2018)
3. Celino, I., et al.: Urbanopoly - A social and location-based game with a purpose to crowdsource your urban data. In: International Conference on Privacy, Security, Risk and Trust (PASSAT), and International Confernece on Social Computing (SocialCom), pp. 910–913. IEEE (2012)
4. Celino, I., et al.: Linking smart cities datasets with human computation – the Case of UrbanMatch. In: Cudré-Mauroux, P. (ed.) ISWC 2012. LNCS, vol. 7650, pp. 34–49. Springer, Heidelberg (2012). https://doi.org/10.1007/978-3-642-35173-0_3
5. Celino, I., Della Valle, E., Gualandris, R.: On the effectiveness of a mobile puzzle game UI to crowdsource linked data management tasks. In: 1st International Workshop on User Interfaces for Crowdsourcing and Human Computation (2014)
6. Celino, I., Fiano, A., Fino, R.: Analysis of a cultural heritage game with a purpose with an educational incentive. In: Bozzon, A., Cudre-Maroux, P., Pautasso, C. (eds.) ICWE 2016. LNCS, vol. 9671, pp. 422–430. Springer, Cham (2016). https://doi.org/10.1007/978-3-319-38791-8_28
7. Chamberlain, J., Poesio, M., Kruschwitz, U.: Phrase detectives: a web-based collaborative annotation game. In: Proceedings of the International Conference on Semantic Systems (I-Semantics 2008), pp. 42–49 (2008)
8. Dong, X., et al.: Knowledge vault: a web-scale approach to probabilistic knowledge fusion. In: Proceedings of the 20th ACM SIGKDD international conference on Knowledge discovery and data mining, pp. 601–610. ACM (2014)
9. Fellegi, I.P., Sunter, A.B.: A theory for record linkage. J. Am. Stat. Assoc. **64**(328), 1183–1210 (1969)
10. Ferrara, A., Nikolov, A., Scharffe, F.: Data linking for the semantic web. In: Semantic Web: Ontology and Knowledge Base Enabled Tools, Services, and Applications, 169 (2013)
11. Fürber, C., Hepp, M.: Using SPARQL and SPIN for data quality management on the semantic web. In: Abramowicz, W., Tolksdorf, R. (eds.) BIS 2010. LNBIP, vol. 47, pp. 35–46. Springer, Heidelberg (2010). https://doi.org/10.1007/978-3-642-12814-1_4

12. Guéret, C., Groth, P., Stadler, C., Lehmann, J.: Assessing linked data mappings using network measures. In: Simperl, E., Cimiano, P., Polleres, A., Corcho, O., Presutti, V. (eds.) ESWC 2012. LNCS, vol. 7295, pp. 87–102. Springer, Heidelberg (2012). https://doi.org/10.1007/978-3-642-30284-8_13
13. Hees, J., Roth-Berghofer, T., Biedert, R., Adrian, B., Dengel, A.: BetterRelations: using a game to rate linked data triples. In: Bach, J., Edelkamp, S. (eds.) KI 2011. LNCS (LNAI), vol. 7006, pp. 134–138. Springer, Heidelberg (2011). https://doi.org/10.1007/978-3-642-24455-1_12
14. Howe, J.: The rise of crowdsourcing. Wired Mag. **14**(6), 1–4 (2006)
15. Law, E., Ahn, L.V.: Human computation. Synth. Lect. Artif. Intell. Mach. Learn. **5**(3), 1–121 (2011)
16. Paulheim, H.: Knowledge graph refinement: a survey of approaches and evaluation methods. Semant. Web **8**(3), 489–508 (2017)
17. Paulheim, H., Bizer, C.: Improving the quality of linked data using statistical distributions. Int. J. Semant. Web Inf. Syst. (IJSWIS) **10**(2), 63–86 (2014)
18. Quinn, A.J., Bederson, B.B.: Human computation: a survey and taxonomy of a growing field. In: Proceedings of the SIGCHI conference on human factors in computing systems, pp. 1403–1412. ACM (2011)
19. Re Calegari, G., Nasi, G., Celino, I.: Human Computation vs. Machine Learning: an Experimental Comparison for Image Classification. Hum. Comput. J. **5**(1), 13–30 (2018). https://doi.org/10.15346/hc.v5i1.2
20. Settles, B.: Active learning. Synth. Lect. Artif. Intell. Mach. Learn. **6**(1), 1–114 (2012)
21. Simperl, E., Norton, B., Vrandečić, D.: Crowdsourcing tasks in linked data management. In: Proceedings of the Second International Conference on Consuming Linked Data, Vol. 782, pp. 61–72. CEUR-WS. org (2011)
22. Siorpaes, K., Hepp, M.: Games with a purpose for the semantic web. IEEE Intell. Syst., 23(3) (2008)
23. Sleeman, J., Finin, T.: Type prediction for efficient coreference resolution in heterogeneous semantic graphs. In: IEEE Seventh International Conference on Semantic Computing (ICSC), pp. 78–85. IEEE (2013)
24. Sleeman, J., Finin, T., Joshi, A.: Topic Modeling for RDF Graphs. In: LD4IE at ISWC, pp. 48–62 (2015)
25. Thaler, S., Simperl, E.P.B., Siorpaes, K.: SpotTheLink: a game for ontology alignment. Wissensmanagement **182**, 246–253 (2011)
26. Ul Hassan, U., O'Riain, S., Curry, E.: Effects of expertise assessment on the quality of task routing in human computation. In: Proceedings of the 2nd International Workshop on Social Media for Crowdsourcing and Human Computation, Paris, France (2013)
27. Von Ahn, L., Dabbish, L.: Designing games with a purpose. Commun. ACM **51**(8), 58–67 (2008)
28. Waitelonis, J., Ludwig, N., Knuth, M., Sack, H.: WhoKnows? evaluating linked data heuristics with a quiz that cleans up DBpedia. Interact. Technol. Smart Educ. **8**(4), 236–248 (2011)
29. Wieser, C., Bry, F., Bérard, A., Lagrange, R.: ARTigo: building an artwork search engine with games and higher-order latent semantic analysis. In: First AAAI Conference on Human Computation and Crowdsourcing (2013)
30. Zaveri, A., Rula, A., Maurino, A., Pietrobon, R., Lehmann, J., Auer, S.: Quality assessment for linked data: a survey. Semant. Web **7**(1), 63–93 (2016)

VoxEL: A Benchmark Dataset for Multilingual Entity Linking

Henry Rosales-Méndez(✉), Aidan Hogan, and Barbara Poblete

IMFD Chile and Department of Computer Science,
University of Chile, Santiago, Chile
{hrosales,ahogan,bpoblete}@dcc.uchile.cl

Abstract. The Entity Linking (EL) task identifies entity mentions in a text corpus and associates them with corresponding entities in a given knowledge base. While traditional EL approaches have largely focused on English texts, current trends are towards language-agnostic or otherwise multilingual approaches that can perform EL over texts in many languages. One of the obstacles to ongoing research on multilingual EL is a scarcity of annotated datasets with the same text in different languages. In this work we thus propose VoxEL: a manually-annotated gold standard for multilingual EL featuring the same text expressed in five European languages. We first motivate and describe the VoxEL dataset, using it to compare the behaviour of state of the art EL (multilingual) systems for five different languages, contrasting these results with those obtained using machine translation to English. Overall, our results identify how five state-of-the-art multilingual EL systems compare for various languages, how the results of different languages compare, and further suggest that machine translation of input text to English is now a competitive alternative to dedicated multilingual EL configurations.

Keywords: Multilingual · Entity linking · Information extraction

1 Introduction

The Entity Linking (EL) task identifies entity mentions in a text corpus and associates them with corresponding entities in a Knowledge Base (KB). In this way, we can leverage the information of publicly available KBs about real-world entities to achieve a better understanding of their semantics and also of natural language. For instance, in the text *"in the world of pop music, there is Michael Jackson and there is everybody else"* quoted from The New York Times, we can link the mention *Michael Jackson* with its corresponding entry in, e.g., the Wikidata KB [35] (`wd:Q2831`), or the DBpedia KB [16] (`dbr:Michael_Jackson`)[1]

Resource type: Dataset.
Permanent URL: https://dx.doi.org/10.6084/m9.figshare.6539675.

[1] Throughout, we use prefixes according to http://prefix.cc.

D. Vrandečić et al. (Eds.): ISWC 2018, LNCS 11137, pp. 170–186, 2018.
https://doi.org/10.1007/978-3-030-00668-6_11

allowing us to leverage, thereafter, the information in the KB about this entity to support semantic search, relationship extraction, text enrichment, entity summarisation, or semantic annotation, amongst other applications.

One of the major driving forces for research on EL has been the development of a variety of ever-expanding KBs that describe a broad selection of notable entities covering various domains (e.g., Wikipedia, DBpedia, Freebase, YAGO, Wikidata). Hence, while traditional Named Entity Recognition (NER) tools focused on identifying mentions of entities of specific types in a text, EL further requires disambiguation of which entity in the KB is being spoken about; this remains a challenging problem. On the one hand, name variations – such as *"Michael Joseph Jackson"*, *"Jackson"*, *"The King of Pop"* – mean that the same KB entity may be referred to in a variety of ways by a given text. On the other hand, ambiguity – where the name *"Michael Jackson"* may refer to various (other) KB entities, such as a journalist (`wd:Q167877`), a football player (`wd:Q6831558`), an actor (`wd:Q6831554`), amongst others – means that an entity mention in a text may have several KB candidates associated with it.

Many research works have addressed these challenges of the EL task down through the years. Most of the early EL systems proposed in the literature were monolingual approaches focusing on texts written in one single language, in most cases English (e.g., [12,18]). These approaches often use resources of a specific language, such as Part-Of-Speech taggers and WordNet[2], which prevent generalisation or adaptation to other languages. Furthermore, most of the labelled datasets available for training and evaluating EL approaches were English only (e.g., AIDA/CoNLL [12], DBpedia Spotlight Corpus [18], KORE 50 [13]).

However, as the EL area has matured, more and more works have begun to focus on languages other than English, including multilingual approaches that are either language agnostic [5,6,8,22] – relying only on the language of labels available in the reference KB – or that can be configured for multiple languages [21,30]. Recognising this trend, a number of multilingual datasets for EL were released, such as for the 2013 TAC KBP challenge[3] and the 2015 SemEval Task 13 challenge[4]. Although such resources are valuable for multilingual EL research – where in previous work [28] we presented an evaluation of EL systems comparing two languages from the SemEval dataset – they have their limitations, key amongst which are their limited availability (participants only[5]), a narrow selection of languages, and differences in text and annotations across languages that makes it difficult to compare the performance in each language. More generally, the EL datasets available in multiple languages – and languages other than English – greatly lags behind what is available for English.

Contributions: In this paper, we propose the VoxEL dataset: a manually-annotated gold standard for EL considering five European languages, namely

[2] https://wordnet.princeton.edu; April 1st, 2018.

[3] https://tac.nist.gov/2013/KBP/; April 1st, 2018.

[4] http://alt.qcri.org/semeval2015/task13/; April 1st, 2018.

[5] We have managed to acquire the SemEval dataset, but unfortunately we were not able to acquire the TAC–KBP dataset: our correspondence was not responded to.

German, English, Spanish, French and Italian. This dataset is based on an online source of multilingual news, where we selected and annotated 15 corresponding news articles for these five languages (75 articles in total). Additionally, we created two versions of VoxEL: a *strict* version where entities correspond to a restricted definition of entity, as a mention of a person, place or organisation (based on traditional MUC/NER definitions), and a *relaxed* version where we considered a broader selection of mentions referring to entities described by Wikipedia. Based on the VoxEL dataset, using the GERBIL evaluation framework [34], we present results for various EL systems, allowing us to compare not only across systems, but also across languages. As an additional contribution, we compare the performance of EL systems configurable for a given language with the analogous results produced by applying state-of-the-art machine translation (Google translate) to English and then applying EL configured for English. Our findings show that most systems perform best for English text. Furthermore, machine translation of input text to English achieves comparable – and often better – performance when compared with dedicated multilingual EL approaches.

2 Preliminaries

We first introduce some preliminaries relating to EL. Let E be a set of entity identifiers in a KB; these are typically IRIs, such as `wd:Q2831`, `dbr:Michael_Jackson`. Given an input text, the EL process can be conceptualised in terms of two main phases. First, Entity Recognition (ER) establishes a set of entity mentions M, where each such mention is a sub-string referring to an entity, annotated with its start position in the input text, e.g., (37, *"Michael Jackon"*). Second, for each mention $m \in M$ recognised by the first phase, Entity Disambiguation (ED) attempts to establish a link between m and the corresponding identifier $e \in E$ for the KB entity to which it refers. The second disambiguation phase can be further broken down into a number of (typical) sub-tasks, described next:

Candidate entity generation: For each mention $m \in M$, this stage selects a subset of the most probable KB entities $E_m \subseteq E$ to which it may refer. There are two high-level approaches by which candidate entities are often generated. The first is a dictionary-based approach, which involves applying keyword or string matching between the mention m and the label of entities from E. The second is an NER-based approach, where traditional NER tools are used to identify entity mentions (potentially) independently of the KB.

Candidate entity ranking: This stage is where the final disambiguation is made: the candidate entities E_m for each mention m are ranked according to some measure indicating their likelihood of being the reference for m. The measures used for ranking each entity $e \in E_m$ may take into account features of the candidate entity e (e.g., centrality), features of the candidate link (m, e) (e.g., string similarity), features involving e and candidates for neighbouring mentions E'_m (e.g., graph distance in the KB), and so forth. Ranking may take the form of an explicit metric that potentially combines several measures, or may be implicit in the use of machine-learning methods that classify candidates, or that compute an optimal assignment of links.

Unlinkable mention prediction: The target KBs considered by EL are often, by their nature, incomplete. In some applications, it may thus be useful to extract entity mentions from the input text that do not (yet) have a corresponding entity in the KB. These are sometimes referred to as *emerging entities*, are typically produced by NER candidate generation (rather than a dictionary approach), and are assigned a label such as *NIL* (Not In Lexicon).

It is important to note that while the above processes provide a functional overview of the operation of most EL systems, not all EL systems follow this linear sequence of steps. Most systems perform recognition first, and once the mentions are identified the disambiguation phase is initiated [18,21]. However, other approaches may instead apply a unified process, building models that create feedback between the recognition and disambiguation steps [7]. In any case, the output of the EL process will be a set of links of the form (m, e), where the mention m in the text is linked to the entity e in the KB, optionally annotated with a confidence score – often called a *support* – for the link.

3 Related Work

We now cover related works in the context of multilingual EL, first discussing approaches and systems, thereafter discussing available datasets.

3.1 Multilingual EL Systems

In theory, any EL system can be applied to any language; as demonstrated in our previous work [28], even a system supporting only English may still be able to correctly recognise and link the name of a person such as *Michael Jackson* in the text of another language, assuming the alphabet remains the same. Hence, the notion of a multilingual EL system can become blurred. For example language-agnostic systems – systems that require no linguistic components or resources specific to a language – can become multilingual simply by virtue of having a reference KB with labels in a different – or multiple different – language(s).

Here we thus focus on EL systems that have published evaluation results over texts from multiple languages[6], thus demonstrating proven multilingual capabilities. We summarise such systems in Table 1, where we provide details on the year of the main publication, the languages evaluated, as well as denoting whether or not entity recognition is supported[7], and whether or not a demo, source code or API is currently available.[8] As expected, a high-level inspection of the table

[6] This excludes systems such as Apache Stanbol, OpenCalais, PoolParty, etc.

[7] Some systems assume that mentions have previously been extracted from the text and are given as input, thereafter focusing only on the disambiguation process.

[8] We presented an earlier version of such a table in previous work [28].

shows that English is the most popularly-evaluated (and thus we surmise supported) language, followed by European languages such as German, Spanish, French, Dutch and Italian. We also highlight that most of the multilingual EL approaches included in the table have emerged since 2010.

Table 1. Overview of multilingual EL approaches; the italicised approaches will be incorporated as part of our experiments.

Name	Year	Evaluated Languages	ER	Demo	Src	API
KIM [25]	2004	EN,FR,ES	✓	✓	✗	✓
TagME [8]	2010	DE,EN,NL	✓	✓	✗	✓
SDA [3]	2011	EN,FR	✓	✗	✗	✗
ualberta [10]	2012	EN,ZH	✓	✗	✗	✗
HITS [7]	2012	EN,ES,ZH	✓	✗	✗	✗
THD [6]	2012	DE,EN,NL	✓	✓	✓	✓
DBpedia Spotlight [5,18]	2013	DA,DE,EN,ES,FR,HU,IT,NL,RU	✓	✓	✓	✓
Wang-Tang [37]	2013	EN,ZH	✓	✗	✗	✗
AGDISTIS [33]	2014	DE,EN	✗	✓	✓	✓
Babelfy [21]	2014	DE,EN,ES,FR,IT	✓	✓	✗	✓
FREME [30]	2016	DE,EN	✓	✗	✓	✓
WikiME [32]	2016	AR,DE,EN,ES,FR,HE,IT,TA,TH, TL,TR,UR,ZH	✓	✓	✗	✗
FEL [23]	2017	EN,ES,ZH	✓	✗	✓	✗
FOX [31]	2017	DE,EN,ES,FR,NL	✓	✓	✓	✓
MAG [22]	2017	DE,EN,ES,FR,IT,JA,NL	✗	✓	✓	✓

We will later conduct experiments using the GERBIL evaluation framework [34], which allows for invoking and integrating the results of a variety of public APIs for EL, generating results according to standard metrics in a consistent manner. Hence, in our later experiments, we shall only consider those systems with a working REST-API made available by the authors of the system. In addition, we will manually label our VoxEL system according to Wikipedia, with which other important KBs such as DBpedia, YAGO, Freebase, Wikidata, etc., can be linked; hence we only include systems that support such a KB linked with Wikipedia. Note that GERBIL automatically takes care of mapping coreferent identifiers across KBs (and even across languages in cases such as DBpedia with different KB identifiers for different languages and cross-language links).

With these criteria in mind, we experiment with the following systems:

TagME (2010) uses analyses of anchor texts in Wikipedia pages to perform EL [8]. The ranking stage is based primarily on two measures: *commonness*, which describes how often an anchor text is associated with a particular Wikipedia entity; and *relatedness*, which is a co-citation measure indicating how frequently candidate entities for different mentions are linked from the same Wikipedia article. TagME is language agnostic: it can take advantage of the Wikipedia Search API to apply the same conceptual process over different language versions of Wikipedia to support multilingual EL.

THD (2012) is based on three measures [6]: *most frequent senses*, which ranks candidates for a mention based on the Wikipedia Search API results for that mention; *co-occurrence*, which is a co-citation measure looking at how often candidate entities for different mentions are linked from the same paragraphs in Wikipedia; and *explicit semantic analysis*, which uses keyword similarity measures to relate mentions with a concept. These methods are language agnostic and applicable to different language versions of Wikipedia.

DBpedia Spotlight (2013) was first proposed to deal with English annotations [18], based on keyword and string matching functions ranked by a probabilistic model based on a variant of a TF–IDF measure. DBpedia Spotlight is largely language-agnostic, where an extended version later proposed by Daiber et al. [5] leverages the multilingual information of the Wikipedia and DBpedia KBs to support multiple languages.

Babelfy (2014) performs EL with respect to a custom multilingual KB BabelNet[9] constructed from Wikipedia and WordNet, using machine translation to bridge the gaps in information available for different language versions of Wikipedia [21]. Recognition is based on POS tagging for different languages, selecting candidate entities by string matching. Ranking is reduced to finding the densest subgraph that relates neighbouring entities and mentions.

FREME (2016) delegates the recognition of entities to the Stanford-NER tool, which is trained over the anchor texts of Wikipedia corpora in different languages. Candidate entities are generated by keyword search over local indexes, which are then ranked based on the number of matching anchor texts in Wikipedia linking to the corresponding article of the candidate entity [30].

With respect to FOX, note that while it meets all of our criteria, at the time of writing, we did not succeed in getting the API to run over VoxEL without error; hence we do not include this system. We also omit AGDISTIS and MAG from our selection because they do not perform recognition, requiring a prior identification of the entities in the input text (finding a suitable NER tool/model is not straightforward for some of the languages in our dataset).

3.2 Multilingual EL Datasets

In order to train and evaluate EL approaches, labelled datasets – annotated with the correct entity mentions and their respective KB links – are essential. In some cases these datasets are labelled manually, while in other cases labels can be derived from existing information, such as anchor texts. In Table 2 we

[9] http://babelnet.org/; April 1st, 2018.

Table 2. Survey of dataset for EL task. For multilingual datasets, the quantities shown refer to the English data available. We present metadata about the relaxed and strict version of our dataset by VoxEL$_R$ and VoxEL$_S$ respectively. (Abbreviations: $|D|$ number of documents, $|S|$ number of sentences, $|E|$ number of entities, **Mn** denotes that all entities were manually annotated.)

Dataset	$\|D\|$	$\|S\|$	$\|E\|$	Mn	Languages
AIDA/CoNLL-Complete [12]	1393	22,137	34,929	✓	EN
KORE50 [13]	50	50	144	✓	EN
IITB [15]	103	1,781	18,308	✓	EN
ACE2004 [26]	57	-	306	✗	EN
AQUAINT [26]	50	533	727	✗	EN
MSNBC [4]	20	668	747	✗	EN
DBpedia Spotlight [18]	10	58	331	✓	EN
N3-RSS 500 [27]	1	500	1000	✓	EN
Reuters 128 [27]	128	-	881	✓	EN
Wes2015 [36]	331	-	28,586	✓	EN
News-100 [27]	100	-	1656	✓	DE
Thibaudet [1]	1	3,807	2,980	✗	FR
Bergson [1]	1	4,280	380	✗	FR
SemEval 2015 Task 13 [20]	4	137	769	✓	EN,ES,IT
DBpedia Abstracts [2]	39,132	-	505,033	✗	DE,EN,ES,FR,IT,JA,NL
MEANTIME [19]	120	597	2,790	✓	EN,ES,IT,NL
VoxEL$_R$	15	94	674	✓	DE,EN,ES,FR,IT
VoxEL$_S$	15	94	204	✓	DE,EN,ES,FR,IT

survey the labelled datasets most frequently used by EL approaches (note that sentence counts were not available for some datasets).

We can see that the majority of datasets provide text in one language only – predominantly English – with the exceptions being as follows:

SemEval 2015 Task 13: is built over a biomedical, math, computer and social domain and is designed to support EL and WSD at the same time, containing annotations to Wikipedia, BabelNet and WordNet [20].

DBpedia Abstracts: provides a large-scale training and evaluation corpora based on the anchor texts extracted from the abstracts (first paragraph) of Wikipedia pages in seven languages [2].[10]

MEANTIME: consists of 120 news articles from WikiNews[11] with manual annotations of entities, events, temporal information and semantic roles [19].[12]

[10] http://wiki-link.nlp2rdf.org/abstracts/; April 1st, 2018.

[11] https://en.wikinews.org/; April 1st, 2018.

[12] http://www.newsreader-project.eu/results/data/wikinews/; April 1st, 2018.

With respect to DBpedia Abstracts, while offering a very large multilingual corpus, the texts across different languages vary, as do the documents available; while such a dataset could be used to compare different systems for the same languages, it could not be used to compare the same systems for different languages. Furthermore, there are no guarantees for the completeness of the annotations since they are anchor texts/links extracted from Wikipedia; hence the dataset is best suited as a large collection of positive (training) examples, in a similar manner to how TagME [8] and FREME [30] use anchor texts.

Unlike DBpedia Abstracts, the SemEval and MEANTIME datasets contain analogous documents translated to different languages (also known as *parallel corpora* [20]). Our VoxEL dataset complements these previous resources but with some added benefits. Primarily, both the SemEval and MEANTIME datasets exhibit slight variations in the annotations across languages, leading to (e.g.) a different number of entity annotations in the text for different languages; for example SemEval [20] reports 1,261 annotations for English, 1,239 for Spanish, and 1,225 for Italian, while MEANTIME [19] reports 2,790 entity mentions for English, 2,729 for Dutch, 2,709 for Italian and 2,704 for Spanish. On the other hand, VoxEL has precisely the same annotations across languages aligned at the sentence level, and also features datasets labelled under two definitions of entity. More generally, we see VoxEL as complementing these other datasets.[13]

4 The VoxEL Dataset

In this section, we describe the VoxEL Dataset that we propose as a gold standard for EL involving five languages: German, English, Spanish, French and Italian. VoxEL is based on 15 news articles sourced from the VoxEurop[14] website: a European newsletter with the same news articles professionally translated to different languages. This source of text thus obviates the need for translation of texts to different languages, and facilitates the consistent identification and annotation of mentions (and their Wikipedia links) across languages. With VoxEL, we thus provide a high-quality resource with which to evaluate the behaviour of EL systems across a variety of European languages.

While the VoxEurop newsletter is a valuable source of professionally translated text in several European languages, there are sometimes natural variations across languages that – although they preserve meaning – may change how the entities are mentioned. A common example is the use of pronouns rather than repeating a person's name to make the text more readable in a given language. Such variations would then lead to different entity annotations across languages, hindering comparability. Hence, in order to achieve the same number of sentences and annotations for each new (document), we applied small manual edits to homogenise the text (e.g., replacing a pronoun by a person's name). On the other hand, sentences that introduce new entities in one particular language,

[13] In previous work we used the SemEval dataset to compare EL systems for English and Spanish texts, where we refer the reader to [28] for more details.

[14] http://www.voxeurop.eu/; April 1st, 2018.

or that deviate too significantly across all languages, are eliminated; fewer than 10% of the sentences from the original source were eliminated.

When labelling entities, we take into consideration the lack of consensus about what is an "*entity*" [14,17,29]: some works conservatively consider only mentions of entities referring to fixed types such as person, organisation and location as entities (similar to the traditional NER/TAC consensus on an entity), while other authors note that a much more diverse set of entities are available in Wikipedia and related KBs for linking, and thus consider any noun-phrase mentioning an entity in Wikipedia to be a valid target for linking [24]. Furthermore, there is a lack of consensus on how overlapping entities – like *New York City Fire Department* – should be treated [14,17]; should *New York City* be annotated as a separate entity or should we only cover maximal entities? Rather than take a stance on such questions – which appear application dependent – we instead create two versions of the data: a *strict* version that considers only maximal entity mentions referring to persons, organisations and locations; and a *relaxed* version that considers any noun phrase mentioning a Wikipedia entity as a mention, including overlapping mentions where applicable. For example, in the sentence *"The European Central Bank released new inflation figures today"* the strict version would only include "*European Central Bank*", while the relaxed version would also include "*Central Bank*" and "*inflation*".

To create the annotation of mentions with corresponding KB identifiers, we implemented a Web tool[15] that allows a user to annotate a text, producing output in the NLP Interchange Format (NIF) [11], as well as offering visualisations of the annotations that facilitate, e.g., revision. For each language, we provide annotated links targeting the English Wikipedia entry, as well as that language's version of Wikipedia (if different from English). In case there was no appropriate Wikipedia entry for a mention of a person, organisation or place, we annotate the mention with a `NotInLexicon` marker. These annotations were created by the first author in English, which were then revised by the other authors according to the two labelling guidelines (*strict* and *relaxed*). The first author then extended these annotations to the other languages using the sentence-level correspondence, thereafter verifying that each language has the same number of annotations and the same set of English Wikipedia identifiers for each sentence.

In summary, VoxEL consists of 15 news articles (documents) from the multilingual newsletter VoxEurop, totalling 94 sentences; the central topic of these documents is politics, particularly at a European level. This text is annotated five times for each language, and two times for the strict and relaxed versions, giving a total of 150 annotated documents and 940 sentences. The same number of annotations is given for each language (including by sentence). For the strict version, each language has 204 annotated mentions, while for the relaxed version, each language has 674 annotated mentions. In the relaxed version, 6.2%, 10.8%, 20.3% and 62.7% of the entries correspond to *persons*, *organisations*, *places* and *others* respectively, while in the strict version the entities that fall in the first three classes constitute 16.9%, 28.7% and 54.4% (*others* are excluded by

[15] https://github.com/henryrosalesmendez/NIFify; April 1st, 2018.

definition under the strict guidelines). Again, this homogeneity of text and annotations across languages was non-trivial to achieve, but facilitates comparison of evaluation results not only across systems, but across languages.

5 Experiments

We now use our proposed VoxEL dataset to conduct experiments in order to explore the behaviour of state-of-the-art EL systems for multilingual settings. In particular, we are interested in the following questions:

- **RQ1**: How does the performance of systems compare for multilingual EL?
- **RQ2**: For which of the five languages are the best results achieved?
- **RQ3**: How would a method based on machine translation to English compare with directly configuring the system for a particular language?

In order to address **RQ1** and **RQ2**, we ran the multilingual EL systems Babelfy, DBpedia Spotlight, FREME, TagME and THD over both versions of VoxEL in all five languages. These experiments were conducted with the GERBIL [34] EL evaluation framework, which provides unified access to the public APIs of multiple EL tools, abstracting different input and output formats using the NIF vocabulary, translating identifiers across KBs, and allowing to apply standard metrics to measure the performance of results with respect to a labelled dataset. GERBIL calls these systems via their REST APIs maintaining default (non-language) parameters, except for the case of Babelfy, for which we analyse two configurations: one that applies a more liberal interpretation of entities to include conceptual entities (Babelfy$_R$), and another configuration that applies a stricter definition of entities (Babelfy$_S$), where the two configurations correspond loosely with the relaxed/strict versions of our dataset.

The results of these experiments are shown in Table 3, where we present micro-measures for Precision (mP), Recall (mR) and F_1 (mF), with all systems, for all languages, in both versions of the dataset.[16] From first impressions, we can observe that two systems – TagME and THD – cannot be configured for all languages, where we leave the corresponding results blank.

With respect to **RQ1**, for the Relaxed version, the highest F_1 scores are obtained by Babelfy$_R$ (0.662: ES) and DBpedia Spotlight (0.650: EN). On the other hand, the highest F_1 scores for the Strict version are TagME (0.857: EN) and Babelfy$_R$ (0.805: ES). In general, the F_1 scores for the Strict version were higher than those for the Relaxed version: investigating further, the GERBIL framework only considers annotations to be false positives when a different annotation is given in the labelled dataset at an overlapping position; hence fewer labels in the Strict dataset will imply fewer false positives overall, which seems to outweigh the effect of the extra true positives that the Relaxed version would generate. Comparing the best Strict/Relaxed results for each system, we can

[16] The GERBIL results are available at https://users.dcc.uchile.cl/~hrosales/ISWC2018_experiment_GERBIL.html.

Table 3. GERBIL Evaluation of EL systems with Micro Recall (*mR*), Precision (*mP*) and F_1 (*mF*). A value "–" indicates that the system does not support the corresponding language. The results in bold are the best for that metric, system and dataset variant comparing across the five languages (i.e., the best in each row, split by Relax/Strict).

		Relaxed					Strict				
		DE	EN	ES	FR	IT	DE	EN	ES	FR	IT
Babelfy_R	*mP*	**0.840**	0.649	0.835	0.824	0.810	**0.932**	0.785	0.929	0.889	0.907
	mR	0.461	0.522	**0.549**	0.488	0.451	0.676	**0.735**	0.710	0.632	0.578
	mF	0.595	0.578	**0.662**	0.613	0.579	0.784	0.759	**0.805**	0.739	0.706
Babelfy_S	*mP*	0.903	0.722	**0.916**	0.912	0.884	**0.942**	0.816	0.923	0.912	0.894
	mR	0.181	**0.219**	0.210	0.200	0.192	0.558	0.524	**0.593**	0.563	0.583
	mF	0.301	0.336	**0.342**	0.328	0.316	0.701	0.638	**0.722**	0.697	0.706
DBspot	*mP*	0.731	**0.745**	0.691	0.658	0.682	0.781	**0.854**	0.690	0.691	0.800
	mR	0.508	**0.577**	0.399	0.360	0.488	0.544	**0.602**	0.382	0.406	0.549
	mF	0.600	**0.650**	0.506	0.466	0.569	0.641	**0.706**	0.492	0.512	0.651
FREME	*mP*	0.762	0.803	0.655	0.737	**0.857**	0.750	**0.871**	0.660	0.739	0.858
	mR	0.161	**0.267**	0.175	0.129	0.213	0.426	**0.764**	0.553	0.416	0.652
	mF	0.266	**0.400**	0.276	0.219	0.342	0.543	**0.814**	0.602	0.532	0.740
TagME	*mP*	0.635	**0.754**	–	–	0.494	0.875	**0.946**	–	–	0.742
	mR	0.232	**0.488**	–	–	0.182	0.652	**0.784**	–	–	0.509
	mF	0.340	**0.592**	–	–	0.266	0.747	**0.857**	–	–	0.604
THD	*mP*	**0.831**	0.806	–	–	–	**0.857**	0.809	–	–	–
	mR	0.109	**0.253**	–	–	–	0.352	**0.647**	–	–	–
	mF	0.194	**0.386**	–	–	–	0.500	**0.719**	–	–	–

see that Babelfy_R, DBpedia Spotlight and FREME have less of a gap between both, meaning that they tend to annotate a broader range of entities; on the other hand, Babelfy_S and THD are more restrictive in the entities they link.

With respect to **RQ2**, considering all systems, we can see a general trend that English had the best results overall, with the best *mF* for DBpedia Spotlight, FREME and TagME. For THD, German had higher precision but much lower recall; a similar result can be seen for FREME in Italian in the Relaxed version. On the other hand, Babelfy generally had best results in German and Spanish, where, in fact, it often had the *lowest* precision in English.

With respect to possible factors that explain such differences across languages, there are variations between languages that may make the EL task easier or harder depending on the features used; for example, systems that rely on capitalisation may perform differently for Spanish, which uses less capitalisation, (e.g., "*Jungla de cristal*": a Spanish movie title in sentence case); and German, where all nouns are capitalised. Furthermore, the quality of EL resources available for different languages – in terms of linguistic components, training sets, contextual corpora, KB meta-data, etc. – may also vary across languages.

Regarding **RQ3**, we present another experiment to address the question of the efficacy of using machine translations. First we note that, although works in related areas – such as cross-lingual ontology matching [9] – have used machine translation to adapt to multilingual settings, to the best of our knowledge, no system listed in Table 1 uses machine translations over the input text (though systems such as Babelfy do use machine translations to enrich the lexical knowledge available in the KB). Hence we check to see if translating a text to English using a state-of-the-art approach – Google Translate[17] – and applying EL over the translated English text would fare better than applying EL directly over the target language; we choose one target language to avoid generating results for a quadratic pairing of languages, and we choose English since it was the only language working for/supported by all systems in Table 3.

A complication for these translation experiments is that while VoxEL contains annotations for the texts in their original five languages, including English, it does not contain annotations for the texts translated to English. While we considered manually annotating such documents produced by Google Translate, we opted against it partly due to the amount of labour it would again involve, but more importantly because it would be specific to one translation service at one point in time: as these translation services improve, these labelled documents would quickly become obsolete. Instead, we apply evaluation on a per-sentence basis, where for each sentence of a text in a non-English language, we translate it and then compare the set of annotations produced against the set of manually-annotated labels from the original English documents; in other words, we check the annotations produced by sentence, rather than by their exact position. This is only possible because in the original VoxEL dataset, we defined a one-to-one correspondence between sentences across the five different languages.

Note that since GERBIL requires labels to have a corresponding position, we thus needed to run these experiments locally outside of the GERBIL framework. Hence, for a sentence s, let A denote the IRIs associated with manual labels for s in the original English text, and let B denote the IRIs annotated by the system for the corresponding sentence of the translated text; we denote true positives by $A \cap B$, false positives by $B - A$, and false negatives by $A - B$.[18]

In Table 4, we show the results of this second experiment, focusing this time on the Micro-F_1 (mF) score obtained for each system over the five languages of VoxEL, again for the relaxed and strict versions. For each system, we consider three experiments: (1) the system is configured for the given language and run over text for the given language, (2) the system is configured for English and run over the text translated from the given language, (3) the system is configured for English and run over the text in the given language without translation. We use the third experiment to establish how the translation to English – rather than the system configuration to English – affects the results. First we note that without using positional information to check false positives (as per GERBIL),

[17] https://translate.google.com/; April 1st, 2018.
[18] To compute Precision, Recall and F_1, we do not require true negatives.

Table 4. Micro F_1 scores for systems performing EL with respect to the VoxEL dataset. For each system and each non-English language, we show the results of three experiments: first, for (_, _) the system is configured for the same language as the input text; second, for (EN,EN$_t$), the system is configured for English and applied to text translated to English from the original language (EN,EN); third, for (EN,_), the system is configured for English and run for the text in the current (original) language. Below the name of each system, we provide the relaxed and strict results for the English text. Underlined results indicate the best of the three configurations for the given system, language and dataset variant (e.g., the best for the columns of three values). The best result for each system across all variations (excluding English input) is bolded.

		Relaxed				Strict			
		DE	ES	FR	IT	DE	ES	FR	IT
Babelfy$_R$ (0.545,0.319)	(_,_)	0.523	**0.541**	0.493	0.504	0.344	0.362	0.309	0.365
	(EN,EN$_t$)	0.507	0.515	0.505	0.501	0.298	0.298	0.314	0.301
	(EN,_)	0.215	0.170	0.195	0.140	0.253	0.239	0.220	0.179
Babelfy$_S$ (0.308,0.567)	(_,_)	0.279	0.325	0.290	0.311	0.572	0.611	0.583	**0.616**
	(EN,EN$_t$)	0.311	0.309	0.322	0.303	0.518	0.523	0.559	0.532
	(EN,_)	0.201	0.179	0.189	0.137	0.376	0.372	0.395	0.258
DBpedia Spotlight (0.466,0.707)	(_,_)	0.400	0.331	0.240	0.342	0.510	0.477	0.481	0.653
	(EN,EN$_t$)	0.441	0.454	0.464	0.449	0.696	0.694	0.721	**0.729**
	(EN,_)	0.209	0.161	0.180	0.188	0.374	0.259	0.326	0.323
FREME (0.407,0.708)	(_,_)	0.282	0.302	0.268	0.373	0.483	0.583	0.479	**0.726**
	(EN,EN$_t$)	0.404	0.403	0.401	0.408	0.701	0.713	0.692	0.711
	(EN,_)	0.166	0.183	0.196	0.222	0.190	0.338	0.342	0.374
TagME (0.462,0.327)	(_,_)	0.414	–	–	–	0.272	–	–	–
	(EN,EN$_t$)	0.431	**0.450**	0.441	0.439	0.330	0.333	0.321	0.336
	(EN,_)	0.188	0.181	0.200	0.148	0.212	0.202	0.197	0.164
THD (0.392,0.625)	(_,_)	0.241	–	–	–	0.546	–	–	–
	(EN,EN$_t$)	0.394	0.392	0.386	0.387	0.597	0.620	0.595	**0.623**
	(EN,_)	0.207	0.175	0.217	0.174	0.251	0.332	0.403	0.352

the results change from those presented in Table 3; more generally, the gap between the Relaxed and Strict version is reduced.

With respect to **RQ3**, in Table 4, for each system, language and dataset variant, we underline which of the three configurations performs best. For example, in DBpedia Spotlight, all values on the (EN,EN$_t$) line – which denotes applying DBpedia Spotlight configured for English over text translated to English – are underlined, meaning that for all languages, prior translation to English outperformed submitting the text in its original language to DBpedia Spotlight configured for that language.[19] In fact, for almost all systems, translating the input text to English generally outperforms using the available language configurations of the respective EL systems, with the exception of Babelfy, where the available multilingual settings generally outperform a prior translation to English (we may recall that in Table 3, Babelfy performed best for texts other than English). We further note that the translation results are generally com-

19 ... it also implies that it outperforms running English EL on text in the original language, though this is hardly surprising and just presented for reference.

petitive with those for the original English text – shown below the name of the system for the Relaxed and Strict datasets – even slightly outperforming those results in some cases. We also observe from the generally poor (EN,_) results that translation is important; in other words, one cannot simply just apply an EL system configured for English over another language and expect good results.

To give a better impression of the results obtained from the second experiment, in Fig. 1, for the selected systems, we show the following aggregations: (1) *Calibrated* (_,_): the mean Micro-F_1 score across the four non-English languages with the EL system configured for that language; (2) *Translation* (EN,EN$_t$): the mean Micro-F_1 score across the four non-English languages with the text translated to English and the EL system configured for English; (3) English (EN,EN): the (single) Micro-F_1 score for the original English text. From this figure, we can see that translation is comparable to native English EL, and that translation often considerably outperforms EL in the original language.

We highlight that using translation to English, the result will be an annotated text in English rather than the original language. However, given that translation is done per-sentence, the EL annotations for the translated English text could potentially be "mapped" back per sentence to the text in the original language; at the very least, the translated English annotations would be a useful reference.

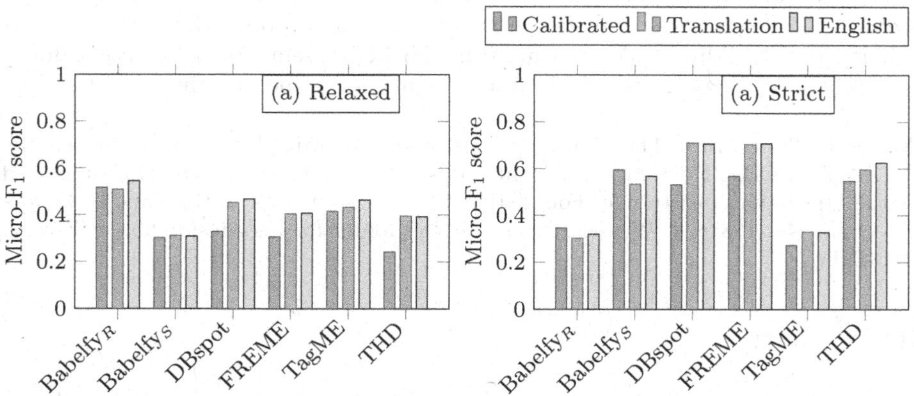

Fig. 1. Summary of the Micro-F_1 results over VoxEL Relaxed/Strict for the translation experiments, comparing mean values for setting the EL system to the language of the text (*Calibrated*), translating the text to English first (*Translation*), and the corresponding F_1 score for EL over the original English text (*English*)

6 Conclusion

While Entity Linking has traditionally focused on processing texts in English, in recent years there has been a growing trend towards developing techniques and systems that can support multiple languages. To support such research, in this paper we have described a new labelled dataset for multilingual EL, which we call VoxEL. The dataset contains 15 new articles in 5 different languages with 2 different criteria for labelling, resulting in a corpus of 150 manually-annotated news articles. In a Strict version of the dataset considering a core of entities, we derive 204 annotated mentions in each language, while in a Relaxed version of the dataset considering a broader range of entities described by Wikipedia, we derive 674 annotated mentions in each language. The VoxEL dataset is distinguished by having a one-to-one correspondence of sentences – and annotated entities per sentence – between languages. The dataset (in NIF) is available online under a CC-BY 4.0 licence: https://dx.doi.org/10.6084/m9.figshare.6539675.

We used the VoxEL dataset to conduct experiments comparing the performance of selected EL systems in a multilingual setting. We found that in general, Babelfy and DBpedia Spotlight performed the most consistently across languages. We also found that with the exception of Babelfy, EL systems performed best over English versions of the text. Next, we compared configuring the multilingual EL system for each non-English language versus applying a machine translation of the text to English and running the system in English; with the exception of Babelfy, we found that the machine translation approach outperformed configuring the system for a non-English language; even in the case of Babelfy, the translation sometimes performed better, while in others it remained competitive. This raises a key issue for research on multilingual EL: state-of-the-art machine translation is now reaching a point where we must ask if it is worth building dedicated multilingual EL systems, or if we should focus on EL for one language to which other languages can be machine translated.

Acknowledgements. The work of Henry Rosales-Méndez was supported by CONICYT-PCHA/Doctorado Nacional/2016-21160017. The work was also supported by the Millennium Institute for Foundational Research on Data (IMFD) and by Fondecyt Grant No. 1181896. We also thank Michael Röder for his considerable help with GERBIL.

References

1. Brando, C., Frontini, F., Ganascia, J.G.: REDEN: Named entity linking in digital literary editions using linked data sets. CSIMQ **7**, 60–80 (2016)
2. Brümmer, M., Dojchinovski, M., Hellmann, S.: DBpedia abstracts: a large-scale, open, Multilingual NLP Training Corpus. In: LREC (2016)
3. Charton, E., Gagnon, M., Ozell, B.: Automatic semantic web annotation of named entities. In: Butz, C., Lingras, P. (eds.) AI 2011. LNCS (LNAI), vol. 6657, pp. 74–85. Springer, Heidelberg (2011). https://doi.org/10.1007/978-3-642-21043-3_10
4. Cucerzan, S.: Large-scale named entity disambiguation based on wikipedia data. In: EMNLP-CoNLL, p. 708 (2007)
5. Daiber, J., Jakob, M., Hokamp, C., Mendes, P.N.: Improving efficiency and accuracy in multilingual entity extraction. In: I-SEMANTICS, pp. 121–124 (2013)
6. Dojchinovski, M., Kliegr, T.: Recognizing, classifying and linking entities with Wikipedia and DBpedia. In: WIKT, pp. 41–44 (2012)

7. Fahrni, A., Göckel, T., Strube, M.: HITS' Monolingual and Cross-lingual Entity Linking System at TAC 2012: A Joint Approach. In: TAC (2012)
8. Ferragina, P., Scaiella, U.: Tagme: on-the-fly annotation of short text fragments (by wikipedia entities). In: CIKM, pp. 1625–1628. ACM (2010)
9. Fu, B., Brennan, R., O'Sullivan, D.: Cross-lingual ontology mapping and its use on the multilingual semantic web. In: Multilingual Semantic Web, pp. 13–20 (2010)
10. Guo, Z., Xu, Y., de Sá Mesquita, F., Barbosa, D., Kondrak, G.: ualberta at TAC-KBP 2012: English and Cross-Lingual Entity Linking. In: TAC (2012)
11. Hellmann, S., Lehmann, J., Auer, S., Brümmer, M.: Integrating NLP Using Linked Data. In: ISWC, pp. 98–113 (2013)
12. Hoffart, J., et al.: Robust disambiguation of named entities in text. In: EMNLP, pp. 782–792. ACL (2011)
13. Hoffart, J., Seufert, S., Nguyen, D.B., Theobald, M., Weikum, G.: KORE: keyphrase overlap relatedness for entity disambiguation. In: CIKM, pp. 545–554 (2012)
14. Jha, K., Röder, M., Ngonga Ngomo, A.-C.: All that glitters is not gold – rule-based curation of reference datasets for named entity recognition and entity linking. In: Blomqvist, E., Maynard, D., Gangemi, A., Hoekstra, R., Hitzler, P., Hartig, O. (eds.) ESWC 2017. LNCS, vol. 10249, pp. 305–320. Springer, Cham (2017). https://doi.org/10.1007/978-3-319-58068-5_19
15. Kulkarni, S., Singh, A., Ramakrishnan, G., Chakrabarti, S.: Collective annotation of Wikipedia entities in web text. In: SIGKDD, pp. 457–466 (2009)
16. Lehmann, J.: DBpedia-a large-scale, multilingual knowledge base extracted from Wikipedia. Semant. Web **6**(2), 167–195 (2015)
17. Ling, X., Singh, S., Weld, D.S.: Design challenges for entity linking. TACL **3**, 315–328 (2015)
18. Mendes, P.N., Jakob, M., García-Silva, A., Bizer, C.: DBpedia spotlight: shedding light on the web of documents. In: I-SEMANTICS, pp. 1–8. ACM (2011)
19. Minard, A., Speranza, M., Urizar, R., Altuna, B., van Erp, M., Schoen, A., van Son, C.: MEANTIME, the NewsReader Multilingual Event and Time Corpus. In: LREC (2016)
20. Moro, A., Navigli, R.: SemEval-2015 Task 13: Multilingual all-words sense disambiguation and entity linking. In: SemEval@ NAACL-HLT, pp. 288–297 (2015)
21. Moro, A., Raganato, A., Navigli, R.: Entity linking meets word sense disambiguation: a unified approach. Trans. ACL **2**, 231–244 (2014)
22. Moussallem, D., et al.: MAG: A multilingual, knowledge-base agnostic and deterministic entity linking approach. In: K-CAP, p. 9 (2017)
23. Pappu, A., Blanco, R., Mehdad, Y., Stent, A., Thadani, K.: Lightweight multilingual entity extraction and linking. In: WSDM, pp. 365–374. ACM (2017)
24. Perera, S., Mendes, P.N., Alex, A., Sheth, A.P., Thirunarayan, K.: Implicit entity linking in Tweets. In: ESWC, pp. 118–132 (2016)
25. Popov, B., et al.: Kim-a semantic platform for information extraction and retrieval. Nat. Lang. Eng. **10**(3–4), 375–392 (2004)
26. Ratinov, L., Roth, D., Downey, D., Anderson, M.: Local and global algorithms for disambiguation to Wikipedia. In: NAACL-HLT, pp. 1375–1384 (2011)
27. Röder, M., et al.: N^3-A collection of datasets for named entity recognition and disambiguation in the NLP interchange format. In: LREC, pp. 3529–3533 (2014)
28. Rosales-Méndez, H., Poblete, B., Hogan, A.: Multilingual entity linking: comparing English and Spanish. In: LD4IE@ISWC, pp. 62–73 (2017)
29. Rosales-Méndez, H., Poblete, B., Hogan, A.: What should entity linking link? In: AMW (2018)

30. Sasaki, F., Dojchinovski, M., Nehring, J.: Chainable and extendable knowledge integration web services. In: ISWC, pp. 89–101 (2016)
31. Speck, R., et al.: Ensemble learning of named entity recognition algorithms using multilayer perceptron for the multilingual web of data. In: K-CAP, p. 26 (2017)
32. Tsai, C.T., Roth, D.: Cross-lingual wikification using multilingual embeddings. In: NAACL-HLT, pp. 589–598 (2016)
33. Usbeck, R., et al.: AGDISTIS - Graph-based disambiguation of named entities using linked data. In: Mika, P., et al. (eds.) ISWC 2014. LNCS, vol. 8796, pp. 457–471. Springer, Cham (2014). https://doi.org/10.1007/978-3-319-11964-9_29
34. Usbeck, R., et al.: GERBIL: General Entity Annotator Benchmarking Framework. In: WWW, pp. 1133–1143 (2015)
35. Vrandečić, D., Krötzsch, M.: Wikidata: a free collaborative knowledgebase. Commun. ACM **57**(10), 78–85 (2014)
36. Waitelonis, J., Exeler, C., Sack, H.: Linked data enabled generalized vector space model to improve document retrieval. In: NLP & DBpedia @ ISWC (2015)
37. Wang, Z., Li, J., Tang, J.: Boosting cross-lingual knowledge linking via concept annotation. In: IJCAI, pp. 2733–2739 (2013)

The Computer Science Ontology: A Large-Scale Taxonomy of Research Areas

Angelo A. Salatino(✉), Thiviyan Thanapalasingam, Andrea Mannocci, Francesco Osborne, and Enrico Motta

Knowledge Media Institute, The Open University, Milton Keynes MK7 6AA, UK
{angelo.salatino, thiviyan.thanapalasingam, andrea.mannocci, francesco.osborne, enrico.motta}@open.ac.uk

Abstract. Ontologies of research areas are important tools for characterising, exploring, and analysing the research landscape. Some fields of research are comprehensively described by large-scale taxonomies, e.g., MeSH in Biology and PhySH in Physics. Conversely, current Computer Science taxonomies are coarse-grained and tend to evolve slowly. For instance, the ACM classification scheme contains only about 2K research topics and the last version dates back to 2012. In this paper, we introduce the Computer Science Ontology (CSO), a large-scale, automatically generated ontology of research areas, which includes about 26K topics and 226K semantic relationships. It was created by applying the Klink-2 algorithm on a very large dataset of 16M scientific articles. CSO presents two main advantages over the alternatives: (i) it includes a very large number of topics that do not appear in other classifications, and (ii) it can be updated automatically by running Klink-2 on recent corpora of publications. CSO powers several tools adopted by the editorial team at Springer Nature and has been used to enable a variety of solutions, such as classifying research publications, detecting research communities, and predicting research trends. To facilitate the uptake of CSO we have developed the *CSO Portal*, a web application that enables users to download, explore, and provide granular feedback on CSO at different levels. Users can use the portal to rate topics and relationships, suggest missing relationships, and visualise sections of the ontology. The portal will support the publication of and access to regular new releases of CSO, with the aim of providing a comprehensive resource to the various communities engaged with scholarly data.

Keywords: Scholarly data · Ontology learning · Bibliographic data
Scholarly ontologies

Resource Type: Dataset, Ontology of Research Areas
Permanent URL: https://cso.kmi.open.ac.uk/
License: CC BY 4.0 International License

D. Vrandečić et al. (Eds.): ISWC 2018, LNCS 11137, pp. 187–205, 2018.
https://doi.org/10.1007/978-3-030-00668-6_12

1 Introduction

Ontologies have proved to be powerful solutions to represent domain knowledge, integrate data from different sources, and support a variety of semantic applications [1–5]. In the scholarly domain, ontologies are often used to facilitate the integration of large datasets of research data [6], the exploration of the academic landscape [7], information extraction from scientific articles [8], and so on. Specifically, ontologies that describe research topics and their relationships are invaluable tools for helping to make sense of research dynamics [7], to classify publications [3], to characterise [9] and identify [10] research communities, and to forecast research trends [11].

Some fields of research are well described by large-scale and up-to-date taxonomies, e.g., MeSH in Biology and PhySH in Physics. Conversely, current Computer Science taxonomies are coarse-grained and tend to evolve slowly. For instance, the current version of ACM classification scheme, containing only about 2K research topics, dates back to 2012 and superseded its 1998 release.

In this paper, we present the *Computer Science Ontology (CSO)*, a large-scale, granular, and automatically generated ontology of research areas which includes about 26K topics and 226K semantic relationships. CSO was created by applying the Klink-2 algorithm on a dataset of 16M scientific articles in the field of Computer Science [12]. CSO presents two main advantages over the alternatives: (i) it includes a very large number of topics that do not appear in other classifications, and (ii) it can be updated automatically by running Klink-2 on recent corpora of publications. Its fine-grained representation of research topics is very useful for characterising the content of research papers at the granular level at which researchers typically operate. For instance, CSO characterises the Semantic Web according to 40 sub-topics, such as Linked Data, Semantic Web Services, Ontology Matching, SPARQL, OWL, SWRL, and many others. Conversely, the ACM classification simply contains three related concepts: "Semantic web description languages", "Resource Description Framework (RDF)", and "Web Ontology Language (OWL)".

CSO was initially created in 2012 and has been continuously updated over the years. During this period, it has supported a range of applications and approaches for community detection, trend forecasting, and paper classification [10, 11, 13]. In particular, CSO powers two tools currently used by the editorial team at Springer Nature (SN): Smart Topic Miner [3] and Smart Book Recommender [14]. The first is a semi-automatic tool for annotating SN books both by means of topics drawn from CSO and tags selected from the internal classification used at Springer Nature. The latter is an ontology-based recommender system that suggests books, journals, and conference proceedings to market at specific venues.

We are now releasing the Computer Science Ontology, so that the relevant communities can take advantage of it and use it as a comprehensive and granular semantic resource to support the development of their own applications. To facilitate its uptake, we have developed the *CSO Portal*, a web application that enables users to download, explore, and provide granular feedback on CSO. The portal offers three different interfaces for exploring the ontology and examining the network of relationships between topics. It also allows users to rate both topics and relationships between topics, as well as suggesting new topics and relationships. The feedback from the community will be considered by an editorial board of domain experts and used to generate new versions of CSO.

We intend to regularly release new versions of CSO that will incorporate user feedback and new knowledge extracted from recent research output. Our aim is to provide a comprehensive solution for describing the Computer Science landscape that will benefit researchers, companies, organisations, and research policy makers.

The paper is structured as follows. In Sect. 2, we discuss the related work, pointing out the existing gaps. In Sect. 3, we describe the Computer Science Ontology, the applications that adopted it, and how it was evaluated. In Sect. 4, we discuss the CSO Portal and the relevant use cases. Finally, in Sect. 5 we summarise the main conclusions and outline future directions of work.

2 Related Work

Ontologies and taxonomies of research topics can support a variety of applications, such as dataset integration, the exploration process in digital libraries, the production of scholarly analytics, and modelling research dynamics [3].

In the field of Computer Science, the best-known taxonomy is the ACM Computing Classification System[1], developed and maintained by the Association for Computing Machinery (ACM). However, this taxonomy suffers from several limitations: in particular, it contains only about 2K research topics and it is developed manually. This is an extremely slow and expensive process and, as a result, its last version dates back to 2012. Hence, while the ACM taxonomy has been adopted by many publishers, in practice it lacks both depth and breadth and releases quickly go out of date.

In the field of Physics and Astronomy, the most popular solution used to be the Physics and Astronomy Classification Scheme (PACS)[2], replaced in 2016 by the Physics Subject Headings (PhySH)[3]. PACS used to associate alphanumerical codes to each subject heading to indicate their position within the hierarchy. However, this setup made its maintenance quite complex and the American Institute of Physics (AIP) discontinued it in 2010. Afterwards, the American Physical Society (APS) developed PhySH, a new classification scheme that has the advantage of being crowdsourced with the support of authors, reviewers, editors and organisers of scientific conferences, so that it is constantly updated with new terms.

The Mathematics Subject Classification (MSC)[4] is the main taxonomy used in the field of Mathematics. This scheme is maintained by Mathematical Reviews and zbMATH and it is adopted by many mathematics journals. It consists of 63 macro-areas classified with two digits: each of them is further refined into over 5K three- and

[1] The ACM Computing Classification System: http://www.acm.org/publications/class-2012.

[2] Physics and Astronomy Classification Scheme: https://publishing.aip.org/publishing/pacs.

[3] PhySH - Physics Subject Headings: https://physh.aps.org/about.

[4] 2010 Mathematics Subject Classification: https://mathscinet.ams.org/msc/msc2010.html.

five-digit classifications representing their sub-areas. The last version dates back to 2010 and typically a new official version is released every ten years.

The Medical Subject Heading (MeSH)[5] [15] is the standard solution in the field of Medicine. It is maintained by the National Library of Medicine of the United States and it is constantly updated by collecting new terms as they appear in the scientific literature.

The JEL[6] classification scheme is the most used classification in the field of Economics. The JEL scheme was created by the Journal of Economic Literature of the American Economic Association. Its last major revision dates back to 1990, but in the last years there have been many incremental changes to reflect the advances in the field [16].

The Library of Congress Classification[7] is a system of library classification that encompasses many areas of science. It was developed by the Library of Congress and it is used to classify books within large academic libraries in USA and several other countries. However, it is much too shallow to support the characterisation of scientific research at a good level of granularity. For instance, the field of Computer Science is covered by only three topics: Electronic computers, Computer science, and Computer software.

A common limitation of most of these taxonomies is that, being manually crafted and maintained by domain experts, they tend to evolve relatively slow and therefore become quickly outdated. To cope with this issue, some institutions (e.g., the American Physical Society) are crowdsourcing their classification scheme. However, the crowdsourcing strategy also suffers from limitations, such as trust and reliability [17].

A complementary strategy is to automatically or semi-automatically generate these classifications using data driven methodologies. In the literature, we can find a variety of approaches for learning taxonomies or ontologies based on natural language processing [18], clustering techniques [19], statistical methods [20], and so on. For instance, Text2Onto [18] is a framework for learning ontologies from a collection of documents. This approach identifies synonyms, sub-/superclass hierarchies, etc. through the application of natural language processing techniques on the sentence structure, where phrases like "such as…" and "and other…" imply a hierarchy between terms. This method presents some similarities with the Klink-2 algorithm [12], but requires the full text of documents. TaxGen [19] is another approach to the automatic generation of a taxonomy from a corpus by means of a hierarchical agglomerative clustering algorithm and text mining techniques. The clustering algorithm first identifies the bottom clusters by observing the linguistic features in the documents, such as co-occurrences of words, names of people, organisations, domain terms and other significant words from the text. Then the clusters are aggregated creating higher-level clusters, which form the hierarchy. This strategy is similar to the one adopted by Klink-

[5] MeSH - Medical Subject Headings: https://www.nlm.nih.gov/mesh.

[6] Journal of Economic Literature: https://www.aeaweb.org/econlit/jelCodes.php.

[7] Library of Congress Classification: https://www.loc.gov/catdir/cpso/lcc.html.

2 for inferring the *relatedEquivalent* relationships. Another approach to automatically create categorisation systems is the subsumption method [20], which computes the conditional probability for a keyword to be associated with another based on their co-occurrence. Given a pair of keywords, this system tries to understand whether there is a subsumption relationship between them, according to certain heuristics. However, this approach is limited to the statistical analysis on the co-occurrence keywords, while Klink-2 goes further by also taking advantage of external sources. It is also possible to combine ontology learning and a crowdsourcing strategy by developing approaches that take in account both statistical measures and user opinions [21, 22]. For instance, Wohlgenannt et al. [21] combine human effort and machine computation by crowd-sourcing the evaluation of an automatically generated ontology with the aim of dynamically validating the extracted relations.

3 The Computer Science Ontology

The Computer Science Ontology is a large-scale ontology of research areas that was automatically generated using the Klink-2 algorithm [12] on the Rexplore dataset [7]. This consists of about 16 million publications, mainly in the field of Computer Science. Some relationships were also refined manually by domain experts during the preparation of two ontology-assisted surveys in the fields of Semantic Web [23] and Software Architecture [13].

The current version of CSO includes 26K topics and 226K semantic relationships. The main root of CSO is Computer Science; however, the ontology includes also a few secondary roots, such as Linguistics, Geometry, Semantics, and so on.

The CSO data model[8] is an extension of the BIBO ontology[9], which in turn builds on SKOS[10]. It includes five semantic relations:

- *relatedEquivalent*, which indicates that two topics can be treated as equivalent for the purpose of exploring research data (e.g., Ontology Matching and Ontology Mapping). For the sake of avoiding technical jargon, in the CSO Portal this predicate is referred to as *alternative label of*.
- *skos:broaderGeneric*, which indicates that a topic is a super-area of another one (e.g., Semantic Web is a super-area of Linked Data). This predicate is referred to as *parent of* in the portal. The inverse relation (*child of*) is instead implicit.
- *contributesTo*, which indicates that the research output of one topic contributes to another. For instance, research in Ontology Engineering contributes to Semantic Web, but arguably Ontology Engineering is not a sub-area of Semantic Web – that is, there is plenty of research in Ontology Engineering outside the Semantic Web area.

8 http://technologies.kmi.open.ac.uk/rexplore/ontologies/BiboExtension.owl.

9 http://purl.org/ontology/bibo/.

10 http://www.w3.org/2004/02/skos.

- *rdf:type*, this relation is used to state that a resource is an instance of a class. For example, a resource in our ontology is an instance of topic.
- *rdfs:label*, this relation is used to provide a human-readable version of a resource's name.

The Computer Science Ontology is available for download in various formats (OWL, Turtle, and CSV) from https://cso.kmi.open.ac.uk/downloads. This ontology is licensed under a Creative Commons Attribution 4.0 International License (CC BY 4.0)[11] meaning that everyone is allowed to:

- copy and redistribute the material in any medium or format;
- remix, transform, and build upon the material for any purpose, even commercially.

In the following subsection, we will discuss the automatic generation of CSO with the Klink-2 algorithm (Sect. 3.1), the applications adopting it (Sect. 3.2), and how it was evaluated (Sect. 3.3).

3.1 CSO Generation

CSO was automatically generated by Klink-2 [12], an algorithm that produces an ontology of research topics by processing scholarly metadata (titles, abstracts, keywords, authors, venues) and external sources (e.g., DBpedia, calls for papers, web pages). Klink-2 can produce a full ontology including all the topics represented in the input dataset or focus on some branches under seed keywords (e.g., "Semantic Web").

In Algorithm 1, we report the pseudocode of Klink-2. The algorithm takes as input a set of keywords and investigates their relationship with the set of their most co-occurring keywords. Klink-2 infers the semantic relationship between keyword x and y by means of three metrics: (i) $H_R(x,y)$, which uses a semantic variation of the subsumption method for measuring the intensity of a hierarchical relationship; (ii) $T_R(x,y)$, which uses temporal information to do the same; and (iii) $S_R(x,y)$, which estimates the similarity between two topics. The first two are used to detect *skos:broaderGeneric* and *contributesTo* relationships, while the latter is used to infer *relatedEquivalent* relationships. Klink-2 then removes loops in the topic network (instruction #9). Finally, it merges keywords linked by a *relatedEquivalent* relationship and splits ambiguous keywords associated to multiple meanings (e.g., "Java"). The keywords produced in this step are added to the initial set of keywords to be further analysed in the next iteration and the while-loop is re-executed until there are no more keywords to be processed. Finally, Klink-2 filters the keywords considered "too generic" or "not academic" according to a set of heuristics (instruction #13) and generates the triples describing the ontology.

---

[11] CC BY 4.0 International License https://creativecommons.org/licenses/by/4.0.

```
Input  : List of keywords keywords, User feedbacks feedbacks
Output: Ontology CSO
relationships={};                              // Initialise an empty set
while some keywords yet to process do
    foreach k1 in keywords do
        candidates = GetCandidates(k1, feedbacks);
        foreach k2 in candidates do
            relationship = InferRelationship(k1, k2, feedback,
             relationships);
        end foreach
    end foreach
    relationships = RemoveLoops(relationships);
    new.keywords = MergeAndSplitKeywords(keywords, relationships);
    keywords = AddNewKeywords(new.keywords);
end while
keywords = FilterTopics(keywords, feedbacks, relationships);
CSO = GenerateSemanticRelationships(relationships);
return(CSO);
```

Algorithm 1. The Klink-2 algorithm used to generate CSO.

Klink-2 was evaluated on the task of generating an ontology of Semantic Web topics using the metadata in the Rexplore dataset in a previous work [12]. For this purpose, we generated with the help of three senior researchers a gold standard ontology[12] including 88 research topics in the field of the Semantic Web. Klink-2 outperformed significantly the alternative algorithms ($p = 0.0005$) yielding a precision of 86% and a recall of 85.5%. For further details about Klink-2 and its evaluation, please refer to [12].

3.2 Applications Using CSO

The Computer Science Ontology has been used to support a variety of applications and algorithms. In this section, we discuss a selection of these systems and how they use the ontology, with the aim of showing the practical value of CSO and inspiring further applications.

Smart Topic Miner [3] (STM)[13] is a tool developed in collaboration with Springer Nature for supporting its editorial team in classifying editorial products according to a taxonomy of research topics drawn from CSO and the Springer Nature internal taxonomy. STM halves the time needed for classifying proceedings from 20–30 to 10–15 min and allows this task to be performed also by assistant editors, thus distributing the load and reducing costs. It is currently used to classify about 800 proceedings books every year, including the ones published in the well-known Lecture Notes in Computer Science (LNCS) series family.

[12] Gold Standard: http://technologies.kmi.open.ac.uk/rexplore/data.

[13] Demo of Smart Topic Miner: http://rexplore.kmi.open.ac.uk/STM_demo.

Smart Book Recommender [14] (SBR)[14] is an ontology-based recommender system that takes as input the proceedings of a conference and suggests books, journals, and other conference proceedings which are likely to be relevant to the attendees of the conference in question. It builds on a dataset of 27K Springer Nature editorial products described with CSO research topics. SBR allows editors to investigate why a certain publication was suggested by means of an interactive view that compares the topics of the suggested publications and those of the input conference.

Augur [11] is an approach that aims to effectively detect the emergence of new research areas by analysing topic networks and identifying clusters associated with an overall increase of the pace of collaboration between research areas. Initially, Augur creates evolutionary networks describing the collaboration between research topics over time. Then it uses a novel clustering algorithm, the Advanced Clique Percolation Method (ACPM), to identify portions of the network that exhibit a significant increase in the pace of collaboration. Each identified clusters of topics represent an area of the network that is nurturing a new research area that should shortly emerge.

Rexplore [7] is a system that leverages novel solutions in large-scale data mining, semantic technologies and visual analytics, to provide an innovative environment for exploring and making sense of scholarly data. Rexplore uses CSO for characterising research papers, authors, and organisations according to their research topics and for producing relevant views. For instance, Rexplore is able to plot the collaboration graph of the top researchers in a field and to visualise researchers in terms of the shifting of their research interests over the years. Rexplore also describes each topic in CSO with a variety of analytics, and allows users to visualise the trends of its sub-topics.

The **Technology-Topic Framework** [24] is an approach that characterises technologies according to their propagation through research topics drawn from CSO, and uses this representation to forecast the future propagation of novel technologies across research fields. The aim is to suggest promising technologies to scholars and accelerate the flow of knowledge from one community to another and the pace of technology propagation. The system was evaluated on a set of 1,118 technologies in the Artificial Intelligence field yielding excellent results.

EDAM [13] is an expert-driven automatic methodology for creating systematic reviews that limits the amount of tedious tasks that have to be performed by human experts. Typically, systematic reviews require domain experts to annotate hundreds of papers manually. EDAM is able to skip this step by (i) characterising the area of interest using an ontology of topics, (ii) asking domain experts to refine this ontology, and (iii) exploiting this knowledge base for classifying relevant papers and producing useful analytics. The first implementation of EDAM adopted CSO for analysing the field of Software Architecture.

3.3 CSO Evaluation

Since its introduction in 2012, the Computer Science Ontology has been used in several studies and proved to effectively support a wide range of tasks such as:

---

[14] Demo of Smart Book Recommender: http://rexplore.kmi.open.ac.uk/SBR_demo.

- forecasting new research topics [11];
- exploration of scholarly data [7];
- automatic annotation of research papers [13];
- detection of research communities [10];
- ontology forecasting [25].

In this section, we will briefly report the results of these studies and highlight the role of CSO.

Forecasting New Research Topics. The evaluation of the Augur system [11] proved that semantically enriching topics networks with CSO yields a significant improvement in performance on the task of predicting the emergence of novel research areas. Table 1 shows precision and recall obtained in the period 1999–2009 by a version of Augur using CSO and by an alternative version that uses only keywords to represent research topics[15].

Exploration of Scholarly Data. The Rexplore system was shown to be able to support users in performing specific tasks more effectively than Microsoft Academic Search (MAS), thanks to its organic representation of research topics [7]. We conducted a user study and asked 26 users to complete three tasks using one of the systems. The users adopting Rexplore completed the task more quickly and with higher success rate, as reported in Table 2.

Table 1. Performance of Augur [11] when characterising topics with keywords or CSO.

	Keywords		CSO	
	Precision	Recall	Precision	Recall
1999	0.68	0.49	**0.86**	**0.76**
2000	0.62	0.39	**0.78**	**0.70**
2001	0.69	0.49	**0.77**	**0.72**
2002	0.65	0.50	**0.82**	**0.80**
2003	0.72	0.54	**0.83**	**0.79**
2004	0.70	0.47	**0.84**	**0.68**
2005	0.62	0.49	**0.71**	**0.66**
2006	0.32	0.32	**0.43**	**0.51**
2007	0.06	0.21	**0.28**	**0.44**
2008	0.06	0.08	**0.15**	**0.33**
2009	0.05	0.59	**0.09**	**0.76**

Table 2. Experimental results (in min:secs) using Rexplore and MAS to perform three different tasks.

Rexplore (CSO) (17 participants)			
	Average time	Standard deviation	Success rate
Task 1	**03:06**	**00:45**	**100%**
Task 2	**08:01**	**02:50**	**94%**
Task 3	**07:51**	**02:32**	**100%**
MAS (no CSO) (9 participants)			
	Average time	Standard deviation	Success rate
Task 1	14:46	00:24	33%
Task 2	13:52	01:35	50%
Task 3	15:00	00:00	0%

[15] The evaluation material of Augur can be found at http://rexplore.kmi.open.ac.uk/JCDL2018.

Automatic Annotation of Research Papers. The aforementioned Expert-Driven Automatic Methodology [13] uses CSO for automatically classifying research papers by categorising under a topic all papers that contain in the title, abstract, or keyword field the label of the topic, its *relatedEquivalent*, or its *skos:narrowerGeneric*. We applied this approach to the field of Software Architecture[16] and found that its performance in classifying papers was not statistically significantly different from those of six senior researchers in the field ($p = 0.77$). Table 3 shows the agreement between the annotators, computed as the ratio of papers which were tagged with the same category by both annotators. The approach adopting CSO yielded the highest average agreement and also obtained the highest agreement with three out of six domain experts.

Table 3. Agreement between annotators (including EDAM) and average agreement of each annotator.

	EDAM (CSO)	User1	User2	User3	User4	User5	User6
EDAM (CSO)	-	56%	68%	64%	64%	76%	64%
User1	**56%**	-	40%	**56%**	36%	48%	44%
User2	68%	40%	-	64%	52%	**76%**	64%
User3	64%	56%	64%	-	52%	64%	**68%**
User4	**64%**	36%	52%	52%	-	**64%**	52%
User5	**76%**	48%	**76%**	64%	64%	-	72%
User6	64%	44%	64%	**68%**	52%	72%	-
AVG	**66%**	45%	58%	59%	51%	63%	60%

Detection of Research Communities. The Temporal Semantic Topic-Based Clustering (TST) is an approach for detecting research communities by clustering researchers according to their research trajectories, defined as distributions of topics over time. We evaluated the full version of TST that characterises the researcher's interests according to CSO against 25 human experts in the fields of Semantic Web and Human Computer Interaction, finding no significant differences ($p > 0.14$). Conversely, an alternative version that simply uses keywords was outperformed by both TST and human experts ($p < 0.0001$).

Ontology Forecasting. The Semantic Innovation Forecast model (SIF) [25] is an approach to predict new concepts of an ontology at time $t + 1$, using only data available at time t. The full version of SIF, learning from concepts in CSO, was able to significantly outperform[17] several variations of LDA [26], as reported in Table 4.

[16] The evaluation material of EDAM can be found at http://rexplore.kmi.open.ac.uk/data/edam.

[17] The evaluation material of SIF can be found at http://technologies.kmi.open.ac.uk/rexplore/ekaw2016/OF.

Table 4. Mean average precision @10 for SIF [25] and other four alternative algorithms based on LDA [26].

YEAR-FORECAST	YEAR-TRAINED	YEAR-PRIOR	SIF (CSO)	LDA	LDA-A	LDA-I	LDA-IA
2000	1999	1997-1999	**0.7031**	0.125	0.4761	0	0.408
2002	2001	1999-2001	**0.875**	0	0.8227	0.6428	0.7486
2004	2003	2001-2003	**0.906**	0	0.5822	0.5726	0.6347
2006	2005	2003-2005	**0.8755**	0.3069	0.7853	0.8385	0.6893
2008	2007	2005-2006	**0.988**	0.398	0.681	0.5661	0.7035
AVG			**0.8695**	0.1659	0.6694	0.524	0.6368

4 The CSO Portal

The CSO Portal is a web application that enables users to download, explore, and provide granular feedback on CSO. It is available at http://cso.kmi.open.ac.uk.

Figure 1 shows an overview of the CSO Portal. We consider three kinds of users: unregistered users, registered users, and members of the editorial board. Unregistered users can download the ontology and browse it by using three alternative interfaces. Registered users are also allowed to post feedback regarding the full ontology or specific topics or relationships. The members of the editorial board have the task of reviewing the user feedback and select the changes to be incorporated in the new releases of CSO.

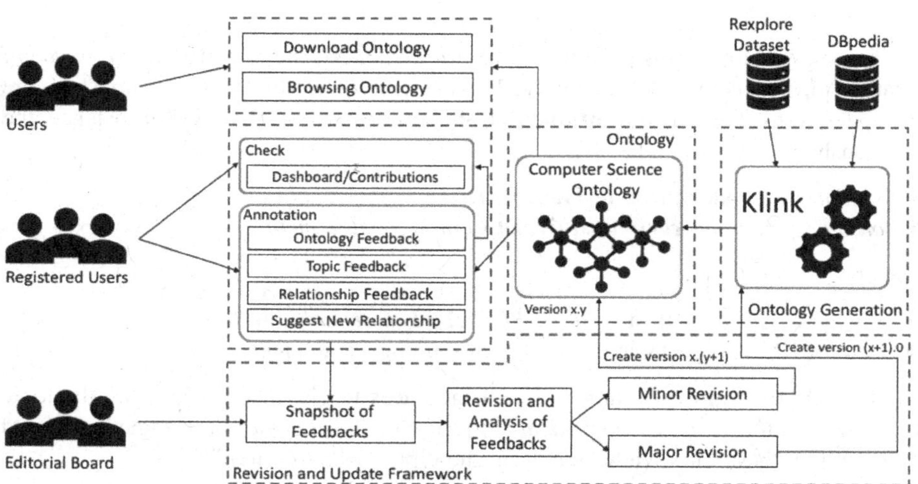

Fig. 1. Overview of the computer science ontology portal.

In the following sections, we will discuss how users can explore CSO and leave feedback at different levels of granularity.

4.1 Exploring CSO

An important functionality of the CSO Portal is the ability to search and navigate the about 26K research topics in CSO. The homepage of the portal (Fig. 2) provides a simple search bar as a starting point. The user can type the label of any topic (e.g., "Semantic Web") and submit it to be redirected to that topic page.

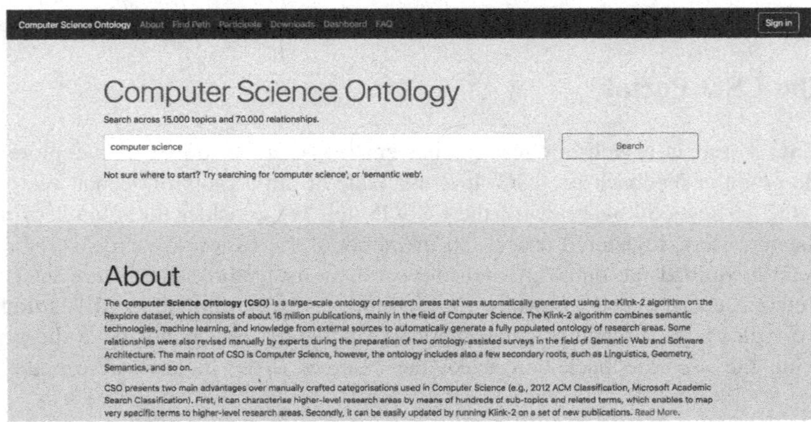

Fig. 2. Homepage of the computer science ontology portal.

For a given topic, this page shows its *skos:broaderGeneric* and *relatedEquivalent* relationships with the relevant topics. For the sake of clarity, these relationships are presented to the users as **parent of/child of** and **alternative label of.** For instance, the relationships:

- *semantic web* **skos:broaderGeneric** *RDF*
- *ontology mapping* **relatedEquivalent** *ontology alignment*

are presented as:

- *semantic web* **parent of** *RDF* or *RDF* **child of** *semantic web*
- *ontology mapping* **alternative label of** *ontology alignment*

The CSO Portal offers three different interfaces to visualise and explore the topic relationships: the *graph view*, the *detailed view*, and the *compact view*. Figures 3, 4 and 5 show how these three views represent the topic "*semantic web*"[18].

[18] http://cso.kmi.open.ac.uk/topics/semantic%20web.

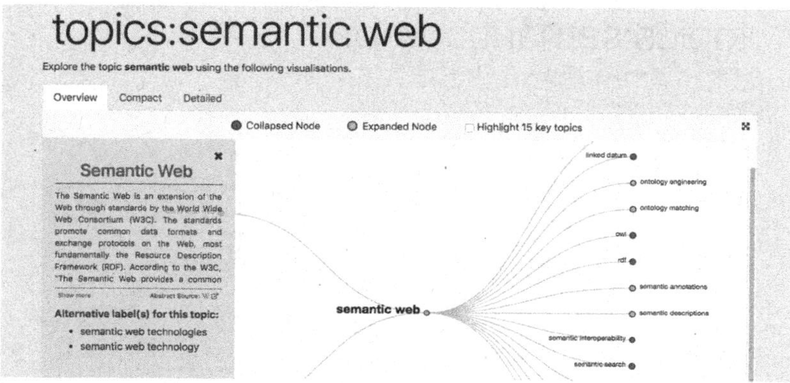

Fig. 3. Screenshot of the resource page related to the topic *"semantic web"* (Overview).

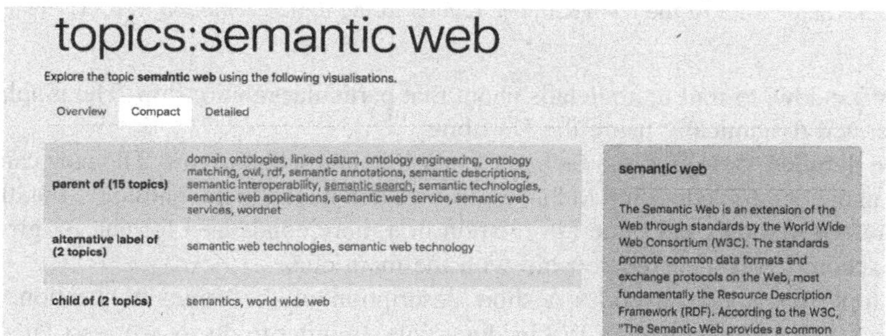

Fig. 4. Screenshot of the resource page related to the topic "*semantic web*" (Compact).

The **graph view** is an interactive interface that allows users to seamlessly navigate the network of topics within CSO. In this view, each topic is represented as a node and the *skos:broaderGeneric* relationships are represented as links. Initially, the view focuses on the topic searched by the user and its direct relationships. The user can also explore the ontology by expanding nodes, hiding unwanted branches, and zooming in and out. The nodes can be expanded or collapsed by left clicking on them. The user can also utilise a checkbox for highlighting the 15 key topics in the branch. This feature allows the user to quickly identify the most significant topics, making use of an approximate count of the relevant papers within the Rexplore dataset [7]. When users right-clicks on a specific node, they are prompted with a menu containing the following two options: (i) *Inspect* – This opens a sidebar window, as shown in Fig. 3, providing more information about the topic (description and equivalent topics), and (ii) *Explore in new page* – This redirects the user to another page where the selected topic is the central node in the graph. The user can also right-click on links, which also opens a

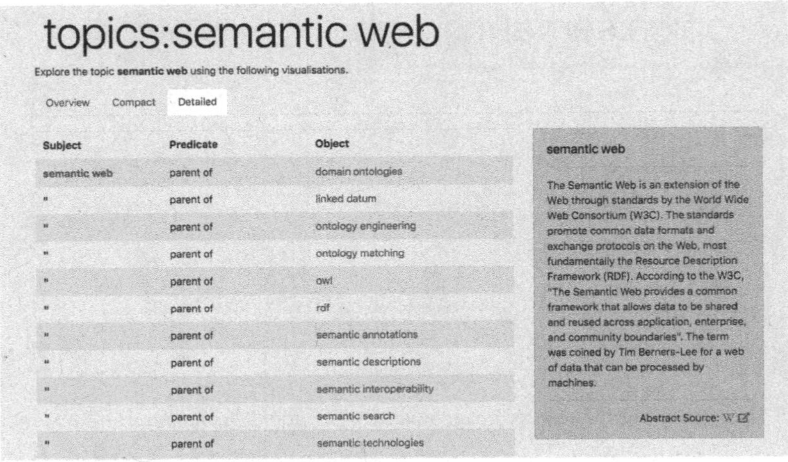

topics:semantic web

Explore the topic **semantic web** using the following visualisations.

Overview Compact Detailed

Subject	Predicate	Object
semantic web	parent of	domain ontologies
"	parent of	linked datum
"	parent of	ontology engineering
"	parent of	ontology matching
"	parent of	owl
"	parent of	rdf
"	parent of	semantic annotations
"	parent of	semantic descriptions
"	parent of	semantic interoperability
"	parent of	semantic search
"	parent of	semantic technologies

semantic web

The Semantic Web is an extension of the Web through standards by the World Wide Web Consortium (W3C). The standards promote common data formats and exchange protocols on the Web, most fundamentally the Resource Description Framework (RDF). According to the W3C, "The Semantic Web provides a common framework that allows data to be shared and reused across application, enterprise, and community boundaries". The term was coined by Tim Berners-Lee for a web of data that can be processed by machines.

Abstract Source: W

Fig. 5. Screenshot of the resource page related to the topic *"semantic web"* (Detailed).

sidebar window, to find more details about that particular relationship. The graph view is generated dynamically using the D3 library[19].

The **detailed view** presents each relevant triple in a separate row. The user can click on the name of a topic to jump to that topic page and navigate the ontology. Finally, the **compact view** shows the same information in a more condensed format, by grouping topics according to their relationship with the main one.

A topic page also provides a short description of the topic in question and a hyperlink to the corresponding Wikipedia article. In order to do so, we associated each topic to the relevant DBpedia entity by feeding a sentence listing the label of the topic and its direct sub and super topics to the DBpedia Spotlight API [27]. The subsequent JSON response contains a list of likely DBpedia pages for the selected topic, each with a number of probability statistics. A data analysis showed that by filtering out candidates with a similarity score less than 1 and an offset value greater than 0 it is possible to identify the correct DBpedia entity with nearly 100% precision. Naturally, not all CSO topics are described in DBpedia.

The portal supports content negotiation and yields different representations of the resources according to the content-type specified in the request. It currently supports '*text/html*', '*application/rdf+xml*', '*text/turtle*', '*application/n-triples*', and '*application/ld+json*'.

4.2 User Feedback

Registered users can provide feedback about the ontology and its relationships in all the alternative views, to be considered for future releases of the Computer Science

[19] D3.js, https://d3js.org.

Ontology (see Sect. 5). In particular, users can offer feedback at (i) ontology level, (ii) topic level, and (iii) relationship level.

The ontology level feedback is a general assessment expressing thoughts and criticisms about CSO. The user can provide it by clicking the feedback tab in the top menu and filling a text form.

Users can give feedback on specific topics by means of a form that can be triggered by clicking an icon near the topic name. Figure 6 shows as example the feedback form for the topic "ontology mapping". Users can rate the topic as "correct", "incorrect" or "is complicated" and comment their rating in a text field. In the same form, users can also suggest one or more relationships that are currently missing from the ontology or a new topic that should be linked by this relationship. Figure 7 shows the form for suggesting new relationships for the topic "ontology mapping". The users can choose the predicate from "parent of", "alternative label of", "and child of". The object could be either a topic that already exists in CSO or a new one.

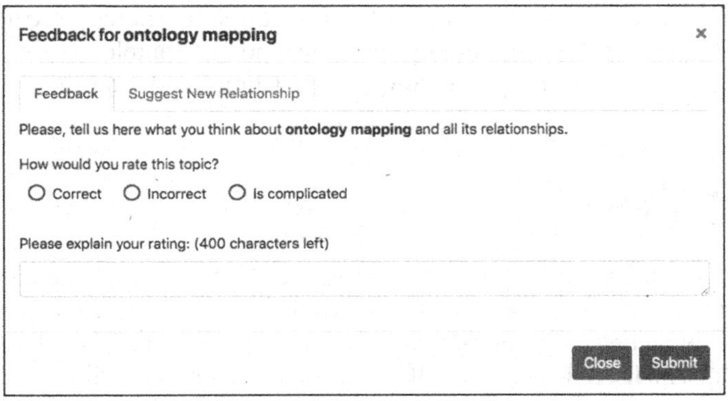

Fig. 6. Form for providing feedback about the topic *"ontology mapping"*.

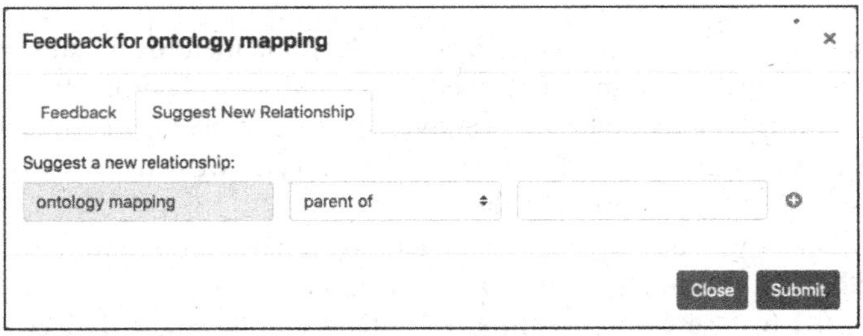

Fig. 7. Form for suggesting new relationships about the topic *"ontology mapping"*.

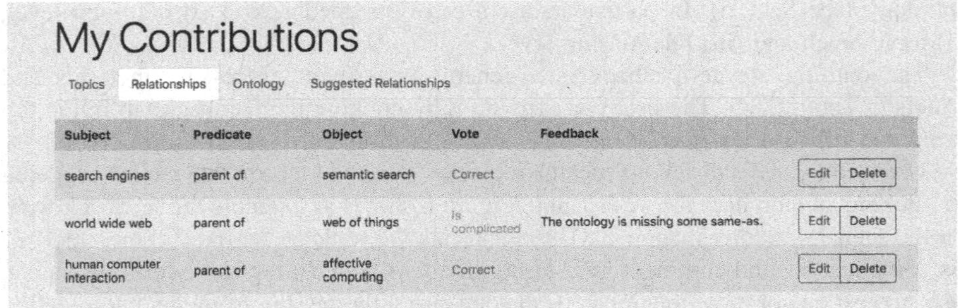

Fig. 8. *My Contribution* page where users can review their own feedback.

Finally, users can offer feedback on specific relationships by means of an alternative form. As in the previous case, they can rate the relationship and add a short comment.

The CSO Portal allows users to review their own feedback entries. In the "*My Contributions*" page (Fig. 8) users can inspect, edit, and delete any previously given suggestion. The feedback entries are organised by typology (ontology level, topic level, relationship level, and recommendation of new relationships), and they can be either retracted or modified.

5 Future Updates

We plan to periodically release new versions of the CSO ontology. The editorial board will supervise this process and review user feedback, distilling a list of recommendations to be implemented in future versions. The composition of the editorial board is currently being finalised. Initially it will comprise a small number of individuals drawn from the Open University and our industrial collaborators (between 4 and 6 people in total). Depending on the success and impact of the initiative we expect that the board will grow significantly in the future and will expand to include representatives of a variety of organizations. Both minor and major revisions will be released on a regular basis.

Minor revisions will be produced by directly implementing in the ontology the changes suggested by users and confirmed by the editorial board. The changes may include: (i) removal of a topic, (ii) removal of a relationship, (iii) inclusion of a relationship, and (iv) inclusion of a topic. In this phase, we will focus on correcting specific errors rather than expanding the ontology.

Major revisions will be produced by generating a new full ontology by feeding the Klink algorithm an up-to-date corpus of publications and the set of "correct relationships" suggested by users and confirmed by the editorial board. Indeed, the current version of Klink-2 is already able to take as input user defined relationships and incorporate them in the automatically generated ontology. The goal is to make sure that major revisions of CSO include all significant research areas that have emerged in the interval since the previous major release.

We aim to produce at least one major revision every year. The timing on the other revisions will depend on the number and quality of feedback entries. For instance, a significant number of negative feedback entries on a certain branch would trigger a comprehensive revision of it. In such a case, we will contact domain experts and invite them to review the associated branch on the CSO Portal. For instance, in a recent study [13], we assessed the CSO branch regarding Software Architecture by generating a spreadsheet representation of it and having it reviewed by three senior researchers. The CSO Portal should make this process simpler and easier to track.

6 Conclusions

In this paper, we presented the Computer Science Ontology (CSO), a large-scale, automatically generated ontology of research areas, which provides a much more comprehensive and granular characterisation of research topics in Computer Science than what is currently available in other state-of-the-art taxonomies of research areas. We discussed its characteristics, briefly introduced several applications which use it, and showed that it successfully supports several useful tasks, such as classifying research papers, exploring scholarly data, forecasting new research topics, detecting research communities, and so on. We also introduced the CSO Portal, a web application that enables users to download, explore, and provide feedback on CSO. We intend to take advantage of the CSO Portal to involve the wider research community in the ontology evolution process, with the aim of periodically releasing up-to-date revisions of CSO and allow members of the community to provide feedback. In this sense, the version of CSO presented in this paper can be considered simply as a starting point.

As future work, we are currently developing a new version of Klink-2 that will consider the quantity and the sentiment of the user feedback on previous versions of the ontology. We also intend to apply our ontology learning techniques to other research fields, such as Biology and Engineering. The ultimate goal is to create a comprehensive set of large-scale and data driven ontologies describing most branches of science.

References

1. Saif, H., He, Y., Alani, H.: Semantic sentiment analysis of twitter. In: Cudré-Mauroux, P., et al. (eds.) ISWC 2012, Part I. LNCS, vol. 7649, pp. 508–524. Springer, Heidelberg (2012). https://doi.org/10.1007/978-3-642-35176-1_32
2. Ding, L., Kolari, P., Ding, Z., Avancha, S.: Using ontologies in the semantic web: a survey. In: Sharman, R., Kishore, R., Ramesh, R. (eds.) Ontologies: A Handbook of Principles, Concepts and Applications in Information Systems, pp. 79–113. Springer, Boston (2007). https://doi.org/10.1007/978-0-387-37022-4_4
3. Osborne, F., Salatino, A., Birukou, A., Motta, E.: Automatic classification of Springer nature proceedings with smart topic miner. In: Groth, P., et al. (eds.) ISWC 2016, Part II. LNCS, vol. 9982, pp. 383–399. Springer, Cham (2016). https://doi.org/10.1007/978-3-319-46547-0_33

4. Middleton, S.E., Roure, D.D., Shadbolt, N.R.: Ontology-based recommender systems. In: Staab, S., Studer, R. (eds.) Handbook on Ontologies. IHIS, pp. 779–796. Springer, Heidelberg (2009). https://doi.org/10.1007/978-3-540-92673-3_35
5. Hotho, A., Staab, S., Stumme, G.: Ontologies improve text document clustering. In: Third IEEE International Conference on Data Mining, pp. 541–544. IEEE Computer Society
6. Livingston, K.M., Bada, M., Baumgartner, W.A., Hunter, L.E.: KaBOB: ontology-based semantic integration of biomedical databases. BMC Bioinform. **16**, 126 (2015)
7. Osborne, F., Motta, E., Mulholland, P.: Exploring scholarly data with rexplore. In: Alani, H., et al. (eds.) ISWC 2013, Part I. LNCS, vol. 8218, pp. 460–477. Springer, Heidelberg (2013). https://doi.org/10.1007/978-3-642-41335-3_29
8. Fathalla, S., Vahdati, S., Auer, S., Lange, C.: Towards a knowledge graph representing research findings by semantifying survey articles. In: Kamps, J., Tsakonas, G., Manolopoulos, Y., Iliadis, L., Karydis, I. (eds.) TPDL 2017. LNCS, vol. 10450, pp. 315–327. Springer, Cham (2017). https://doi.org/10.1007/978-3-319-67008-9_25
9. Bettencourt, L.M.A., Kaiser, D.I., Kaur, J.: Scientific discovery and topological transitions in collaboration networks. J. Informetr. **3**, 210–221 (2009)
10. Osborne, F., Scavo, G., Motta, E.: Identifying diachronic topic-based research communities by clustering shared research trajectories. In: Presutti, V., d'Amato, C., Gandon, F., d'Aquin, M., Staab, S., Tordai, A. (eds.) ESWC 2014. LNCS, vol. 8465, pp. 114–129. Springer, Cham (2014). https://doi.org/10.1007/978-3-319-07443-6_9
11. Salatino, A.A., Osborne, F., Motta, E.: AUGUR: forecasting the emergence of new research topics. In: Joint Conference on Digital Libraries 2018, Fort Worth, Texas, pp. 1–10 (2018)
12. Osborne, F., Motta, E.: Klink-2: integrating multiple web sources to generate semantic topic networks. In: Arenas, M., et al. (eds.) ISWC 2015. LNCS, vol. 9366, pp. 408–424. Springer, Cham (2015). https://doi.org/10.1007/978-3-319-25007-6_24
13. Osborne, F., Muccini, H., Lago, P., Motta, E.: Reducing the Effort for Systematic Reviews in Software Engineering Pre-Print: https://bit.ly/2sobCkI
14. Thanapalasingam, T., Osborne, F., Birukou, A., Motta, E.: Ontology-based recommendation of editorial products. In: International Semantic Web Conference 2018, Monterey, CA, USA (2018)
15. Lipscomb, C.E.: Medical subject headings (MeSH). Bull. Med. Libr. Assoc. **88**, 265–266 (2000)
16. Cherrier, B.: Classifying economics: a history of the JEL codes. J. Econ. Lit. **55**, 545–579 (2017)
17. Clough, P., Sanderson, M., Gollins, T.: Examining the limits of crowdsourcing for relevance assessment. IEEE Internet Comput. **17**, 32–38 (2013)
18. Cimiano, P., Völker, J.: Text2Onto. In: Montoyo, A., Muñoz, R., Métais, E. (eds.) NLDB 2005. LNCS, vol. 3513, pp. 227–238. Springer, Heidelberg (2005). https://doi.org/10.1007/11428817_21
19. Muller, A., Dorre, J., Gerstl, P., Seiffert, R.: The TaxGen framework: automating the generation of a taxonomy for a large document collection. In: Proceedings of the 32nd Annual Hawaii International Conference on Systems Sciences, HICSS-32. Abstracts and CD-ROM of Full Papers, p. 9. IEEE Computer Society (1999)
20. Sanderson, M., Croft, B.: Deriving concept hierarchies from text. In: Proceedings of the 22nd Annual International ACM SIGIR Conference on Research and Development in Information Retrieval - SIGIR 1999, pp. 206–213. ACM Press, New York (1999)
21. Wohlgenannt, G., Weichselbraun, A., Scharl, A., Sabou, M.: Dynamic integration of multiple evidence sources for ontology learning. J. Inf. Data Manag. **3**, 243–254 (2012)

22. Mortensen, J.M., Musen, M.A., Noy, N.F.: Crowdsourcing the verification of relationships in biomedical ontologies. In: AMIA Annual Symposium Proceedings 2013, pp. 1020–1029 (2013)
23. Kirrane, S., et al.: A decade of semantic web research through the lenses of a mixed methods approach. Semant. Web J. - Prepr. (2018)
24. Osborne, F., Mannocci, A., Motta, E.: Forecasting the spreading of technologies in research communities. In: Proceedings of the Knowledge Capture Conference (2017)
25. Cano-Basave, A.E., Osborne, F., Salatino, A.A.: Ontology forecasting in scientific literature: semantic concepts prediction based on innovation-adoption priors. In: Blomqvist, E., Ciancarini, P., Poggi, F., Vitali, F. (eds.) EKAW 2016. LNCS (LNAI), vol. 10024, pp. 51–67. Springer, Cham (2016). https://doi.org/10.1007/978-3-319-49004-5_4
26. Blei, D.M., Edu, B.B., Ng, A.Y., Edu, A.S., Jordan, M.I., Edu, J.B.: Latent Dirichlet allocation. J. Mach. Learn. Res. **3**, 993–1022 (2003)
27. Daiber, J., Jakob, M., Hokamp, C., Mendes, P.N.: Improving efficiency and accuracy in multilingual entity extraction. In: Proceedings of the 9th International Conference on Semantic Systems - I-SEMANTICS 2013, p. 121. ACM Press, New York (2013)

DistLODStats: Distributed Computation of RDF Dataset Statistics

Gezim Sejdiu[1(✉)], Ivan Ermilov[2], Jens Lehmann[1,3], and Mohamed Nadjib Mami[1,3]

[1] Smart Data Analytics, University of Bonn, Bonn, Germany
{sejdiu,jens.lehmann,mami}@cs.uni-bonn.de
[2] Department of Computer Science, University of Leipzig, 04109 Leipzig, Germany
iermilov@informatik.uni-leipzig.de
[3] Fraunhofer IAIS, Sankt Augustin, Germany
{jens.lehmann,mohamed.nadjib.mami}@iais.fraunhofer.de

Abstract. Over the last years, the Semantic Web has been growing steadily. Today, we count more than 10,000 datasets made available online following Semantic Web standards. Nevertheless, many applications, such as data integration, search, and interlinking, may not take the full advantage of the data without having a priori statistical information about its internal structure and coverage. In fact, there are already a number of tools, which offer such statistics, providing basic information about RDF datasets and vocabularies. However, those usually show severe deficiencies in terms of performance once the dataset size grows beyond the capabilities of a single machine. In this paper, we introduce a software component for statistical calculations of large RDF datasets, which scales out to clusters of machines. More specifically, we describe the first distributed in-memory approach for computing 32 different statistical criteria for RDF datasets using Apache Spark. The preliminary results show that our distributed approach improves upon a previous centralized approach we compare against and provides approximately linear horizontal scale-up. The criteria are extensible beyond the 32 default criteria, is integrated into the larger SANSA framework and employed in at least four major usage scenarios beyond the SANSA community.

1 Introduction

Over the last two decades, the Semantic Web has grown from a mere idea for modeling data in the web, into an established field of study driven by a wide range of standards and protocols for data consumption, publication and exchange on the Web. For the record, today we count more than 10,000 datasets openly available online using Semantic Web standards[1]. Thanks to such standards, large datasets

Resource type: Software Framework.
Website: http://sansa-stack.net/distlodstats/.
Permanent URL: https://doi.org/10.6084/m9.figshare.6080711.

[1] http://lodstats.aksw.org/.

D. Vrandečić et al. (Eds.): ISWC 2018, LNCS 11137, pp. 206–222, 2018.
https://doi.org/10.1007/978-3-030-00668-6_13

became machine-readable [13]. Nevertheless, many applications such as data integration, search, and interlinking may not take full advantage of the data without having *a priori* statistical information about its internal structure and coverage. RDF dataset statistics can be beneficial in many ways, for example: (1) Vocabulary reuse (suggesting frequently used similar vocabulary terms in other datasets during dataset creation), (2) Quality analysis (analysis of incoming and outcoming links in RDF datasets to establish hubs similar to what pagerank has achieved in the traditional web), (3) Coverage analysis (verifying whether frequent dataset properties cover all similar entities and other related tasks), (4) privacy analysis (checking whether property combinations may allow to uniquely identify persons in a dataset) and (5) link target analysis (finding datasets with similar characteristics, e.g. similar frequent properties) for interlinking candidates.

A number of solutions have been conceived to offer users such statistics about RDF vocabularies [17] and datasets [7,9]. However, those efforts showed severe deficiencies in terms of performance when the dataset size goes beyond the main memory size of a single machine. This limits their capabilities to medium-sized datasets only, which paralyzes the role of applications in embracing the increasing volumes of the available datasets.

As the memory limitation was the main shortcoming in the existing works, we investigated parallel approaches that distribute the workload among several separate memories. One solution that gained traction over the past years is the concept of *Resilient Distributed Dataset (RDDs)*, initially suggested at [18], which are in-memory data structures. Using RDDs, we are able to perform operations on the whole dataset stored in a significantly enlarged distributed memory.

Apache Spark[2] is an implementation of the concept of RDDs. It allows performing coarse-grained operations over voluminous datasets in a distributed manner in parallel. It extends earlier efforts in the area such as Hadoop MapReduce.

In this paper, we introduce a software component "DistLODStats" for statistical evaluation of large RDF datasets, which scales out to clusters of multiple machines. We extend the approach proposed in [5] for computing 32 different statistical criteria for RDF datasets. Our contributions can be summarized as follows:

- We propose an algorithm for computing RDF dataset statistics and implement it using an efficient framework for large-scale, distributed and in-memory computations: Apache Spark.
- We perform an analysis of the complexity of the computational steps and the data exchange between nodes in the cluster.
- We evaluate our approach and demonstrate empirically its superiority over a previous centralized approach.
- We integrated the approach into the SANSA framework, where it is actively maintained and re-uses the community infrastructure (mailing list, issues trackers, website etc.).
- We briefly describe four usage scenarios for DistLODStats.

[2] http://spark.apache.org.

The paper is structured as follows: Our approach for the computation of RDF dataset statistics is detailed in Sect. 2 and evaluated in Sect. 3. Related work on the computation of RDF statistics is discussed in Sect. 5. Finally, we conclude and suggest planned extensions of our approach in Sect. 6.

2 Approach

In this paper, we adopted the 32 statistical criteria proposed in [5]. In contrast to [5], we perform the computation in a large-scale distributed environment using Spark and the concept of RDDs. Instead of processing the input RDF dataset directly, this approach requires the conversion to an RDD that is composed of three elements: *Subject*, *Property* and *Object*. We name such an RDD a *main dataset.*

The statistical criteria proposed in [5] are formalized as a triple (F, D, P) consisting of a filter condition F, a derived dataset D and a post processing operation P. In our approach, we adapt the definition of those elements to be applicable to RDDs.

Definition 1 (Statistical criterion). *A statistical criterion* $\mathcal{C}$ *is a triple* $\mathcal{C} = (F, D, P)$*, where:*

- F *is a SPARQL filter condition.*
- D *is a derived dataset from the main dataset (RDD of triples) after applying* F.
- P *is a post-processing filter operating on the data structure* D.

F acts as a filter operation, which determines whether a specific criterion is matched against a triple in the *main dataset.* D is the result of applying the criterion on the *main dataset.* P is an operation applied to D to (optionally) perform further computational steps. If no extra computation are needed, P just returns exactly the results from the intermediate dataset D.

2.1 Main Dataset Data Structure

The *main dataset* is based on an RDD data structure which is a basic building block of the Spark framework. RDDs are in-memory collections of records that can be operated in parallel on large clusters. By using RDDs, Spark abstracts away the differences of the underlying data sources. RDDs during their lifecycle are kept in-memory, which enables efficient reuse of RDDs during several consequent transformations. Spark provides fault-tolerance by keeping a lineage information (a Directed Acyclic Graph (DAG) of transformations) for each RDD. This way any RDD can be reconstructed in case of node failure by tracing back the lineage. Spark enables full control over the persistence state and partitioning of the RDDs in the cluster. Thus, we can further improve computational efficiency of statistical criteria by planning a suitable storage strategy (i.e. alternating between memory and disk). For example, we can precisely determine which RDDs will be reused, and manage the degree of parallelism by specifying how an RDD is partitioned across the available resources.

Definition 2 (Basic Operations). *All the statistical criteria can be represented in our approach using the following basic operations: map, filter, reduce-by, and group-by. These operations can be formalized as follows:*

- *map* : $I \rightarrow O$, *where* I *is an input RDD and* O *is an output RDD. Map transforms each value from an input RDD into another value, following a specified rule.*
- *filter* : $I \rightarrow O$, *where* I *is an input RDD and* O *is an output RDD, which contains only the elements that satisfy a condition.*
- *reduce* : $I \rightarrow O$, *where* I *is an input RDD of key-value* (K, V) *pairs and* O *is an output RDD of* $(K, list(V))$ *pairs.*
- *group-by* : $(I, F) \rightarrow O$, *where* I *is an input RDD of pairs* $(K, list(V))$, F *is a grouping function (e.g., count, avg), and* O *is an output RDD containing the values in* $list(V)$ *from* I *aggregated using the grouping function.*

2.2 Distributed LODStats Architecture

The computation of statistical criteria is performed as depicted in Fig. 1. Our approach consists of three steps: (1) saving RDF data in scalable storage, (2) parsing and mapping the RDF data into the *main dataset*, and (3) performing statistical criteria evaluation on the *main dataset* and generating results.

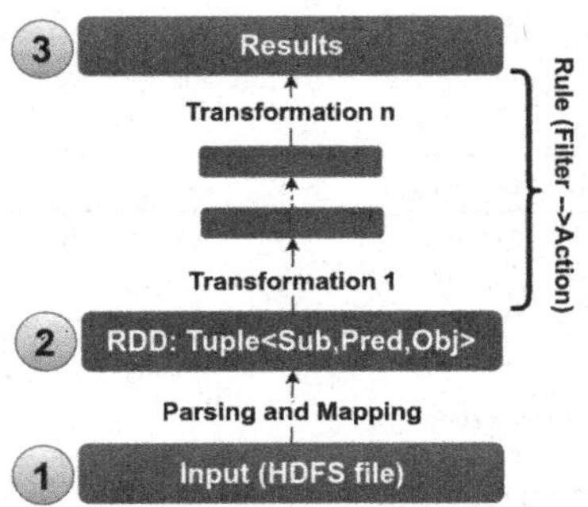

Fig. 1. RDD lineage of a criterion execution.

Fetching the RDF Data (Step 1): RDF data needs first to be loaded into a large-scale storage that Spark can efficiently read from. For this purpose, we use HDFS (Hadoop Distributed File-System)[3]. HDFS is able to accommodate any type of data in its raw format, horizontally scale to arbitrary number of nodes, and replicate data among the cluster nodes for fault tolerance. In such a distributed environment, Spark adopts different data locality strategies to try to perform computations as close to the needed data as possible in HDFS and thus avoid data transfer overhead.

[3] https://hadoop.apache.org/docs/r1.2.1/hdfs_design.html.

Parsing and Mapping RDF into the Main Dataset (Step 2): In the course of Spark execution, data is parsed into triples and loaded into an RDD of the following format: *Triple<Subj,Pred,Obj>* (by using the Spark *map* transformation).

Statistical Criteria Evaluation (Step 3): For each criterion, Spark generates an execution plan, which is composed of one or more of the following Spark transformations: *map*, *filter*, *reduce* and *group-by*.

2.3 Algorithm

The DistLODStats algorithm (see Algorithm 1) constructs the *main dataset* from an RDF file (line 1). Afterwards, the algorithm iterates over the criteria defined inside the DistLODStats framework and evaluates them (lines 4, 6 and 8).

To define a statistical criterion inside the DistLODStats framework, one must specify *filter*, *action*, and *postProc* methods. The evaluation of the criterion then starts first by the *filter* method (line 4) that is used to apply the rule filters of the criterion (Rule Filter in Table 1). Applied on a *main dataset*, this latter will return a new RDD with a subset of the triples. Next, the *action* method is used to apply the criterion's rule action (Rule Action in Table 1). Applied on the filtered RDD, this either computes statistics directly or reorganizes the RDD so statistics can be computed in the next step. At the end, the *postProc* method is used as an optional operation to perform further statistical computations (e.g. average after count or sort).

Algorithm 1. DistLODStats.

```
  input : RDF: an RDF dataset, C: a list of criterion.
1 RDD mainDataset = RDF.toRDD < Triple > ()
2 mainDataset.cache()
3 foreach c ∈ C do
4     triples ← c.filter(mainDataset)
5     triples.cache()
6     triples ← c.action(triples)
7     if c.hasPostProc then
8         triples ← c.postProc(triples)
```

In our work, we make use of Spark caching techniques. Basically, if an RDD is constructed from a data source e.g. file, or through a lineage of RDDs, and then cached, there is no need to construct the RDD again the next time it is needed. We have used two different approaches for caching: (1) caching the *main dataset* entirely (line 2), and (2) caching a derived RDD after applying the criteria *filter* on the *main dataset* (line 5). In the first approach, the RDD is constructed from the RDF source during the first criteria computation, so the next criteria do not need to fetch it again. In the second approach, the RDD resulting from executing the *filter* of one criterion is cached and used by any other criterion sharing the same *filter* pattern.

2.4 Complexity Analysis

The performance of criteria computation depends on two factors mainly:

- **Data Shuffling and Filtering.** In general, the computation can be expensive if there is data movement involved during the distributed execution, which is also known as shuffling. This generally happens when there is a data reduction (in the map-reduce sense). This entails cases like grouping together similar data or applying aggregation functions for SUM, AVG, COUNT, etc. Another factor influencing the performance of criteria computation are filters. The more data is filtered in early stages, the less processing is required in subsequent steps.
- **Data Scanning.** To execute the criterion filter on the same data, data is scanned only once for all criteria. However if data changes state, for example is mapped to another form with new columns added, then another scan of the new state is needed. Finally, if data is shuffled across cluster nodes, then a new scan is needed as well.

Per-criterion Complexity Analysis. Based on the two previous factors, we performed a complexity analysis of each statistical criterion. The results are reported in Table 2. We deem the complexity is mostly linear corresponding to cases where only one or limited number of scans is required. However there are situations where the complexity can increase when there are iterative executions, like the case of data sorting or graph-based computations (e.g. finding cycles or getting the path between two edges).

Below we give an overview of complexity analysis for our most operators used through our approach.

The complexity of $map()$ and $filter()$ itself is linear with respect to the number of input triples. The overall complexity depends on the functions passed to them. Consider an RDD as a single data structure on memory, any other operations (such as map and filter) are linear, or $O(n)$. The subsequent step is to split this RDD between s nodes, the complexity on each node then becomes $O(n/s)$. Let be f a function with complexity $O(f)$, then its complexity will be $O(n/s * O(f))$. As evident from the formula $O(n/s * O(f))$, the runtime increases linearly when the size of RDD increases and *decreases* linearly with the number of nodes in the cluster in case of a function f with with $O(f) = O(1)$.

The complexity of the $sortBy$ operation according to Spark[4] is a sampled $O(n)$, which means only the unique sample keys m (with $m \leq n$) are sorted and lead to a complexity of $O(m * log(m))$ plus the ranges of key sets. Afterwords, the data is shuffled around in $O(n)$ which is costly as sorting needs to be applied internally for the range of keys collected on a given partition p, i.e. $O(p * log(p))$ time is required.

[4] https://github.com/apache/spark/blob/d5b1d5fc80153571c308130833d0c0774de62c92/core/src/main/scala/org/apache/spark/Partitioner.scala#L101.

Table 1. Definition of Spark rules (using Scala notation) per criterion.

	Criterion	Rule (Filter → Action)		Postproc.
1	used classes	p=RDF_TYPE && o.isURI()	→ map(_.o)	–
2	class usage count	p=RDF_TYPE && o.isURI()	→ map(o => (o, 1)).reduceByKey(_ + _)	take(100)
3	classes defined	p=RDF_TYPE && s.isURI()&& (o=RDFS_CLASS||o=OWL_CLASS)	→ map(_.s)	–
4	class hierarchy depth	p=RDFS_SUBCLASS_OF && s.isIRI() && o.isIRI()	→ G += (?s,?o)	depth(G)
5	property usage		→ map(f => (f.pred, 1)) .reduceByKey(_ + _)	take(100)
6	property usage distinct per subj.		→ groupBy(_.subj) .reduceByKey(_ + _)	count
7	property usage distinct per obj.		→ groupBy(_.obj) .reduceByKey(_ + _)	count
8	properties distinct per subj.		→ groupBy(_.subj) .combineByKey(_ + _)	sum/count
9	properties distinct per obj.		→ groupBy(_.obj) .combineByKey(_ + _)	sum/count
10	outdegree		→ map(_.s).map(f => (f, 1)) .combineByKey(_ + _)	sum/count
11	indegree		→ map(_.o).map(f => (f, 1)) .combineByKey(_ + _)	sum/count
12	property hierarchy depth	p=RDFS_SUBPROPERTY_OF && s.isIRI() && o.isIRI()	→ G += (?s,?o)	depth(G)
13	subclass usage	p=RDFS_SUBPROPERTY_OF	→ count()	–
14	triples		→ count()	–
15	entities mentioned		→ map(f=>(s.isURI(), p.isURI(),o.isURI())).count	–
16	distinct entities		→ map(f=>(s.isURI(), p.isURI(),o.isURI())).distinct	–
17	literals	o.isLiteral()	→ count()	–
18	blanks as subj.	s.isBlank()	→ count()	–
19	blanks as obj.	o.isBlank()	→ count()	–
20	datatypes	o.isLiteral()	→ map(o => (o.dataType(), 1)) .reduceByKey(_ + _)	–
21	languages	o.isLiteral()	→ map(o => (o.languageTag(), 1)) .reduceByKey(_ + _)	–
22	average typed string length	o.isLiteral() && obj .getDatatype()=XSD_STRING)	→ count(); len+=o.length()	len/count
23	average untyped string length	o.isLiteral() && o.getDatatype().isEmpty()	→ count(); len+=o.length()	len/count
24	typed subject	p=RDF_TYPE	→ count()	–
25	labeled subject	p=RDFS_LABEL	→ count()	–
26	sameAs	p=OWL_SAME_AS	→ count()	–
27	links	!s.getNS() =(o.getNS())	→ map(f => (s.getNS()+ o.getNS())) .map(f=> (f, 1)).reduceByKey(_ + _)	–
28	max per property {int,float,time}	o.getDatatype()={XSD_INT | XSD_float | XSD_datetime}	→ map(f => (f.p, f.o)) .maxBy(_._2)	–
29	average per property {int,float,time}	o.getDatatype()={XSD_INT | XSD_float | XSD_datetime}	→ m1=>map(_.o).count m2=>map(_.p).count	m1/m2
30	subj. vocabularies		→ map(f => (f.s.getNS())) .map(f => (f, 1)).reduceByKey(_ + _)	–
31	pred. vocabularies		→ map(f => (f.p.getNS())) .map(f => (f, 1)).reduceByKey(_ + _)	–
32	obj. vocabularies		→ map(f => (f.o.getNS())) .map(f => (f, 1)).reduceByKey(_ + _)	–

Table 2. Complexity and data shuffling breakdown by statistical criterion. Notation conventions: n = number of triples; V = number of vertices; E = number of edges.

Criterion	Runtime complexity	Data shuffling and data scanning
(1, 3)	O(n)	Data is filtered locally and returned, i.e. no data exchange is needed
(2, 5)	As sorting is required to retrieve the top 100 results, i.e. the complexity depends on the sorting algorithm used	This operation can be implemented in a map-reduce fashion: classes initially are distributed across the cluster, so calculating their counts requires data to be shuffled and then reduced. The sorting in post-processing requires moving the data. Currently, data is sorted in each node and the union of the datasets is subsequently sorted as well
(6, 7, 8, 9)	O(n)	Following a map-reduce approach, the data is first mapped to <subject,property> pairs and then reduced by subject, so data needs to be shuffled prior to the grouping. De-duplication (distinct) is automatically achieved by the reduce function
(4, 12)	O(V+E)	The best representation of this criterion is a graph where data is already connected, and only linear traversal is required so no data transfer is needed
(10, 11, 20, 21)	O(n)	Following a map-reduce approach, data is first mapped to <subject,1> and then reduced by subject counting the 1s, so data needs to be shuffled prior to the grouping
(13, 14)	O(n)	The count is performed locally and the individual counts are summed up for the cluster, i.e. no data movement is needed
(15)	O(n)	Counting of entities with mentioned s, p and o is done in parallel, so the overall count uses individual counts and sums them. Hence, no data transfer is needed
(16)	O(n)	This is similar to 15, but instead of counting, just returning the triples, so data is saved directly after checking isURI and saved back, i.e. no data is moved
(17, 18, 19, 24, 25, 26, 27, 30, 31, 32)	O(n)	Data is filtered and then counted in each node, the overall count can be obtained by summing up individual counts, so no data movement
(23, 23)	O(n)	The computation requires to project out the objects only and map them to the length of themselves, then the average is computed by summing up the length dividing by the size of each map. The AVG count is done in parallel in each node and then the AVG of all AVGs is a matter of getting single values from each node, so no data movement is needed
(28)	O(n)	Obtaining the maximum per property requires also reducing data distributed in the cluster, so data movement needed
(29)	O(n)	The data here is also reduced by property, so the sum and the count, thus the average, can happen in the same time. Either way, data needs to be moved across the cluster

2.5 Implementation

DistLODStats comprises three main phases depicted in Fig. 2 and explained previously. The output of the *Computing* phase will be the statistical results represented in a human-readable format e.g. VoID, or row data. We expressed the three phases of the 32 criteria using the basic operations defined in Definition 2. Next, those have been mapped to Spark transformations and actions in Table 1, where: *map* is mapped directly to Spark `Map()`, *reduce* is mapped to `groupByKey()`, and *group-by* is mapped to `reduceByKey()`. Exceptions of this general strategy were done for the implementation of the post processing steps

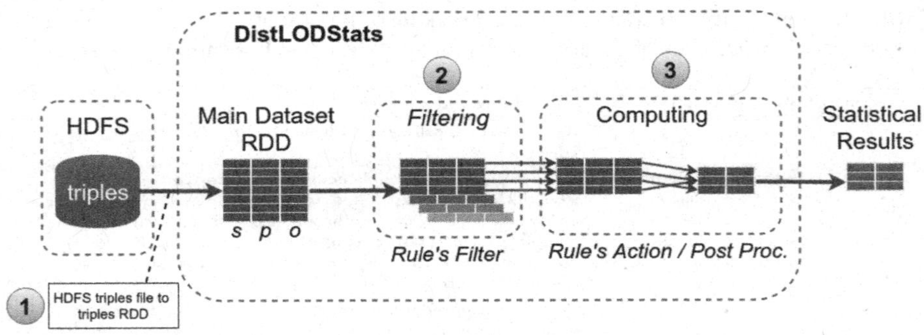

Fig. 2. Overview of DistLODStats's abstract architecture.

of Criteria 4 and 12, where we use a Spark GraphX[5], which is more suitable for this particular case of graph-oriented criterion computation.

Furthermore, we provide a Docker image of the system[6] available under *Apache License 2.0*, integrated within the BDE platform[7] - an open source Big Data Processing Platform allowing users to install numerous big data processing tools and frameworks and create working data flow applications.

We implemented DistLODStats using Spark-2.2.0, Scala 2.11.11 and Java 8. DistLODStats has meanwhile been integrated into SANSA [6,11], an open source[8] *data flow processing engine* for performing distributed computation over large-scale RDF datasets. It provides data distribution, communication, and fault tolerance for manipulating large RDF graphs and applying machine learning algorithms on the data at scale. Via this integration, DistLODStats can also leverage the developer and user community as well as infrastructure behind the SANSA project. This also ensure the sustainability of DistLODStats given that SANSA is backed by several grants until at least 2021.

3 Evaluation

The aim of our evaluation is to see how well our approach can perform against non-distributed approaches as well as analysing the scalability of the distributed approach. In particular, we addressed the following questions: (Q_1): How does the runtime of the algorithm change when more nodes in the cluster are added? (Q_2): How does the algorithm scale to larger datasets? (Q_3): How does the algorithm scale to a larger number of datasets?

In the following, we present our experimental setup including the datasets used. Thereafter, we give an overview of our results, which we subsequently discuss in the final part of this section.

---

[5] https://spark.apache.org/docs/latest/graphx-programming-guide.html.
[6] https://github.com/SANSA-Stack/Spark-RDF-Statistics.
[7] https://github.com/big-data-europe.
[8] https://github.com/SANSA-Stack.

3.1 Experimental Setup

We used one synthetic and two real world datasets for our experiments:

1. We chose the geospatial dataset LinkedGeoData [16] which offers a spatial RDF knowledge base derived from OpenStreetMap.
2. As a cross domain dataset, we selected *DBpedia* [10] (v 3.9). DBpedia is a knowledge base with a large ontology.
3. As a synthetic dataset, we chose to use the Berlin SPARQL Benchmark (BSBM) [2]. It is based on an e-commerce use case which is built around a set of products that are offered by different vendors. The benchmark provides a data generator, which can be used to create sets of connected triples of any particular size.

Properties of these datasets are given in Table 3.

Table 3. Dataset summary information (nt format).

⟶		DBpedia			BSBM		
	LinkedGeoData	en	de	fr	2 GB	20 GB	200 GB
#nr. of triples	1,292,933,812	812,545,486	336,714,883	340,849,556	8,289,484	81,980,472	817,774,057
Size (GB)	191.17	114.4	48.6	49.77	2	20	200

For the evaluation, all data is stored on the same HDFS cluster using Hadoop 2.8.0. All experiments were carried out on a 6 nodes cluster (1 master, 5 workers): Intel(R) Xeon(R) CPU E5-2620 v4 @ 2.10 GHz (32 Cores), 128 GB RAM, 12 TB SATA RAID-5. The experiments on a local mode are all performed on a single instance of the cluster. We ran two centralized versions of LODStats (explained below at Sect. 3.2) for comparison. The machines were connected via a Gigabit network. All experiments were executed three times and the average value is reported.

3.2 Results

We evaluate our approach using the above datasets to compare it against the original LODStats. We carried out two sets of experiments. First, we evaluate the execution time of our distributed approach against the original approach. Second, we evaluate the horizontal scalability via increasing nodes (machines) in the cluster. Results of the experiments are presented in Table 4, Figs. 3, 4 and 5.

Distributed Processing on Large-Scale Datasets

To address Q_1, we started our experiments by evaluating the *speedup* gained by adopting a distributed implementation of LODStats criteria using our approach, and compare it against the original centralized version. We run the experiments on four datasets ($DBpedia_{en}$, $DBpedia_{de}$, $DBpedia_{fr}$, and *LinkedGeoData*) in

Table 4. Distributed Processing on Large-Scale Datasets.

⟶	Runtime (h) (mean/std)				
	LODStats		DistLODStats		
	(a) files	(b) bigfile	(c) local	(d) cluster	(e) speedup ratio
LinkedGeoData	n/a	n/a	36.65/0.13	**4.37**/0.15	7.4x
$M^{en}_{DBpedia}$	**24.63**/0.57	fail	25.34/0.11	**2.97**/0.08	7.6x
$M^{de}_{DBpedia}$	n/a	n/a	10.34/0.06	**1.2**/0.0	7.3x
$M^{fr}_{DBpedia}$	n/a	n/a	10.49/0.09	**1.27**/0.04	7.3x

a local environment on a single instance with two configurations: (1) files of the dataset are considered separately, and (2) one big file–all files concatenated.

Table 4 shows the performance of two algorithms applied to the four datasets. The column LODStats $^{(a)}$ reports on the performance of LODStats on files separately (considering each file as a sequence of execution), the next columns LODStats$^{(b)}$ reports on the performance of LODStats using a single big file by concatenating each file, and the last columns reports on the performance of DistLODStats on the same case as previously i.e. the performance for one big dataset in local mode (*c*) and cluster mode (*d*). We observe that the execution in DistLODStats$^{(c),(d)}$ finishes with all the datasets (see Fig. 3). However, for LODStats$^{(a),(b)}$ the execution often fails at different stages of the execution. In particular, *n/a* indicates parser exceptions and *fail* out of memory exceptions. The only case where the execution finishes and actually slightly outperforms DistLODStats$^{(c)}$ on a single node is executing LODStats on the dataset $DBpedia^{en}$ split into files (25.34 h for DistLODStats$^{(c)}$ vs 24.63 h in LODStats$^{(a)}$). This is because the DistLODStats$^{(c)}$ considers the input dataset as a big file instead of evaluation it on each file separately. LODStats streams the criteria one by one, so having a large dataset streamed that way would lead to very high processing times. However, with small data as input, the processing can finish in short amount of time, but the results can be very inaccurate.

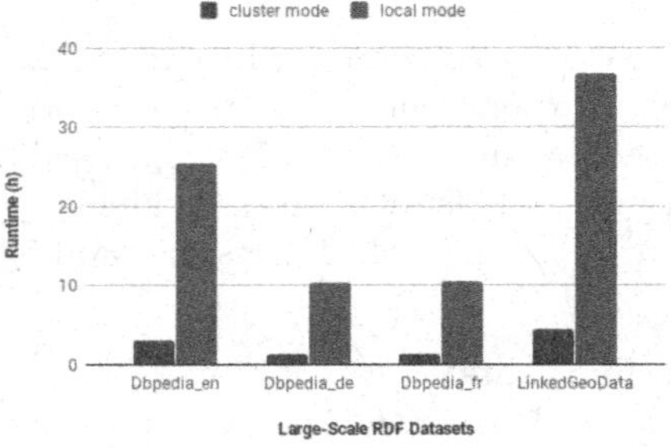

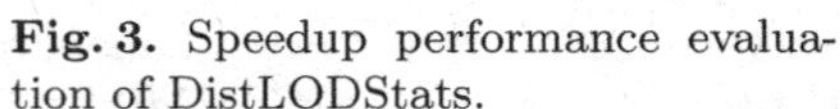
Fig. 3. Speedup performance evaluation of DistLODStats.

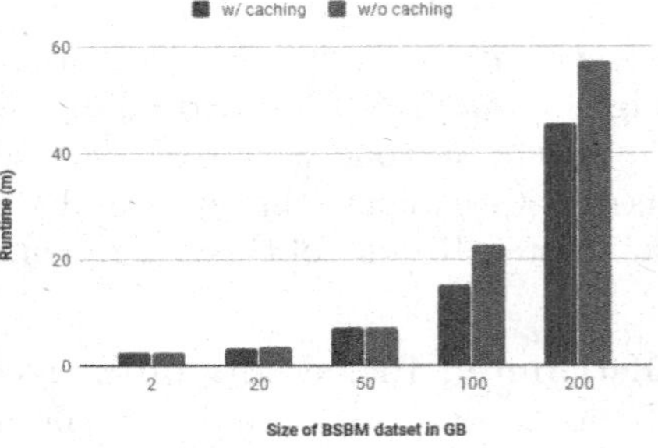

Fig. 4. Sizeup performance evaluation of DistLODStats.

Figure 3 shows the speedup performance evaluation for large-scale RDF Datasets for DistLODStats on local mode and cluster mode, respectively. All results illustrate consistent improvement for each dataset when running on a cluster. The maximum speedup is 7.6x and the geometric mean of the speedup is 7.4x.

For example, on $DBpedia_{en}$, the time on cluster mode is about 2.97 h which is 7.6 times faster than evaluating DistLODStats on local mode (about 25.34 h). The reason why the time spent on local mode extremely decreases is that the size of the working directory of worker processes is too large and Spark uses threads for distributing the tasks.

Scalability

Sizeup Scalability. To measure the performance of *size-up* i.e. scalability of our approach, we run experiments on three different sizes. This analysis keeps the number of nodes in a cluster constant, we fix the number of workers (nodes) to 5 and grow the size of datasets to measure whether a given algorithm can deal with larger datasets. Since real-world datasets are considered to be unique in the size and also on other aspects e.g. number of unique terms, we chose the BSBM benchmark tool to generate artificial datasets of different sizes. We started by generating a dataset of 2 GB. Then we iteratively increased the size of datasets by one order of magnitude.

On each dataset, we ran the distributed algorithm and the runtime is reported on Fig. 4. The x-axis is a generated BSBM dataset per each order of 10x magnitude.

By comparing the runtime (see Fig. 4), we note that the execution time cost grows linearly and is near-constant when the size of the dataset increases. As expected, it stays near-constant as long as the data fits in memory. This demonstrates one of the advantages of utilizing an in-memory approach in performing the statistics computation. The overall time spent in data read/write and network communication found in disk-based approaches is no present in distributed in-memory computing. The performance only starts to degrade when substantial amounts of data need to be written to disk due to memory overflows. The results show scalability of our algorithm in context of sizeup, which answers question Q_2.

Node Scalability. In order to measure node scalability, we use variations of the number of the workers on our cluster. The number of workers varies from 1, 2, 3 and 4 to 5.

Let T_N be the time required to complete the task on N workers. The speedup S is the ratio $S = \frac{T_L}{T_N}$, where T_L is the execution time of the algorithm on local mode. Efficiency measures the processing power being used (i.e speedup per worker). It is defined as the time to run the algorithm on N workers compared to the time to run algorithm on local mode: $E = \frac{S}{N} = \frac{T_L}{NT_N}$.

Figure 5 shows the speedup for $BSBM_{50GB}$. We can see that as the number of workers increases, the execution time cost is super-linear. As depicted in Fig. 6, the speedup performance trend is consistent as the number of workers increases.

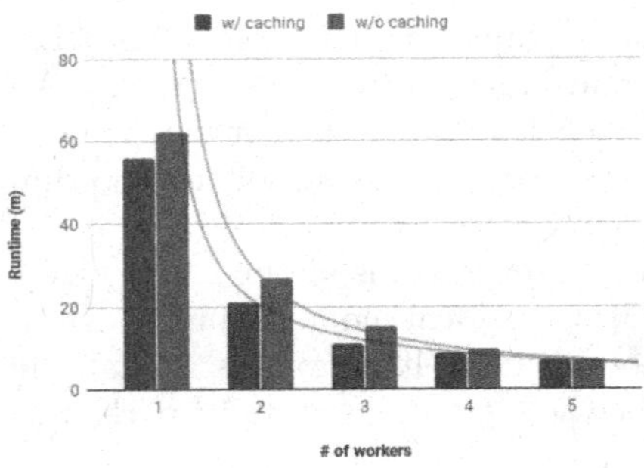

Fig. 5. Scalability performance evaluation on DistLODStats.

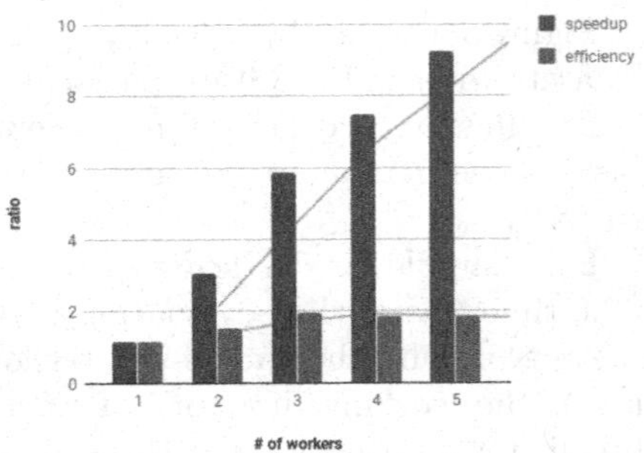

Fig. 6. Speedup ratio and efficiency of DistLODStats.

In contrast, as the number of workers was increased from 1 to 5, efficiency increased only up to the 4th worker for $BSBM_{50GB}$ dataset. This implies that the tasks generated from the given dataset were covered with almost 4 nodes. The results imply that DistLODStats can achieve near linear or even super linear scalability in performance, which answers question Q_3.

Breakdown by Criterion

Now we analyze the overall runtime of criteria execution. Fig. 7 reports on the runtime of each criterion on both $BSBM_{20GB}$ and $BSBM_{200GB}$ datasets.

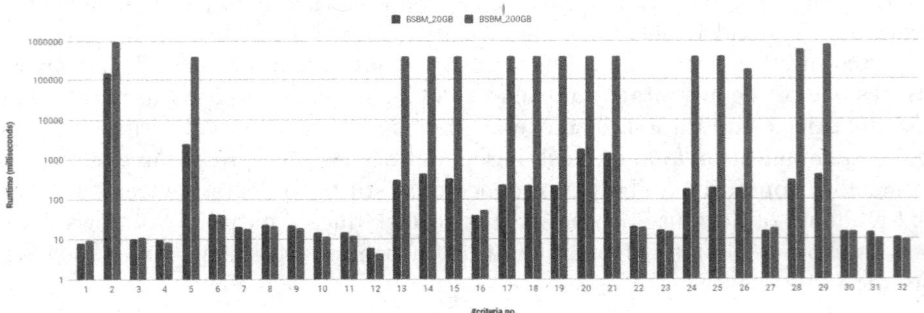

Fig. 7. Overall breakdown by criterion analysis (log scale).

Discussion. DistLODStats consists of 32 predefined criteria most of which have a runtime complexity of $O(n)$ where n is the number of input triples. The breakdown for BSBM with two instances is shown in Fig. 7. The results obtained confirm to a large extent the pre-analysis made in Subsect. 2.4. The execution is longer when there is data movement in the cluster compared to when data is processed without movement e.g. Criterion 2, 3 and 4. There are some criteria that are quite efficient to compute even with data movement e.g. 22, 23. This is because data is largely filtered before the movement. Criterion 2 and 28 are the most expensive ones in terms of time of execution. This is most probably because of the sorting and maximum algorithm used by Spark. Criteria 20 and 21 are particularly expensive because of the extra overhead caused by extracting the data type and language for each particular object of type Literal. Criteria like 14 and 15 do not require movement of data, but yet are inefficient in execution. This is because the data is not filtered previously. The last three criteria do include data movement but are among the most efficient ones. This is because the low number of namespaces the chosen datasets have.

Overall, the evaluation study conducted demonstrates that parallel and distributed computation of the different statistical values is scalable, i.e. the execution finishes in reasonable time relative to the high volume of datasets.

4 Use Cases

DistLODStats is a generic tool for horizontally scalable statistics evaluation. We are aware of the following major users of the tool:

Comprehensive Statistics – LODStats. LODStats[9] is a project, which has crawled RDF data from metadata portals for the past seven years. It interacts with the CKAN dataset metadata registry to obtain a comprehensive picture of the current state of the Data Web. The drawback of the previous engine for LODStats is its inability to horizontally scale out, which naturally limited its scope to small and medium size datasets. For this reason, statistical criteria for several large-scale datasets were not reflected in the project website. Meanwhile, DistLODStats is used as underlying engine overcoming the previous limitations and generating statistical descriptions, including e.g. VoID, for large parts of the Linked Open Data Cloud.

Big Data Platform – BDE. Big Data Europe (BDE)[10] [1] is an open source big data processing platform allowing users to deploy Big Data processing tools and frameworks. Those tools and frameworks usually generate large amounts of log data. DistLODStats is used for computing statistics over those logs within the BDE platform. BDE uses the Mu Swarm Logger service[11] for detecting docker events and convert their representation to RDF. In order to generate visualisations of log statistics, BDE then calls DistLODStats from SANSA-Notebooks [6].

Blockchain – Alethio Use Case. Alethio is building an Ethereum analytics platform that strives to provide transparency over the transaction pool of the Ethereum p2p network. Their 5 billion triple data set contains large scale blockchain transaction data modelled as RDF according to the structure of the Ethereum ontology[12]. Alethio is using SANSA in general and DistLODStats

[9] http://lodstats.aksw.org/.
[10] https://github.com/big-data-europe.
[11] https://github.com/big-data-europe/mu-swarm-logger-service.
[12] https://github.com/ConsenSys/EthOn.

specifically in order to perform large-scale batch analytics, e.g. computing the asset turnover for sets of accounts, computing attack pattern frequencies and Opcode usage statistics. DistLODStats was run on a 100 node cluster with 400 cores to compute those statistics.

LOD Summaries – ABSTAT. ABSTAT[13][14] is a framework that aims to provide a better understanding of linked data sets. It implements an ontology-driven linked data summarization approach. DistLODStats is used for data set summarisation of large-scale RDF datasets in this context.

5 Related Work

In this section, we provide an overview of related work regarding RDF dataset statistics calculation. To the best of our knowledge, all but one existing approaches use small to medium scale datasets and do not horizontally scale. A dataset is large-scale w.r.t. a particular task in the scope of this article if the main memory on commodity hardware is insufficient to perform the task (without swapping to disk). We mention here, for example RDF_{Pro} [3], which offers a suite of stream-oriented, highly optimized processors for common tasks, such as data filtering, RDFS inference, smushing, as well as statistics extraction. The second related approach we are aware of is Aether [12], which is an application for generating, viewing and comparing extended VoID statistical descriptions of RDF datasets. The tool is useful, for example, in getting to know a newly encountered dataset, in comparing the different versions of a dataset, and in detecting outliers and errors. Luzzu [4] is a quality assessment framework for linked data. Its Quality Metric Language (LQML), is a domain specific language (DSL) that enables knowledge engineers to declaratively define quality metrics whose definitions can be understood more easily. LQML offers notations, abstractions and expressive power, focusing on the representation of quality metrics. However, only one work we came across that provided a distributed framework for RDF statistics computation: LODOP [8]. LODOP adopts a MapReduce approach for computing, optimizing, and benchmarking data profiling techniques. It uses Apache Pig as the underlying computation engine (Hadoop-based). LODOP implements 15 data profiling tasks comparing to 32 in our work. Because of the usage of MapReduce, the framework has a significant drawback: materialization of intermediate results between Map and Reduce and between two subsequent jobs is done on disk. DistLODStats does not use the disk-based MapReduce framework (Hadoop), but rather bases its computation mainly in-memory, so runtime performance is presumably better [15]. Unfortunately, we were unable to run LODOP for comparison. This is due to technical problems encountered, despite the very significant effort we devoted to deploy and run it. To the best of our knowledge, DistLODStats is the first software component for in-memory distributed computation of RDF dataset statistics.

[13] http://abstat.disco.unimib.it/.

6 Conclusions and Future Work

For obtaining an overview over the Web of Data as well as evaluating the quality of individual datasets, it is important to gather statistical information describing characteristics of the internal structure of datasets. However, this process is both data-intensive and computing-intensive and it is a challenge to develop fast and efficient algorithms that can handle large scale RDF datasets.

In this paper, we presented DistLODStats, a novel software component for distributed in-memory computation of RDF Datasets statistics implemented using the Spark framework. DistLODStats is maintained and has an active community due to its integration in SANSA. Our definition of statistical criteria provides a framework reducing the implementation effort required for adding further statistical criteria. We showed that our approach improves upon a previous centralized approach we compare against. Since Spark RDDs are designed to scale horizontally, cluster sizes can be adapted to dataset sizes accordingly. Although we achieved reasonable results in terms of scalability, we plan to further improve time efficiency by persisting the data to an even higher extend more in memory and perform load balancing.

Acknowledgment. This work was partly supported by the EU Horizon2020 projects BigDataEurope (GA no. 644564), QROWD (GA no. 723088), WDAqua (GA no. 642795) and BigDataOcean (GA no. 732310).

References

1. Auer, S., et al.: The BigDataEurope platform - supporting the variety dimension of big data. In: 17th International Conference on Web Engineering (2017)
2. Bizer, C., Schultz, A.: The berlin SPARQL benchmark. Int. J. Semant. Web Inf. Syst. **5**, 1–24 (2009)
3. Corcoglioniti, F., Rospocher, M., Mostarda, M., Amadori, M.: Processing billions of RDF triples on a single machine using streaming and sorting. In: Proceedings of the 30th Annual ACM Symposium on Applied Computing, pp. 368–375. ACM (2015)
4. Debattista, J., Auer, S., Lange, C.: Luzzu – a methodology and framework for linked data quality assessment. J. Data Inf. Qual. (JDIQ) **8**(1), 4 (2016)
5. Auer, S., Demter, J., Martin, M., Lehmann, J.: LODStats — an extensible framework for high-performance dataset analytics. In: ten Teije A., et al. (eds.) Knowledge Engineering and Knowledge Management, EKAW 2012, Lecture Notes in Computer Science, vol. 7603, pp. 353–362. Springer, Heidelberg (2012). https://doi.org/10.1007/978-3-642-33876-2_31
6. Ermilov, I., et al.: The Tale of Sansa Spark. In: Proceedings of 16th International Semantic Web Conference, Poster and Demos (2017)
7. Ermilov, I., Martin, M., Lehmann, J., Auer, S.: Linked open data statistics: collection and exploitation. In: Proceedings of the 4th Conference on Knowledge Engineering and Semantic Web (2013)
8. Forchhammer, B., Jentzsch, A., Naumann, F.: LODOP - multi-query optimization for linked data profiling queries. In: International Workshop on Dataset PROFIling and fEderated Search for Linked Data (PROFILES), Heraklion, Greece (2014)

9. Langegger, A., Wöß, W.: RDFstats - an extensible RDF statistics generator and library. In: DEXA Workshops, pp. 79–83. IEEE Computer Society (2009)
10. Lehmann, J., et al.: DBpedia - a large-scale, multilingual knowledge base extracted from wikipedia. Semant. Web J. **6**(2), 167–195 (2015)
11. Lehmann, J., et al.: Distributed semantic analytics using the SANSA stack. In: Proceedings of 16th International Semantic Web Conference (2017)
12. Mäkelä, E.: Aether – generating and viewing extended VoID statistical descriptions of RDF datasets. In: Presutti, V., Blomqvist, E., Troncy, R., Sack, H., Papadakis, I., Tordai, A. (eds.) ESWC 2014. LNCS, vol. 8798, pp. 429–433. Springer, Cham (2014). https://doi.org/10.1007/978-3-319-11955-7_61
13. Ngonga Ngomo, A.-C., Auer, S., Lehmann, J., Zaveri, A.: Introduction to linked data and its lifecycle on the web. In: Reasoning Web (2014)
14. Palmonari, M., Rula, A., Porrini, R., Maurino, A., Spahiu, B., Ferme, V.: ABSTAT: linked data summaries with ABstraction and STATistics. In: Gandon, F., Guéret, C., Villata, S., Breslin, J., Faron-Zucker, C., Zimmermann, A. (eds.) ESWC 2015. LNCS, vol. 9341, pp. 128–132. Springer, Cham (2015). https://doi.org/10.1007/978-3-319-25639-9_25
15. Shi, J., et al.: Clash of the titans: mapreduce vs. spark for large scale data analytics. Proc. VLDB Endow. **8**(13), 2110–2121 (2015)
16. Stadler, C., Lehmann, J., Höffner, K., Auer, S.: LinkedGeoData: a core for a web of spatial open data. Semant. Web J. **3**(4), 333–354 (2012)
17. Vandenbussche, P.-Y., Atemezing, G.A., Poveda-Villalón, M., Vatant, B.: Linked open vocabularies (LOV): a gateway to reusable semantic vocabularies on the web. Semant. Web, 1–16 (2015). Preprint(Preprint)
18. Zaharia, M., et al.: Resilient distributed datasets: a fault-tolerant abstraction for in-memory cluster computing. In: Proceedings of the 9th USENIX Conference on Networked Systems Design and Implementation (2012)

Knowledge Integration for Disease Characterization: A Breast Cancer Example

Oshani Seneviratne(✉), Sabbir M. Rashid(✉), Shruthi Chari(✉), Jamie P. McCusker(✉), Kristin P. Bennett(✉), James A. Hendler(✉), and Deborah L. McGuinness(✉)

Rensselaer Polytechnic Institute, Troy, NY 12080, USA
{senevo,rashis2,charis,mccusj2,bennek}@rpi.edu,
{hendler,dlm}@cs.rpi.edu

Abstract. With the rapid advancements in cancer research, the information that is useful for characterizing disease, staging tumors, and creating treatment and survivorship plans has been changing at a pace that creates challenges when physicians try to remain current. One example involves increasing usage of biomarkers when characterizing the pathologic prognostic stage of a breast tumor. We present our semantic technology approach to support cancer characterization and demonstrate it in our end-to-end prototype system that collects the newest breast cancer staging criteria from authoritative oncology manuals to construct an ontology for breast cancer. Using a tool we developed that utilizes this ontology, physician-facing applications can be used to quickly stage a new patient to support identifying risks, treatment options, and monitoring plans based on authoritative and best practice guidelines. Physicians can also re-stage existing patients or patient populations, allowing them to find patients whose stage has changed in a given patient cohort. As new guidelines emerge, using our proposed mechanism, which is grounded by semantic technologies for ingesting new data from staging manuals, we have created an enriched cancer staging ontology that integrates relevant data from several sources with very little human intervention.

Keywords: Ontologies · Knowledge integration
Deductive inference · Automatic extraction
Cancer characterization · Cancer staging guidelines

Resource: https://cancer-staging-ontology.github.io.

1 Introduction

Our goal is to improve health knowledge infrastructures by use of semantic technologies to support data integration in an environment of quickly evolving medical information. We present a prototype system that uses semantic technologies to integrate medical information relevant for characterizing breast cancer. Our

D. Vrandečić et al. (Eds.): ISWC 2018, LNCS 11137, pp. 223–238, 2018.
https://doi.org/10.1007/978-3-030-00668-6_14

system can automatically parse the guidelines from the cancer staging manual and construct OWL axioms [3] that can be used to infer recommended personalized options for patients. These inferences are made using the data related to the treatment and monitoring of the disease that are represented in RDF [19].

1.1 Background

The authoritative staging system is published by the American Joint Committee on Cancer (AJCC). As the inaugural authors of the cancer staging manuals have stated in [2]:

> "Staging of cancer is not an exact science. As new information becomes available about etiology and various methods of diagnosis and treatment, the classification and staging of cancer will change."

Since the inception of the cancer staging manual in 1977, there have been eight editions. The latest AJCC Cancer Staging Manual, Eighth Edition (AJCC 8th Edition) [1], makes a tangible effort to incorporate biologic and molecular markers to create a more contemporary personalized approach using pathologic prognostic staging. This has increased the complexity of the staging criteria.

In order to stage tumors, many physicians rely on cancer staging manuals, or compact 'cheat sheets' derived from the contents of these manuals. However, since the new staging guideline incorporates additional data streams, the physicians have to traverse increasingly tedious decision trees.

In terms of discovering relevant treatment and monitoring options based on the stage, or more broadly the characterization of the disease, physicians usually refer to the National Comprehensive Cancer Network (NCCN) Guidelines [18]. Navigating these guidelines also is often a tedious process. Furthermore, in order to keep up with the growing and rapidly changing body of knowledge, physicians may also use subscription services such as *UpToDate*[1], which has articles on many of the state of the art topics in medicine, including cancer. However, physicians may not have enough time to sift through these articles and ascertain the information that is relevant for the case at hand.

1.2 Related Work

Initial work related to an ontology that captured cancer staging information is available in Massicano et al. [21] for the AJCC 6th edition [27]. Boeker et al. [5] have also created an ontology for the same guideline in which they focus on tumors in the colon and rectum. The biggest difference between the previous ontologies and our cancer staging ontology is the inclusion of additional biomarkers as per the AJCC 8th edition staging criteria, which were not available in the previous staging editions. These biomarkers used in the new edition significantly increased the complexity of the criteria required to stage a tumor. Additionally, the previous ontologies do not model real-world representations of the tumor

[1] *UpToDate* - a clinical decision support resource: http://www.uptodate.com.

concepts in their axioms nor specify those in the comments. In those ontologies, the tumor is of a certain *rdf:type* *T* (class representing severity of tumor size: *T0-T4*), *N* (class representing the severity of the spread to the lymph nodes: *N0-N3*) and *M* (class representing whether the cancer has metastasized: *M0-M1*). In the real world, representation for tumor size T has a value in millimeters (or centimeters) that is used to derive a T value of 0–4. Similarly, N has a value for the number of lymph nodes affected that is used to derive the severity rating from 0–3. Thus, their approach of representing the cancer characterization using just the *rdf:type* to the corresponding T, N, M classes is problematic because when any of these derived classifications change as per a new guideline, the RDF graph has to change with it, representing the new classification. In our knowledge graph, these values are encoded as attributes to give them temporal extent, avoiding this problem.

Furthermore, in addition to including classes for all cancer stages for the respective guideline, we also map the breast cancer terms to community-accepted terms from the National Cancer Institute thesaurus (NCIt) [12], and incorporate recommended tests and treatment plans from the openly reusable Clinical Interpretations of Variants in Cancer (CIViC) [14] data that can be used to provide stage specific recommendations. Furthermore, our ontology includes terms that are not included in NCIt or AJCC, such as more specific subclasses of tumor characteristics (*T1*, *T1_as*, *T1_am*, *T1NOS*, etc.) that are available in the Surveillance, Epidemiology, and End Results (SEER) dataset [16].

1.3 Overview of the Knowledge Integration System for Breast Cancer Characterization

We developed our prototype primarily to address the issue of rapidly changing information in characterizing disease, specifically breast cancer. Since manual

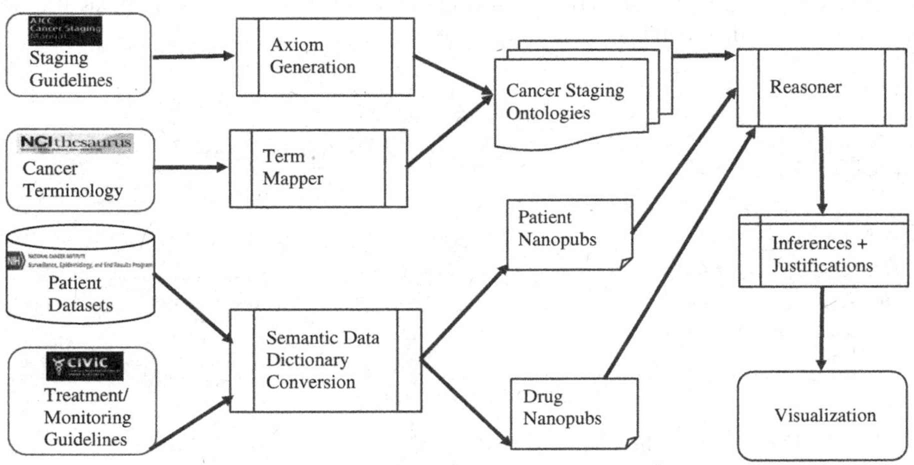

Fig. 1. Knowledge integration architecture for breast cancer characterization

look-up of the breast cancer staging criteria is prone to human error, our system was designed to support automated navigation through the tedious decision trees to minimize any look up errors. We also provide support for integration of data from various sources. Figure 1 depicts the overall knowledge integration architecture that will be explained in detail in the following sections.

2 Development of the Cancer Staging Ontologies

As mentioned in related work (Sect. 1.2) the last known staging ontologies were created for the AJCC 6th edition. There are no ontologies for the AJCC 7th and 8th editions to the best of our knowledge. We describe the process we followed when constructing these new staging ontologies, accounting for the complexity of the data streams in the new guideline.

2.1 Cancer Staging Terms

The previous breast cancer staging guidelines (i.e. AJCC 7th edition [10] and earlier) only considered anatomical features such as the size of the tumor (T), the number of lymph nodes affected (N), and whether the cancer has metastasized (M). Additionally considered in the new staging guidelines [1] are biomarkers including human epidermal growth factor receptor 2 (HER2), estrogen (ER) and progesterone (PR) receptor statuses and tumor grade (Grade). This addition has led to a more complex set of rules for staging criteria using the classes corresponding to the specific stages in the AJCC 7th and 8th editions that we incorporated into our Cancer Staging Terms (CST) ontology.

Figure 2 depicts the 8th staging edition staging class hierarchy. Each stage class includes the properties *cst:hasRecommendedTest*, *cst:hasTreatmentOption*, and *rdfs:subClassOf* assertions where applicable. We added the *rdfs:comments* to better describe the concepts in the ontology based on the descriptions available in the medical literature and to support explanation.

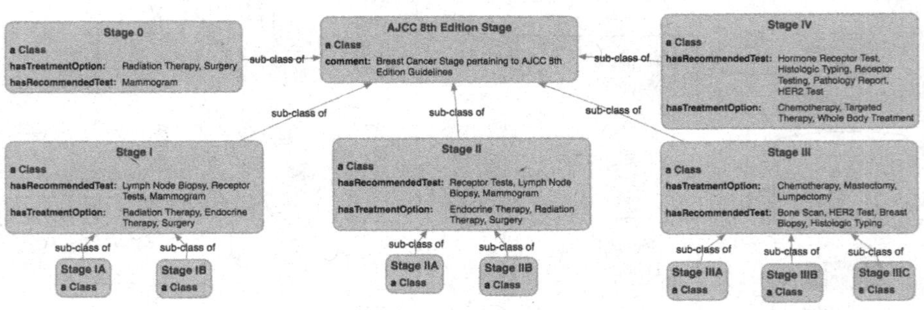

Fig. 2. Stage hierarchy of the AJCC cancer staging 8th Edition

Furthermore, in the AJCC staging manuals, and in the data we ingested from other sources, we observed different subclasses for the broader classification of the features considered, i.e. T, N, M, HER2, ER, PR, and Grade in the ontology. Figure 3 depicts a small subset of these classifications, which includes various Tumor size (T) classes. Similarly, there are other subclass assertions, and mappings to the NCIt classes for N, M, HER2, ER, PR, and Grade. We augmented these classes with the *rdfs:comment*, *rdfs:label*s, and the *owl:equivalentClass* obtained from NCIt [12]. These *rdfs:comment*s and *rdfs:label*s are used to explain a particular conclusion resulting from the application of a reasoner utilizing the ontology explained in detail in Sect. 5.

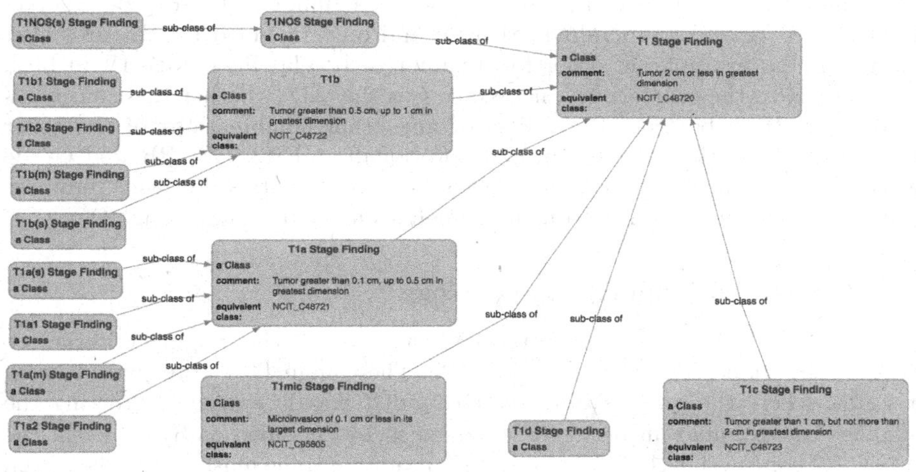

Fig. 3. Hierarchy of the Tumor Size (T) Classes in Our Integrated Ontology. *We created the tumor size classes in the left two columns to support our integration and reasoning. These classes reflect content in SEER* [16] *and not AJCC* [1].

2.2 Translating Staging Criteria into Structured Mappings

We extracted 19 criteria from AJCC 7th edition, and 407 criteria for clinical prognostic stage grouping from AJCC 8th edition. A script was necessary for the 8th edition since the complexity of the staging guideline has increased with the addition of the biomarkers. Table 1 illustrates the number of different combinations for staging criteria observed in the two staging guidelines. The non-linear expansion of the number of combinations is due to the complex interaction of the additional biomarkers HER2, ER, PR and Tumor Grade.

For each of the two staging guidelines, we created corresponding 'map files' to represent the conditions required for a tumor to be classified a certain stage from 0–IV. We created 18 such map files for the two guidelines (AJCC 7th and 8th editions), with 9 map files representing each stage from 0, IA, IB, IIA, IIB,

Table 1. Number of feature combinations for determining stage

Stage	0	IA	IB	IIA	IIB	IIIA	IIIB	IIIC	IV
AJCC 7th **Edition**	1	1	2	3	2	5	3	1	1
AJCC 8th **Edition**	1	57	33	77	39	82	92	25	1

IIIA, IIIB, IIIC, and IV. Each line in the map file in a 7th edition stage contains the set of possible T, N and M combinations that would result in that stage being assigned to the tumor. The map files for the 8th edition followed a similar form, but also included the additional features HER2, ER, PR and Grade.

If any of the features can be ***any*** value for a tumor to be staged, the map file omitted those corresponding features, and only included the features that mattered. For example, in order for a tumor to be classified stage IV in both the guidelines, the only criteria necessary was the 'M' (whether the cancer has metastasized) to be true. Regardless of any other combinations of the other features T, N in the 7th edition, and additionally HER2, ER, PR and Grade in the 8th edition, the tumor will always be classified stage IV, thus only one combination is available for both the guidelines for determining stage IV.

2.3 Structured Mappings to Ontology

In order to automatically generate OWL axioms for the staging criteria, we utilized the map files created in Sect. 2.2. These map files were parsed using a script, where the property *owl:intersectionOf* was leveraged in creating the axioms. For example, in order for a tumor to be classified as Stage IA in the AJCC 7th edition (i.e. *AJCC7_Stage_IA*), a tumor profile must satisfy the axiom in Listing 2.1. However, for the same tumor to be classified as Stage IA in the AJCC 8th edition (i.e. *AJCC8_Stage_IA*), only one of the 57 axioms must be satisfied (Listing 2.2 demonstrates one such axiom). We developed the breast cancer staging ontology for the AJCC 7th edition (BCS7) and the ontology for the AJCC 8th edition (BCS8) using the above-mentioned procedure to codify all the axioms related to classifying tumors.

```
@prefix cst: <http://idea.tw.rpi.edu/cancer_staging_terms.owl#> .
[] a owl:Class; rdfs:subClassOf cst:AJCC7_Stage_IA;
    owl:intersectionOf ( cst:T1 cst:N0 cst:M0 ).
```

Listing 2.1. The Only OWL Axiom for a Tumor to be Classified as *Stage IA* in the AJCC 7th Edition

```
[] a owl:Class; rdfs:subClassOf cst:AJCC8_Stage_IA;
    owl:intersectionOf ( cst:T1 cst:N0 cst:M0 cst:Grade1
    cst:HER2_Neg cst:ER_Neg cst:PR_Pos ).
...
```

Listing 2.2. One of the Many OWL Axioms for a Tumor to be Classified as *Stage IA* in the AJCC 8th Edition

3 Integrated Cancer Knowledge Graph

We chose RDF [19] as the underlying knowledge representation model to handle heterogeneous data while providing interoperable representations. The CST, BCS7 and BCS8 developed in Sect. 2, are part of our Integrated Cancer Knowledge Graph. Additionally, we extracted data from crowd sourced, open source, reusable cancer resources to augment the knowledge graph with treatment and monitoring options based on the stage inferred using the cancer staging ontologies developed.

3.1 Integrating Data from Other Cancer Data Sources

There are many services that provide vast collections of data that may be useful and relevant in a cancer knowledge graph. Some of these services include CIViC [14], OncoKB [6], MyCancerGenome [23] and Integrative Onco Genomics [13].

As a proof of concept, we incorporated data from CIViC [14], which has crowd sourced, open source and reusable data that identifies drugs that may interact with biomarkers. Additionally from the CIViC data dumps, related articles and their trust ratings captured in the form of provenance were also incorporated. Nanopublications [24] were created for this data using a semantic annotation approach called Semantic Data Dictionaries (SDDs) [25], which simplifies the ability to express the full semantics of a dataset. The SDD process [25] allowed us to link the data concepts with each other, as well as reference implicit entities in the data, and link the corresponding data elements as characteristics of these entities. The concepts contained in the data records needed to be mapped with related terms from domain specific ontologies such as NCI thesaurus (NCIt) [12] and Uniprot [7], as well as general purpose ontologies such as Semanticscience Integrated Ontology (SIO) [9].

The dictionary mapping table of the SDD that was used for CIViC maps 14 different features in the dataset such as *Drugs*, *Status*, *Evidence ID*, *Evidence Level*, *Gene*, *Variant*, *Disease* and the *Trust Rating* to the respective classes available in SIO and NCIt. These classes are used for type assignment when creating a knowledge graph from the data. For example, the *Drug* column in the dataset is mapped to *sio:Drug*, *Gene* column to *sio:Gene*, etc. Furthermore, the classes specified in the *attributeOf*, *inRelationTo* and *wasDerivedFrom* are used in semantically modeling relationships in the generated nanopublications.

A codebook was used to map over 200 specific values found in the CIViC data to the corresponding terms in existing ontologies. The *Disease* types that were found in CIViC were mapped to concepts in the Human Disease Ontology (DOID) [26], Experimental Factor Ontology (EFO) [20], and NCIt [12]. For example, the concept for the *HER2-receptor Positive Breast Cancer* in our knowledge graph is mapped to concepts such as *efo:1000294*, *doid:0060079*, *ncit:C53556*[2]. Similarly, drugs were mapped to concepts from the Drug Bank [29], the Drug Ontology (DRON) [15], and/or Chemical Entities of Biological Interest (ChEBI) Ontology [8], genes were mapped to terms in Uniprot [7], etc.

[2] These specific mappings were looked up using Ontobee (http://www.ontobee.org).

4 Converting Patient Records to RDF

In order to evaluate our cancer staging ontology, we needed cancer patient data that included the characteristics of the tumor in RDF, ideally in the nanopublications format [24]. The SEER datasets [16] contained the desired data which included demographic information, tumor stage as per the older AJCC 6th edition, and the survival status of patients treated from 1980–2012. We browsed the datasets using the statistical software, SEER*Stat[3], and downloaded a subset of the data to create the patient nanopublications.

Due to the anonymity and privacy constraints on the medical data, the SEER patient records lacked any identifying information like the patient name. However, for our use case, i.e. to model a patient, we needed an identifying attribute, so we annotated the patient records with names from Python's Natural Language Toolkit (NLTK) name corpus [4] to assign a name to each patient record. The patient data was then fed through the SDD pipeline [25] to generate knowledge graphs that included nanopublications that captured the attributes of a patient and where that information came from within an assertion in the patient graph. Utilization of the SDD approach allowed us to semantically represent relationships such as the age of patient at diagnosis (i.e. the attribute *sio:Age* as *sio:attributeOf* the patient which *sio:existsAt* the time of diagnosis). We mapped 29 such features for a patient record in the SEER dataset in the data dictionary, and the codebook contains 100+ mappings to terms in NCIt.

Since some of the values occurring in SEER did not match existing terms in the ontologies, we leveraged our Cancer Staging Terms (CST) ontology, introduced in Sect. 2. A codebook mapping corresponding to SEER was defined that would generate standard values and map commonly occurring terms to their ontology equivalents. As the structured format of the data is insufficient to capture the implicit linkages within the attributes of the dataset, a SEER dictionary mapping was defined that established the entity-attribute mappings to facilitate the conversion of the data to the named graphs with nanopublications.

5 Inference Agent

We developed a deductive inference agent on the Whyis knowledge graph framework [22] to infer the stage of a tumor, and the corresponding treatment/-monitoring plans. Whyis provides an environment for automated generalized inference over changes to the knowledge graph, supporting the generation of derived knowledge. The framework enables knowledge curation using a Semantic Extract, Transform, and Load tool for creating RDF from tabular sources, as well as automated mapping of external linked data knowledge sources. Furthermore, developers can create custom views for visualizing the data in the underlying knowledge graph.

The Whyis inference agent was built to reason over the nanopublications pertaining to the patient data records constructed in Sect. 4, using the cancer

[3] SEER*Stat: https://seer.cancer.gov/seerstat.

staging ontologies CST, BCS7 and BCS8 introduced in Sect. 2. While the SDD process [25] allowed us to model the data easily, it resulted in some challenges in terms of writing inference rules in OWL, such as finding appropriate paths between entities or attributes specific to the nanopublications, as well as inference over individuals rather than just classes. To address these issues, we decided to take a route similar to SPARQL DL [28] and built SPARQL templates for different OWL reasoning profiles, as well as custom inference rules based on the SDD files, to be consumed by the inference agent.

An example configuration for an OWL inference rule is shown in Listing 5.1, and an example custom rule, auto-generated with the utilization of files generated by the SDD process is shown in Listing 5.2.

```
"Class Subsumption Closure": (
    where = "?resource rdfs:subClassOf ?class .
    ?class rdfs:subClassOf+ ?superClass .",
    construct="?resource rdfs:subClassOf ?superClass .",
    explanation="Since {{class}} is a subclass of {{superClass}},
    any class that is a subclass of {{class}} is also a subclass
    of {{superClass}}. Therefore, {{resource}} is a subclass of
    {{superClass}}.")
```

Listing 5.1. Example Configuration for an OWL Inference Rule (Class Subsumption)

```
"AJCC8 Stage IIIA": (
    resource="?Tumor",
    prefixes="...",
    construct="?Tumor cst:hasAJCCStage cst:AJCC8_Stage_IIIA .",
    where=tnm_where +
      ?T rdf:type cst:T3 .
      ?N rdf:type cst:N3 .
      ?M rdf:type cst:M0 .
      ?Grade rdf:type cst:Grade1 .
      ?HER2 rdf:type cst:HER2_Pos .
      ?ER rdf:type cst:ER_Pos .
      ?PR rdf:type cst:PR_Pos .)
```

Listing 5.2. Example Configuration for a Custom Inference Rule (One of the Criteria for a Tumor to be Classified as Stage IIIA in the AJCC 8th Edition)

These configurations are used to instantiate the following variables in the inference agent: `resource`, `prefixes`, `where`, `construct` and `explanation`. The `prefixes` and `where` variables are used in a SPARQL query that selects relevant URIs from the triple store. The `resource` variable is used to refer to which element returned by the query will be appended new triples. The form of the new triples that will be added is specified in the `construct` variable. An explanation for the rule creating this new knowledge is specified in the `explanation` variable.

Generating Explanations

Our data conversion process captures the provenance of the various sources, which we convert to nanopublications, as well as the explanations behind why specific assertions were inferred. These natural language explanations make it easier for a non-technical user, who might not have an in-depth knowledge of the staging rules, to understand why a certain stage was inferred. When an explicit `explanation` is not provided in the rule, it is derived from the `where` clause used to create the assertion corresponding to the inference. The `explanation` is then associated with that assertion on the new inferred stage using the *prov:used* property. As an example, when the custom inference rule specified in Listing 5.2 is fired on *Patient D*, whose tumor satisfies the criteria given in the `where` clause in that rule, an explanation similar to the one shown in Listing 5.3 will be generated. For better readability of the explanation, the *rdfs:label* or *rdfs:comment* of the values that get bound to variables such as `?T,?N,?M`, etc. (i.e. 'Primary Tumor size', 'Degree of spread to lymph nodes', 'Presence of distant metastasis', etc.) are used instead of the actual class names.

```
Patient D's tumor was found to be AJCC8 Stage IIIA since the
following are true:
- Primary Tumor size is T3 .
- Degree of spread to lymph nodes is N3 .
- Presence of distant metastasis is M0 .
- Tumor Grade is Grade3 .
- Human Epidermal growth factor Receptor 2 (HER2) is Positive.
- Estrogen Receptor (ER) is Positive.
- Progesterone Receptor (PR) is Positive.
```

Listing 5.3. Example of Explanation for Inferring a Stage in the AJCC 8th Edition

Using a similar strategy, we are able to identify possible drug treatment plans from the cancer database CIViC by equating the disease type that the drugs target, to the inferred cancer stage of the patient. To achieve this, we generated custom inference rules from the CIViC SDD files, and once the inferencer runs these rules on the patient nanopublications, the corresponding explanations were generated and attached to the stage assertion nanopublications.

6 Visualization of the Cancer Characterization

In order to demonstrate the integrated cancer knowledge graph and the reasoning capabilities of the Whyis inference agent, we built a visualization tool that displays different treatment paths and guideline impacts to a patient in the form of interactive reports as introduced by Kennedy et al. [17]. The visualization is built on the Whyis knowledge graph framework (introduced in Sect. 5). When a user, say a physician, selects a patient record, they are presented with information that helps enhance their diagnostic process, and in some cases, eliminates the manual labor of walking through the decision trees in the guidelines to support cancer staging decisions. As can be seen in the Figs. 4 and 5, the view is divided into four sections: (1) *Patient Details*, (2) *Biomarker and Staging*, (3) *Treatment Plan*, and (4) *Suggested Drugs*.

In this visualization tool, it is possible to choose between the three latest AJCC staging guidelines, i.e. AJCC 6th, 7th, and 8th editions. Once a guideline

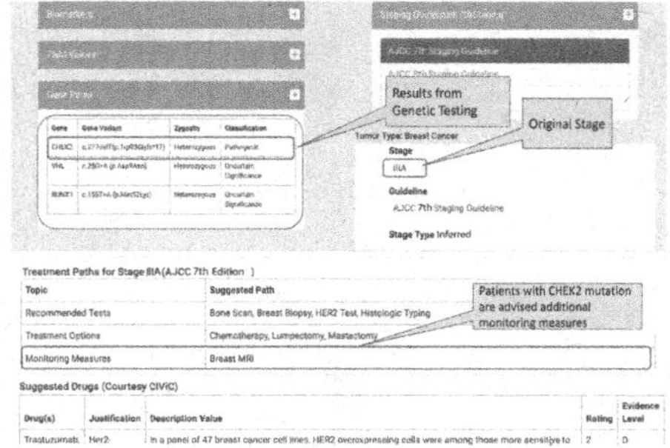

Fig. 4. AJCC 7th edition staging characterization

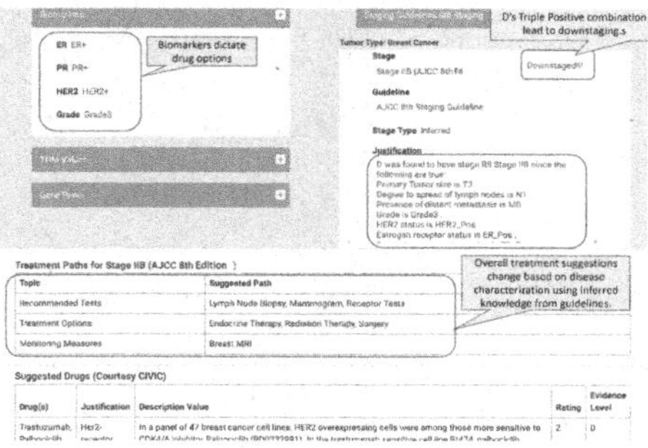

Fig. 5. AJCC 8th edition staging characterization

is selected, the view dynamically loads newly derived knowledge using asynchronous JavaScript SPARQL POST requests. The derived knowledge includes the inferred stage, whether this is an up-stage/down-stage/no-change, and the explanations behind the inferred stage. Based on the inferred stage for the guideline selected, the corresponding treatment and monitoring options available in our integrated cancer knowledge graph (i.e. CIViC drug nanopublication records) are also queried and presented to the user.

A screenshot of a patient's report as per the older 7th edition is shown in Fig. 4, and the same patient's report according to the newer 8th edition is shown in Fig. 5. Note the differences in the inferred stage–the patient is down-staged from IIIA in the 7th edition to IIB in the 8th edition. There are also some changes to the treatment and monitoring options based on this new inferred stage.

7 Evaluation

We used our cancer staging ontologies and the inference agent on 250 randomly selected SEER patient records to estimate prevalence of stage changes between different staging guidelines. We anticipated a number of changes given that the latest AJCC 8th edition utilizes additional biomarkers to determine stage. These SEER patient records were first transformed into nanopublications using the SDD process [25] as explained in Sect. 4, after which our inference agent was applied to determine the stage as per the two guidelines.

The aggregated view of these stage transitions from the AJCC 7th to the 8th edition is shown in Fig. 6. As can be seen in the figure, a majority of the patients' stage did not change, but a statistically significant percentage of patients were either up-staged or down-staged. For example, out of the patients who were assigned to have stage IIB cancer according to the 7th edition (19% of the population), 15% were down-staged to IB, 30% were reclassified to IIA, 38% remained in stage IIB, and 13% and 5% were up-staged to IIIA and IIIB respectively.

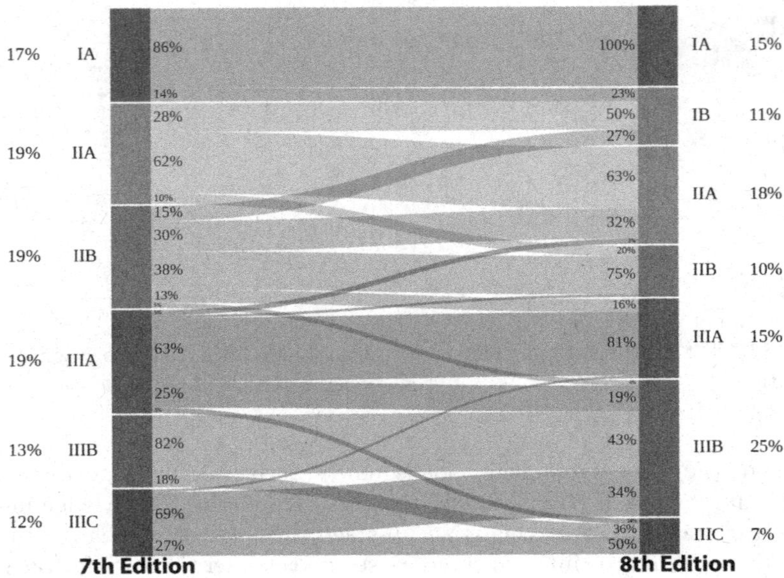

Fig. 6. Stage transitions of 250 patient records from SEER

This indicates that there is a strong need for re-characterizing breast cancer according to the new guideline. Our ontologies and the supporting tools provide the first step in this process.

8 Discussion

We have utilized semantic technologies for all aspects of our system: from characterizing breast cancer and representing synthetic patient data to loading structured and unstructured treatment and monitoring data into a knowledge graph.

For the integrated cancer knowledge graph generation, we mapped concepts in several datasets using a codebook and modeled a structure amongst the attributes using the dictionary mapping table. The deductive inference agent we developed leverages SPARQL DL reasoning, where queries are used to select existing triples and construct new triples. This was done over common inference rules including class subsumption and class or property equivalence closures, as well as custom rules pertaining to the cancer staging. The inference was applied to heterogeneous data sources in our integrated cancer knowledge graph to seamlessly derive new knowledge by applying the inference rules. The visualization we created is able to react to changes in the triple store that results in automatic updates to what the user is seeing. The information in our cancer knowledge graph is kept current with periodic semantic extract transform load updates. Our system allows one to consider a multitude of parameters related to tumor biology as well as standard pathology simultaneously and can easily updated to support new classification criteria.

Resource Contributions
We expect the following publicly available artifacts, along with the applicable documentation, to be useful resources for anyone interested in analyzing breast cancer data according to the new and the old cancer staging guidelines.

1. Ontologies:
   (a) Cancer Staging Terms (CST)
   (b) Breast Cancer Staging Ontology for the AJCC 7th Edition (BCS7)
   (c) Breast Cancer Staging Ontology for the AJCC 8th Edition (BCS8)
2. Semantic Annotations:
   (a) Semantic Data Dictionaries
   (b) Code Books
   (for SEER and CIViC)
3. Source Code:
   (a) AJCC Guideline Extractor
   (b) OWL Axiom Generator
   (c) Whyis Inference Agent
   (d) Custom Inference Rules
   (e) Visualization
4. Data:
   (a) SEER Nanopublications
   (b) CIViC Nanopublications

9 Future Work

There are many online resources with rapidly changing information from clinical trials, as well as data from basic science research with useful cancer data that can be leveraged to augment the cancer knowledge graph. However, when multiple data streams are combined, especially drug information, there may be inconsistent or ambiguous information. Therefore, we will need to resolve such issues using a combination of provenance, data integrity, and trust in the source and/or the methodology.

The inference agent we developed can be used to identify treatment paths based on a patient's cancer stage. However, the CIViC data [14] we used for this purpose defines treatment paths for the broader stages (i.e. stage II as opposed to AJCC's narrower IIA or IIB stages). Therefore, we plan to ascertain the correct treatment paths for all the narrower stages and add those in to the cancer knowledge graph. We plan to incorporate additional data sources such as the NCCN clinical practice guidelines in oncology [18], which is the authoritative source for physicians in identifying suitable cancer treatment and monitoring plans. This will allow our inference agent to output the precise treatment paths, in addition to the ones that are obtained by linking the patient's inferred narrower AJCC tumor stage to the broader stage with ontological properties such as *rdfs:subClassOf* relationships. We also expect the future ontologies to be built using the AJCC API as a resource for all valid values on stage permutations.

We plan to expand the inference capabilities, which are currently restricted to class, instance, property subsumption, equivalence, and inversion closures, to other techniques that will help derive even more relevant knowledge. For example, we believe it would be useful to infer 'patients like me' using instance matching and identify alternate treatment paths that have worked in the past for similar patients, and predict response to a treatment path using temporal reasoning.

As new guidelines will infer new staging results, updates or fixes to the patient data or existing guidelines are needed. The Whyis framework provides an effective mechanism to 'retire' old inferences and trigger computation of new ones, as long as the nanopublication has the same URI. The framework tracks a nanopublication URI when a new version is added, removing older versions, as well as any inferences that are made on them. We opted to create different classes for the stage based on the guideline, so that we can switch between different guidelines easily. For example, we have *AJCC7_Stage_IIA* and *AJCC8_Stage_IIA*, as opposed to a generic *Stage_IIA*. Therefore, encoding the information about which guideline the staging criteria is from, in the provenance assertion for that triple, without having to make that explicit class, and utilizing the provenance information in the inference to determine the stage per the selected guideline is a useful addition. This change requires versioning of assertions in our integrated cancer knowledge graph, and some changes to the custom inference rules.

10 Conclusion

We have presented a prototype knowledge integration system that can be used to encapsulate the breadth of information required to characterize disease. The specific domain problem we address is characterizing breast cancer, which today is predominantly done by manually looking up cancer staging guidelines. In fact, oncology is moving towards adopting the concept of *Precision Oncology*, in which the treatment plans and therapies are driven by data from personalized genetic markers independent of cancer type [11]. In the future, new guidelines for cancer staging are expected to incorporate genomic test results analyzed in the context of the patient's history, which will further increase the complexity of the staging criteria, requiring automated mechanisms similar to the techniques illustrated in this paper. Therefore, it is our expectation that the resources contributed in this paper and the methodologies to ingest rapidly changing information, will be useful to application designers who are aiming to support next generation precision medicine assistant tools.

Acknowledgements. This work is partially supported by IBM Research AI through the AI Horizons Network. We thank our colleagues from IBM (Amar Das, Ching-Hua Chen) and RPI (John Erickson, Alexander New, Rebecca Cowan) who provided insight and expertise that greatly assisted the research.

References

1. Amin, M.B., et al.: The Eighth Edition AJCC Cancer Staging Manual: Continuing to build a bridge from a population-based to a more personalized approach to cancer staging. CA: Cancer J. Clin. **67**(2), 93–99 (2017)
2. Beahrs, O.H., Henson, D.E., Hutter, R.V., Myers, M.H.: Manual for staging of cancer. Am. J. Clin. Oncol. **11**(6), 686 (1988)
3. Bechhofer, S.: OWL: Web ontology language. In: Encyclopedia of Database Systems, pp. 2008–2009. Springer (2009)
4. Bird, S., Loper, E.: NLTK: the natural language toolkit. In: Proceedings of the ACL 2004 on Interactive Poster and Demonstration Sessions, p. 31. Association for Computational Linguistics (2004)
5. Boeker, M., França, F., Bronsert, P., Schulz, S.: TNM-O: ontology support for staging of malignant tumours. J. Biomed. Semant. **7**(1), 64 (2016)
6. Chakravarty, D., et al.: OncoKB: a precision oncology knowledge base. JCO Precis. Oncol. **1**, 1–16 (2017)
7. Consortium, U.: UniProt: a hub for protein information. Nucleic Acids Res. **43**(D1), D204–D212 (2014)
8. Degtyarenko, K., et al.: ChEBI: a database and ontology for chemical entities of biological interest. Nucleic Acids Res. **36**(Suppl. 1), D344–D350 (2007)
9. Dumontier, M., Baker, C.J., Baran, J., Callahan, A., Chepelev, L., Cruz-Toledo, J., Del Rio, N.R., Duck, G., Furlong, L.I., Keath, N.: The Semanticscience Integrated Ontology (SIO) for biomedical research and knowledge discovery. J. Biomed. Semant. **5**(1), 14 (2014)
10. Edge, S.B., Compton, C.C.: The American Joint Committee on Cancer: the 7th edition of the AJCC cancer staging manual and the future of TNM. Ann. Surg. Oncol. **17**(6), 1471–1474 (2010)
11. Garraway, L.A., Verweij, J., Ballman, K.V.: Precision oncology: an overview. J. Clin. Oncol. **31**(15), 1803–1805 (2013)
12. Kumar, A., Smith, B.: Oncology ontology in the NCI thesaurus. In: Miksch, S., Hunter, J., Keravnou, E.T. (eds.) AIME 2005. LNCS (LNAI), vol. 3581, pp. 213–220. Springer, Heidelberg (2005). https://doi.org/10.1007/11527770_30
13. Gonzalez-Perez, A., et al.: IntOGen-mutations identifies cancer drivers across tumor types. Nature Methods **10**(11), 1081 (2013)
14. Griffith, M., et al.: CIViC is a community knowledgebase for expert crowdsourcing the clinical interpretation of variants in cancer. Nat. Genet. **49**(2), 170 (2017)
15. Hanna, J., Joseph, E., Brochhausen, M., Hogan, W.R.: Building a drug ontology based on RxNorm and other sources. J. Biomed. Semant. **4**(1), 44 (2013)
16. Hayat, M.J., Howlader, N., Reichman, M.E., Edwards, B.K.: Cancer statistics, trends, and multiple primary cancer analyses from the surveillance, epidemiology, and end results (SEER) program. Oncologist **12**(1), 20–37 (2007)
17. Kennedy, B.M., Oren, L.G., Buehring Jr., W.J.: Interactive report generation system and method of operation, uS Patent 5,937,155, 10 August 1999
18. Kim, H.L., Puymon, M.R., Qin, M., Guru, K., Mohler, J.L.: NCCN clinical practice guidelines in oncology^{TM}. J. Natl. Compr. Cancer Netw. (2013)
19. Klyne, G., Carroll, J.J.: Resource Description Framework (RDF): Concepts and Abstract Syntax. Technical report. World Wide Web Consortium (2006). http://www.w3.org/TR/rdf-concepts/
20. Malone, J., et al.: Modeling sample variables with an Experimental Factor Ontology. Bioinformatics **26**(8), 1112–1118 (2010)

21. Massicano, F., et al.: An ontology for TNM clinical stage inference. In: ONTOBRAS (2015)
22. McCusker, J.P.: Whyis: nano-scale knowledge graph publishing, management, and analysis framework (2018). https://github.com/tetherless-world/whyis/
23. Micheel, C.M., Lovly, C.M., Levy, M.A.: My cancer genome. Cancer Genet. **207**(6), 289 (2014)
24. Mons, B., Velterop, J.: Nano-Publication in the e-science era. In: Workshop on Semantic Web Applications in Scientific Discourse (SWASD 2009), pp. 14–15 (2009)
25. Rashid, S.M., Chastain, K., Stingone, J.A., McGuinness, D.L., McCusker, J.P.: The semantic data dictionary approach to data annotation & integration. In: Proceedings of the First Workshop on Enabling Open Semantic Science (SemSci), pp. 47–54 (2017)
26. Schriml, L.M., Arze, C., Nadendla, S., Chang, Y.W.W., Mazaitis, M., Felix, V., Feng, G., Kibbe, W.A.: Disease Ontology: a backbone for disease semantic integration. Nucleic Acids Res. **40**(D1), D940–D946 (2011)
27. Singletary, S.E., Greene, F.L., Sobin, L.H.: Classification of isolated tumor cells. Cancer **98**(12), 2740–2741 (2003)
28. Sirin, E., Parsia, B.: SPARQL-DL: SPARQL Query for OWL-DL. In: OWLED, vol. 258 (2007)
29. Wishart, D.S., et al.: DrugBank: a comprehensive resource for in silico drug discovery and exploration. Nucleic Acids Res. **34**(Suppl. 1), D668–D672 (2006)

Comunica: A Modular SPARQL Query Engine for the Web

Ruben Taelman(✉), Joachim Van Herwegen, Miel Vander Sande, and Ruben Verborgh

Ghent University – imec – IDLab, Ghent, Belgium
{ruben.taelman,joachim.herwegen,miel.vandersande,ruben.verborgh}@ugent.be

Abstract. Query evaluation over Linked Data sources has become a complex story, given the multitude of algorithms and techniques for single- and multi-source querying, as well as the heterogeneity of Web interfaces through which data is published online. Today's query processors are insufficiently adaptable to test multiple query engine aspects in combination, such as evaluating the performance of a certain join algorithm over a federation of heterogeneous interfaces. The Semantic Web research community is in need of a flexible query engine that allows plugging in new components such as different algorithms, new or experimental SPARQL features, and support for new Web interfaces. We designed and developed a Web-friendly and modular meta query engine called Comunica that meets these specifications. In this article, we introduce this query engine and explain the architectural choices behind its design. We show how its modular nature makes it an ideal research platform for investigating new kinds of Linked Data interfaces and querying algorithms. Comunica facilitates the development, testing, and evaluation of new query processing capabilities, both in isolation and in combination with others.

1 Introduction

Linked Data on the Web exists in many shapes and forms—and so do the processors we use to query data from one or multiple sources. For instance, engines that query RDF data using the SPARQL language [1] employ different algorithms [2,3] and support different language extensions [4,5]. Furthermore, Linked Data is increasingly published through different Web interfaces, such as data dumps, Linked Data documents [6], SPARQL endpoints [7] and Triple Pattern Fragments (TPF) interfaces [8]. This has led to entirely different query evaluation strategies, such as server-side [7], link-traversal-based [9], shared client-server query processing [8], and client-side (by downloading data dumps and loading them locally).

The resulting variety of implementations suffers from two main problems: a lack of sustainability and a lack of comparability. Alternative query algorithms and features are typically either implemented as forks of existing software packages [10–12] or as independent engines [13]. This

D. Vrandečić et al. (Eds.): ISWC 2018, LNCS 11137, pp. 239–255, 2018.
https://doi.org/10.1007/978-3-030-00668-6_15

practice has limited sustainability: forks are often not merged into the main software distribution and hence become abandoned; independent implementations require a considerable upfront cost and also risk abandonment more than established engines. Comparability is also limited: forks based on older versions of an engine cannot meaningfully be evaluated against newer forks, and evaluating combinations of cross-implementation features—such as different algorithms on different interfaces—is not possible without code adaptation. As a result, many interesting comparisons are never performed because they are too costly to implement and maintain. For example, it is currently unknown how the Linked Data Eddies algorithm [13] performs over a federation [8] of brTPF interfaces [14]. Another example is that the effects of various optimizations and extensions for TPF interfaces [10–17] have only been evaluated in isolation, whereas certain combinations will likely prove complementary.

In order to handle the increasing heterogeneity of Linked Data on the Web, as well as various solutions for querying it, there is a need for a flexible and modular query engine to experiment with all of these techniques—both separately and in combination. In this article, we introduce Comunica to realize this vision. It is a highly modular meta engine for federated SPARQL query evaluation over heterogeneous interfaces, including TPF interfaces, SPARQL endpoints, and data dumps. Comunica aims to serve as a flexible research platform for designing, implementing, and evaluating new and existing Linked Data querying and publication techniques.

Comunica differs from existing query processors on different levels:

1. The **modularity** of the Comunica meta query engine allows for extensions and customization of algorithms and functionality. Users can build and fine-tune a concrete engine by wiring the required modules through an RDF configuration document. By publishing this document, experiments can repeated and adapted by others.
2. Within Comunica, multiple **heterogeneous interfaces** are first-class citizens. This enables federated querying over heterogeneous sources and makes it for example possible to evaluate queries over any combination of SPARQL endpoints, TPF interfaces, datadumps, or other types of interfaces.
3. Comunica is implemented using **Web-based technologies** in JavaScript, which enables usage through browsers, the command line, the SPARQL protocol [7], or any Web or JavaScript application.

Comunica and its default modules are publicly available on GitHub and the npm package manager under the open-source MIT license (canonical citation: https://zenodo.org/record/1202509#.Wq9GZhNuaHo).

This article is structured as follows. In the next section, we discuss the related work, followed by the main features of Comunica in Sect. 3. After that, we introduce the architecture of Comunica in Sect. 4, and its implementation in Sect. 5. Next, we compare the performance of different Comunica configurations with the TPF Client in Sect. 6. Finally, Sect. 7 concludes and discusses future work.

2 Related Work

In this section, we illustrate the many possible degrees of freedom for SPARQL query evaluation, and show that they are hard to combine, which is the problem we aim to solve with Comunica. We first discuss the SPARQL query language, its engines, and algorithms. After that, we discuss alternative Linked Data publishing interfaces, and their connection to querying. Finally, we discuss the software design patterns that are essential in the architecture of Comunica.

2.1 The Different Facets of SPARQL

SPARQL [1] is the W3C-recommended RDF query language. The traditional way to implement a SPARQL query processor is to use it as an interface to an underlying database, resulting in a so-called SPARQL endpoint [7]. This is similar to how an SQL interface provides access to a relation database. The internal storage can either be a native RDF store, e.g., AllegroGraph [18] and Blazegraph [19], or a non-RDF store, e.g., Virtuoso [20] uses a object-relational database management system.

Various algorithms have been proposed for optimized SPARQL query evaluation. Some algorithms for example use the concept of query rewriting [2] based on algebraic equivalent query operations, others have proposed the optimization of Basic Graph Pattern evaluation [3] using selectivity estimation of triple patterns.

In order to evaluate SPARQL queries over datasets of different storage types, SPARQL query frameworks were developed, such as Jena (ARQ) [21], RDFLib [22], rdflib.js [23] and rdfstore-js [24]. Jena is a Java framework, RDFLib is a python package, and rdflib.js and rdfstore-js are JavaScript modules. Jena—or more specifically the ARQ API—and RDFLib are fully SPARQL 1.1 [1] compliant. rdflib.js and rdfstore-js both support a subset of SPARQL 1.1. These SPARQL engines support in-memory models or other sources, such as Jena TDB in the case of ARQ. Most of the query algorithms are tightly coupled to these frameworks, which makes swapping out query algorithms for specific query operators hard or sometimes even impossible. Furthermore, complex things such as federated querying over heterogeneous interfaces are difficult to implement using these frameworks, as they are not supported out-of-the-box. This issue of modularity and heterogeneity are two of the main problems we aim to solve within Comunica. The differences between Comunica and existing frameworks will be explained in more detail in Sect. 3.

The Triple Pattern Fragments client [8] (also known as Client.js or `ldf-client`) is a client-side SPARQL engine that retrieves data over HTTP through Triple Pattern Fragments (TPF) interfaces [8]. Different algorithms [10,16,17] for this client and TPF interface extensions [11,12,14,15] have been proposed to reduce effort of server or client in some way. All of these efforts are however implemented and evaluated in isolation. Furthermore, the implementations are tied to TPF interface, which makes it impossible to use them for other types of datasources and interfaces. With Comunica, we aim to solve this by modularizing query operation implementations into separate modules, so that they can be plugged in and combined in different ways, on top of different datasources and interfaces.

With Semantic Web technologies providing the capability to integrate data from different sources, federated query processing has been an active area of research. However, most of the existing frameworks require SPARQL endpoints on every source. The TPF Client instead federates over TPF interfaces, and achieves similar performance compared to the state of the art [8] despite its usage of a more lightweight interface. However, no frameworks exist that enable federation over heterogeneous interfaces, such as the federation over any combination of SPARQL endpoints and TPF interfaces. With Comunica, we aim to fill this gap. In addition dataset-centric approaches, alternative methods such as link-traversal-based query evaluation [9] exist to query a web of Linked Data documents.

2.2 Linked Data Fragments

In order to formally capture the heterogeneity of different Web interfaces to publish RDF data, the Linked Data Fragment [8] (LDF) conceptual framework uniformly characterizes responses of Web interfaces to RDF-based knowledge graphs. The simplest type of LDF is a data dump—it is the response of a single HTTP requests for a complete RDF dataset. Other types of LDFs includes responses of SPARQL endpoints, TPF interfaces, and Linked Data documents.

Existing LDF research highlights that, when it comes to publishing datasets on the Web, there is no silver bullet: no single interface works well in all situations, as each one involves trade-offs [8]. As such, data publishers must choose the type of interface that matches their intended use case, target audience and infrastructure. This however complicates client-side engines that need to retrieve data from the resulting heterogeneity of interfaces. As shown by the TPF approach, interfaces can be self-descriptive and expose one or more features [25], to describe their functionality using a common vocabulary [26,27]. This allows clients without prior knowledge of the exact inputs and outputs of an interface to discover its usage at runtime.

A design goal of Comunica is to facilitate interaction with any current and future interface within the LDF framework, both in single-source and federated scenarios.

2.3 Software Design Patterns

In the following, we discuss three software design patterns that are relevant to the modular design of the Comunica engine.

Publish-Subscribe Pattern. The publish-subscribe [28] design pattern involves passing messages between publishers and subscribers. Instead of programming publishers to send messages directly to subscribers, they are programmed to publish messages to certain categories. Subscribers can subscribe to these categories which will cause them to receive these published messages, without requiring prior knowledge of the publishers. This pattern is useful for decoupling software components from each other, and only requiring prior knowledge of message categories. We use this pattern in Comunica for allowing different implementations of certain tasks to subscribe to task-specific buses.

Actor Model. The actor model [29] was designed as a way to achieve highly parallel systems consisting of many independent agents communicating using messages, similar to the publish–subscribe pattern. An actor is a computational unit that performs a specific task, acts on messages, and can send messages to other actors. The main advantages of the actor model are that actors can be independently made to implement certain specific tasks based on messages, and that these can be handled asynchronously. These characteristics are highly beneficial to the modularity that we want to achieve with Comunica. That is why we use this pattern in combination with the publish-subscribe pattern to let each implementation of a certain task correspond to a separate actor.

Mediator Pattern. The mediator [30] pattern is able to reduce coupling between software components that interact with each other, and to easily change the interaction if needed. This can be achieved by encapsulating the interaction between software components in a mediator component. Instead of the components having to interact with each other directly, they now interact through the mediator. These components therefore do not require prior knowledge of each other, and different implementations of these mediators can lead to different interaction results. In Comunica, we use this pattern to handle actions when multiple actors are able to solve the same task, by for example choosing the best actor for a task, or by combining the solutions of all actors.

3 Requirement Analysis

In this section, we discuss the main requirements and features of the Comunica framework as a research platform for SPARQL query evaluation. Furthermore, we discuss each feature based on the availability in related work. The main feature requirements of Comunica are the following:

SPARQL query evaluation The engine should be able to interpret, process and output results for SPARQL queries.

Modularity Different independent modules should contain the implementation of specific tasks, and they should be combinable in a flexible framework. The configurations should be describable in RDF.

Heterogeneous interfaces Different types of datasource interfaces should be supported, and it should be possible to add new types independently.

Federation The engine should support federated querying over different interfaces.

Web-based The engine should run in Web browsers using native Web technologies.

In Table 1, we summarize the availability of these features in similar works.

Table 1. Comparison of the availability of the main features of Comunica in similar works. (1) A subset of SPARQL 1.1 is implemented. (2) Querying over SPARQL endpoints, other types require implementing an internal storage interface. (3) Downloading of dumps. (4) Federation only over SPARQL endpoints using the SERVICE keyword.

Feature	TPF Client	ARQ	RDFLib	rdflib.js	rdfstore-js	Comunica
SPARQL	✓(1)	✓	✓	✓(1)	✓(1)	✓(1)
Modularity						✓
Heterogeneous interfaces		✓(2,3)	✓(2,3)	✓(3)	✓(3)	✓
Federation	✓	✓(4)	✓(4)			✓
Web-based	✓			✓	✓	✓

3.1 SPARQL Query Evaluation

The recommended way of querying within RDF data, is using the SPARQL query language. All of the discussed frameworks support at least the parsing and execution of SPARQL queries, and reporting of results.

3.2 Modularity

Adding new functionality or changing certain operations in Comunica should require minimal to no changes to existing code. Furthermore, the Comunica environment should be developer-friendly, including well documented APIs and auto-generation of stub code. In order to take full advantage of the Linked Data stack, modules in Comunica must be describable, configurable and wireable in RDF. By registering or excluding modules from a configuration file, the user is free to choose how heavy or lightweight the query engine will be. Comunica's modular architecture will be explained in Sect. 4. ARQ, RDFLib, rdflib.js and rdfstore-js only support customization by implementing a custom query engine programmatically to handle operators. They do not allow plugging in or out certain modules.

3.3 Heterogeneous Interfaces

Due to the existence of different types of Linked Data Fragments for exposing Linked Datasets, Comunica should support heterogeneous interfaces types, including self-descriptive Linked Data interfaces such as TPF. This TPF interface is the only interface that is supported by the TPF Client. Additionally, Comunica should also enable querying over other sources, such as SPARQL endpoints and data dumps in RDF serializations. The existing SPARQL frameworks mostly support querying against SPARQL endpoints, local graphs, and specific storage types using an internal storage adapter.

3.4 Federation

Next to the different type of Linked Data Fragments for exposing Linked Datasets, data on the Web is typically spread over different datasets, at different locations. As mentioned in Sect. 2, federated query processing is a way to query over the combination of such datasets, without having to download the complete datasets and querying over them locally. The TPF client supports federated query evaluation over its single supported interface type, i.e., TPF interfaces. ARQ and RDFLib only support federation over SPARQL endpoints using the SERVICE keyword. Comunica should enable combined federated querying over its supported heterogeneous interfaces.

3.5 Web-Based

Comunica must be built using native Web technologies, such as JavaScript and RDF configuration documents. This allows Comunica to run in different kinds of environments, including Web browsers, local (JavaScript) runtime engines and command-line interfaces, just like the TPF-client, rdflib.js and rdfstore-js. ARQ and RDFLib are able to run in their language's runtime and via a command-line interface, but not from within Web browsers. ARQ would be able to run in browsers using a custom Java applet, which is not a native Web technology.

4 Architecture

In this section, we discuss the design and architecture of the Comunica meta engine, and show how it conforms to the modularity feature requirement. In summary, Comunica is collection of small modules that, when wired together, are able to perform a certain task, such as evaluating SPARQL queries. We first discuss the customizability of Comunica at design-time, followed by the flexibility of Comunica at run-time. Finally, we give an overview of all modules.

4.1 Customizable Wiring at Design-Time Through Dependency Injection

There is no such thing as the Comunica engine, instead, Comunica is a meta engine that can be instantiated into different engines based on different configurations. Comunica achieves this customizability at design-time using the concept of dependency injection [31]. Using a configuration file, which is created before an engine is started, components for an engine can be selected, configured and combined. For this, we use the Components.js [32] JavaScript dependency injection framework, This framework is based on semantic module descriptions and configuration files using the Object-Oriented Components ontology [33].

Description of Individual Software Components. In order to refer to Comunica components from within configuration files, we semantically describe all Comunica components using the Components.js framework in JSON-LD [34]. Listing 1 shows an example of the semantic description of an RDF parser.

Description of Complex Software Configurations. A specific instance of a Comunica engine can be initialized using Components.js configuration files that describe the wiring between components. For example, Listing 2 shows a configuration file of an engine that is able to parse N3 and JSON-LD-based documents. This example shows that, due to its high degree of modularity, Comunica can be used for other purposes than a query engine, such as building a custom RDF parser.

Since many different configurations can be created, it is important to know which one was used for a specific use case or evaluation. For that purpose, the RDF documents that are used to instantiate a Comunica engine can be published as Linked Data [33]. They can then serve as provenance and as the basis for derived set-ups or evaluations.

```
{
  "@context": [ ... ],
  "@id": "npmd:@comunica/actor-rdf-parse-n3",
  "components": [
    {
      "@id":              "crpn3:Actor/RdfParse/N3",
      "@type":            "Class",
      "extends":          "cbrp:Actor/RdfParse",
      "requireElement": "ActorRdfParseN3",
      "comment":          "An actor that parses Turtle-like RDF",
      "parameters": [
        {
          "@id": "caam:Actor/AbstractMediaTypedFixed/mediaType",
          "default": [ "text/turtle", "application/n-triples" ]
        }
      ]
    }
  ]
}
```

Listing 1: Semantic description of a component that is able to parse N3-based RDF serializations. This component has a single parameter that allows media types to be registered that this parser is able to handle. In this case, the component has four default media types.

```
{
  "@context": [ ... ],
  "@id": "http://example.org/myrdfparser",
  "@type": "Runner",
  "actors": [
```

```
    { "@type": "ActorInitRdfParse",
      "mediatorRdfParse": {
        "@type": "MediatorRace",
        "cc:Mediator/bus": { "@id": "cbrp:Bus/RdfParse" }
      } },
    { "@type": "ActorRdfParseN3",
      "cc:Actor/bus": "cbrp:Actor/RdfParse" },
    { "@type": "ActorRdfParseJsonLd",
      "cc:Actor/bus": "cbrp:Actor/RdfParse" },
  ]
}
```

Listing 2: Comunica configuration of `ActorInitRdfParse` for parsing an RDF document in an unknown serialization. This actor is linked to a mediator with a bus containing two RDF parsers for specific serializations.

4.2 Flexibility at Run-Time Using the Actor-Mediator-Bus Pattern

Once a Comunica engine has been configured and initialized, components can interact with each other in a flexible way using the actor [29], mediator [30], and publish-subscribe [28] patterns. Any number of actor, mediator and bus modules can be created, where each actor interacts with mediators, that in turn invoke other actors that are registered to a certain bus.

Fig. 1 shows an example logic flow between actors through a mediator and a bus. The relation between these components, their phases and the chaining of them will be explained hereafter.

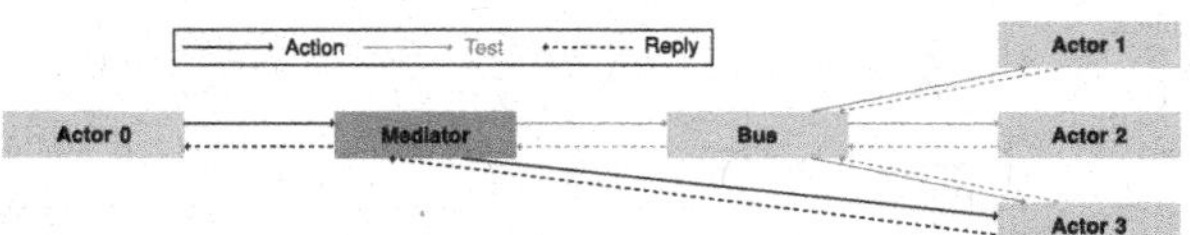

Fig. 1. Example logic flow where Actor 0 requires an action to be performed. This is done by sending the action to the Mediator, which sends a test action to Actors 1, 2 and 3 via the Bus. The Bus then sends all test replies to the Mediator, which chooses the best actor for the action, in this case Actor 3. Finally, the Mediator sends the original action to Actor 3, and returns its response to Actor 0.

Relation Between Actors and Buses. Actors are the main computational units in Comunica, and buses and mediators form the glue that ties them together and makes them interactable. Actors are responsible for being able to accept certain messages via the bus to which they are subscribed, and for responding with an answer. In order to avoid a single high-traffic bus for all message types which could cause performance issues, separate buses exist for different message types. Fig. 2 shows an example of how actors can be registered to buses.

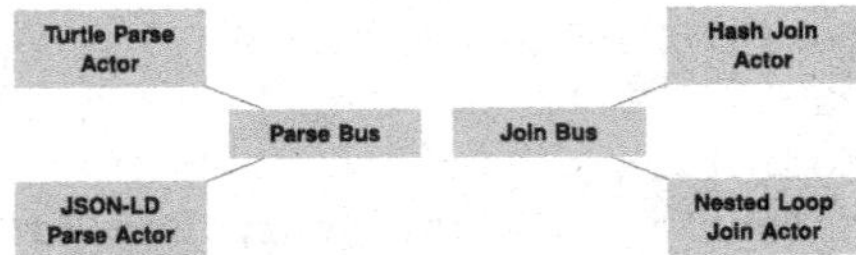

Fig. 2. An example of two different buses each having two subscribed actors. The left bus has different actors for parsing triples in a certain RDF serialization to triple objects. The right bus has actors that join query bindings streams together in a certain way.

Mediators Handle Actor Run and Test Phases. Each mediator is connected to a single bus, and its goal is to determine and invoke the best actor for a certain task. The definition of 'best' depends on the mediator, and different implementations can lead to different choices in different scenarios. A mediator works in two phases: the test phase and the run phase. The test phase is used to check under which conditions the action can be performed in each actor on the bus. This phase must always come before the run phase, and is used to select which actor is best suited to perform a certain task under certain conditions. If such an actor is determined, the run phase of a single actor is initiated. This run phase takes this same type of message, and requires to effectively act on this message, and return the result of this action. Fig. 3 shows an example of a mediator invoking a run and test phase.

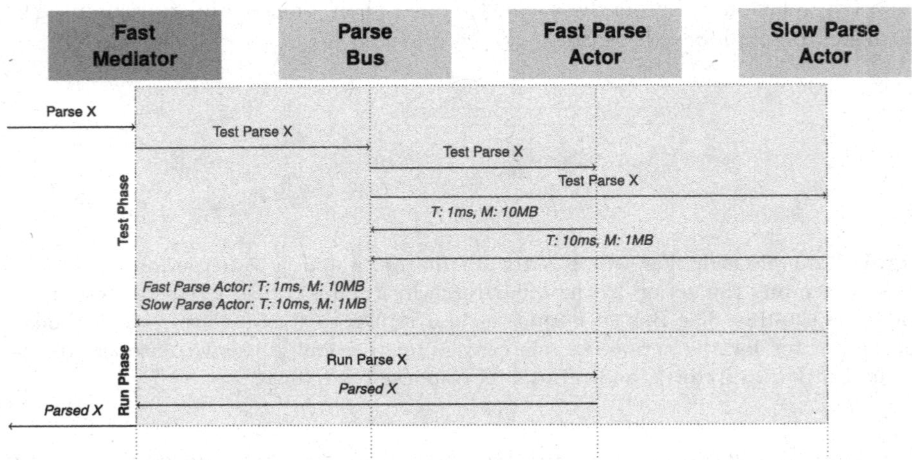

Fig. 3. Example sequence diagram of a mediator that chooses the fastest actor on a parse bus with two subscribed actors. The first parser is very fast but requires a lot of memory, while the second parser is slower, but requires less memory. Which one is best, depends on the use case and is determined by the Mediator. The mediator first calls the tests the actors for the action, and then runs the action using the best actor.

4.3 Modules

At the time of writing, Comunica consists of 79 different modules. This consists of 13 buses, 3 mediator types, 57 actors and 6 other modules. In this section, we will only discuss the most important actors and their interactions.

The main bus in Comunica is the query operation bus, which consists of 19 different actors that provide at least one possible implementation of the typical SPARQL operations such as quad patterns, basic graph patterns (BGPs), unions, projects, ... These actors interact with each other using streams of quad or solution mappings, and act on a query plan expressed in in SPARQL algebra [1].

In order to enable heterogeneous sources to be queried in a federated way, we allow a list of sources, annotated by type, to be passed when a query is initiated. These sources are passed down through the chain of query operation actors, until the quad pattern level is reached. At this level, different actors exist for handling a single source of a certain type, such as TPF interfaces, SPARQL endpoints, local or remote datadumps. In the case of multiple sources, one actor exists that implements a federation algorithm defined for TPF [8], but instead of federating over different TPF interfaces, it federates over different single-source quad pattern actors.

At the end of the pipeline, different actors are available for serializing the results of a query in different ways. For instance, there are actors for serializing the results according to the SPARQL JSON [35] and XML [36] result specifications, but actors with more visual and developer-friendly formats are available as well.

5 Implementation

Comunica is implemented in TypeScript/JavaScript as a collection of Node modules, which are able to run in Web browsers using native Web technologies. Comunica is available under an open license on GitHub and on the NPM package manager. The 79 Comunica modules are tested thoroughly, with more than 1,200 unit tests reaching a test coverage of 100%. In order to be compatible with existing JavaScript RDF libraries, Comunica follows the JavaScript API specification by the RDFJS community group, and will actively be further aligned within this community. In order to encourage collaboration within the community, we extensively use the GitHub issue tracker for planned features, bugs and other issues. Finally, we publish detailed documentation for the usage and development of Comunica.

We provide a default Linked Data-based configuration file with all available actors for evaluating federated SPARQL queries over heterogeneous sources. This allows SPARQL queries to be evaluated using a command-line tool, from a Web service implementing the SPARQL protocol [7], within a JavaScript application, or within the browser. We fully implemented SPARQL 1.0 [37] and a subset of SPARQL 1.1 [1] at the time of writing. In future work, we intend to implement additional actors for supporting SPARQL 1.1 completely.

Comunica currently supports querying over the following types of heterogeneous datasources and interfaces:

- Triple Pattern Fragments interfaces [8]
- Quad Pattern Fragments interfaces (an experimental extension of TPF with a fourth graph element)
- SPARQL endpoints [7]
- Local and remote dataset dumps in RDF serializations.
- HDT datasets [38]
- Versioned OSTRICH datasets [39]

In order to demonstrate Comunica's ability to evaluate federated query evaluation over heterogeneous sources, the following guide shows how you can try this out in Comunica yourself.

Support for new algorithms, query operators and interfaces can be implemented in an external module, without having to create a custom fork of the engine. The module can then be plugged into existing or new engines that are identified by RDF configuration files.

In the future, we will also look into adding support for other interfaces such as brTPF [14] for more efficient join operations and VTPF [15] for queries over versioned datasets.

6 Performance Analysis

One of the goals of Comunica is to replace the TPF Client as a more flexible and modular alternative, with at least the same functionality and similar performance. The fact that Comunica supports multiple heterogeneous interfaces and sources as shown in the previous section validates this flexibility and modularity, as the TPF Client only supports querying over TPF interfaces.

Next to a functional completeness, it is also desired that Comunica achieves similar performance compared to the TPF Client. The higher modularity of Comunica is however expected to cause performance overhead, due to the additional bus and mediator communication, which does not exist in the TPF Client. Hereafter, we compare the performance of the TPF Client and Comunica and discover that Comunica has similar performance to the TPF Client. As the main goal of Comunica is modularity, and not absolute performance, we do not compare with similar frameworks such as ARQ and RDFLib. Instead, relative performance of evaluations using the same engine under different configurations is key for comparisons, which will be demonstrated using Comunica hereafter.

For the setup of this evaluation we used a single machine (Intel Core i5-3230M CPU at 2.60 GHz with 8 GB of RAM), running the Linked Data Fragments server with a HDT-backend [38] and the TPF Client or Comunica, for which the exact versions and configurations will be linked in the following workflow. The main goal of this evaluation is to determine the performance impact of Comunica, while keeping all other variables constant.

In order to illustrate the benefit of modularity within Comunica, we evaluate using two different configurations of Comunica. The first configuration (Comunica-sort) implements a BGP algorithm that is similar to that of the original TPF Client: it sorts triple patterns based on their estimated counts and evaluates and joins them in that order. The second configuration (Comunica-smallest) implements a simplified version of this BGP algorithm that does not sort all triple patterns in a BGP, but merely picks the triple pattern with the smallest estimated count to evaluate on each recursive call, leading to slightly different query plans.

We used the following evaluation workflow:

1. Generate a WatDiv [40] dataset with scale factor=100.
2. Generate the corresponding default WatDiv queries with query-count=5.
3. Install the server software configuration, implementing the TPF specification, with its dependencies.
4. Install the TPF Client software, implementing the SPARQL 1.1 protocol, with its dependencies.
5. Execute the generated WatDiv queries 3 times on the TPF Client, after doing a warmup run, and record the execution times results.
6. Install the Comunica software configuration, implementing the SPARQL 1.1 protocol, with its dependencies, using the Comunica-sort algorithm.
7. Execute the generated WatDiv queries 3 times on the Comunica client, after doing a warmup run, and record the execution times.
8. Update the Comunica installation to use a new configuration supporting the Comunica-smallest algorithm.
9. Execute the generated WatDiv queries 3 times on the Comunica client, after doing a warmup run, and record the execution times.

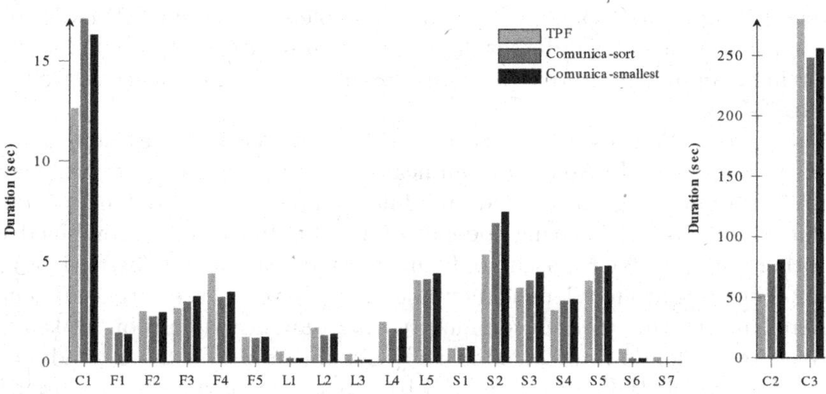

Fig. 4. Average query evaluation times for the TPF Client, Comunica-sort, and Comunica-smallest for all queries (shorter is better). C2 and C3 are shown separately because of their higher evaluation times.

The results from Fig. 4 show that Comunica is able to achieve similar performance compared to the TPF Client. Concretely, both Comunica variants are faster for 11 queries, and slower for 9 queries. However, the difference in evaluation times is in most cases very small, and are caused by implementation details, as the implemented algorithms are equivalent. Contrary to our expectations, the performance overhead of Comunica's modularity is negligible. Comunica therefore improves upon the TPF Client in terms of modularity and functionality, and achieves similar performance.

These results also illustrate the simplicity of comparing different algorithms inside Comunica. In this case, we compared an algorithm that is similar to that of the original TPF Client with a simplified variant. The results show that the performance is very similar, but the original algorithm (Comunica-sort) is faster in most of the cases. It is however not always faster, as illustrated by query C1, where Comunica-sort is almost a second slower than Comunica-smallest. In this case, the heuristic algorithm of the latter was able to come up with a slightly better query plan. Our goal with this result is to show that Comunica can easily be used to compare such different algorithms, where future work can focus on smart mediator algorithms to choose the best BGP actor in each case.

7 Conclusions

In this work, we introduced Comunica as a highly modular meta engine for federated SPARQL query evaluation over heterogeneous interfaces. Comunica is thereby the first system that accomplishes the Linked Data Fragments vision of a client that is able to query over heterogeneous interfaces. Not only can Comunica be used as a client-side SPARQL engine, it can also be customized to become a more lightweight engine and perform more specific tasks, such as for example only evaluating BGPs over Turtle files, evaluating the efficiency of different join operators, or even serve as a complete server-side SPARQL query endpoint that aggregates different datasources. In future work, we will look into supporting supporting alternative (non-semantic) query languages as well, such as GraphQL [41].

If you are a Web researcher, then Comunica is the ideal research platform for investigating new Linked Data publication interfaces, and for experimenting with different query algorithms. New modules can be implemented independently without having to fork existing codebases. The modules can be combined with each other using an RDF-based configuration file that can be instantiated into an actual engine through dependency injection. However, the target audience is broader than just the research community. As Comunica is built on Linked Data and Web technologies, and is extensively documented and has a ready-to-use API, developers of RDF-consuming (Web) applications can also make use of the platform. In the future, we will continue maintaining and developing Comunica and intend to support and collaborate with future researchers on this platform.

The introduction of Comunica will trigger a new generation of Web querying research. Due to its flexibility and modularity, existing areas can be combined and evaluated in more detail, and new promising areas that remained covered so far will be exposed.

Acknowledgements. The described research activities were funded by Ghent University, imec, Flanders Innovation & Entrepreneurship (AIO), and the European Union. Ruben Verborgh is a postdoctoral fellow of the Research Foundation - Flanders.

References

1. Harris, S., Seaborne, A., Prud'hommeaux, E.: SPARQL 1.1 Query Language. W3C (2013). https://www.w3.org/TR/2013/REC-sparql11-query-20130321/
2. Schmidt, M., Meier, M., Lausen, G.: Foundations of SPARQL query optimization. In: Proceedings of the 13th International Conference on Database Theory, pp. 4–33 (2010)
3. Stocker, M., Seaborne, A., Bernstein, A., Kiefer, C., Reynolds, D.: SPARQL basic graph pattern optimization using selectivity estimation. In: Proceedings of the 17th International Conference on World Wide Web, pp. 595–604 (2008)
4. Erling, O., Mikhailov, I.: RDF support in the virtuoso DBMS. In: Pellegrini, T., Auer, S., Tochtermann, K., Schaffert, S. (eds.) Networked Knowledge - Networked Media: Integrating Knowledge Management, New Media Technologies and Semantic Systems, vol. 221, pp. 7–24. Springer, Heidelberg (2009). https://doi.org/10.1007/978-3-642-02184-8_2
5. Cheng, J., Ma, Z.M., Yan, L.: f-SPARQL: a flexible extension of SPARQL. In: Bringas, P.G., Hameurlain, A., Quirchmayr, G. (eds.) DEXA 2010. LNCS, vol. 6261, pp. 487–494. Springer, Heidelberg (2010). https://doi.org/10.1007/978-3-642-15364-8_41
6. Berners-Lee, T.: Linked Data (2009). https://www.w3.org/DesignIssues/LinkedData.html
7. Feigenbaum, L., Todd Williams, G., Grant Clark, K., Torres, E.: SPARQL 1.1 Protocol. W3C (2013). https://www.w3.org/TR/2013/REC-sparql11-protocol-20130321/
8. Verborgh, R., et al.: Triple pattern fragments: a low-cost knowledge graph interface for the web. J. Web Semantics **37**(38), 184–206 (2016)
9. Hartig, O.: An overview on execution strategies for Linked Data queries. Datenbank-Spektrum **13**, 89–99 (2013)
10. Van Herwegen, J., Verborgh, R., Mannens, E., Van de Walle, R.: Query execution optimization for clients of triple pattern fragments. In: Gandon, F., Sabou, M., Sack, H., d'Amato, C., Cudré-Mauroux, P., Zimmermann, A. (eds.) ESWC 2015. LNCS, vol. 9088, pp. 302–318. Springer, Cham (2015). https://doi.org/10.1007/978-3-319-18818-8_19
11. Vander Sande, M., Verborgh, R., Van Herwegen, J., Mannens, E., Van de Walle, R.: Opportunistic linked data querying through approximate membership metadata. In: Arenas, M., et al. (eds.) ISWC 2015. LNCS, vol. 9366, pp. 92–110. Springer, Cham (2015). https://doi.org/10.1007/978-3-319-25007-6_6
12. Van Herwegen, J., De Vocht, L., Verborgh, R., Mannens, E., Van de Walle, R.: Substring filtering for low-cost linked data interfaces. In: Arenas, M., et al. (eds.) ISWC 2015. LNCS, vol. 9366, pp. 128–143. Springer, Cham (2015). https://doi.org/10.1007/978-3-319-25007-6_8

13. Acosta, M., Vidal, M.-E.: Networks of linked data eddies: an adaptive web query processing engine for RDF data. In: Arenas, M., et al. (eds.) ISWC 2015. LNCS, vol. 9366, pp. 111–127. Springer, Cham (2015). https://doi.org/10.1007/978-3-319-25007-6_7
14. Hartig, O., Buil-Aranda, C.: Bindings-restricted triple pattern fragments. In: Debruyne, C. (ed.) OTM 2016. LNCS, vol. 10033, pp. 762–779. Springer, Cham (2016). https://doi.org/10.1007/978-3-319-48472-3_48
15. Taelman, R., Vander Sande, M., Verborgh, R., Mannens, E.: Versioned triple pattern fragments: a low-cost linked data interface feature for web archives. In: Proceedings of the 3rd Workshop on Managing the Evolution and Preservation of the Data Web (2017)
16. Folz, P., Skaf-Molli, H., Molli, P.: CyCLaDEs: a decentralized cache for triple pattern fragments. In: Sack, H., Blomqvist, E., d'Aquin, M., Ghidini, C., Ponzetto, S.P., Lange, C. (eds.) ESWC 2016. LNCS, vol. 9678, pp. 455–469. Springer, Cham (2016). https://doi.org/10.1007/978-3-319-34129-3_28
17. Taelman, R., Verborgh, R., Colpaert, P., Mannens, E.: Continuous client-side query evaluation over dynamic Linked Data. In: Sack, H., Rizzo, G., Steinmetz, N., Mladenić, D., Auer, S., Lange, C. (eds.) ESWC 2016. LNCS, vol. 9989, pp. 273–289. Springer, Cham (2016). https://doi.org/10.1007/978-3-319-47602-5_44
18. Aasman, J.: AllegroGraph: RDF Triple Database, vol. 17. Oakland Franz Incorporated, Cidade (2006)
19. Thompson, B.B., Personick, M., Cutcher, M.: The Bigdata® RDF graph database. In: Linked Data Management, pp. 193–237 (2014)
20. Erling, O., Mikhailov, I.: Virtuoso: RDF support in a native RDBMS. In: de Virgilio, R., Giunchiglia, F., Tanca, L. (eds.) Semantic Web Information Management, pp. 501–519. Springer, Heidelberg (2010). https://doi.org/10.1007/978-3-642-04329-1_21
21. Apache Jena. https://jena.apache.org/
22. RDFLib. https://rdflib.readthedocs.io/en/stable/
23. rdflib.js. https://github.com/linkeddata/rdflib.js
24. rdfstore-js. https://github.com/antoniogarrote/rdfstore-js
25. Verborgh, R., Dumontier, M.: A Web API ecosystem through feature-based reuse. CoRR. abs/1609.07108 (2016)
26. Lanthaler, M., Gütl, C.: Hydra: a vocabulary for hypermedia-driven web APIs. In: LDOW, vol. 996 (2013)
27. Taelman, R., Verborgh, R.: Declaratively describing responses of hypermedia-driven web APIs. In: Proceedings of the 9th International Conference on Knowledge Capture (2017)
28. Birman, K., Joseph, T.: Exploiting virtual synchrony in distributed systems. ACM (1987)
29. Hewitt, C., Bishop, P., Steiger, R.: Session 8 formalisms for artificial intelligence a universal modular actor formalism for artificial intelligence. In: Advance Papers of the Conference, p. 235. Stanford Research Institute (1973)
30. Gamma, E.: Design Patterns: Elements of Reusable Object-oriented Software. Pearson Education India, Delhi (1995)
31. Fowler, M.: Inversion of Control Containers and the Dependency Injection pattern (2004). https://martinfowler.com/articles/injection.html
32. Taelman, R.: Components.js. http://componentsjs.readthedocs.io/en/latest/
33. Van Herwegen, J., Taelman, R., Capadisli, S., Verborgh, R.: Describing configurations of software experiments as Linked Data. In: Proceedings of the 1st Workshop on Enabling Open Semantic Science (2017)

34. World Wide Web Consortium, et al.: JSON-LD 1.0: a JSON-based serialization for linked data (2014)
35. Grant Clark, K., Feigenbaum, L., Torres, E.: SPARQL 1.1 Query Results JSON Format. W3C (2013). https://www.w3.org/TR/2013/REC-sparql11-results-json-20130321/
36. Hawke, S.: SPARQL Query Results XML Format (Second Edition). W3C (2013). https://www.w3.org/TR/rdf-sparql-XMLres/
37. Prud'hommeaux, E., Seaborne, A.: SPARQL Query Language for RDF. W3C (2008). https://www.w3.org/TR/2008/REC-rdf-sparql-query-20080115/
38. Fernández, J.D., Martínez-Prieto, M.A., Gutiérrez, C., Polleres, A., Arias, M.: Binary RDF representation for publication and exchange (HDT). Web Semant. Sci. Serv. Agents World Wide Web **19**, 22–41 (2013)
39. Taelman, R., Vander Sande, M., Verborgh, R.: OSTRICH: versioned random-access triple store. In: Proceedings of the 27th International Conference Companion on World Wide Web (2018)
40. Aluç, G., Hartig, O., Özsu, M.T., Daudjee, K.: Diversified stress testing of RDF data management systems. In: Mika, P., et al. (eds.) ISWC 2014. LNCS, vol. 8796, pp. 197–212. Springer, Cham (2014). https://doi.org/10.1007/978-3-319-11964-9_13
41. Facebook, Inc.: GraphQL. Working Draft, October 2016. http://facebook.github.io/graphql/October2016/

VoCaLS: Vocabulary and Catalog of Linked Streams

Riccardo Tommasini[1(✉)], Yehia Abo Sedira[1], Daniele Dell'Aglio[2], Marco Balduini[1], Muhammad Intizar Ali[3], Danh Le Phuoc[4], Emanuele Della Valle[1], and Jean-Paul Calbimonte[5]

[1] Politecnico di Milano, DEIB, Milan, Italy
{riccardo.tommasini,marco.balduini,emanuele.valle}@polimi.it,
yehiamohamed.abosedera@mail.polimi.it
[2] University of Zurich, Zurich, Switzerland
dellaglio@ifi.uzh.ch
[3] Insight Center for Data Analytics, National University of Ireland, Galway, Ireland
ali.intizar@insight-centre.org
[4] Technichal University of Berlin, Berlin, Germany
danh.lephuoc@tu-berlin.de
[5] University of Applied Sciences and Arts Western Switzerland, Sierre, Switzerland
jean-paul.calbimonte@hevs.ch

Abstract. The nature of Web data is changing. The popularity of news feeds and social media, the rise of the Web of Things, and the adoption of sensor technologies are examples of streaming data that reached the Web scale. The different nature of streaming data calls for specific solutions to problems like data integration and analytics. There is a need for streaming-specific Web resources: new vocabularies to describe, find and select streaming data sources, and systems that can cooperate dynamically to solve stream processing tasks. To foster interoperability between these streaming services on the Web, we propose the Vocabulary & Catalog of Linked Streams (VoCaLS). VoCaLS is a three-module ontology to (i) publish streaming data following Linked Data principles, (ii) describe streaming services and (iii) track the provenance of stream processing.

1 Introduction

Streams have become increasingly more relevant in several scenarios, including sensor data analytics, social networks, or the Internet of Things. Handling the variety and velocity dimensions together has proven to be hard, and the Semantic Web community has answered to these challenges producing languages, models, systems, and benchmarks under the *Stream Reasoning* umbrella [12]. Despite the progress that Stream Reasoning efforts constitute, the interest in exploring stream publication and consumption mechanisms on the Web has only recently gained attention [14,23]. As opposed to traditional static and stored RDF data, streams are produced and consumed in a different way, focusing on the liveness and dynamics of the data, and often requiring alternative protocols and

D. Vrandečić et al. (Eds.): ISWC 2018, LNCS 11137, pp. 256–272, 2018.
https://doi.org/10.1007/978-3-030-00668-6_16

mechanisms for dealing with data velocity. Systems that consume streams for processing (e.g., reasoning, filtering, learning, event detection) require standards for interchanging data about the streams, including endpoint information, processing capabilities, data structure, pull and push retrieval options, and querying specificities.

Different studies have partially tackled these problems in the past [4,24], although there is still no general agreement on a shared set of principles, as it is the case with *static* Linked Data. A set of challenges and requirements regarding the availability of streams on the Web has been presented in [14], providing a road-map towards a Web of Data Streams. Further examples in this scope include generic RESTful service interfaces for streaming data, such as in [6,24]; or the RSP Service Interface [3], providing a programming API for continuous query engines. Other approaches adopted a Linked Data-based publishing strategy [20], although in practice they show the inadequacy of static data publishing for this purpose. Also, systems like TripleWave [22] allow live provisioning of RDF stream data through push and pull mechanisms, thus generating live RDF stream endpoints. Nevertheless, in all these cases, metadata about the streams and their access points and methods have used ad-hoc description vocabularies or project specific ontologies.

This paper presents a (i) set of requirements that take into account recent challenges and issues; (ii) it describes a **Vo**cabulary for **Ca**taloging and **L**inking **S**treams and streaming services on the web (VoCaLS[1]). Last but not least, the paper (iii) draws a road map towards the creation of a catalog that would make streams discoverable, accessible, and reusable. VoCaLS is an emerging resource that standardizes the mechanisms to publish and consume semantic streams on the Web. This includes not only the publication of streams but also the consumption and processing, regardless of implementations details and design choices of different RDF Stream Processing (RSP) and Stream Reasoning systems and languages. This vocabulary constitutes a foundational step towards the long-term goal of allowing Web-centered interactions among RDF Stream processing services. VoCaLS has been engineered as a collaborative effort, following the discussions and results of the work of the W3C RSP Community Group[2]. The vocabulary has been made openly available through a permanent URI, it has been submitted to the Linked Open Vocabularies (LOV) repository [26], it is published under a CC-BY 4.0 license, and its documentation is made available through the Widoco toolset [17]. Furthermore, the ontology itself has been designed in a generic manner, so that it can be reused and combined with domain-specific and technology-specific vocabularies.

The remainder of the paper is structured as follows: we in Sect. 2, we discuss some motivating use-cases. We present the requirements analysis in Sect. 3. We show reused vocabularies Sect. 4 before introducing the VoCaLS modules and how to combine them with other vocabularies in Sect. 5. Section 6 describes the related work, while Sect. 7 concludes the paper and presents the roadmap.

[1] VoCaLS URI: https://w3id.org/rsp/vocals#.
[2] https://www.w3.org/community/rsp/.

2 Use-Cases

In this section, we describe three use-cases that motivate the design and the adoption of a vocabulary for describing streams and streaming services.

2.1 Stream Discovery and Selection in Smart Cities

Smart cities are one of the early adopters of IoT technologies and subsequently of stream processing. Since many publisher and consumers coexist in city sensor networks, middlewares and semantic technologies are commonly used to enable automated discovery and integration [16,18]. The publishers purpose is making the stream findable to the middleware, while the consumers intent is finding and selecting the proper streams that solve a given information need. Therefore, urban data streams are enriched with semantic annotations; their selection is automated using technologies that interpret descriptions.

The most significant hindrance to this approach is centralization. Middlewares often rely on a central repository of stream descriptions, because interoperability requires a standardized interfaces. With the adoption of a shared vocabulary, the decentralized automated discovery would be possible at the Web scale. As for Linked Datasets, stream provisioning services will be able to exchange descriptions to agents requests on-demand, reducing the middleware load.

2.2 Streaming Service Discovery and Federation

Federating a query to a remote stream endpoint is a desirable feature for a situation where data are naturally distributed. This is particularly true for streaming sources, where the time required to gather the data might overcome the responsiveness requirements posed by an information-need.

To make RDF Stream Processing (RSP) federation work in practice, we should follow the example of the Linked Data community. SPARQL query federation relies on protocols and dataset descriptions such as VoID [1] and DCat [21]. To support interactions between query engines on the Web, we need (i) standardizing the language, (ii) fixing the protocols and, last but not least, (iii) agreeing on the vocabulary to describe various resources.

Looking at the RSP state-of-the-art, (i) Dell'Aglio et al. [13] provided a reference model that unifies RSP languages and reconcile the execution semantics of existing RSP engines [5,19]. (ii) Balduini et al. [3] designed a set of RESTful APIs that regulate how to interact with an RDF Stream Processing (RSP) engine in a declarative way.

The missing building block is (iii) the adoption of a shared vocabulary to publish the streams and describe the services. Languages, APIs, and vocabularies together would foster interoperability between different implementation and instances of RSP Services, which up to this point has not happened yet.

2.3 Reproducibility of Empirical Research

Benchmarking is a relevant research topic within the Stream Processing community. Ontologies, datasets, and queries were proposed so far as benchmarks to evaluate engines performance and processing capabilities. Recent efforts tried to formalize an experimental environment that could make empirical research systematic and foster experiment reproducibility [25]. In this context, cataloging available streams, profiling the features of the engine, and tracking the provenance of the experiment as used queries and obtained results, would improve the research outcomes, fostering reproducibility and repeatability.

3 Requirements Analysis

In this section, we present our requirement analysis for VoCaLS. Building on our previous results [13,14,23], we have identified a series of challenges that have to be addressed to comply with the needs of stream processing scenarios. This analysis also takes into account the use cases discussed in the W3C RSP Community Group, as well as the general requirements[3]. We organized the challenges in three main topics: Publication & discovery, Access & processing, and Provenance & licensing.

Publication & Discovery. This aspect refers to the description of streams and streaming services, shared according to the Linked Data principles [7], for the creation of catalogs and discovery endpoints. In particular, a *stream description* should (C1) characterize the contents of a (RDF) stream content and (C2) describe the characteristics of the stream source. Moreover, a *streaming service description* should (C3) describe available endpoints from which streams can be accessed/processed/generated. In this scope, such a vocabulary should be able to answer to questions such as:

- What is the identifier/address of a stream?
- Who created, maintains and/or publishes the stream?
- How frequently is the stream content produced?
- Where is located the vocabulary describing the stream content?

Access & Processing. These challenges focus on descriptions of protocols and APIs to obtain data from the streams, communicate with the streaming services and manipulate the data. To serve the necessity of managing streaming data in real-time, it is crucial to (C4) describe the capabilities of streaming services, such as stream processing engines and reasoners, in terms of their features (e.g. available operators, entailment regimes, etc.). It is also needed to (C5) maintain the order of elements in the stream. Moreover, it is important to (C6) allow the selection of stream partitions and windows, which can be dumped, transmitted or filtered, enabling time-series analysis and replay. Questions relevant to this scope include:

[3] https://w3id.org/rsp/requirements.

- Where can I access the live stream?
- Where can I access the stream as a static dataset?
- Where can I access the history of the historical stream as a static dataset?
- Where can I access the stream starting from a point in the past?
- What is the preceding element of a given stream element?

Provenance & Licensing. These challenges refer to tracking the transformations that involve streaming data, and those that occur on the streams, as well as contracts that regulate data access by actors involved in such transformations (C7). Questions in this scope include:

- How can we describe the process that generated the stream?
- How can we describe the process that generated the stream windows?
- Which datasets or streams were used to derive the stream?

To support our requirement analysis, we investigated how the community perceives the challenges, we identified. Therefore, we designed a survey[4] that aims at gathering more precise information about the perception of stream metadata needs, and the relation between Streaming Data and Linked Data. The hypothesis we started from was that current vocabularies for static/stored (Linked) Data are not enough to satisfy scenarios involving streaming data. We therefore formulated 18 questions (not counting those with multiple options), which aim at (i) investigating the potential impact of a vocabulary resource (3 questions), (ii) probing the relevance of specific challenges (8 questions), and (iii) quantifying the knowledge of the survey respondents on the indicated themes (7 questions).

We collected 34 answers[5] mainly from the Stream Reasoning and Linked Data communities. We asked the participants to self-evaluate their competences on Linked Data, Stream Processing, and Stream Reasoning. Moreover, we cross-checked their answers with simple technical questions. As a result, the survey respondents showed an equally distributed knowledge of the two domains (Linked Data and Stream Reasoning). Although only 11% of them declared to be confident with vocabularies like DCAT, VoID, and DCTerms, they all acknowledged the Linked Data principles.

From the investigation, it emerges that for 55% of the respondents Streaming Data are highly relevant research-wise (5 points on a maximum of 5). Moreover, 35% of the respondents have a high interest in Linked Data. This suggests that being the respondents equally distributed between the two communities, Streaming Data is relevant to the Linked Data community too.

The survey also shows that 51% respondents evaluated the challenges we presented at least as *important* or *crucial*. Nevertheless, most of the challenges resulted to be unsatisfiable or not entirely addressed by the most popular Linked Data vocabularies, e.g., VoID. Indeed, excepting C3 and C7, for which respectively 45% and 51% of the respondents positively answered, all the remaining challenges collected mostly uncertain answers (i.e., Maybe/I Don't Know). All the results are reported in Table 1.

[4] https://ysedira.github.io/vocals/survey.md.
[5] https://goo.gl/zsEJXe.

Table 1. Challenges Relevance and VoID adequacy to solve them. Legend: U:Useless; M:Marginal; IDK:I Don't Know; I:Important; C:Crucial.

	Relevance					VoID Adequacy		
Challenges	U	M	IDK	I	C	Yes	No	IDK/Maybe
C1	0.00%	6.06%	3.03%	30.30%	60.61%	18.18%	15.15%	66.67%
C2	0.00%	9.09%	21.21%	54.55%	15.15%	27.27%	18.18%	54.55%
C3	0.00%	6.06%	6.06%	54.55%	33.33%	45.45%	9.09%	45.45%
C4	0.00%	18.18%	6.06%	33.33%	42.42%	3.03%	45.45%	51.52%
C5	0.00%	9.09%	12.12%	63.64%	15.15%	18.18%	18.18%	63.64%
C6	6.06%	9.09%	27.27%	48.48%	9.09%	33.33%	9.09%	57.58%
C7	0.00%	21.21%	27.27%	45.45%	6.06%	51.52%	9.09%	39.39%

Last but not least, we considered that one of the main differences between streaming and static data is related to the protocols required to access them [23]. Our idea was confirmed by 61% of the survey respondents that agreed (29%) or strongly agreed (32%) with our statement.

Furthermore, we specifically asked our respondents to evaluate with a score from 1 to 5 the nature of streaming data as pull-based (1) or push-based (5). 21 % of the respondents consider streaming data as naturally push-based against the 9% that consider them as pull-based. 35% percent of the respondents are inclined to push-based (4) against the 6% (2). The remaining 29% of the respondents expressed a neutral vote (3).

Supported by the results we acquired, we formulate the following requirements for our vocabulary to satisfy. VoCaLS must:

R1 enable the description of streams, i.e., characterizing their content, relevant statistics, and the license of use;
R2 enable the description of streaming services, i.e., characterizing their capabilities, their APIs, and the license of use;
R3 enable historical stream processing/analysis and replay, i.e., allowing stream storage and dumping of stream samples;
R4 enable provenance tracking at any level, i.e., characterizing stream (a) creation, (b) publication, and (c) storage; but also denoting manipulation and management concerning to existing theoretical frameworks;
R5 tame velocity for streaming data management, i.e., prioritize push-based content provisioning to pull-based one, and encouraging the adoption of an active stream processing paradigm;
R6 tame variety for streaming data management, i.e., do not bind the specification to any domain specific vocabulary, e.g., SSN [11] for IoT or SIOC [8] for Social Media, and to any specific data models, e.g., RDF Streams.

In general, the survey results show that the requirements we collected justify the introduction of a new vocabulary dedicated to describing the different aspects that are only partially covered by existing vocabularies. The limitations of current vocabularies are related to the fact that streaming data requires different (potentially multiple) access methods, going beyond pull and one-off query mechanisms. It is also evident that unlike traditional Linked Data, RDF streams cannot be just de-referenced. Indeed, data items have to be recovered in a streaming fashion or at least partitioned in windows. Finally, RDF stream services often have different features and operators that may result in different types of streaming results. Differences among systems in this respect have been studied in the past [13], although there are no vocabularies available that represent this type of information.

4 Background and Vocabulary Reuse

In this section, we describe related vocabularies that VoCaLS reuses, and those that inspired part of its design.

Dataset description vocabularies were designed primarily with static and stored (linked) data in mind. However, they provide metadata descriptions for any sort of datasets published on the Web. Indeed, they have found a wide use not only within the Semantic Web community but also the wider Open Data movements all over the world.

The **Data Catalog Vocabulary (DCAT)** [21] is an RDF vocabulary designed to foster interoperability among data web published catalogs. It focuses on describing how datasets are accessible and distributed. From DCAT, we extended the notions of *Distribution*, *Dataset* and *Catalog*.

The **Dublin Core Terms (DCterms)**[6] is the first vocabulary made to describe both physical and Web resources, and provides fifteen generic terms to ad dataset metadata. The properties *title, creator, subject, description* can be used in combination with VoCaLS, since they do not directly refer to datasets.

The **Vocabulary of Interlinked Datasets (VoID)** [1] aims at describing RDF datasets and cross-dataset links. VoID's use-cases comprise dataset discovery, selection and query optimization. From VoID, properties related to *data-dumps*, *features*, and specific static resources can be used.

The **SPARQL Service Description (SD)**[7] is a W3C recommendation that contains the necessary terminology to describe SPARQL endpoints and, thus, it is relevant w.r.t. VoCaLS Service Description. VoCaLS does not extend SPARQL-related terms from SD on purpose, since we consider more appropriate to maintain the specification unbiased. However, SD can be reused in several streaming scenarios, due to the similarity between some SPARQL and RSP operations. Indeed, SD properties like entailment regime, and supported languages can be used with VoCaLS.

The **Provenance Ontology (PROV-O)** allows to describe agent-entity-activity relations that captures the semantics of transformations. VoCaLS Prove-

[6] http://dublincore.org/documents/dcmi-terms/.

[7] https://www.w3.org/TR/sparql11-service-description/#.

nance modules extends PROV-O adding some minimal nomenclature that simplifies the usage of this pattern in the context of streaming data processing.

We discuss now two vocabulary drafts that were then merged into VoCaLS. In fact, we decided to re-design the whole vocabulary, considering the limitations of these two early attempts:

The **Vocabulary of Interlinked Streams (VoIS)** [23] extends VoID around four challenges—Discovery, Access, Recall, and Provenance—to model stream interlinking. VoIS is an ontology that provides the classes to publish streams attaching a static description, and allows defining several access methods (e.g., WebSockets, RSP engine) that can be attached to the description. The **Web Stream Processing (WeSP)** [14] refines the idea of SLD, and is currently implemented in a set of systems, such as TripleWave [22] and CQELS. WeSP comprises a vocabulary to exchange RDF streams on the Web that is built on top of DCAT and the SPARQL SD. It allows describing the stream; some models for stream serializations, and communication protocols.

Given that both VoIS and WeSP have still little or no adoption, and considering their compatibility issues with other vocabularies such as DCAT, we preferred not to reuse terms from them and design VoCaLS from scratch.

5 Vocabulary of Linked Streams

In this section we present the design of VoCaLS. The vocabulary is organized in three modules: VoCaLS Core, which describes the core elements of the vocabulary, VoCaLS Service Description, which describes RDF stream service descriptions, and VoCaLS Provenance, focused on streaming data transformation and manipulation. We will introduce each module separately, along with illustrative examples.

5.1 Core Vocabulary

VoCaLS Core concepts are based on an extension of DCAT to represent streams on the Web. As depicted in Fig. 1 and presented in Listing 1.1, the model introduces the basic abstractions to represent streams. A (i) `vocals:StreamDescriptor` is a document accessible via HTTP that holds metadata about the stream and its contents. A (ii) `vocals:Stream` represents a Web stream, i.e., an unbounded sequence of time-varying data elements [13] that

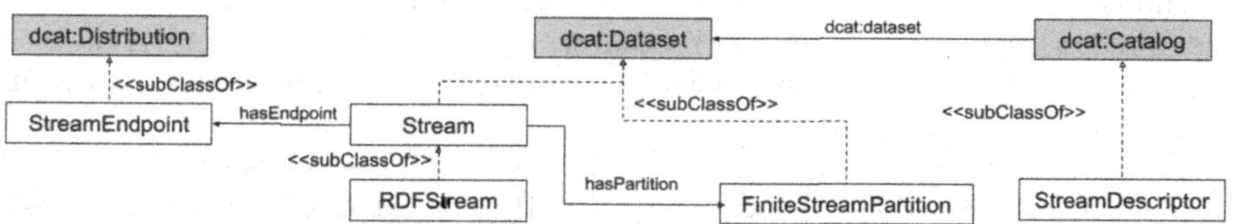

Fig. 1. VoCaLS Core module

might be findable and accessible on the Web, and which can be consumed via a (iii) `vocals:StreamEndpoint`. An example of Stream Endpoint is available in Listing 1.1, line 8. Finally, a (iv) `vocals:FiniteStreamPartition` is a portion of the stream available for regular Linked Data services to access and process its content.

```
1  <> a vocals:StreamDescriptor , vsd:CatalogService ;
2      dcat:dataset :MilanTrafficStream .
3  :MilanTrafficStream a vocals:RDFStream ;
4     vocals:hasEndpoint :MilanTrafficStreamEndpoint ;
5     dct:title "Milan Traffic Stream"^^xsd:string ;
6     dct:publisher <www.3cixty.eu>;
7     dct:description "Aggregated stream produced by traffic sensors in
          Milan"^^xsd:string .
8  :MilanTrafficStreamEndpoint a vocals:StreamEndpoint ;
9     dct:license <https://creativecommons.org/licenses/by-nc/4.0/> ;
10    dct:format frmt:JSON-LD ;
11    dcat:accessURL "ws://example.org/traffic/milan".
```

Listing 1.1. An RDF Stream and Endpoint descriptions using VoCaLS. Prefixes have been omitted.

VoCaLS Core module enables stream producers to publish metadata to describe streams (R1). It also provides a way to represent and describe finite stream partitions that facilitate historical stream processing (R3). With such metadata available on the Web, consumers can discover and select streams relevant to their tasks: for instance, a consumer can retrieve all available endpoints for a given stream, as per Listing 1.2.

```
SELECT ?endpoint
WHERE {
 :traffic_stream a vocals:Stream ;
    vocals:hasEndpoint ?endpoint .
}
```

Listing 1.2. SPARQL query retrieving a StreamEndpoint.

5.2 Streaming Service Description

VoCaLS Service Description focuses on offering a way to publish metadata related to various streaming services and their capabilities, enabling consumers to discover and select services suitable to their needs. The `vsd:StreamingService` is an abstraction to represent a service that deals with data streams of any type. Continuous query engines, stream reasoners, and RDF stream publishers are valid examples.

As depicted in Fig. 2, three classes of RDF streaming services were identified, although others could be added if needed: (i) `vsd:CatalogService`, a service that may provide metadata about streams, their content, query endpoints and more. (ii) `vsd:PublishingService`, which represents a service that publishes

RDF streams (e.g. TripleWave in Listing 1.3), possibly following a Linked Data compliant scheme, and (iii) `vsd:ProcessingService`, which models a stream processing service that performs any kind of transformation on streaming data, e.g. querying, reasoning, filtering, as in Listing 1.5. These services include the possibility of specifying push-based publishing paradigms (R5).

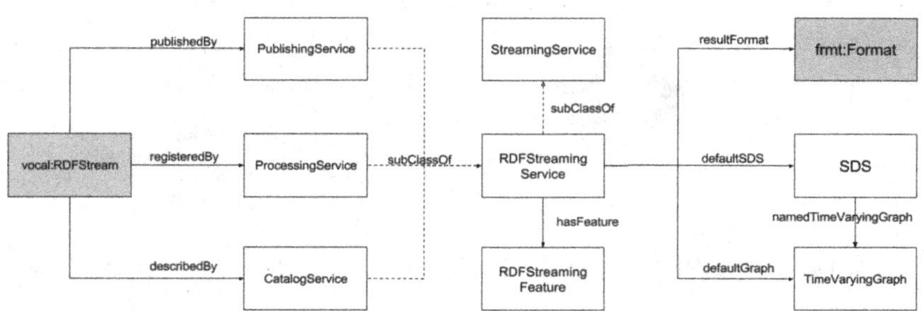

Fig. 2. VoCaLS Service Description classes subset describing RDF streaming services.

```
:trplwv1 a vsd:PublishingService ;
  vsd:hasFeature vsd:replaying ;
  vsd:hasFeature vsd:filtering ;
  vsd:resultFormat frmt:JSON-LD .
```

Listing 1.3. RSP Publishing description using VoCaLS. Prefixes omitted.

Figure 3 shows how VoCaLS Service Description can be used to describe different services, associating each service to the various `vsd:RDF StreamingFeature` it provides, such as what is the reporting policy [13] used by the query engine, which type of time is control is applied, and what the timestamp associated with each stream element represents.

```
SELECT ?sv
WHERE {
    ?sv vsd:hasFeature vsd:filtering  ;
        vsd:registeredStream :RDF_S1 .
}
```

Listing 1.4. SPARQL query to retrieve service having vsd:filtering capabilities and stream :RDF_S1 registered

VoCaLS Service Description makes streaming services annotation straightforward. For instance, Listing 1.5 shows an example of Service Description that uses VoCaLS to describe an instance of the C-SPARQL engine. The running engine can perform *vsd:windowing* and *vsd:filtering*, and it currently registered one RDF Stream. Another example, the RSP engine in Fig. 3 has RDF_S1 stream registered and can perform filtering and windowing operations. By using VoCaLS Service Description, these services and their features can be published on the

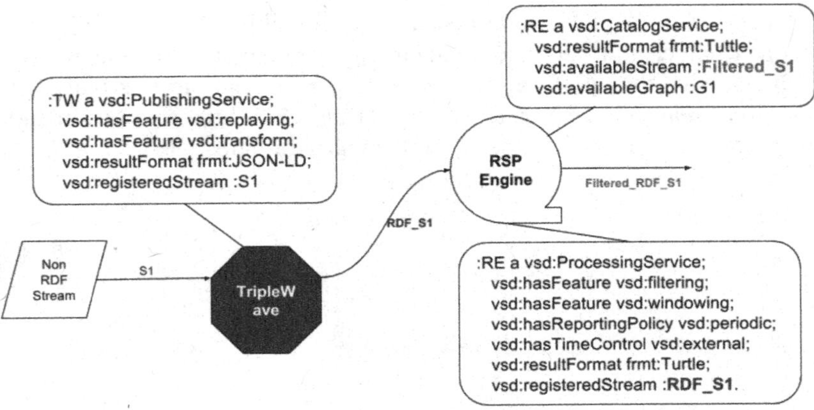

Fig. 3. Describing RDF streaming services using VoCaLS Service Description

web (R2), thus allowing consumers to access service descriptions and select services suitable for their needs, as in the query in Listing 1.4 Moreover, VoCaLS is extensible and, thus, service-specific extensions are possible, e.g., in Listing 1.5, line 7 a custom feature describes a C-SPARQL timestamp function.

```
1 :csparql a vsd:ProcessingService ;
2     vsd:hasFeature vsd:windowing, :timestamp_function;
3     vsd:availableGraphs [ a vsd:TimeVaryingGraph ] ;
4     vsd:hasRegisteredStreams [ a vocals:RDFStream ] .
5 :timestamp_function a vsd:RDFStreamingRSFeature ;
6     dcterms:description "Takes an RDF Triple and returns its
          timestamp."
7 vsd:windowing a vsd:RDFStreamingFeature .
```

Listing 1.5. RSP Engine description using VoCaLS. Prefixes have been omitted.

5.3 Stream Transformation Provenance

VoCaLS Provenance module focuses on tracking the provenance of stream processing services, i.e., tracing the consequences of operations performed over the streams. The module defines four main classes: (i) `vprov:R2ROperator` refers to operators that produce RDF mappings (relations) from other RDF mappings [2], e.g., sum and count. (ii) `vprov:R2SOperator` represents operators that produce a stream from a relation [2], for instance replaying a static dataset as a stream. (iii) `vprov:S2ROperator` refers to operators that produce relations from streams, e.g., windowing (Fig. 4).

Finally, (iv) `vprov:S2SOperator` allows describing operators that produce a stream from another stream. To represent the most common operators, VoCaLS Provenance already contains several subclasses of the four generic ones presented above, e.g., `vprov:WidowOperator`, or `vprov:FilterOperator`.

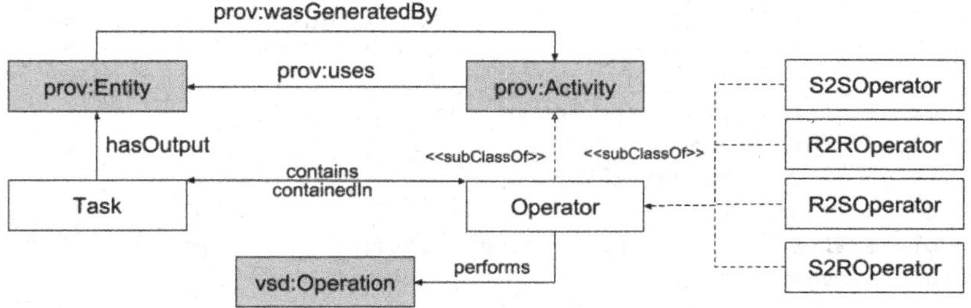

Fig. 4. VoCaLS Provenance subset for RDF stream processing operators.

```
SELECT ?res ?op
WHERE { ?op a vprov:Operator; prov:uses ?res.
        ?str prov:wasGeneratedBy :traffic_stream }
```

Listing 1.6. SPARQL query sources and operations generating a stream.

Through VoCaLS Provenance, the provenance of a stream can be modeled (R4), enabling queries that indicate resources contributed to the stream or which operations were executed, as shown in Listing 1.6.

Listing 1.7 illustrates how to track the provenance of a stream that was generated as the result of `vprov:Task` which contains a sequence of `vprov:Operators`. The execution order of a set of operators is represented in a linked list by `vprov:followedBy` and `vprov:precededBy`

```
:t1 a vprov:Task; vprov:contains :op1,:op2,op3 ; vprov:hasOutput :out
    .
:op1 a vprov:S2ROperator; prov:uses :in_stream.
:op2 a vprov:R2ROperator; prov:uses :w1; vprov:precededBy :op1.
:op3 a vprov:R2SOperator; prov:uses :r1; vprov:precededBy :op2.
:w1 a vsd:Window; prov:wasGeneratedBy :op1.
:r1 a vocals:FiniteStatePartition; prov:wasGeneratedBy :op2.
:out a vocals:RDFStream; prov:wasGeneratedBy :op3.
```

Listing 1.7. Example of RSP operations using VoCaLS Provenance.

5.4 Combining VoCaLS with Other Vocabularies

In the following, we show how VoCaLS can be combined with existing vocabularies (R6). This is especially useful to describe the stream content. In fact, VoCaLS only focus is annotating the stream and streaming services metadata, rather then modeling the annotations within the streams. On the other hand, other ontologies such as SSN [11] and SAO [18] are important resources to describe what is streaming on. In Listing 1.8, we used these to vocabularies to enrich the stream description of Listing 1.1. The SSN ontology is used to represent the source device, and the observation data (event:Event) it produces. The SAO ontology is used to characterize information about the output of a stream observation (Stream Event).

```
:CadornaTrafficStream a ssn:Output, vocals:Stream .
:TrafficFlowSensing a ssn:sensing, sao:StreamEvent ;
  prov:used :CadornaTrafficFlow ;
  ssn:hasOutput :CadornaTrafficStream.
:CadornaTrafficSensor a ssn:SensingDevice ;
  ssn:observes  :TrafficFlow  ;
  ssn:implements :TrafficFlowSensing .
:CadornaTrafficFlow a ssn:ObservationValue, sao:StreamData ;
  prov:wasDerivedFrom :CTObservation .
:CTObservation a ssn:Observation, vsd:TimeVaryingGraph, event:Event ;
  ssn:observedProperty :TrafficFlow ;
  ssn:observationResult :CTSensorOutput ;
  event:time [a time:Instant ;
              time:inXSDDateTime "2013-01-01T00:00:00"^^xsd:dateTime ]
              .
```

Listing 1.8. VoCaLS with SAO and SSN Ontologies. Prefixes omitted.

6 Related Work

In this section, we position VoCaLS within the state of the art. We discuss how existing solutions already addressed some of the challenges that we presented in Sect. 3. Moreover, we use our requirement analysis to highlight differences and commonalities between them. Table 2 summarizes the comparison.

Table 2. Requirements vs State-of-the-Art. Symbol Legend: Empty cell, i.e. not covered; ≃, i.e. partially covered; ✓, i.e. covered.

Requirements	R1	R2	R3	R4.a	R4.b	R4.c	R5	R6
LSD	✓	≃		✓				✓
SLD	✓		✓	✓			✓	
LDN					≃	≃	≃	≃
SAO	≃		≃	≃	✓			✓
CES		✓				✓	✓	✓
VoCaLS	✓	✓	✓	✓	✓	✓	✓	✓

Linked Stream Data (LSD) [24]. This work proposed a mechanism to identify and access data streams coming from sensor networks. LSD takes into account temporal and spatial aspects and it enables discovery using URIs that models sensors, time, space and their combinations. Authors did not propose any protocol extensions w.r.t. Linked Data. Therefore, LSD supports access through streaming protocols, i.e., active paradigm. The problem of storing relevant portions of the stream is not discussed, while provenance tracking is possible for data sources using descriptions but not for stream publishing or transformations.

Streaming Linked Data (SLD) [6]. Barbieri et al. proposed SLD to publish Streaming Linked Data using an RSP engine. The approach is based on two concepts: the Stream Graph (or S-Graph) and Instantaneous Graph (or I-Graph). The S-Graph is a document that refers to stream elements and contains other relevant information. Indeed, it enables stream discovery and can partially track the provenance of the RSP engine activity. The I-Graph identifies an element of the stream. Although SLD does not propose any protocol extension, the RSP engine privileges the active paradigm to publish the stream. Nevertheless, they do not discuss how to dump stream portions nor how to track the provenance of the transformation.

Linked Data Notifications (LDN) [10] is a W3C recommendation[8] that aims at making Web Notifications de-referenceable, persistent, and reusable, i.e., compliant with Linked Data principles. Such a protocol orchestrates the communication between *senders*, *receivers* and *consumers*. Accessing the stream contents might be possible using LDN since it is not bounded to any specific transmission protocols, although the communication methods between consumer and receiver are RESTful. LDN enables the tracking of the provenance of the involved actors, but not specifically stream transformations. Although a first attempt to specialize LDN for RDF streams was presented in [9] neither discovery of streams and services nor stream finite portions are within the scope of the work, which targets communication/sharing between actors rather than exploration and querying.

Stream Annotation Ontology (SAO) [18]. SAO can be used to express the features of stream elements, i.e., *StreamEvents*, but not Streaming services. SAO allows publishing derived data about IoT streams, and it deals with representation of aggregated data. However, the vocabulary does not aim at describing transformations in depth. Provenance tracking is possible to some extent for temporal relations. SAO can be combined with other vocabularies to enrich the description of sensor data.

Complex Event Services (CES) [15]. Gao et al. proposed CES to support smart cities applications' development using semantic technologies. The ontology extends OWL-S to support automated discovery and integration of sensor streams. It was designed to describe event services and requests, therefore it can be used to annotate streaming services. However, there is no distinction between streams publisher and consumers. Provenance tracking is possible at the level of transformation by distinguishing primitive and complex event services. The ontology is designed to be used in combination with the ACEIS middleware and other vocabularies (SSN).

7 Conclusion

In this paper, we introduced VoCaLS, a vocabulary for describing RDF streams, streaming services, and stream transformations. VoCaLS is designed to allow the

[8] https://linkedresearch.org/ldn/.

publication, discovery, consumption, and provision of RDF streams. It includes the capability of describing streaming services; the operations and features that they support, and the workflows that detail how streaming data is generated or processed.

The design of VoCaLS has followed a community-driven approach, starting from the W3C RSP Community group results, and a requirement analysis described in Sect. 3 The challenges presented in this work and the potential impact of this ontology have been supported by the survey described in the paper, which added to the analysis of the state of the art, show the timeliness and adequacy of VoCaLS. The proposed ontology has been designed as a generic resource, which can be combined with domain-specific vocabularies (e.g. for IoT), and reuses and inherits elements from widely used vocabularies such as DCAT and VoID. Furthermore, VoCaLS is also the result of the evolution and merge of two independent preliminary works [14,23] in this area.

VoCaLS has been published following well-principled practices for the publication of the vocabulary, including the set up of permanent URIs, the availability of full open documentation using Widoco, the availability of sources in Github[9], its inclusion in the LOV repository, and the setup of redirects for serving different ontology formats.

Road Map: Regarding the adoption and sustainability plans for VoCaLS, several steps have been taken in this direction. First, given that the establishment of a common vocabulary is one of the main goals of the W3C RSP Community Group, we have started the process of elevating this vocabulary as an official Group Note. The adoption and support from the authors, as a relevant part of this community, will contribute positively to this endeavor. Once this step is achieved, the RSP Community Group is expected to take the responsibility for maintenance, updates and dissemination.

Second, an important goal is to foster the adoption of VoCaLS within relevant communities. For this purpose we initiated the creation of a catalog of streams descriptions[10]. We started annotating all the historical streams that where published for benchmarking purposes. Moreover, we developed a simple utility[11] to support the annotation of new streams. Finally, in order to lead by example, we have launched the integration of VoCaLS within relevant services and software available for the RSP community: the RSP Services, RSPLab, and TripleWave.

References

1. Alexander, K., Cyganiak, R., Hausenblas, M., Zhao, J.: Describing linked datasets. In: Workshop on Linked Data on the Web, LDOW (2009)
2. Arasu, A., Babu, S., Widom, J.: The CQL continuous query language: semantic foundations and query execution. VLDB J. **15**(2), 121–142 (2006)

[9] https://github.com/ysedira/vocals.
[10] https://github.com/ysedira/vocals/tree/master/catalog.
[11] https://github.com/ysedira/stream-annotation-tool.

3. Balduini, M., Della Valle, E.: A restful interface for RDF stream processors. In: ISWC (Posters & Demos), vol. 1035, pp. 209–212. CEUR-WS.org (2013)
4. Balduini, M., Della Valle, E., Dell'Aglio, D., Tsytsarau, M., Palpanas, T., Confalonieri, C.: Social listening of city scale events using the streaming linked data framework. In: Alani, H., et al. (eds.) ISWC 2013. LNCS, vol. 8219, pp. 1–16. Springer, Heidelberg (2013). https://doi.org/10.1007/978-3-642-41338-4_1
5. Barbieri, D.F., Braga, D., Ceri, S., Della Valle, E., Grossniklaus, M.: C-SPARQL: a continuous query language for RDF data streams. Intl. J. Semant. Comput. **4**(01), 3–25 (2010)
6. Barbieri, D.F., Della Valle, E.: A proposal for publishing data streams as linked data - a position paper. In: LDOW (2010)
7. Berners-Lee, T., Bizer, C., Heath, T.: Linked data-the story so far. IJSWIS **5**(3), 1–22 (2009)
8. Breslin, J.G., Decker, S., Harth, A., Bojars, U.: SIOC: an approach to connect web-based communities. Intl. J. Web Based Commun. **2**(2), 133–142 (2006)
9. Calbimonte, J.P.: Linked data notifications for rdf streams. In: Proceedings of the Web Stream Processing (WSP) Workshop at ISWC, pp. 66–73 (2017)
10. Capadisli, S., Guy, A., Lange, C., Auer, S., Sambra, A., Berners-Lee, T.: Linked data notifications: a resource-centric communication protocol. In: Blomqvist, E., Maynard, D., Gangemi, A., Hoekstra, R., Hitzler, P., Hartig, O. (eds.) ESWC 2017. LNCS, vol. 10249, pp. 537–553. Springer, Cham (2017). https://doi.org/10.1007/978-3-319-58068-5_33
11. Compton, M., et al.: The SSN ontology of the W3C semantic sensor network incubator group. J. Web Semant. **17**, 25–32 (2012)
12. Della Valle, E., Ceri, S., Van Harmelen, F., Fensel, D.: It's a streaming world! reasoning upon rapidly changing information. IEEE Intell. Syst. **6**, 83–89 (2009)
13. Dell'Aglio, D., Della Valle, E., Calbimonte, J.P., Corcho, O.: RSP-QL semantics: a unifying query model to explain heterogeneity of RDF stream processing systems. Intl. J. Semant. Web Inf. Syst. (IJSWIS) **10**(4), 17–44 (2014)
14. Dell'Aglio, D., Le Phuoc, D., Le-Tuan, A., Ali, M.I., Calbimonte, J.P.: On a web of data streams. In: ISWC DeSemWeb (2017)
15. Gao, F., Ali, M.I., Curry, E., Mileo, A.: Automated discovery and integration of semantic urban data streams: the ACEIS middleware. Future Gener. Comp. Syst. **76**, 561–581 (2017)
16. Gao, S., Scharrenbach, T., Bernstein, A.: The CLOCK data-aware eviction approach: towards processing linked data streams with limited resources. In: Presutti, V., d'Amato, C., Gandon, F., d'Aquin, M., Staab, S., Tordai, A. (eds.) ESWC 2014. LNCS, vol. 8465, pp. 6–20. Springer, Cham (2014). https://doi.org/10.1007/978-3-319-07443-6_2
17. Garijo, D.: WIDOCO: a wizard for documenting ontologies. In: d'Amato, C., et al. (eds.) ISWC 2017. LNCS, vol. 10588, pp. 94–102. Springer, Cham (2017). https://doi.org/10.1007/978-3-319-68204-4_9
18. Kolozali, S., Bermúdez-Edo, M., Puschmann, D., Ganz, F., Barnaghi, P.M.: A knowledge-based approach for real-time IoT data stream annotation and processing. In: 2014 IEEE International Conference on Internet of Things, Taipei, Taiwan, 1–3 September 2014, pp. 215–222 (2014)
19. Le-Phuoc, D., Dao-Tran, M., Xavier Parreira, J., Hauswirth, M.: A native and adaptive approach for unified processing of linked streams and linked data. In: Aroyo, L., et al. (eds.) ISWC 2011. LNCS, vol. 7031, pp. 370–388. Springer, Heidelberg (2011). https://doi.org/10.1007/978-3-642-25073-6_24

20. Le-Phuoc, D., Nguyen-Mau, H.Q., Parreira, J.X., Hauswirth, M.: A middleware framework for scalable management of linked streams. J. Web Semant. **16**, 42–51 (2012)
21. Maali, F., Cyganiak, R., Peristeras, V.: Enabling interoperability of government data catalogues. In: Wimmer, M.A., Chappelet, J.-L., Janssen, M., Scholl, H.J. (eds.) EGOV 2010. LNCS, vol. 6228, pp. 339–350. Springer, Heidelberg (2010). https://doi.org/10.1007/978-3-642-14799-9_29
22. Mauri, A., et al.: TripleWave: Spreading RDF Streams on the Web. In: Groth, P., et al. (eds.) ISWC 2016. LNCS, vol. 9982, pp. 140–149. Springer, Cham (2016). https://doi.org/10.1007/978-3-319-46547-0_15
23. Sedira, Y.A., Tommasini, R., Della Valle, E.: Towards vois: a vocabulary of interlinked streams. In: ISWC DeSemWeb (2017)
24. Sequeda, J.F., Corcho, O.: Linked stream data: a position paper. In: SSN, pp. 148–157. CEUR-WS. org (2009)
25. Tommasini, R., Della Valle, E., Mauri, A., Brambilla, M.: RSPLab: RDF stream processing benchmarking made easy. In: d'Amato, C., et al. (eds.) ISWC 2017. LNCS, vol. 10588, pp. 202–209. Springer, Cham (2017). https://doi.org/10.1007/978-3-319-68204-4_21
26. Vandenbussche, P.Y., Atemezing, G.A., Poveda-Villalón, M., Vatant, B.: Linked open vocabularies (LOV): a gateway to reusable semantic vocabularies on the web. Semant. Web **8**(3), 437–452 (2017)

A Complex Alignment Benchmark: GeoLink Dataset

Lu Zhou[1(✉)], Michelle Cheatham[1(✉)], Adila Krisnadhi[2(✉)], and Pascal Hitzler[1(✉)]

[1] DaSe Lab, Wright State University, Dayton, OH 45435, USA
{zhou.34,michelle.cheatham,pascal.hitzler}@wright.edu
[2] Faculty of Computer Science, Universitas Indonesia, Depok, Jawa Barat 16424, Indonesia
adila@cs.ui.ac.id

Abstract. Ontology alignment has been studied for over a decade, and over that time many alignment systems and methods have been developed by researchers in order to find simple 1-to-1 equivalence matches between two ontologies. However, very few alignment systems focus on finding complex correspondences. One reason for this limitation may be that there are no widely accepted alignment benchmarks that contain such complex relationships. In this paper, we propose a real-world dataset from the GeoLink project as a potential complex alignment benchmark. The dataset consists of two ontologies, the GeoLink Base Ontology (GBO) and the GeoLink Modular Ontology (GMO), as well as a manually created reference alignment, that was developed in consultation with domain experts from different institutions. The alignment includes 1:1, 1:n, and m:n equivalence and subsumption correspondences, and is available in both Expressive and Declarative Ontology Alignment Language (EDOAL) and rule syntax.

1 Introduction

Ontology alignment is an important step in enabling computers to query and reason across the many linked datasets on the semantic web. This is a difficult challenge because the ontologies underlying different linked datasets can vary in terms of subject area coverage, level of abstraction, ontology modeling philosophy, and even language. Due to the importance and difficulty of the ontology alignment problem, it has been an active area of research for over a decade [12].

Ideally, alignment systems should be able to uncover any entity relationships across two ontologies that can exist within a single ontology. Such relationships have a wide range of complexity, from basic 1-to-1 equivalence, such as a Person in one ontology being equivalent to a Human in another ontology, to arbitrary m-to-n relationships, such as a Professor with a hasRank property value of "Assistant" in one ontology being a subclass of the union of the Faculty and TenureTrack classes in another. Unfortunately, the majority of the research activities in the field of ontology alignment remains focus on the simplest end

D. Vrandečić et al. (Eds.): ISWC 2018, LNCS 11137, pp. 273–288, 2018.
https://doi.org/10.1007/978-3-030-00668-6_17

of this scale – finding 1-to-1 equivalence relations between ontologies. Part of the reason for this may be that there are no widely used and accepted ontology alignment benchmarks that involve complex relations.

This paper seeks to take a step in that direction by proposing a complex alignment benchmark based on two ontologies which were developed by domain experts jointly with the reference alignment, and which in fact were developed for deployment on major ocean science data repository platforms, i.e., without the actual intention to develop an alignment benchmark. For this reason, the benchmark, including the reference alignment, can be considered to be (a) objective, in that it was created for deployment and not for benchmarking, (b) realistic, in that it captures an application use case developed for deployment, and (c) a valid ground truth alignment, in that the two ontologies and the reference alignment were developed together, by domain experts. We argue that it is therefore of rather unique nature and will inform complex ontology alignment research from a practical and applied perspective, rather than artificial laboratory-like. The benchmark, coincidently, as this was the requirement of the use case, has a particular focus on relationships involving properties, which is particularly interesting because those have been shown to be rather difficult to handle for current alignment approaches [1].

The main contributions of this paper are therefore the following:

- Presentation of two ontologies to support data representation, sharing, integration, and discovery for the geoscience research domain.
- Creation of an alignment between these two ontologies that includes 1:1, 1:n, and m:n correspondences, and given the creation history and usage of the alignment, it is fair to say that the alignment constitutes a gold-standard reference.
- Publication of the benchmark alignment in both rule syntax and EDOAL format[1] at a persistent URL[2] under a CC-BY license.

In addition, we have analyzed and categorized the mapping rules constituting the alignment. We found several which had not been classified or discussed previously, and we will present and discuss our analysis.

This paper is organized as follows. Section 2 discusses the few existing ontology alignment benchmarks that involve relationships other than 1-to-1 equivalence. Section 3 gives further background on the GeoLink modeling process, including why two different but related ontologies were developed. Section 4 discusses the alignment between the two GeoLink ontologies, along with some descriptive statistics and an analysis of the types of mapping rules constituting the alignment. Section 5 concludes with a discussion of potential future work in this area.

1 http://alignapi.gforge.inria.fr/edoal.html.
2 https://doi.org/10.6084/m9.figshare.5907172.

2 Related Work

Most work associated with evaluating the performance of ontology alignment systems has been done in conjunction with the Ontology Alignment Evaluation Initiative (OAEI)[3]. These yearly events allow developers to test their alignment systems on various tracks that evaluate performance on different facets of the problem such as instance matching, large ontology matching, and interactive matching, among others. Currently, most of these tracks involve the identification of 1-to-1 equivalence relationships, such as a Participant being equivalent to an Attendee. In 2009, the OAEI ran an "oriented" matching track that challenged systems to find subsumption relationships such as a Book is a subclass of a Publication. However, this track was abandoned after one year. Some system developers complained that the quality of the reference alignment was low [2]. This frustrated system developers and hindered participation. A discussion at the last two Ontology Matching workshops[4] made it clear that the community is interested in complex alignment, but that lack of applicable benchmarks is hindering progress. Our proposed benchmark seeks to address this concern by providing a reference alignment as a benchmark, and by addressing the quality issue of the previous benchmark by the fact that the process leading to the reference alignment guarantees its high quality.

In addition to using the OAEI benchmark, alignment systems that attempt to identify subsumption relations have sometimes used their own manually developed (and sometimes unpublished) reference alignments [5]. Other subsumption systems have evaluated the precision of their approach by manually validating relations produced by their system, while foregoing an assessment of recall [13]. Other related work has centered on developing a benchmark for compound alignments, which the authors define as mappings between class or property expressions involving more than two ontologies [10]. Their first step in this direction was to create a set of reference alignments containing relations of the form $< X, Y, Z, R, M >$, where X, Y and Z are classes from three different ontologies and R is a relation between Y and Z that results in a class expression that is related to X by the relation M. For example, a DisabledVeteran (X) is equivalent to (M) the intersection (R) of Veteran (Y) and Disabled (Z). This benchmark is based on cross-products among the Open Biomedical Ontologies (OBO) Foundry[5], which have been manually validated by at least two experts.

The work presented herein differs from these approaches by considering a wider range of relationship types (beyond subsumption and the type of ternary relation described in [10]), as they naturally arose out of the application from which the reference alignment was taken.

More related work is currently being undertaken by Thieblin and her colleagues, who are creating a complex alignment benchmark using the Conference track ontologies within the OAEI [14]. This work is partially completed, and at

[3] http://oaei.ontologymatching.org.
[4] http://www.ontologymatching.org/.
[5] http://www.obofoundry.org/.

the time of this writing it covers three of the seven ontologies. In addition, we are collaborating with them (under their direction) to complete the dataset and prepare a new task in OAEI to evaluate complex alignment systems. The reference alignment we describe herein differs from the effort by Thieblin et al. in that the GeoLink ontologies and alignment constitute real-world datasets designed for practice and applied by geoscientists, rather than being an artificial artifact designed solely for alignment benchmarking. Furthermore, data from seven geoscience repositories have been published according to the GeoLink schema and they are available online[6]. This instance data can in the future be used by alignment systems that employ extensional matching techniques. In contrast to this, significant instance data is not readily available for most of the OAEI Conference Track ontologies.

3 The GeoLink Modeling Process

Benchmarks come in at least two varieties. On the one hand, there are artificial benchmarks that provide a kind-of laboratory setting for evaluation. On the other hand, there are benchmarks created from data as it is used in realistic use cases or even deployed scenarios. Both of these types are important, and they cover different aspects of the spectrum, and may have different advantages. Artificial benchmarks can be made to be balanced, or to focus on certain aspects of a problem, and sometimes they can be used to test scalability issues more easily as different versions of the same benchmark set may be easily producible. Natural benchmarks, on the other hand, may expose issues arising in practice which may easily be overlooked by designers of artificial benchmarks, in particular in a young field such as complex ontology alignment. Natural benchmarks also may come with an independently verified gold standard baseline, as in our case.

The project that this benchmark arose from is called GeoLink [15] and was funded under the U.S. National Science Foundation's EarthCube initiative. This planned decade-long endeavor is a recognition that oftentimes the most innovative and useful discoveries come at the intersection of traditional fields of research. This is particularly true in the geosciences, which often bring together disparate groups of researchers such as geologists, meteorologists, climatologists, ecologists, archaeologists, and so on. For its part, GeoLink employs semantic web technologies to support data representation, sharing, integration, and discovery [9]. In particular, seven diverse geoscience datasets have been brought together into a single data repository.

At the beginning of the project, some providers' data resided in relational databases while others' had been published as RDF triples and exposed via a SPARQL endpoint. Because each provider had their own schema, the first step in the GeoLink project was to develop a unified schema according to which all data providers could publish their data [9]. Creating a unified schema for independently developed datasets is sometimes difficult, and the final product often ends up requiring providers to shoehorn their data into a schema that does

[6] http://data.geolink.org.

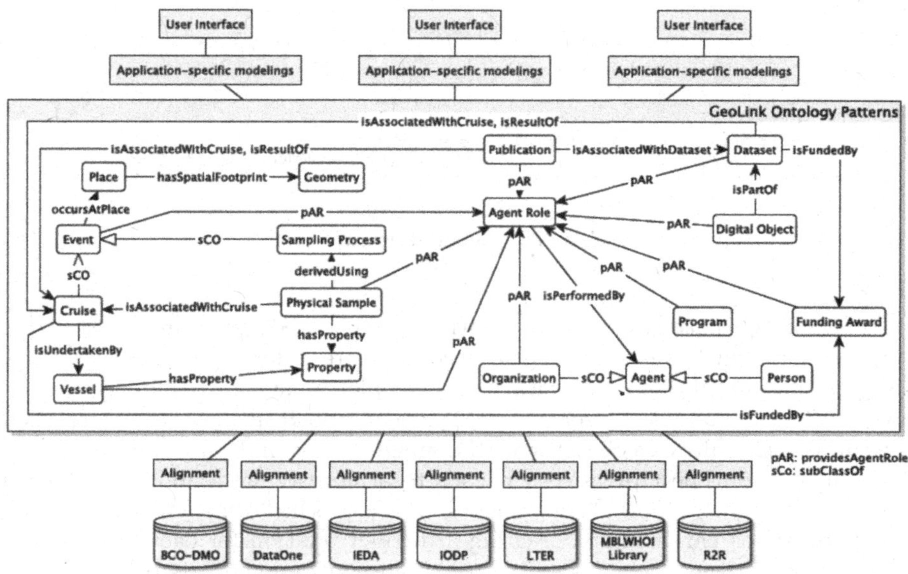

Fig. 1. Intended usage of the GMO

not quite fit. GeoLink uses an approach that relies on ontology design patterns (ODP) in an attempt to avoid this issue [4]. An ODP represents a reusable solution to a recurring modeling problem and generally encodes a specific abstract notion, such as a process, event, and agent, etc. These are frequently the small areas of semantic overlap that exist between datasets from different subfields of the same high-level domain. ODPs provide a structured and application-neutral representation of the key concepts within a domain. Throughout the first year of the project, geoscientists, data providers and ontologists worked together to identify and model the important concepts within the geosciences that recurred across two or more datasets. The result of this were what we call ontology modules, based on ODPs, and eventually they were stitched together to form the GMO [7].

As shown in Fig. 1, the GMO allows data providers to publish only those aspects of their data modeled by the GMO according to that schema. Any data that the provider has, which is not covered by that schema, can be published using the provider's own schema, since no other providers have similar content. For example, in Fig. 1, the provider R2R has data related mostly to the cruise and vessel modules in the lower left of the Fig. 1, and so, it publishes its related data using that terminology. R2R also has data not modeled by the GMO, and so, it uses its own terminology when publishing that information. This freedom is intended to make the publishing process easier. However, some problems still remained.

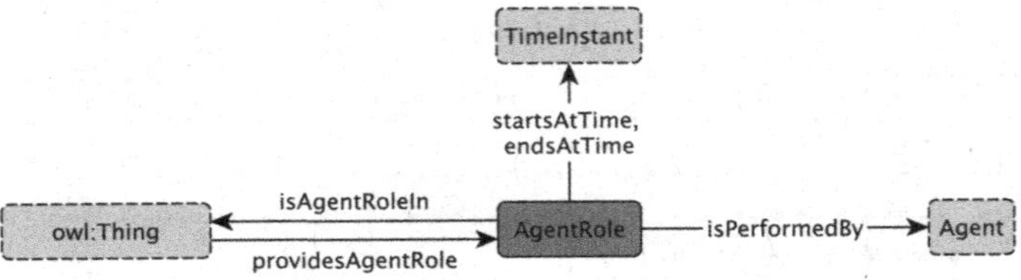

Fig. 2. The Agent Role pattern

Some of the patterns contain a rather complicated structure, mostly due to reification, which was employed to accommodate different perspectives (e.g., based on granularity) on the data. For example, many of the data providers have information about the sponsor of a project, and R2R has a native relation in their schema called hasSponsor with domain Award and range Organization. However following best practices, it leads to a more versatile model if being a sponsor is recognized (and thus modeled) as a role which an agent (in this case an organization) can assume. Creating a distinct relation for each type of role on a project (sponsor, chief scientist, research assistant, etc.) is brittle, in the sense that if new roles will be added later, potentially due to the inclusion of a new dataset, then the schema will need to be edited by adding new vocabulary for new roles together with (possibly complex) role relationships. Another issue with using a relation such as hasSponsor is that a more fine-grained data repository may have additional temporal information related to the sponsor role, and then it is not clear how to add this temporal information to the hasSponsor model without punning. Essentially, hasSponsor should better be expressed as a ternary relation between award, organization, and the type of relation (in this case, being a sponsor) expressed using an individual which can be reused in all sponsor relationships. In terms of ODPs, this is realized by reusing the Agent Role pattern, shown in abstract form in Fig. 2. This approach both allows new roles to be added easily (by subclassing AgentRole) and supports temporal queries if desired.

Unfortunately, while the data providers recognized the utility of this modeling approach, they found it cumbersome to map their data to it. Looking at their own schemas, they found nothing equivalent to AgentRole, and looking at the GMO, they found no obvious way to model the Sponsor field in their database. Additionally, reification led to the generation of blank nodes and the need to create and maintain many URIs. A simpler interface for the data providers was therefore requested.

To accommodate this, a second ontology, together with a manual alignment between this ontology and the GMO, was created to bridge the gap via an intermediate schema that is "flatter" than the patterns and closer to the data providers' own schemas, but still easy to align to the GMO modules because it has been developed directly out of the GMO. This ontology is referred to as the GBO. The providers publish their data according to the GBO and then SPARQL construct queries which encode the alignment can be used to map data to the GMO. From the very beginning, it was intended that the data integration process would be based on manual, and thus high-quality, mappings between different schemas. As a consequence, ontology alignment systems were not employed to make these mappings, not even to inform human decisions. All mappings were established as a collaborative effort between the data repository providers, the domain experts, and the ontology engineers involved in the modeling and deployment process. Because the GBO was manually engineered directly from the GMO in order to serve this particular purpose, the alignment is guaranteed to be precisely the one intended by the developers. I.e. the alignment is guaranteed to contain all of the relations necessary to solve this real-world alignment problem and no superfluous relations have been included. We argue that this characteristic makes the GeoLink ontologies a good example of a complex ontology alignment problem that can be used as a benchmark for systems that attempt to automate such alignment processes: While it is not a synthetic benchmark, it reflects complex alignment issues encountered in practice.

The example below illustrates the use of the GBO and its alignment to the GMO. In the GBO, there is a relation called hasSponsor with a domain that includes Award and range Organization. This mirrors many of the providers' existing schemas. Providers publish triples either according to the GMO schema (e.g., if they have temporal information), or according to the GBO schema.

```
x:award1 a                view:Award ;
         view:hasSponsor x:org1 .
x:org1   a                view:Organization .
```

Then, the GBO-oriented triples are converted into the GMO schema using this SPARQL construct:

```
PREFIX view: <http://schema.geolink.org/dev/view#>
CONSTRUCT {
  ?x     a                    :FundingAward ;
         :providesAgentRole _:bn1 .
  _:bn1 :isPerformedBy        ?y ;
         a                    :SponsorRole .
  ?y     a                    :Organization .
} WHERE {
  ?x     a                    view:Award ;
         view:hasSponsor      ?y .
  ?y     a                    view:Organization
}
```

Let us look at this by means of a schema diagram. In Fig. 3, the three nodes and the two solid arrows indicate the graph pattern used to express the sponsoring organization role in the GMO. The dashed arrow is sometimes called a *shortcut* [8]. This shortcut (which is not part of the GMO) "flattens" this part of the GMO, and in the GBO, the `:SponsorRole` node is removed, but the shortcut

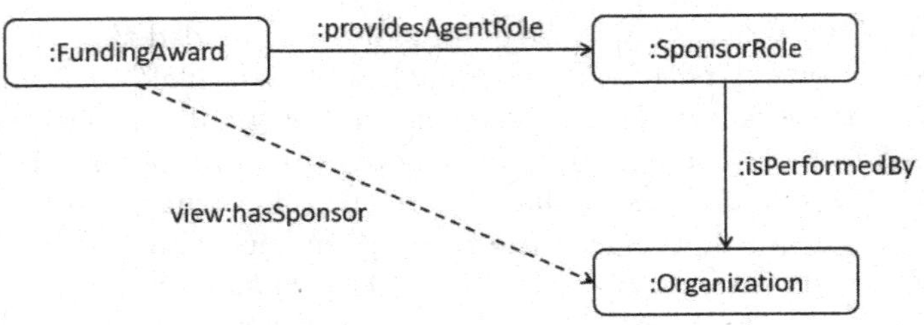

Fig. 3. A schema diagram to explain an example alignment

is added (and `:FundingAward` and `:Organization` have been replaced by the local `view:Award` and `view:Organization`, respectively).

Note that there is no doubt here about the intended alignment between the corresponding parts of the GBO and the GMO: `view:Award` and `:FundingAward` should be mapped to each other (as equivalence), as should `view:Organization` and `:Organization`. It is also clear that that the relation `view:hasSponsor` between an `view:Award` and an `view:Organization` should be aligned (as equivalence) to the concatenation of `:providesAgentRole` and `:isPerformedBy`, provided the entity shared by the two relation expressions is of type `:SponsorRole`, and the chain starts at a `:FundingAward` and ends at a `:Organization`. I.e. a complex alignment is required to express this very natural relationship between these two ontology snippets. Below we will give more examples of complex alignments arising from our setting, when we discuss the different alignment patterns we have identified. The example above is a "Typed Property Chain Equivalence" in our classification, and below we discuss this example further.

More information about the GMO and the project is available from [6] and from the project website[7].

4 The GeoLink Complex Alignment Benchmark

4.1 Dataset

In order to prepare the GeoLink ontologies for use as a complex alignment benchmark, some changes to the namespaces were required. As we introduced in the previous section, several ODPs and modules were created to represent the frequently recurring concepts in the GeoLink datasets, and these were stitched together to form the GMO. During this process, the namespace of some entities was changed from one that reflected its originating pattern to the namespace of the GMO, which is `http://gmo#`. For example, the class FundingAward was originally in the fundingaward pattern, with the namespace http://schema.geolink.org/1.0/pattern/fundingaward#. After merging these modules, the namespace of the class FundingAward became `http://gmo#`. This

[7] http://www.geolink.org/.

Table 1. The number of classes, object properties, and data properties in both GeoLink ontologies

Ontology	Classes	Object Properties	Data Properties
GeoLink Base Ontology	40	149	49
GeoLink Modular Ontology	156	124	46

has been applied to all entities except those that are imported from other ontologies, which retain their original namespace. For example, the namespace of the class Instant, which is imported from http://www.w3.org/2006/time#, remains unchanged. Additionally, the namespace of entities in the GBO has been changed from http://schema.geolink.org/1.0/base/main# to `http://gbo#`.

Table 1 shows the number of classes and properties in both ontologies. Both ontologies are comparable in size to ontologies currently used by the OAEI, meaning that they are within the capabilities of most current ontology alignment systems to handle.

4.2 Simple and Complex Correspondences

In order to understand the correspondences in the benchmark, we give the formal definition of simple and complex correspondences.

Simple Correspondence. Simple correspondence refers to basic 1-to-1 simple alignment between two ontologies, including class and property. It not only includes 1-to-1 equivalence, but also contains 1-to-1 subsumption, and 1-to-1 disjointness.

Complex Correspondence. Complex correspondence refers to more complex patterns, such as 1-to-n equivalence, 1-to-n subsumption, m-to-n equivalence, m-to-n subsumption, and m-to-n arbitrary relationship.

We have identified 12 different kinds of simple and complex correspondence patterns in the GeoLink complex alignment benchmark. Table 2 presents these different patterns and the corresponding number and category in the whole dataset. As the table shows, the alignment consists predominantly of complex relationships. In the following, we explain these alignment types, from simple 1-to-1 correspondence to complex m-to-n correspondence, with a formal pattern and example each.

Class Equivalence. The first pattern is just simple 1-to-1 class equivalence. Classes C_1 and C_2 are from ontology O_1 and ontology O_2, respectively.

Formal Pattern: $C_1(x) \leftrightarrow C_2(x)$
Example: $\mathsf{Award}(x) \leftrightarrow \mathsf{FundingAward}(x)$

Class Subsumption. This pattern is very similar to the first pattern. But, instead of class equivalence, this pattern describes simple 1-to-1 class subsumption.

Table 2. The alignment pattern types found in the GeoLink complex alignment benchmark, along with the number of times each occurs and the type of relation.

Pattern	Occurrences	Category
Class Equivalence	10	1:1
Class Subsumption	2	1:1
Property Equivalence	7	1:1
Property Equivalence Inverse	5	1:1
Class Typecasting Equivalence	4	1:n
Class Typecasting Subsumption	1	1:n
Property Typecasting Subsumption	5	1:n
Property Typecasting Subsumption Inverse	5	1:n
Typed Property Chain Equivalence	26	m:n
Typed Property Chain Equivalence Inverse	17	m:n
Typed Property Chain Subsumption	17	m:n
Typed Property Chain Subsumption Inverse	12	m:n

Formal Pattern: $C_1(x) \rightarrow C_2(x)$
Example: $\mathsf{GeoFeature}(x) \rightarrow \mathsf{Place}(x)$

Property Equivalence. Property alignment is also an important part of ontology alignment research [8]. This pattern captures simple 1-to-1 property equivalence. Property p_1 and property p_2 are from ontology O_1 and ontology O_2, respectively. The property can be either a data property or an object property.

Formal Pattern: $p_1(x, y) \leftrightarrow p_2(x, y)$
Example: $\mathsf{hasAward}(x, y) \leftrightarrow \mathsf{fundedBy}(x, y)$

Property Equivalence Inverse. This pattern is similar to the previous one, just that the domain and range values of a property are switched when it aligns to a property in another ontology.

Formal Pattern: $p_1(x, y) \leftrightarrow p_2(y, x)$
Example: $\mathsf{isAwardOf}(x, y) \leftrightarrow \mathsf{fundedBy}(y, x)$

Class Typecasting Equivalence. This pattern is more specific than the previous ones. The idea of typecasting, and why it is important in ontology modeling, is formally introduced and discussed in [8]. The pattern indicates that individuals of type C_1 in one ontology are cast into a subclass of C_2 in the other ontology. Note that punning is employed here – x is treated as an individual on the left hand side of the rule and as a class on the right hand side. For example, an instance of PlaceType in the GBO might be 'ocean'. This is cast into a subclass of Place in the GMO. The reverse is also true: if the GMO has a subclass of Place called Island, then 'island' is an instance of the class PlaceType in the GBO.

Formal Pattern: $C_1(x) \leftrightarrow \mathsf{rdfs{:}subclassOf}(x, C_2)$
Example: $\mathsf{PlaceType}(x) \leftrightarrow \mathsf{rdfs{:}subclassOf}(x, \mathsf{Place})$

Class Typecasting Subsumption. This pattern is almost identical to the one above, except that the rule only holds in one direction. In the example, a GeoFeatureType (which comes from the General Bathymetric Chart of the Oceans[8] vocabulary) is always a type of Place, but there are types of Places that are not GeoFeatureType.

Formal Pattern: $C_1(x) \rightarrow \mathsf{rdfs{:}subclassOf}(x, C_2)$
Example: $\mathsf{GeoFeatureType}(x) \rightarrow \mathsf{rdfs{:}subclassOf}(x, \mathsf{Place})$

Property Typecasting Subsumption. This pattern is similar in spirit to the Class Typecasting patterns mentioned above. However in this case, a property is cast into a class assignment statement. In a sense, this alignment drops information, as y does not occur on the right hand side

Formal Pattern: $p_1(x, y) \rightarrow \mathsf{rdf{:}type}(x, C_2)$
Example: $\mathsf{hasPlaceType}(x, y) \rightarrow \mathsf{rdf{:}type}(x, \mathsf{Place})$

We note here that some rules that fall under this category are not exact translations of the underlying SPARQL queries, due to expressibility constraints in EDOAL (see Sect. 4.3 below). For instance, instead the example above, which states that the hasPlaceType object property is subsumed by an rdf:type statement with the range value of Place, we would actually like to state the following, which reflects the SPARQL query:

Formal Pattern: $p_1(x, y) \leftrightarrow \mathsf{rdf{:}type}(x, y) \wedge \mathsf{rdfs{:}subclassOf}(y, C_2)$
Example: $\mathsf{hasPlaceType}(x, y) \leftrightarrow \mathsf{rdf{:}type}(x, y) \wedge \mathsf{rdfs{:}subclassOf}(y, \mathsf{Place})$

For instance, we would like a rule that implies that the GBO statement hasPlaceType(Honolulu,Island) is equivalent to stating that Honolulu is a type of Island *and* that Island is a subclass of Place in the GMO. In other words, one of the individuals occurring as a property filler on the GBO side is cast into a class on the GMO side. At the same time, the other property filler on the GBO side is asserted to be an instance of this class. However, this is not possible because the statement requires a variable (y), and that is not supported

[8] https://www.gebco.net.

by the core EDOAL language. The EDOAL specification does mention a pattern language that might enable this type of statement, but it does not appear to be fully supported at this time.

Property Typecasting Subsumption Inverse. This pattern is the same as the one above, except that the property fillers are flipped.

Formal Pattern: $p_1(x, y) \rightarrow \mathsf{rdf{:}type}(y, C_2)$
Example: $\mathsf{isPlaceTypeOf}(x, y) \rightarrow \mathsf{rdf{:}type}(y, \mathsf{Place})$

Again, in some cases we would actually like to state the following, which cannot be fully expressed in EDOAL, to the best of our knowledge:

Formal Pattern: $p_1(x, y) \rightarrow \mathsf{rdf{:}type}(y, x) \wedge \mathsf{rdfs{:}subclassOf}(x, C_2)$
Example: $\mathsf{isGeoFeatureTypeOf}(x, y) \rightarrow \mathsf{rdf{:}type}(y, x) \wedge \mathsf{rdfs{:}subclassOf}(x, \mathsf{Place})$

Typed Property Chain Equivalence. A property chain is a classical complex pattern that was introduced in [11]. This pattern captures the situation related to the hasSponsor property discussed in detail in Sect. 3. The pattern applies when a property, together with a type restriction on one or both of its fillers, in one ontology have been used to "flatten" the structure of the other ontology by short-cutting a property chain in that ontology. The pattern also ensures that the types of the property fillers involved in the property chain are typed appropriately in the other ontology. The formal pattern and example are shown below. The classes D_i and property r are from ontology O_1, and classes C_i and properties p_i are from ontology O_2.

Formal Pattern:

$$D_1(x_1) \wedge r(x_1, x_{n+1}) \wedge D_2(x_{n+1}) \leftrightarrow C_1(x_1) \wedge p_1(x_1, x_2) \wedge C_2(x_2) \wedge \cdots \wedge p_n(x_n, x_{n+1}) \wedge C_{n+1}(x_{n+1})$$

Example[9]:

$$\mathsf{Award}(x) \wedge \mathsf{hasSponsor}(x, z) \leftrightarrow \mathsf{FundingAward}(x) \wedge \mathsf{providesAgentRole}(x, y) \wedge \mathsf{SponsorRole}(y) \wedge \mathsf{performedBy}(y, z)$$

Note that in this and all following patterns, any of the D_i or C_i may be omitted (in which case they are essentially $\top$). Also, for the left-to-right direction, we assume that $x_2, \ldots x_n$ are existentially quantified variables.

[9] In contrast to the example discussed in Fig. 3, we leave out `:Organization` and `view:Organization`, because it is possible, in principle, that a non-organization agent (e.g., an individual) may sponsor.

Typed Property Chain Equivalence Inverse. This pattern is the same as the one above, except that the property fillers are flipped.

Formal Pattern:

$$D_1(x_1) \wedge r(x_1, x_{n+1}) \wedge D_2(x_{n+1}) \leftrightarrow C_1(x_{n+1}) \wedge p_1(x_{n+1}, x_n) \wedge C_2(x_n) \wedge \cdots \wedge p_n(x_2, x_1) \wedge C_{n+1}(x_1)$$

Example:

$$\mathsf{Award}(z) \wedge \mathsf{isSponsorOf}(x, z) \leftrightarrow \mathsf{FundingAward}(z) \wedge \mathsf{provideAgentRole}(z, y) \wedge \mathsf{SponsorRole}(y) \wedge \mathsf{performedBy}(y, x)$$

Typed Property Chain Subsumption. This is identical to the Typed Property Chain Equivalence pattern except that the relationship only holds in one direction.

Formal Pattern:

$$D_1(x_1) \wedge r(x_1, x_{n+1}) \wedge D_2(x_{n+1}) \rightarrow C_1(x_1) \wedge p_1(x_1, x_2) \wedge C_2(x_2) \wedge \cdots \wedge p_n(x_n, x_{n+1}) \wedge C_{n+1}(x_{n+1})$$

Example:

$$\mathsf{Cruise}(x) \wedge \mathsf{hasChiefScientist}(x, z) \rightarrow \mathsf{Cruise}(x) \wedge \mathsf{providesAgentRole}(x, y) \wedge \mathsf{AgentRole}(y) \wedge \mathsf{performedBy}(y, z)$$

Typed Property Chain Subsumption Inverse. This pattern is the same as the one above, except that the property fillers are flipped.

Formal Pattern:

$$D_1(x_1) \wedge r(x_1, x_{n+1}) \wedge D_2(x_{n+1}) \rightarrow C_1(x_{n+1}) \wedge p_1(x_{n+1}, x_n) \wedge C_2(x_n) \wedge \cdots \wedge p_n(x_2, x_1) \wedge C_{n+1}(x_1)$$

Example:

$$\mathsf{Cruise}(z) \wedge \mathsf{isChiefScientistOf}(x, z) \rightarrow \mathsf{Cruise}(z) \wedge \mathsf{providesAgentRole}(z, y) \wedge \mathsf{AgentRole}(y) \wedge \mathsf{performedBy}(y, x)$$

In [11], four alignment types were identified, some of which are subsumed by ours. We do not at all claim that our classification above is exhaustive, but we consider it a refinement of the ones listed in [11]. We conjecture that there are many more important ones of relevance to other use cases. Mapping out the space of complex alignment types is, in our understanding, helpful for further research into complex alignment algorithms.

4.3 Format in EDOAL and Rule Syntax

As mentioned previously, SPARQL construct queries are used to convert data published by the data providers according to the GBO into the schema described in the GMO, because the GMO employs modeling practices that enhance extensibility and facilitate reasoning. However, most ontology alignment benchmarks are not formatted in SPARQL but rather according to the format provided by the Alignment API [5]. The standard alignment format is not expressive enough to capture complex relations. However, the Alignment API also provides a format called EDOAL that can be used to express these types of relations. This format can be read and manipulated programmatically using the Alignment API, and is therefore very convenient for ontology alignment researchers. In addition, EDOAL is already accepted by the ontology alignment community. It has been used by others when proposing new alignment benchmarks [10] and [14], and we continue that approach here. Because EDOAL can be difficult for humans to parse quickly, we have also expressed the alignments in using a naive rule syntax. The rule presentation is not intended for programmatic manipulation, but rather to make it easier for humans to read and understand the alignments. Both versions of the alignment, along with the GBO and GMO ontologies, can be downloaded from https://doi.org/10.6084/m9.figshare.5907172 under a CC-BY License. We apply HermiT [3] reasoning to the ontologies independently to check satisfiability, since some EODAL mappings which are part of our benchmark do not seem to be expressible in OWL DL. The GeoLink website[10] contains detailed documentation of the dataset and provides users with more insights about the resource, such as all entities, patterns, and relationships between them in both ontologies.

5 Conclusion

Complex alignment has been discussed for a long time, but relatively little work has been done to advance the state of the art of complex ontology alignment. The lack of an available complex alignment benchmark may be a primary reason for the slow speed of development. In addition, most current alignment benchmarks have been created by humans for the sole purpose of evaluating alignment systems, and they may not always represent real-world cases. In this paper, we have proposed a complex alignment benchmark based on the real-world GeoLink dataset. The two ontologies and the reference alignment were designed and created by ontologists and geoscience domain experts to support data representation, sharing, integration and discovery. We take advantage of these ontologies to create a complex alignment benchmark. In our dataset, the alignments not only cover 1:1 simple correspondences, but also contain 1:n and m:n complex relations. All correspondences required to convert between the two ontologies (a key goal of ontology alignment) are guaranteed to be present, because one ontology was consciously created from the other, with SPARQL queries to mitigate

[10] http://schema.geolink.org/.

each change. In addition, the alignment has been evaluated by domain experts from different organizations to ensure high quality. Moreover, the ontologies and alignments in both rule and EDOAL syntax have been published in FigShare with an open access license for reusability.

As future work in this area, we plan to put forth this alignment problem as a potential new track within the OAEI. To be specific, we will create three sub-tasks related to our benchmark to simplify for benchmarking and in order to help researchers to work towards full complex alignment. The three sub-tasks we have in mind are entity identification, relationship identification, and full complex alignment identification. In entity identification, the researchers are asked to identify the entities involved in a complex alignment, including classes and properties. In relationship identification, the system will endeavor to find the concrete relationships, such as equivalence, subsumption, intersection, value restriction, and so on., that hold between the entities when we give it the involved entities generated by entity identification as input. Full complex alignment identification is a combination of the former two steps. With respect to the maintenance of the benchmark, our intention is to remain actively involved for years to come in the OAEI complex alignment benchmarking track, and to also develop corresponding alignment methods. We thus have an intrinsic interest in keeping the benchmark maintained and usable, which would, e.g., mean that we are prepared to transfer it to new benchmarking framework if required in the future. At the same time, based on participants' feedback, we will modify the reference alignment if necessary to perfect the benchmark by making it more convenient to use. This may involve, for example, making the alignment available in additional formats. Furthermore, we also plan to create an automated alignment system to tackle the alignment problem set forth by this benchmark.

Acknowledgments. We would like to thank the geosciences data providers for sharing the data, and the domain experts for helping understand the concepts to create the ontologies and evaluate the reference alignment. In addition, we would also like to show our gratitude to Jerome Euzenat for providing advice regarding the conversion of rules to EDOAL.

References

1. Cheatham, M., Hitzler, P.: The properties of property alignment. In: Proceedings of the 9th International Workshop on Ontology Matching collocated with the 13th International Semantic Web Conference (ISWC 2014), Riva del Garda, Trentino, Italy, October 20, 2014, pp. 13–24 (2014)
2. David, J.: AROMA results for OAEI 2009. In: Proceedings of the 4th International Workshop on Ontology Matching (OM-2009) Collocated with the 8th International Semantic Web Conference (ISWC-2009) Chantilly, USA, October 25 (2009)
3. Glimm, B., Horrocks, I., Motik, B., Stoilos, G., Wang, Z.: Hermit: an OWL 2 reasoner. J. Autom. Reason. **53**(3), 245–269 (2014)

4. Hitzler, P., Gangemi, A., Janowicz, K., Krisnadhi, A., Presutti, V. (eds.): Ontology Engineering with Ontology Design Patterns - Foundations and Applications, Studies on the Semantic Web, vol. 25. IOS Press, Netherlands (2016)
5. Jain, P., Hitzler, P., Sheth, A.P., Verma, K., Yeh, P.Z.: Ontology alignment for linked open data. In: Patel-Schneider, P.F., et al. (eds.) ISWC 2010. LNCS, vol. 6496, pp. 402–417. Springer, Heidelberg (2010). https://doi.org/10.1007/978-3-642-17746-0_26
6. Krisnadhi, A.: Ontology Pattern-Based Data Integration. Ph.D. thesis, Wright State University (2015)
7. Krisnadhi, A., et al.: The GeoLink modular oceanography ontology. In: Arenas, M., et al. (eds.) ISWC 2015. LNCS, vol. 9367, pp. 301–309. Springer, Cham (2015). https://doi.org/10.1007/978-3-319-25010-6_19
8. Krisnadhi, A.A., Hitzler, P., Janowicz, K.: On the capabilities and limitations of OWL regarding typecasting and ontology design pattern views. In: Tamma, V., Dragoni, M., Gonçalves, R., Ławrynowicz, A. (eds.) OWLED 2015. LNCS, vol. 9557, pp. 105–116. Springer, Cham (2016). https://doi.org/10.1007/978-3-319-33245-1_11
9. Krisnadhi, A.A., et al.: The geolink framework for pattern-based linked data integration. In: Proceedings of the ISWC 2015 Posters & Demonstrations Track co-located with the 14th International Semantic Web Conference (ISWC-2015), Bethlehem, PA, USA, October 11, 2015 (2015)
10. Pesquita, C., Cheatham, M., Faria, D., Barros, J., Santos, E., Couto, F.M.: Building reference alignments for compound matching of multiple ontologies using OBO cross-products. In: Proceedings of the 9th International Workshop on Ontology Matching collocated with the 13th International Semantic Web Conference (ISWC 2014), Riva del Garda, Trentino, Italy, October 20, 2014, pp. 172–173 (2014)
11. Ritze, D., Meilicke, C., Sváb-Zamazal, O., Stuckenschmidt, H.: A pattern-based ontology matching approach for detecting complex correspondences. In: Proceedings of the 4th International Workshop on Ontology Matching (OM-2009) collocated with the 8th International Semantic Web Conference (ISWC-2009) Chantilly, USA, October 25, 2009 (2009)
12. Shvaiko, P., Euzenat, J.: Ontology matching: state of the art and future challenges. IEEE Trans. Knowl. Data Eng. **25**(1), 158–176 (2013)
13. Suchanek, F.M., Abiteboul, S., Senellart, P.: PARIS: probabilistic alignment of relations, instances, and schema. PVLDB **5**(3), 157–168 (2011)
14. Thiéblin, É., Haemmerlé, O., Hernandez, N., dos Santos, C.T.: Towards a complex alignment evaluation dataset. In: Proceedings of the 12th International Workshop on Ontology Matching co-located with the 16th International Semantic Web Conference (ISWC 2017), Vienna, Austria, October 21, 2017, pp. 217–218 (2017)
15. You, J.: Geoscientists aim to magnify specialized web searching (2015)

In-Use Track

Supporting Digital Healthcare Services Using Semantic Web Technologies

Gintaras Barisevičius, Martin Coste, David Geleta, Damir Juric, Mohammad Khodadadi, Giorgos Stoilos(✉), and Ilya Zaihrayeu

Babylon Health, London SW3 3DD, UK
{gintaras.barisevicius,martin.coste,david.geleta,damir.juric, mohammad.khodadadi,giorgos.stoilos,ilya.zaihrayeu}@babylonhealth.com

Abstract. We report on our efforts and faced challenges in using Semantic Web technologies for the purposes of supporting healthcare services provided by Babylon Health. First, we created a large medical Linked Data Graph (LDG) which integrates many publicly available (bio)medical data sources as well as several country specific ones for which we had to build custom RDF-converters. Even for data sources already distributed in RDF format a conversion process had to be applied in order to unify their schemata, simplify their structure and adapt them to the Babylon data model. Another important issue in maintaining and managing the LDG was versioning and updating with new releases of data sources. After creating the LDG, various services were built on top in order to provide an abstraction layer for non-expert end-users like doctors and software engineers which need to interact with it. Finally, we report on one of the key use cases built in Babylon, namely an AI-based chatbot which can be used by users to determine if they are in need of immediate medical treatment or they can follow a conservative treatment at home. To match user text to our internal AI-models an NLP-based knowledge extraction and logic-based reasoning approach was implemented and evaluation provided with encouraging results.

1 Introduction

The use of Semantic Web technologies such as Linked Data have started to be used extensively in many real-world applications [5,8,15]. Especially in the biomedical domain, OWL has been adopted since the early days of the Semantic Web and used to create a large number of medical ontologies [19], prominent examples of which are SNOMED [22], FMA [11], NCI [10], the Disease ontology [20], and many more. Many of these ontologies cover different and complementary topics such as genes, human phenotypes, proteins, and so on, and can be quite heterogeneous making it hard to retrieve information in a uniform way. Linking them under a homogeneous data model over which applications can be built would be beneficial [5,18].

Semantic technologies have also been adopted within Babylon Health.[1] Babylon offers healthcare services through a mobile application. Users can register to

[1] https://www.babylonhealth.com/.

D. Vrandečić et al. (Eds.): ISWC 2018, LNCS 11137, pp. 291–306, 2018.
https://doi.org/10.1007/978-3-030-00668-6_18

the app and have video consultations with doctors and healthcare professionals. The service also allows doctors to prescribe drugs to patients which can receive them from pharmacies of their choice. Moreover, patients can also receive referrals to health specialists or book lab exams with nearby facilities. Besides consultation with doctors, Babylon has been developing an AI-based doctor accessible through a chatbot, which can be used for symptom checking and triaging—that is, determining if the conditions that a user is entering in the chatbot are critical and he/she is in need of immediate medical attention or he/she can follow a conservative treatment at home.

Various services within Babylon generate, exchange, and consume clinical and health data and knowledge. For example, information extraction and text annotation services have been developed in order to process patient entered text and recognise the relevant medical terms that are entered. These terms may need to be compared with symptoms and risk factors in our symptom checking and triaging engines or with past diagnosis stored in the user profiles. Various other services in Babylon like drug prescribing or billing also deal with medical data like drugs, their side effects, contraindications and more.

To support the above services a medical Linked Data Graph (LDG) has been created by converting various medical data sources into RDF. Hence, all data within Babylon (diagnosis, drugs, etc.) are encoded using codes from medical ontologies like SNOMED, NCI, ICD-10, and more. Standards are heavily used in order to represent complex medical conditions and reasoning services have been implemented in order to achieve high degree of interoperability and intercommunication between the services. Some of the challenges faced in realising this use case are the following:

- Biomedical data sources are highly heterogeneous and custom converters had to be implemented in order to unify and harmonise them.
- Although efforts like BioPortal [19] and Bio2RDF [2] already offer a very large number of medical ontologies in RDF, several country specific clinical data sources are missing.
- OWL often exhibits complex structure (e.g., complex and/or nested class expressions) which would be impossible to be interpreted by our non-expert end-users (mostly doctors and software engineers). Consequently, even data sources distributed in OWL had to be converted to our simplified model.
- Updating our LDG with new releases of the data sources is a non-trivial issue since services already operate over the existing schema and changes may alter their behaviour.
- It would not be possible for our end-users to interact with triple-stores and use SPARQL, hence abstraction layers, services, and browsers had to be implemented in order help them use and feel comfortable with Semantic Web technologies and the LDG.
- Comparing OWL classes with existing reasoners is too strict in a real-world setting where one has to deal with language ambiguity and variability.

In the following, we first present our efforts in creating a medical Linked Data Graph and show how we addressed the above challenges. Next, we present several of the services we built around it in order to make the content accessible and easy to use by our non-experts. In more detail, we built a middle-ware service, called ClinicalKnowledge, whose purpose is to provide an abstraction layer to the LDG through a set of REST services. In addition, we have also implemented a web-browsing tool which can be used to search for classes and see their content like relations to other classes, direct super-classes, and more. Although, our LDG does not store complex OWL class expressions, such expressions are used in other components and services within Babylon like the triaging engine or patient profiles where complex medical conditions are formed by combining IRIs from the LDG and using OWL constructors. These expressions need to be compared with each other in order for services to exchange knowledge and interoperate and for these purposes a custom (hybrid) reasoner was implemented. Finally, we report on the triaging use-case built in Babylon and the role of our hybrid reasoner in matching user text to the internal triaging and symptom checking models. Preliminary evaluation of our Semantic Web-based (NLP plus logic) solution provided with encouraging results.

2 Building a Medical Linked Data Graph

The overall architecture of our platform is depicted in Fig. 1. The pipeline currently supports structured (RDF/OWL) and semi-structured (XML, CSV/TSV) data sources. All sources (even those already distributed in RDF) undergo a conversion process in which their schema and structure is processed in order to adapt it to the RDF model used in Babylon and reconcile their differences as much as possible. This conversion process also links the sources to an *Upper Level Ontology* which consists of an abstract medical model via which access is realised. Since data sources often feature overlaps, ontology alignment algorithms [21] are also used in order to establish mappings between the various sources and improve the level of integration. All converted data sources as well as the computed mappings are loaded into GraphDB.[2] The pipeline also supports the integration of information extracted from unstructured (web) data sources via Machine Reading [9] and crawling but the description of this pipeline is out of the scope of this paper. On top of the LDG a set of services is provided for outside clients to interact with the LDG. As it can be seen in the architecture, the LDG is continuously updated with new data sources as these are released. In the following sections we present further technical details about the aforementioned components.

2.1 Data Sources

Today a wealth of medical knowledge and data sources are available on-line. Several of these are already distributed in OWL and/or RDF, prominent examples of which are SNOMED, NCI, FMA, the ontologies in BioPortal, and many more. The UMLS project also consists of a continuous effort towards integrating

[2] https://ontotext.com/products/graphdb/.

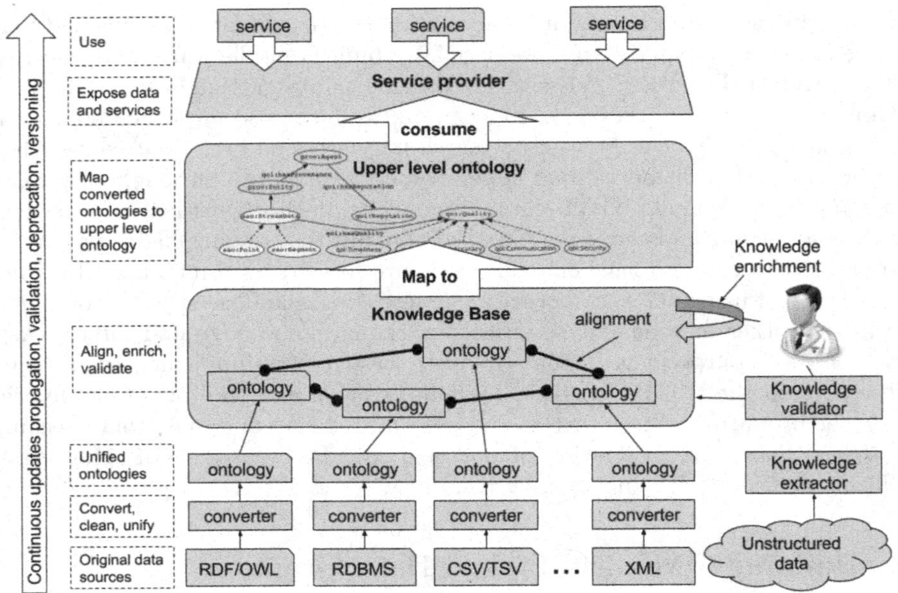

Fig. 1. Architecture of Babylon's linked data graph generation pipeline

and linking biomedical ontologies under a common vocabulary and providing homogeneous access through the UMLS semantic network [17]. The Babylon LDG uses several ontologies from UMLS, however, two issues were identified: (i) UMLS does not contain the most up to date releases of data sources, e.g., new releases of SNOMED are integrated with a six months delay, and (ii) it is missing some country specific data sources like dm+d (which is actually updated weekly), the Canadian Clinical Drug Dataset, and more.

Other prominent efforts in integrating biomedical data sources under a common RDF-based model is the Bio2RDF effort [2,4]. Bio2RDF could be a potential fit for the Babylon use case, however, it mostly focuses on Genes, Proteins, Genomics, and so forth, while Babylon mostly focuses on clinical services like Diseases, Risk Factors, Symptoms, and Drugs, so data sources like SNOMED and NCI which are missing from Bio2RDF are critical. In addition, most importantly the schema and structure of the generated Linked Data Graph had to be under the control of Babylon in order to be easily customisable and adaptable to internal service requirements. For these reasons several custom converters had to be implemented and an LDG was constructed from scratch.

The LDG was built by integrating almost 300 structured as well as semi-structured data sources. We have used most of the UMLS, several ontologies from BioPortal, latest versions of well-known ontologies like FMA, SNOMED, NCI, coding systems like ICD-10(pcs) and ReadCodes, as well as several country specific sources and extensions like dm+d, RxNorm, and more. Table 1 presents

Table 1. List of Important Data sources in our LDG.

Ontology	s/o IRIs	properties	$\sharp\{iri_1\ p\ iri_2\ .\}$	$\sharp\{iri_1\ p\ \mathsf{Lt}\ .\}$	$\sqsubseteq$
General Medical Vocabularies					
SNOMED	326 k	78	926 k	1,1 m	486 k
NCI	133 k	204	298 k	1,8 m	147 k
CHV	57 k	8	0	247 k	0
MeSH	2 m	44	3,6 m	9,6 m	187 k
MedDRA	26 k	8	0	885 k	34 k
Drug Ontologies					
dm+d	309 k	33	444 k	1,6 m	0
SIDER	1 m	28	2,6 m	8,4 m	0
Drugbank	10 m	36	74 m	195 k	0
RxNorm	114 k	41	989 k	1,12 m	200 k
DailyMed	10 k	31	57 k	80 k	0
Coding Systems					
OPCS-4	11 k	6	0	22 k	9 k
ICD-10	11 k	12	11 k	35 k	11 k
ICD-10pcs	190 k	9	0 k	708 k	190 k
CTV3	322 k	97	679 k	868 k	278 k
Read2	89 k	9	0	355 k	89 k

statistics about some of the sources integrated in our LDG; it shows the number of classes and individuals (i.e., IRIs in the subject or object position of triples), the number of properties, the number of triples of the form $s\ p\ o$. where both s and o are IRIs, and the number of such triples where o is an `owl:Literal` and the number of subClassOf axioms ($\sqsubseteq$). For readability we have rounded up numbers and used "k" to indicate thousands and "m" to indicate millions. In total the Babylon medical LDG consists of about 280 million triples which were loaded in GraphDB.

2.2 Schema and Data Model

Since the LDG is stored in a triple-store all data sources are serialised to triples. This creates problems when the original source contains complex class expressions that require the use of blank nodes in order to be serialised. For example, the OWL axiom $\mathsf{Malaria} \sqsubseteq \exists \mathsf{mayHaveFinding.Fever}$ in the NCI ontology is serialised into the following set of triples, where _ : x is a blank node:

```
Malaria  rdfs:subClassOf      _:x .
    _:x  rdf:type             owl:Restriction .
    _:x  owl:onProperty       mayHaveFinding .
    _:x  owl:someValuesFrom   Fever .
```

Clearly, it will be cumbersome for services to operate over such structures as well as problematic for non-expert end-users to browse over them. Hence, our converters perform some level of simplification and normalisation whenever this is possible. For example, the above OWL axiom is serialised into the following triple:

`Malaria mayHaveFinding Fever .`

Other examples of performed normalisation are axioms of the form $A \equiv C \sqcap D$ which are normalised into $C \sqcap D \sqsubseteq A, A \sqsubseteq C$, and $A \sqsubseteq D$. From them only those that can be translated to triples without the use of blank nodes are added to our LDG (i.e., the latter two in the above case) while the others are saved in OWL format for potential future use like reasoning.

Another issue is that authors of different data sources choose different ways and names to represent the same information. For example, at least the following properties have been found in various ontologies for encoding synonym labels:

http://snomed.info/field/Description.term.synonym
http://www.geneontology.org/formats/oboInOwl#hasExactSynonym
http://www.w3.org/2004/02/skos/core#altLabel

In order to unify the model, our converters are replacing these labels with the label `skos:altLabel`. Every class has zero or more `skos:altLabel` properties attached to it and exactly one `skos:prefLabel` property per language tag.

Besides schema harmonisation and simplification, the converted ontologies need to also satisfy some logical and structural constraints. First, every converted ontology $\mathcal{O}$ contains exactly one top-level "root" class—that is, a class O_{root} such that for every atomic class $A \in \mathsf{Sig}(\mathcal{O})$ we have $\mathcal{O} \models A \sqsubseteq O_{\mathsf{root}}$ and $O_{\mathsf{root}} \sqsubseteq A \notin \mathcal{O}$. Second, $\mathcal{O}$ must not contain any cluster of equivalent classes—that is, no list of classes $A_1, \ldots, A_n$ should exist such that $\{A_1 \sqsubseteq A_2, A_2 \sqsubseteq A_3, \ldots, A_n \sqsubseteq A_1\} \subseteq \mathcal{O}$. From a semantic point of view such "loops" are not problematic [1], however, these complicate implementations of graph algorithms like, traversing the subClassOf hierarchy, computing paths between two entities, defining the depth of an ontology, and so one, hence it was decided that loops in imported ontologies would be eliminated. This is done using a depth-first search algorithm which detects them and removes the last subClassOf link.

2.3 Source Updates and Version Management

The medical domain is a very dynamic one and sources are updated very frequently. For example, a new version of SNOMED and UMLS is released every six months while dm+d is released every week. It is critical that updates are imported in the Babylon system as soon as possible since these can provide data on new or retracted drugs, risk factors, diseases, etc. Updates, however, bring the issue of ontology versioning and management. Note that obsolete content cannot be simply removed from the LDG as it may be in use by some services, thus a careful and controlled migration plan is needed. We explored two approaches to ontology versioning:

1. Encode source version in class IRIs. An advantage of this is that the content associated with IRIs never changes and services always receive the same content for the same set of input IRIs. A disadvantage is that, as some services cache IRIs, IRI migration to new versions is required and is a complicated process.
2. Keep IRIs version unaware and manage content updates behind the scenes. An advantage is that, migration is not required, however, when a source is updated services may receive unexpected or different results for the same IRI as content has changed.

From the above we are currently following approach 1. and the main motivation is to ensure that services have a fixed behaviour which does not change with source updates. To implement this approach our converters make IRIs version aware by using the following scheme:

```
bblPrefix:{ontology_id}-{ontology_version}/{resource_id}
```

For example, the class Malaria in NCI versions 17.07e and 17.12 is represented as follows:

http://kb.babylonhealth.com/nci_thesaurus-17.07e/C34797
http://kb.babylonhealth.com/nci_thesaurus-17.12/C34797

This way two different versions of the same data source can co-exist in the LDG. When a new version is integrated, owners of services within Babylon are notified and they start to migrate gradually to the new IRIs. This process can be asynchronous as different services may migrate at different points in time, so the two versions of the same data source may co-exist for even up to six months. After migration is successfully completed the old converted data source is removed from the LDG.

2.4 Upper Level Ontology

Different ontologies and datasets may use different vocabulary and structure in order to represent the same real-world medical concept. For example, to represent medical conditions one ontology may use the class Disease another the class Disorder while another the class ClinicalFinding. For that purposes the use of an upper-level-ontology (ULO) which will provide uniform and source independent access to the underlying LDG has been advocated [7, 12]. In the above example, the ULO can contain one class `ulo:Disease` and all the aforementioned classes can be declared to be subclasses of it. The classes and properties in ULO are those entities that are exposed to the services that are using the LDG and are called *Semantic Types*. Every class in the LDG is associated with at least one semantic type.

In Babylon we adopted the UMLS semantic network (SN) [3] as a starting point for our ULO, however, this was subsequently extended or altered when seemed necessary. More precisely, if a data source contained a top-level class that we feel could be interesting to be exposed to the services, then we created a new class in the ULO with the same label and linked the imported ontology class with the new one. For example, UMLS SN does not contain a class for rare diseases whereas the Babylon LDG includes the RareDisease ontology from BioPortal which provide such a grouping of diseases.

Ontologies that we import from UMLS are already associated with entities in ULO (since our ULO originates from the UMLS SN), however, ontologies which we convert using custom converters are not, hence these have to be assigned one. Given an ontology $\mathcal{O}$ and our ULO $\mathcal{O}_u$ the process of assigning Semantic Types to the classes in $\mathcal{O}$ is the following:

1. Identify "top-level" classes in $\mathcal{O}$ that represent a similar real-world notion as some class in $\mathcal{O}_u$. By similar notion we mean that both $\mathcal{O}$ and $\mathcal{O}_u$ contain classes with the same or similar labels; e.g., both contain classes with label "Disease" (same label), or one contains a class with label "Physiological Process" and the other a class with label "Physiological Function" (similar label). In essence this is a manual alignment process.
2. The relations identified in the previous step between a class C in $\mathcal{O}$ and a class D in $\mathcal{O}_u$ are recorded in the converter configuration in the form $\langle C, D\rangle$.
3. The converter for $\mathcal{O}$ reads the configuration and creates for every link $\langle C, D\rangle$ and every $\mathcal{O} \models C' \sqsubseteq C$ the triple C `bbl:hasSTY` D . assigning the Semantic Type D to every "descendant" class of C in $\mathcal{O}$.

A similar approach is followed for properties, however, properties of an imported ontology are linked to properties from ULO using subPropertyOf axioms. For example, "part of" relations in SNOMED, FMA, and the Alzheimer's Disease ontologies are declared to be subPropertyOf the ULO partOf relation while six "has ingredient" relations from different data sources are also linked under the ULO hasIngredient property.

Besides accessing the data through a unique abstract model, the ULO can also be used for checking consistency of the underlying LDG [12]. Some of the top-level classes of the ULO have been declared to be disjoint, e.g., Organism and ManufacturedObject. Then, the following query checks if the LDG contains classes that that are sub-classes of both of them.

```
ask where { ?x rdfs:subClassOf :Organism, :ManufacturedObject . }
```

which should return false.

In total our ULO contains 817 classes, 349 properties, 816 subClassOf axioms and 332 subPropertyOf axioms.

2.5 Source Alignment

An important part of data integration is discovering links (mappings) between the entities of the various different data sources. For example, class Malaria appears in at least 15 different sources in UMLS, as well as in SNOMED, NCI, and more. In each of them complementary information may be described about this condition, e.g., one source may describe its symptoms, another drugs to treat it and so on. It would be beneficial if we can establish mappings between these classes as then, when one queries for information about this disease data from all sources would be returned.

For that purpose a large effort was spent towards pair-wise aligning the imported data sources. Alignments were build by mostly using the mappings included in UMLS (which are used to build the silver standard in the OAEI campaign [13]); e.g., SNOMED Malaria is mapped to the NCI one if they share the same UMLS code. Mappings were initially stored in GraphDB as equivalence axioms, e.g., if class A is mapped to class B then axiom $A \equiv B$ is added. However, it quickly became clear that this approach does not scale as these axioms cause a combinatorial explosion of the inferred statements computed by GraphDB during loading (GraphDB materialises inferences at loading); more precisely, all ancestors and descendants of A (resp. B) become ancestors and descendants of B (resp. A). Our partial solution was a bit unconventional. Instead, we encoded mappings using `owl:sameAs` and used GraphDB's `owl:sameAs` optimisation[3] to significantly reduce the number of inferred statements.

We were able to load about 3 million mappings between entities of the LDG, however, it has become apparent that we have reached the limits of the capabilities of state-of-the-art triple-stores. Loading the LDG with these mappings takes about 36 hours and our attempts to load all computed mappings (about 4.5 million) have failed.[4] Another issue that has been raised is that although mappings help us complement the information that each source contains for each class it also causes duplication and redundancy. If all Malaria classes have a `skos:definition` in all these 15 sources and all of them are linked with `owl:sameAs`, then after reasoning every such class will contain 15 such definitions. Moreover, the ancestors of each class will be the union of the ancestor of each of these classes creating a blow-up in the number of ancestors of a class. To alleviate this issue one needs to built post-processing filtering mechanism on top of the LDG [5,18]. Such mechanisms were implemented, however, initial results show that they do not scale well in practice.

3 Data Usage and Querying

In the current section we provide some details about mid-level services have been built on top of our LDG. These services provide a form of abstraction layer for accessing and browsing the LDG or for comparing classes w.r.t. the knowledge stored in the LDG.

[3] http://graphdb.ontotext.com/documentation/standard/sameas-optimisation.html.

[4] Alternative triple-stores have also been investigated. We have also tried non-materialisation-based systems which although much faster in loading (as they don't perform reasoning) they fail at query time when they perform backward-chaining reasoning.

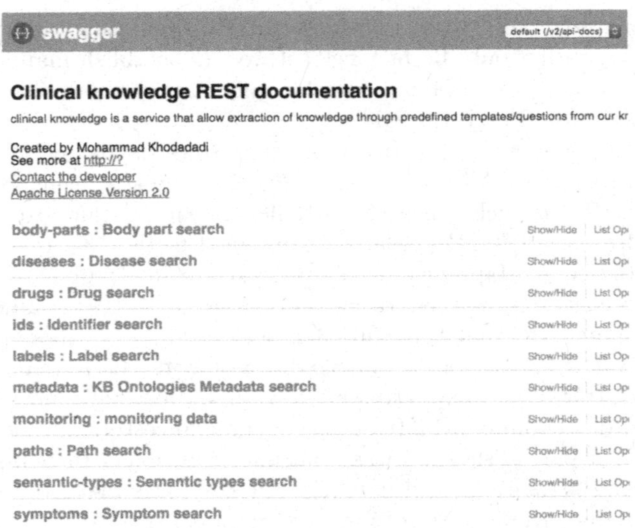

Fig. 2. Class browsing page.

3.1 Querying with Clinical Knowledge

To abstract away from the SPARQL syntax and provide our end-users with the ability to easily retrieve data from the LDG we have built a data access service, called ClinicalKnowledge (CK). CK is using the ULO which, as stated, is the abstract unified model of our medical LDG. CK provides 50 REST services which implement simple or more complex SPARQL queries over our LDG possibly with post-processing on the returned results. Those 50 services are grouped into 10 categories which can be depicted in Fig. 2. The most important services are the following:

- **Disease:** provides services to retrieve information about diseases, e.g., associated diseases, causes of a disease, associated drugs, symptoms, etc.
- **Label:** provides label-based search services, e.g., given some word return a list of IRIs having this word as `skos:prefLabel` or `skos:altLabel`.
- **Semantic Types:** given an IRI, services of this category can be used to retrieve its assigned Semantic Type (i.e., ULO classes), or given a Semantic Type to list all classes annotated with it, and more.
- **Paths:** Given an IRI, services of this group return paths of classes (w.r.t. subClassOf) from that IRI to the top-level or to leaf classes of the LDG.

3.2 Browsing with KB-Explorer

Besides retrieving data in a tabular format using CK, an important part for building services in Babylon or debugging existing ones is to browse through the stored data and understand them. For these purposes an in-house web-browser

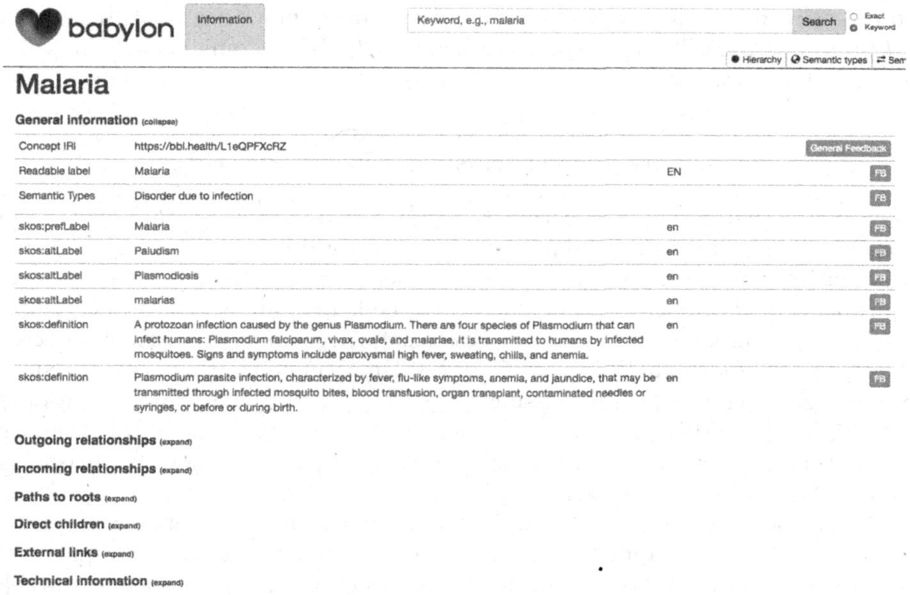

Fig. 3. Class browsing page.

for our LDG was implemented, called KB-explorer. A screen-shot of the class page of the KB-explorer for the class Malaria is depicted in Fig. 3.

The following information can be depicted for each class (some of this information is concealed in order not to overwhelm users):

- Preferred and alternative labels of the class with their language tags.
- The semantic types from ULO that have been assigned to this class, in this case DisorderDueToInfection.
- The `skos:definition` associated to the class (in this case there are two definitions coming from different sources).
- The paths in the hierarchy from the current class to top-level classes.
- Direct super/sub-classes of a class as well as its relations with other classes
- External links to DBpedia and/or Wikidata.

Through KB-explorer users can also browse the classes and properties of ULO as well as annotate medical terms in a text using an annotator that has been built in Babylon for experimentation purposes.

3.3 Class Comparison with a Hybrid Lightweight Reasoner

Babylon services produce and consume classes which are constructed using IRIs from the LDG. In several cases, a class expression may have to be formed in order to capture the medical condition of a patient. For example, the LDG contains no pre-defined class for the notion of a "recent injury in left leg" but this can be captured using OWL class expression using SNOMED atomic classes in either of the following ways:

$$C_1 = \mathsf{RecentInjury} \sqcap \exists \mathsf{findingSite.LeftLeg}$$
$$C_2 = \mathsf{Injury} \sqcap \exists \mathsf{occurred.Recent} \sqcap \exists \mathsf{findingSite.LeftLeg}$$

Inevitably, different services within Babylon will use either of the above (and possibly even more) ways to represent this real-world notion. It would be beneficial for interoperability purposes if we can determine that these two classes are equivalent. In theory, this can be done with the help of an OWL reasoner. If our LDG ($\mathcal{O}_{ldg}$) contained the axiom $ax := \mathsf{RecentInjury} \equiv \mathsf{Injury} \sqcap \exists \mathsf{occurred.Recent}$ then we would indeed have $\mathcal{O}_{ldg} \models C_1 \equiv C_2$. Unfortunately, in a very large number of cases such axioms are missing from $\mathcal{O}_{ldg}$ (in fact, the above axiom does not exist in SNOMED and hence neither in $\mathcal{O}_{ldg}$) and, moreover, the vast size of $\mathcal{O}_{ldg}$ makes it at least challenging to use any of the existing OWL reasoners to perform sub-class reasoning.

For these reasons a custom lightweight approximate reasoner was implemented on top of GraphDB. Since we are mostly dealing with class expressions containing existential quantifiers over which GraphDB is incomplete [6], the reasoner is using some of the consequence-based techniques presented in the literature [14] to improve the inference capabilities of GraphDB. However, it does not implement a complete $\mathcal{EL}$ calculus due to scalability reasons. In addition, in order to tackle the issue with the lack of axioms for classes (like the one mentioned above) it is also using NLP-based knowledge extraction techniques to extract (possibly missing) axioms from class labels. For example, consider class RecentInjury with preferred label "Recent Injury". Dependency parsing [16] is applied on the label in order to break it into word "Injury" with *modifier* "Recent". Then, Named Entity Disambiguation is applied to associate IRIs from the LDG to each of these words; assume that we successfully pick the IRIs of classes Injury and Recent. Finally, a relation from ULO is selected in an attempt to ideally build the expression $\mathsf{Injury} \sqcap \exists \mathsf{occurred.Recent}$. This is then used to replace class RecentInjury in C_1 building C_1' which is essentially the same as C_2 hence being able to determine that the two classes are equivalent.

4 Babylon Chatbot and Triaging

Patient triaging is one of the central automated services offered by Babylon through app's chatbot. Triaging is the process of sorting patients into groups based on their need for immediate medical treatment and can be used in hospital emergency rooms when limited medical resources are only available.

In addition to triaging, Babylon's chatbot also supports general purpose queries, like "Get me info for Malaria" or"What are the symptoms of flu". Figure 4 presents snapshots from Babylon's chatbot. Figure 4a depicts the initial screen prompting the user to enter some text, Fig. 4b depicts a triaging interaction with a user and Fig. 4c an information retrieval one. In order to determine

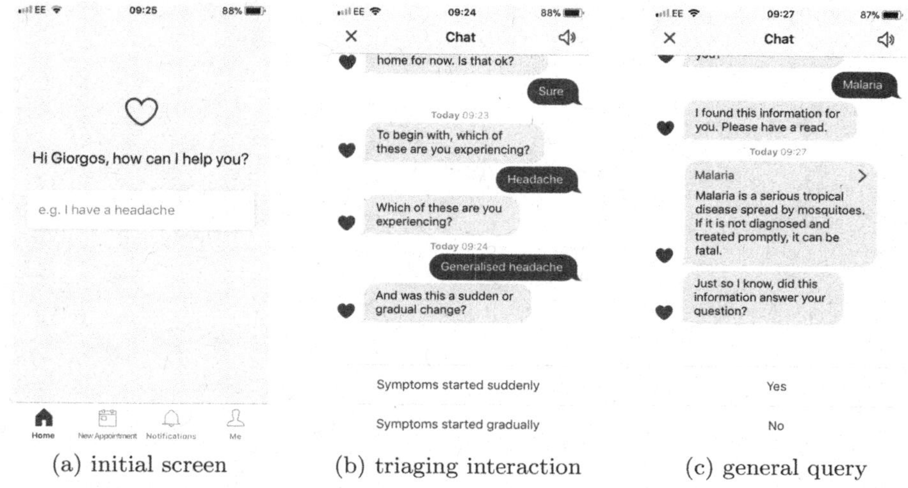

(a) initial screen (b) triaging interaction (c) general query

Fig. 4. Babylon chatbot

which type of interaction to initiate the original user text goes through a classifier which has been trained in Babylon. In Fig. 4b the user initially entered the phrase "My head hurts" hence initiating a triaging interaction. In contrast, in Fig. 4c the user entered the word "Malaria" and the classifier determined that the user entered a general information retrieval query.

Triaging is implemented with the notion of a *flow*. A flow is a directed graph where every node is a multiple choice question to be asked to the user and the answers determine the subsequent node and question to be asked. Every answer in a node is associated with an OWL class expression which is built using entities from the LDG and possibly also OWL constructors. For example, a node can contain the question "Right, do you have any pain?" with the following potential answers and associated classes:

- "ansId": 1, "txt": "No", None.
- "ansId": 2, "txt": "Yes, in one part of my head", $\mathsf{Headache} \sqcap \exists \mathsf{fSite}.\mathsf{HeadPart}$.
- "ansId": 3, "txt": "Yes, a general headache", $\mathsf{GeneralizedHeadache}$.

Flows are created by in-house doctors using a platform developed in Babylon. OWL constructors that have been used so far are $\neg$, $\sqcap$, $\sqcup$ and $\exists$. For example, the answer "Painful to touch scalp or temples" is associated to the class expression $\mathsf{Tenderness} \sqcap \exists \mathsf{fSite}.(\mathsf{Scalp} \sqcup \mathsf{Temples})$. Doctors have so far created flows for nineteen body parts or medical conditions some of which are, Fever, Chest, Pregnancy, Foot, Mouth, Head, Abdomen, and more. A flow can contain more than 50 nodes and each path in this graph encodes a potential interaction with the user until a conclusion is reached. So far more than 1,000 possible interactions have been encoded in the form of flows. The head-flow used to produce the triaging interaction in Fig. 4b is depicted in Fig. 5.

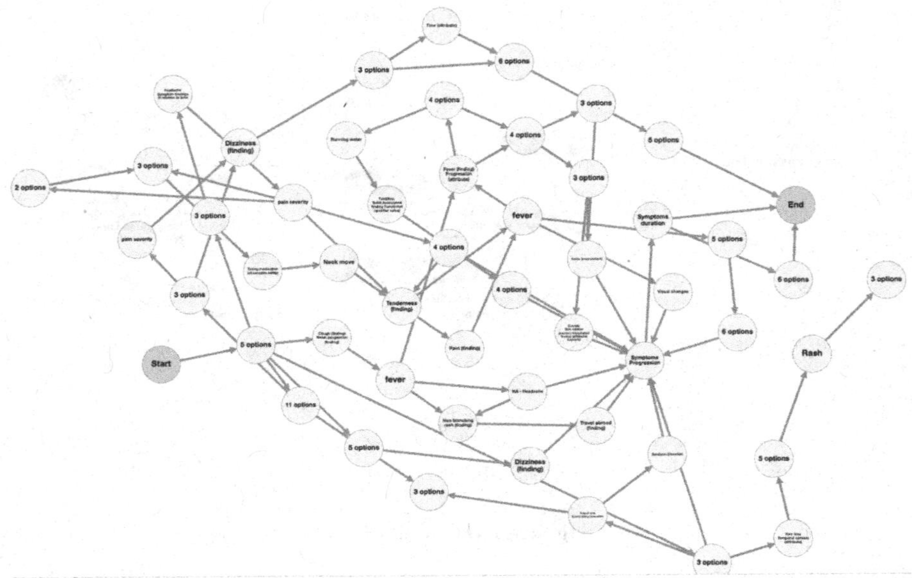

Fig. 5. Head flow

Which flows to activate given some user text is accomplished using our Knowledge-Extraction (KE) and hybrid reasoner methods. From a user text like "I feel a pain around my heart" our KE method extracts the class expression $\mathsf{Pain} \sqcap \exists \mathsf{fSite.Heart}$. Then, the reasoner is used to find all flows which contain nodes that are superclasses or this class; e.g., flows that contain nodes annotated with ChestPain. To evaluate our approach in-house doctors created 680 natural language queries mimicking the text that users would type into the chatbot as well as the list of expected superclasses from the LDG. For example, this test set includes text like"I cut my finger" or "My lower back hurts" while the expected LDG classes are HurtFinger and LowerBackPain, respectively. We measured 0.967 precision and 0.799 recall. The approach was compared against a Machine Learning one based on sentence embeddings and showed better precision and recall, hence the current setting uses the KE+hybrid reasoner while the sentence embedder is used as a fall-back solution.

So far the Babylon app has been downloaded about 600 K times within the UK. Moreover, 392 in-house or contracted doctors have conducted about 135 K video consultations while there have been about 326 K completed conversations with the chatbot.

5 Lessons Learnt and Conclusions

We have presented our efforts, challenges, design decisions, and solutions to problems faced while trying to use Semantic Web technologies in an industrial-strength healthcare application in Babylon Health. Our main lesson learned is that indeed "Linked Data Is Merely More Data" [12]. Following the best practices described in [12]—that is, an upper-level-ontology, integrity constraints to check validity of data, building abstraction layers on top of SPARQL, etc., does improve usability of the LDG. However, still its vast size and the heterogeneity of the sources makes it hard to maintain a consistent structure, comprehend and work with the data, link them using alignment, and build intelligent applications on top. Even if links are discovered, inferring equivalence and unifying (merging) the linked entities cannot be realised at this scale. Finally, comparing class expressions using traditional reasoning systems is very restrictive and the lack of a complete set of axioms in the background knowledge provides a limited recall.

On the positive side Semantic Technologies help at least in the following aspects. Complex class expressions allow us to dynamically represent almost any medical notion (condition) without the need to pre-define all of them (a task clearly impossible). The use of formal semantics for comparing classes helped us achieve a very high precision while the integration of NLP-based techniques also quite good recall improving on previous purely ML-based approaches. Formal semantics, integrity constraints, and the ULO also allowed us to ensure further data quality and consistency while the release of sources using standards relatively easy to create our LDG. Last but not least, materialisation in triples-stores is helping us infer new knowledge and also execute hierarchy traversal queries in a scalable way.

Regarding future plans we are currently in the process of re-building our LDG from scratch. We will follow a more conservative approach starting with few sources as a seed and enriching them with information from other sources [23]. We are also in the process of further improving our KE and hybrid reasoning approaches as well as enriching medical data sources with new knowledge.

References

1. Baader, F.: Terminological cycles in KL-ONE-based knowledge representation languages. In: Proceedings of the 8th National Conference on Artificial Intelligence, pp. 621–626 (1990)
2. Belleau, F., Nolin, M., Tourigny, N., Rigault, P., Morissette, J.: Bio2RDF: towards a mashup to build bioinformatics knowledge systems. J. Biomed. Inform. **41**(5), 706–716 (2008)
3. Bodenreider, O., McCray, A.T.: Exploring semantic groups through visual approaches. J. Biomed. Inform. **36**(6), 414–432 (2003)
4. Callahan, A., Cruz-Toledo, J., Ansell, P., Dumontier, M.: Bio2RDF Release 2: improved coverage, interoperability and provenance of life science linked data. In: Cimiano, P., Corcho, O., Presutti, V., Hollink, L., Rudolph, S. (eds.) ESWC 2013. LNCS, vol. 7882, pp. 200–212. Springer, Heidelberg (2013). https://doi.org/10.1007/978-3-642-38288-8_14
5. Chen, B., Dong, X., Jiao, D., Wang, H., Zhu, Q., Ding, Y., Wild, D.J.: Chem2Bio2RDF: a semantic framework for linking and data mining chemogenomic and systems chemical biology data. BMC Bioinform. **11**, 255 (2010)

6. Cuenca Grau, B., Motik, B., Stoilos, G., Horrocks, I.: Completeness guarantees for incomplete ontology reasoners: theory and practice. J. Artif. Intell. Res. **43**, 419–476 (2012)
7. Dumontier, M., Baker, C.J.O., Baran, J., Callahan, A., Chepelev, L.L., Cruz-Toledo, J., Rio, N.R.D., Duck, G., Furlong, L.I., Keath, N., et al.: The semantic-science integrated ontology (SIO) for biomedical research and knowledge discovery. J. Biomed. Semant. **5**, 14 (2014)
8. Elmer, S., Jrad, F., Liebig, T., ul Mehdi, A., Opitz, M., Stauß, T., Weidig, D.: Ontologies and reasoning to capture product complexity in automation industry. In: Proceedings of the 16th International Semantic Web Conference Posters & Demonstrations and Industry Tracks (2017)
9. Etzioni, O., Banko, M., Cafarella, M.J.: Machine reading. In: Proceedings of the 21st National Conference on Artificial Intelligence (AAAI), pp. 1517–1519 (2006)
10. Golbeck, J., Fragoso, G., Hartel, F.W., Hendler, J.A., Oberthaler, J., Parsia, B.: The national cancer institute's thésaurus and ontology. J. Web Semant. **1**(1), 75–80 (2003)
11. Golbreich, C., Grosjean, J., Darmoni, S.J.: The foundational model of anatomy in OWL 2 and its use. Artif. Intell. Med. **57**(2), 119–132 (2013)
12. Jain, P., Hitzler, P., Yeh, P.Z., Verma, K., Sheth, A.P.: Linked data is merely more data. In: Linked Data Meets Artificial Intelligence, Papers from the 2010 AAAI Spring Symposium, Technical Report SS-10-07, Stanford, California, USA (2010)
13. Jiménez-Ruiz, E., Grau, B.C., Horrocks, I.: Exploiting the UMLS metathesaurus in the ontology alignment evaluation initiative. In: Proceedings of the 2nd International Workshop on Exploiting Large Knowledge Repositories (2012)
14. Kazakov, Y.: Consequence-driven reasoning for horn SHIQ ontologies. In: Proceedings of the 21st International Joint Conference on Artificial Intelligence (IJCAI), pp. 2040–2045 (2009)
15. Knoblock, C.A., et al.: Lessons learned in building linked data for the American art collaborative. In: d'Amato, C., et al. (eds.) ISWC 2017. LNCS, vol. 10588, pp. 263–279. Springer, Cham (2017). https://doi.org/10.1007/978-3-319-68204-4_26
16. Kübler, S., McDonald, R.T., Nivre, J.: Dependency Parsing. Synthesis Lectures on Human Language Technologies. Morgan & Claypool Publishers, San Rafael (2009)
17. McCray, A., Nelson, S.: The representation of meaning in the UMLS. Meth. Inform. Med. **34**, 193–201 (1995)
18. Nolin, M., Dumontier, M., Belleau, F., Corbeil, J.: Building an HIV data mashup using Bio2RDF. Brief. Bioinform. **13**(1), 98–106 (2012)
19. Salvadores, M., Alexander, P.R., Musen, M.A., Noy, N.F.: Bioportal as a dataset of linked biomedical ontologies and terminologies in RDF. Semant. Web **4**(3), 277–284 (2013)
20. Schriml, L.M., et al.: Disease ontology: a backbone for disease semantic integration. Nucleic Acids Research **40(Database–Issue)**, 940–946 (2012)
21. Shvaiko, P., Euzenat, J.: Ontology matching: state of the art and future challenges. IEEE Trans. Knowl. Data Eng. **25**(1), 158–176 (2013)
22. Spackman, K.A., Campbell, K.E., Côté, R.A.: SNOMED RT: a reference terminology for health care. In: American Medical Informatics Association Annual Symposium, AMIA 1997, Nashville, TN, USA, October 25–29, 1997 (1997)
23. Stoilos, G., Geleta, D., Shamdasani, J., Khodadadi, M.: A novel approach and practical algorithms for ontology integration. In: Proceedings of 17th International Semantic Web Conference (ISWC) (2018)

Semantic Technologies for Healthy Lifestyle Monitoring

Mauro Dragoni(✉), Marco Rospocher, Tania Bailoni, Rosa Maimone, and Claudio Eccher

Fondazione Bruno Kessler, Trento, Italy
{dragoni,rospocher,tbailoni,rmaimone,cleccher}@fbk.eu

Abstract. People are nowadays well aware that adopting healthy lifestyles, i.e., a combination of correct diet and adequate physical activity, may significantly contribute to the prevention of chronic diseases. We present the use of Semantic Web technologies to build a system for supporting and motivating people in following healthy lifestyles. Semantic technologies are used for modeling all relevant information, and for fostering reasoning activities by combining real-time user-generated data and domain expert knowledge. The proposed solution is validated in a realistic scenario and lessons learned from this experience are reported.

1 Introduction

In the last decades, health care systems in many countries have invested substantial effort in informing people about the benefits of adopting healthy behaviors in their lives, including the prevention of several medical problems such as cognitive decline, obesity, disability, and death from major chronic diseases [1]. Given the increasing popularity of mobile and personalized applications and devices (e.g., smart watches), a natural follow up of this effort is the development of platforms capable of providing user tailored advices, motivating people to adopt healthy behaviors. Although Internet-based and mobile technologies allow to collect data from personal devices, off-the-shelf wearable sensors, and external sources, exploiting these data to generate effective personalized recommendations and to engage people in developing and maintaining healthier patterns of living, is a challenging task. To carry out this task, a system providing personalized healthy lifestyle support has to take into account and reason on a considerable amount of knowledge from different domains (e.g. user attitudes, preferences and environmental conditions, etc.), in order to generate effective personalized recommendations, and to adapt the message in response to the environment and the user status. Examples of such knowledge include food content and nutrients, physical activities accompanied by information concerning their categorization and effort, user attitudes and preferences, linguistic knowledge, and smart environment information (places, weather, etc.) Moreover, the creation, update and maintenance of the knowledge bases of the system by domain experts must be

D. Vrandečić et al. (Eds.): ISWC 2018, LNCS 11137, pp. 307–324, 2018.
https://doi.org/10.1007/978-3-030-00668-6_19

facilitated in order to obtain a flexible system that can be deployed in different domains, for different target users, and able to evolve in response to the continuous progress of the domain knowledge.

In this paper, we present the knowledge-based solution implemented into the PerKApp platform, a system providing personalized healthy lifestyle recommendations and advices. This work, follows the preliminary description of PerKApp [2], where the abstract architecture of the overall platform has been presented. Here, we put emphasis on presenting (i) the key knowledge component, responsible for generating healthy lifestyle advices and recommendations, (ii) the collaborative environment used by domain experts and knowledge engineers to model the knowledge exploited by the system, and (iii) the observation about how the overall platform, and in particular the knowledge part, behaves in a real-world environment. Then, we notably stress the role of the rule-based reasoner that is responsible of analyzing the compliance between data provided by users and prescription rules modeled by experts, and of providing feedback when the compliance is violated.

The rest of the paper is structured as follows. Section 2 provides a brief overview of available platforms for the monitoring of people's lifestyles and on the use of rule-based reasoners in real-world scenario. In Sect. 3 we discuss the challenges we want to address. While, in Sect. 4 we present the scenario where the PerKApp platform was used: namely the *Key to Health* project. Then, Sect. 5 presents in more detail the knowledge component. In Sect. 6 we exemplify PerKApp in a concrete scenario, while an evaluation of reasoning performances is reported in Sect. 7 together with the lessons learned from this experience. Finally, Sect. 8 concludes.

2 Related Work

The promotion of healthy lifestyle is a recent topic and the availability of systems working in this direction is limited. Nevertheless, some approaches based on Semantic Web technologies have been previously proposed.

In [3,4], intelligent systems recommending exercises tailored to the user profile are presented. Both exploit an ontology for representing information about the patient profile, goals and health data, and about the exercises and their effects. Inference rules are then combined with ontological data for providing health advices, i.e., for suggesting exercises to perform.

A further proposal is the Medical Decision Support System discussed in [5]. The presented solution allows (i) the collection of patients' relevant information via a mobile application prompting questions related to the patient's medical background, and (ii) the creation of customized advices based on the information previously collected and on the changes of patient's lifestyle. The system exploits a number of ontologies, including: (i) the Patient Profile Ontology describing general information, medical information, and medical measurements; (ii) the Questionnaire Ontology, formalizing concepts representing generic components of a questionnaire, sub-questionnaires, questions, potential answers, etc.; (iii) the Lifestyle Ontology; and, (iv) the Guidelines Ontology.

Finally, [6] presented an approach for designing a semantic reasoning engine for supporting coaching profiles. This system uses a web-based interface for collecting user data and an ontology for analyzing and processing such data. This way, created profiles can be used to optimize the coaching activities of professionals.

More in general, reasoning-based systems have been already adopted in the health-care domain [7–9].

With respect to the works presented in the literature, we provide a fully-fledged Semantic Web based solution supporting (i) the modeling and storing of all the data required to provide personalized healthy lifestyle support, as well as (ii) the definition and execution, via a reasoning engine, of a dynamic sets of rules performing real-time monitoring of people lifestyles. The output of the reasoning task is then used for suggesting people to change their habits in order to follow healthier behaviors. Our proposal fits in the context of ontology-centric decision support systems [10], as all the data processed (e.g., user profile, meals) and produced (e.g., violations) by the system are stored in an ontology-based knowledge base. To the best of our knowledge, our contribution is innovative with respect to the state of the art due the capability of integrating a multiple actor modeling environment with the possibility of performing a rule-based reasoning activities, at a very alimentary fine-grained level, for monitoring people behaviors in the healthcare context.

3 Personalized Healthy Lifestyle Support

Systems for personalized healthy lifestyle recommendations and advices fall in the broad area of decision support. The goal of these systems is to help and guide users in taking healthy-informed decisions about their lifestyle, on aspects such as food consumption or everyday physical activities. Such systems have to take a decision (e.g., suggesting some physical exercises or conscious food consumption), similarly as a human expert would do, based on available data (e.g., nutrients ingested in the last meals, user health conditions), and to communicate these decisions to the users according to their preferred means and modalities.

Developing a healthy lifestyle personal assistant requires addressing several challenges from the knowledge perspective:

- C1: to capture and model how experts work ("expert's knowledge") in assessing adherence to healthy lifestyle recommendations: among others, this involves representing different/heterogeneous kinds of data (e.g., food, nutrients, physical activities, user pathologies or needs) and how they interplay in defining a healthy lifestyle (e.g., the best practices an expert would recommend);
- C2: to develop effective and efficient techniques implementing expert's knowledge: these techniques apply expert's knowledge on real-time users' data, to assess how healthy is a user's lifestyle and identify violations of healthy lifestyle recommendations;

C3: to embed these techniques in an efficient/scalable/flexible decision support system: the system has to acquire, process, apply expert's knowledge to provide motivational messages, on the data of a (possibly large) community of monitored users;

C4: to provide facilities supporting experts in specifying and revising the prescriptions for a healthy lifestyle applied by the systems: as healthy lifestyle best practices are not immutable, and sometimes not even universally shared, but they continuously evolve (e.g., new prescriptions for new typologies of users), there is a need to facilitate experts in defining and revising the prescription used in the systems.

In the next sections, we show how the use of semantic technologies is particularly suited to address these challenges, and can foster further research and development in the area of personalized health lifestyle support.

4 The PerKApp Platform

The PerKApp platform,[1] schematically summarized in Fig. 1, is a personalized healthy lifestyle recommendations and advices system implementing the 3-layer architecture described in Sect. 3.

The Input/Output layer consists in a mobile application[2] supporting the acquisition of user information about consumed foods and physical activities. Concerning foods, users are able to select them from a list of almost 6,000 basic foods and dishes; while for physical activities, we connected the PerKApp platform with the APIs of the most popular services and wearable devices (Google Fit, Misfit, Polar, and Garmin), or users can specify what they did among the more than 800 activities available. The Persuasive Layer consists in the development of a natural language generation component integrating a set of persuasive strategies supporting users in behavioral change. The analysis of the challenges linked with these layers (i.e. the effort necessary for providing data rather than the methodology for communicating with users effectively) and how they have been addressed is out of the scope of this paper.

Semantic Technologies in PerKApp. Ontologies and rules are used to address Challenge C1. Ontologies enables to represent in a connected and comprehensive way all the content relevant for the given domain. More precisely, the HeLiS ontology [11], comprises different interconnected modules formally representing expert's knowledge and the data needed by experts to assess the adherence of a user to healthy lifestyle best practices. The choice of adopting an ontological representation enables both to easily reuse existing resources, and to make available to other initiatives the results of the modeling performed in PerKApp.

---

[1] http://perkapp.fbk.eu/.

[2] The application is available at https://play.google.com/store/apps/details?id=eu.fbk.trec.lifestyle. The current version is only in Italian. We are in the process of translating it in English and German.

For instance, for the food domain PerKApp integrates content from the OpenFoodFacts database, which provides information about foods through an RDF dump or an API service. At the same time, as readily reusable resources about physical activities are currently lacking, their ontological modeling in PerKApp can be reused in other initiatives needing structured information about them.

Semantic rules are used to capture the checks performed by healthy life-style experts, to detect user's violations of healthy lifestyle best practices, and to decide whether some recommendation or motivational message has to be notified to the user (the rules do not decide on the way or terminology to be used in the message, a task addressed in the Persuasion Layer). The collection, modeling, and revision of such rules (Challenge C4) by domain experts is supported by the integration of specific user facilities and informal templates into MoKi [12], a collaborative tool that has already been used for supporting the work of domain experts in the health domain [13].

The application and evaluation of these rules on real-time user data (Challenge C2) is performed by an inference engine module, powered by RDFpro [14], a tool enabling several RDF and Named Graph manipulation tasks, including rule-based reasoning. In particular, RDFpro has been configured with specific entailment rules featuring custom SPARQL functions to support the specific requirements of the presented scenario.

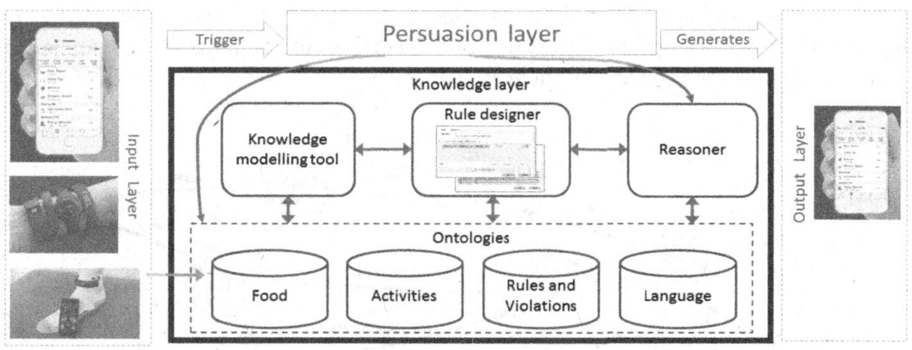

Fig. 1. The overview of the architecture implemented for the PerKApp project.

Finally (Challenge C3), all the data management part is supported by state-of-the-art triplestores, using named graphs to efficiently organize data (e.g., by user) and efficiently tuned SPARQL queries for all data access, manipulation and analysis tasks. As a result, PerKApp provides an evaluation testbed for the joint use of these technologies at scale and, as mentioned in Sect. 2, it contributes filling the current gap about the realization of resources and services integrating knowledge and reasoning capabilities exploiting food, physical activity, and user personalized information in synergy.

Key to Health. As a specific case study, PerKApp has been deployed and evaluated in the context of the project *Key to Health* on workplace health promotion (WHP) inside our institution (FBK). WHP, defined as "the combined efforts of employers, employees, and society to improve the mental and physical health and well-being of people at work",[3] aims at preventing the onset of chronic diseases related to an incorrect lifestyle through organizational interventions directed to workers. Actions can concern the promotion of correct diet, physical activity, and social and individual well-being, as well as the discouragement of bad habits, such as smoking and alcohol consumption. Within the Key to Health project, PerKApp has been used by 92 FBK's workers (both researchers and employers) as a tool to persuade and motivate them to follow WHP recommendations. This study represented a first step within the "Trentino Salute 4.0" digital health initiative aiming at extending the availability of the PerKApp platform at the whole Trentino Province before the end of 2018 and at Italian National level during the two-year period 2019–2020.

5 Knowledge Layer Components

In this Section, we focus on the components that are mainly involved in the Knowledge Layer of PerKApp. We show how the platform supports the domain experts in defining monitoring rules (Sect. 5.1). Then, we describe how reasoning is implemented to evaluate these rules (Sect. 5.2). Here, we skip a presentation of the underlying ontology which details are described in [15]. However, we report a schema including the main concepts in Fig. 2.

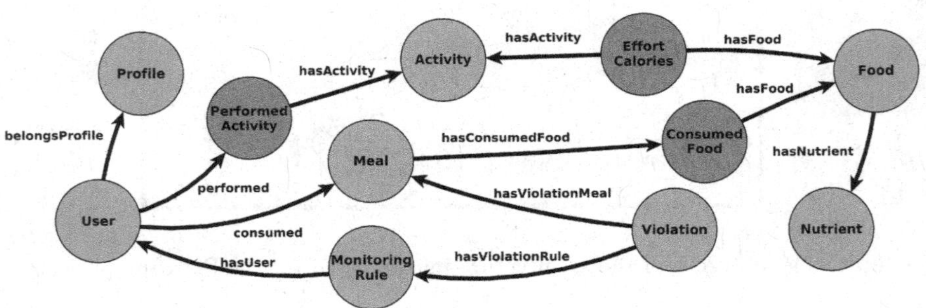

Fig. 2. The HeLiS ontology.

[3] Luxembourg Declaration on Workplace health promotion in the European Union. 1997.

5.1 Experts Support Facilities

As mentioned in Sect. 3, one of the challenges we wanted to address was the development of flexible facilities supporting domain experts in defining rules for monitoring user behaviors. Moreover, one of the requirements is to support the collaborative work of domain experts. Indeed, the definition of dietary rules may lead to disagreement between experts in deciding, for example, the amount of a specific food that a person has to eat during a day and the weekly frequency.

Domain experts' facilities have been implemented as an extension of MoKi [12]. The modeling of concepts and individuals related to the `Food`, `Activity`, and `User` branches is supported by views and forms enabling create, read, update, and delete (CRUD) operations. MoKi provides domain experts with several collaboration facilities (e.g., discussion pages, revision history), whose effectiveness was empirically confirmed [16]. Here, we want to focus on two flexible facilities offered by the tool: the definition of rules and the monitoring of user's behavior.

The definition of rules is supported by an interface that permits domain experts to create, view, or revise rules in a collaborative way, covering both the dietary and activity domains. Each form element is associated with one of the properties of a `MonitoringRule` instance and can be populated with actual values supplied by expert to instantiate the rule template, avoiding the need to learn a formal language for expressing rules. Within the interface, domain experts are able to select the timing dimension that the rule refers to, the type of check that the rule has to perform on user data (e.g., the presence of a specific food or nutrient), the type of entity that the rule has to monitor (i.e., `Food`, `Nutrient`, or `Activity`), and the name identifying that specific entity. Then, a comparison operator has to be selected together with the type of operation to be performed on the acquired data. Finally, other information, like a priority used for reporting, can be associated with the rule.

The second facility we want to highlight is the real-time monitoring of users' behaviors where experts can explore the reports generated by the tool. In this particular case, experts can observe, for a specific meal, the number of associated violations and their related details. The usefulness of this report is two-fold: on the one hand, experts can observe how much a user is respecting the rules assigned; on the other hand, the provided information is useful also for the experts themselves. Indeed, they can analyze if the evaluation of the modeled rules in a real-world context is correct or not. This is a critical aspect in a collaborative environment and, in particular, in the medical domain due to possible disagreements between experts in validating the results of the monitoring activity. In Sect. 7.4, we report a brief discussion about this point by presenting the results of a comparative analysis between experts on the output provided by the reasoner in four different real-world scenarios.

5.2 Rule-Based Reasoning

Reasoning in PerKApp has the goal of verifying that a user's lifestyle is consistent with the monitoring rules defined by domain experts, detecting and possibly materializing violations in the knowledge base, upon which further actions may be taken. Reasoning can be implemented via the fixed point, forward chaining evaluation of IF-THEN *entailment rules* (cf. *monitoring rules*, which are RDF individuals in the HeLiS ontology) that implement the semantics of OWL 2 RL (to account for TBox declarations in the ontology) and of monitoring rules by matching non-conforming patterns in RDF data and asserting corresponding `Violation` individuals.

Aspects of Interest. To effectively implement monitoring rules via rule-based reasoning, the following aspects need to be taken into accounts:

- *Non-monotonic reasoning.* The evaluation of monitoring rules involves aggregations (e.g., total calories of a meals) and negation-as-failure (e.g., a food is not consumed if there is no `ConsumedFood` for it), which make reasoning a non-monotonic process where acquiring new data may invalidate previous inferences.
- *Generation of new individuals.* The materialization of detected violations implies the creation of new individuals in the knowledge base, whose identity is tightly linked to the one of existing individuals (e.g., a meal for a violation affecting it).
- *Interplay with RDFS/OWL reasoning.* The evaluation of monitoring rules (e.g., that a meal must contain cereals) relies on the prior computation of OWL 2 inferences (e.g., to determine that `RiceFlakes` are `Cereals`), while at the same time resulting violation assertions can be the subject of further RDFS/OWL inferences.
- *Static/dynamic data.* The *Food*, *Activity*, and *Monitoring* branches of the ontology are largely static and are available up front, allowing to pre-compute inferences over this data and/or to optimize the entailment rules used real-time on user data.
- *Per-user reasoning.* As monitoring rules operate on a single user, reasoning can be done on a per-user basis achieving scalability via the parallel/distributed processing of multiple users at the same time.

Reasoning Workflow. We perform reasoning in PerKApp using RDFpro [14], a tool that allows us to consider the aforementioned aspects by supporting out-of-the-box OWL 2 RL and the fixed point evaluation of `INSERT... WHERE...` SPARQL-like entailment rules that leverage the full expressivity of SPARQL (e.g., `GROUP BY` aggregation, negation via `FILTER NOT EXISTS`, derivation of RDF nodes via `BIND`). We organize reasoning in two *offline* and *online* phases as shown in Fig. 3.

Offline, a one-time processing of the *static* part of the ontology (*Food*, *Activity*, *Monitoring* branches) is performed to materialize its deductive closure, based on OWL 2 RL and some additional pre-processing rules that identify the most specific types of each `Nutrient` individual (this information greatly helps in aggregating their amounts). The resulting *closed ontology* is then used to *expand* a set of PerKApp-specific entailment rules that evaluate monitoring rules on the *dynamic* user data described by the *User* branch of the ontology. The expanded rules are responsible of: (i) computing the OWL 2 RL deductive closure of user data; (ii) assert `Violation` individuals; and (iii) enrich these individuals with various triples according to the `Violation` model in the ontology. The expansion process consists in pre-computing the matches of specific rule body atoms that can only appear in the static ontology (e.g., `rdfs:subClassOf` statements), producing variable bindings that are replaced in entailment rules to derive a new (larger) set of simpler rules that can be evaluated more efficiently.

Online, each time the reasoning is triggered (e.g., a new meal or performed activity is entered), the user data is merged with the closed ontology and its deductive closure based on the *expanded rules* is computed. This process can occur both on a per-user basis or globally on the whole knowledge base. The resulting `Violation` individuals and their RDF descriptions are stored back in the knowledge base.

Rule Example. We report below an example of PerKApp entailment rule responsible of detecting violations of *DAY-Rules* constraining the daily amount of calories consumed.

```
INSERT { ?v :hasViolationRule ?rule; :hasViolationQuantity ?cal;
            :hasViolationUser ?user; :hasTimestamp ?checkTs. }
WHERE { { SELECT  ?rule ?user ?timestamp (SUM(?a / ?e * ?fc) AS ?cal)
          WHERE { ?rule a :MonitoringRule; :timing :Day; :command "hasCalories";
                        :appliesTo/:containsUser? ?user.
                  ?meal :hasUser ?user; :hasTimestamp ?ts; :hasConsumedFood ?cf.
                  ?cf :amountFood ?a; :hasFood [:amountCalories ?fc; :ediblePart ?e]
                  BIND (:computeTimestamp(:Day, ?ts) AS ?checkTs) }
          GROUP BY ?rule ?user ?checkTs }
        ?rule :hasOperator ?op.
        OPTIONAL { ?rule :hasMonitoredValue ?val }
        OPTIONAL { ?rule :hasMonitoredValueInterval ?interval.
                   ?interval :lowerBound ?lower; :upperBound ?upper. }
        FILTER (?op = "equal" && ?cal != ?val    || ?op = "less" && ?cal >= ?val ||
                ?op = "greater" && ?cal <= ?val || ?cal < ?lower || ?cal > ?upper)
        BIND (:mintViolation(?rule, ?user, ?checkTs) AS ?v) }
```

The nested `SELECT` query takes a monitoring rule and groups affected meals by user and by the *check* timestamp where daily rules for that meal should be evaluated (computed via a SPARQL custom function implemented in Java). The sum of calories is computed for each monitoring rule, user, and check timestamp. The rest of the `WHERE` clause evaluates the monitoring rule operator against the specified target value or value interval. If a violation is detected, a URI for a

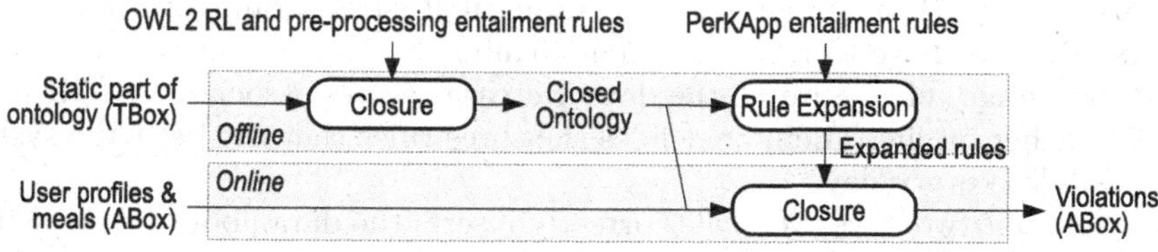

Fig. 3. Reasoning workflow in PerKApp.

new `Violation` individual is minted and that individual and its core violation data are asserted by the `INSERT` clause.

Scheduling. We leverage the distinction among *UB-Rules*, *DAY-Rules*, and *WEEK-Rules* to schedule the online reasoning phase at different times to improve efficiency:

- *UB-Rules* are evaluated in real-time on a meal or performed action immediately after they are received on the server, to enable the possibility of providing an immediate feedback to users. This kind of reasoning suffers of the possibility of high concurrency due to the amount of people providing their data during a small time interval. Hence, by restricting to evaluate only the necessary *UB-Rules* after each meal or activity, we are able to manage potential bottlenecks in elaborating data.
- *DAY-Rules* and *WEEK-Rules* are evaluated in background, respectively on a daily or weekly basis. Their evaluation implies the collection and aggregation of relevant amounts of data, requiring significant time for being analyzed. Therefore, the evaluation of these rules is scheduled at night or in other time slots when the system is relatively idle, so to avoid affecting the performance of the whole system.

6 The PerKApp Platform in Action

In order to explain in a clearer way which are the structures of the monitoring rules generated through the actions of the domain experts and the data acquired by the users, we describe here a simple scenario that will help the comprehension of how the reasoner, described in Sect. 5.2, works.

Let us consider the following scenario: Michelle is a sales agent of an important company. Her working days are very stressful especially in the morning, where she often has to drive for long distances for reaching her customers. After her workday, she uses to run as relaxing practice. In the last period, she started to have some dizziness during the morning and, sometimes even while driving. After a colloquium with her doctor, he noticed that she is used to have very light and quick breakfast and that the level of physical activity she does is maybe too much with respect to the diet she is following. For this reason, her doctor asked her to use the PerKApp application for monitoring what she eats in the morning and which is the actual level of physical activity she does. The doctor ask her to follow three rules: (i) the amount of calories for breakfast has to be higher than 250; (ii) the breakfast always has to contain at least 80 grams of cereals; and, (iii) she does not have to run more than 45 min.

At first stage (cf. Figure 4), the doctor creates a user associated to Michelle, and details her profile. Then, he defines the three rules that the PerKApp system has to validate every day.

For the first two days, Michelle correctly inserts the data about her breakfast and physical activity as shown in Fig. 5.

Based on this data, the ontology integrated within the PerKApp platform and the RDF encoding of the configured monitoring rules, the PerKApp reasoner

```
:MorningDizziness a :Profile.

:Michelle a :User; :hasUserId "4953"^^xsd:int; :belongsProfile :MorningDizziness.

:MR1 a :MonitoringRule; :appliesTo :MorningDizziness; :timing :Breakfast;
   :monitoredEntityType :Food; :monitoredEntity :Cereal; :command "contains";
   :hasOperator ">"; :hasMonitoredValue "80"^^xsd:int; ...
:MR2 a :MonitoringRule; :appliesTo :MorningDizziness; :timing Breakfast;
   :command "hasCalories"; :hasOperator ">"; :hasMonitoredValue "250"^^xsd:int; ...
:MR3 a :MonitoringRule; :appliesTo :MorningDizziness; :timing Day;
   :monitoredEntityType :Activity; :monitoredEntity :Running; :command "contains";
   :hasOperator "<"; :hasMonitoredValue "45"^^xsd:int; ...
```

Fig. 4. Example of a user profile with the description of the associated rules.

```
:Michelle :consumed :Breakfast-4953-1, :Breakfast-4953-2.
:Michelle :performed :Running-4953-1, :Running-4953-2.

:Breakfast-4953-1 a :Breakfast; :hasTimestamp "2017-04-14T07:19:00Z"^^xsd:dateTime;
   :hasConsumedFood [ :hasFood :GoatMilk; :amountFood "250"^^xsd:int ],
                    [ :hasFood :RiceFlakes; :amountFood "100"^^xsd:int ].
:Breakfast-4953-2 a :Breakfast; :hasTimestamp "2017-04-15T07:34:00Z^^xsd:dateTime";
   :hasConsumedFood [ :hasFood :TeaInCups; :amountFood "300"^^xsd:int ],
                    [ :hasFood :Apple; :amountFood "150"^^xsd:int ].

:Running-4953-1 :hasTimestamp "2017-04-14T17:45:00Z^^xsd:dateTime";
    a :PerformedActivity; :hasActivity :Running; :activityDuration "42"^^xsd:int .
:Running-4953-2 :hasTimestamp "2017-04-15T18:20:00Z^^xsd:dateTime";
    a :PerformedActivity; :hasActivity :Running; :activityDuration "38"^^xsd:int .
```

Fig. 5. Example of data provided by users and transformed in their structured representation.

determines that the amount of calories of the two breakfasts and the amount of performed physical activity satisfy the monitoring rules `MR1` and `MR3`, respectively. However, the reasoner determines that no cereals where consumed in the second breakfast, leading to a violation of the cereals rule (`MR1`). The `Violation` individual shown in Fig. 6 is thus asserted in the knowledge base.

```
:violation-4953-1-20170415 a :Violation;
   :hasViolationRule :MR1; :hasTimestamp "2017-04-15T07:34:00Z"^^xsd:dateTime;
   :hasViolationUser :Michelle; :hasViolationMeal :Breakfast-4953-2;
   :hasViolationQuantity "0"^^xsd:int; :hasViolationExpectedQuantity "80"^^xsd:int.
```

Fig. 6. Example of violation (excerpt).

As a result, Michelle did not receive any feedback from the system after the first breakfast as she respected all the monitoring rules provided by the doctor. The same happened also after the registration of her physical activities as she did not exceed the amount of minutes suggested by the doctor. Whereas, she receives a textual feedback from the PerKApp application after the second breakfast, communicating that she did not consume the suggested quantity of cereals. This kind of feedback aims to remember to Michelle that she has to follow doctor's suggestions for avoiding the arise of dizziness during the morning. At the same time, the doctor is able to see the violations of Michelle. This way, by combining the generated violations and Michelle's feedback about her health status after using the PerKApp application for a certain period of time, the doctor can verify if his hypothesis were true.

7 Evaluation

In this Section, we report the evaluation activities we performed on the PerKApp platform with a particular emphasis on the scalability aspect. Indeed, our main goal is to validate the suitability, and potential drawbacks, concerning the deployment of the PerKApp platform in bigger scenarios. We start by presenting an analysis of the data we collected during the first forty-five days of the *Key to Health* project (Sect. 7.1) and by extending our observations to a synthetic scenario simulating wider contexts (Sect. 7.2). Then, we report some lessons learned from the development and deployment of the PerKApp platform (Sect. 7.4).

7.1 Analysis of Real Data

Within the *Key to Health* project, PerKApp has been used in a period of 45 days by 92 users , 49 of which have reported data about their meals on a regular basis, and about their physical activities only occasionally. We thus focus on the meal data provided by those 49 users, analyzing it as well as the reasoning process evaluating monitoring rules.

Figure 7a provides an average characterization of the meals inserted by users based on their type (breakfast, lunch, snack, dinner), in terms of number of composing foods, calories and main nutrients (carbs, lipids, proteins), and the number of triples necessary to encode this meal data in the triplestore; a daily per-user aggregate is also reported. The data give evidence (on average) of a 1500 Kcal daily diet, although users may have omitted some consumed food or underestimated its amount, either unintentionally (i.e., they forget to enter a meal) or intentionally (e.g., to "hide" the consumption of unhealthy foods). The number of triples needed for a meal is in the order of few tens, suggesting that the representation of meals in the ontology is compact and thus makes easier to store and manipulate large numbers of meals. On average, 76 triples per user per day are currently needed, meaning that a small PerKApp deployment supporting 1B triples would manage to store one month worth of meal data for over 400 K users.

The box plots in Fig. 7c summarize the distributions of reasoning time and the number of output violations for the three different times reasoning is performed:[4] after each single meal to check *UB-Rules*; at the end of each day for *DAY-Rules*; and at the end of each week for *WEEK-Rules*. Both reasoning time and the number of violations increase moving from a single meal to a whole week, as more input data is being processed and more rules can be violated. Each

[4] Deployment machine: 2 CPUs Intel Xeon X5690 3.47GHz, 8GB RAM, 160GB HDD.

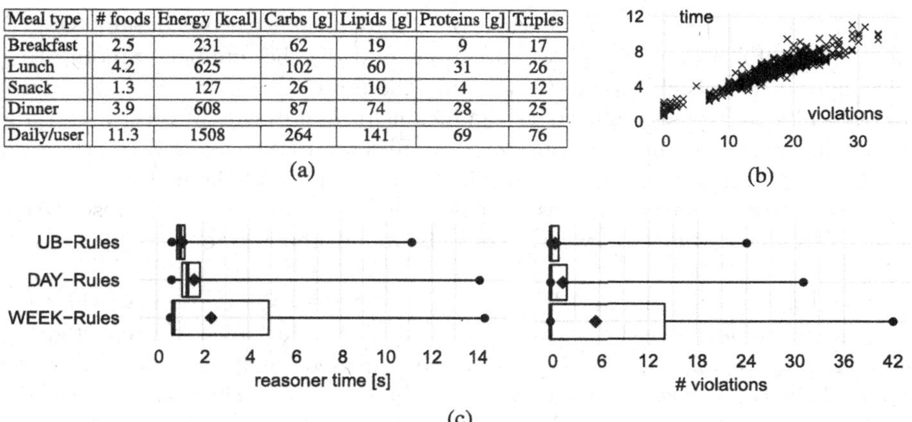

Meal type	# foods	Energy [kcal]	Carbs [g]	Lipids [g]	Proteins [g]	Triples
Breakfast	2.5	231	62	19	9	17
Lunch	4.2	625	102	60	31	26
Snack	1.3	127	26	10	4	12
Dinner	3.9	608	87	74	28	25
Daily/user	11.3	1508	264	141	69	76

Fig. 7. Evaluation on *Key to Health* scenario: (a) Consumed foods, nutrients and RDF triples for user-supplied meals, by type and aggregated per user/day; (b) Correlation between reasoning time and output violations (WEEK-Rules); and (c) Reasoning time and output violations distributions.

violation corresponds on average to the assertion of 18 triples in the triplestore. Reasoning takes around 1 s in half of the cases (medians in box plots), but other cases require much more time (up to 14 s for *WEEK-Rules*) and lead to an increase in the average reasoning time (diamond means in box plots). This variability is the result of the interplay of different factors. In particular, we observe a significant correlation between reasoning time and the number of produced violations, shown in the scatter plot of Fig. 7b, that may be partly justified by the fact that additional entailment rules are triggered to populate detected violations.

7.2 Assessing Reasoning Performance

We performed some scalability tests to assess the reasoning performance under different settings. We focused our tests on two aspects. Firstly, we wanted to observe the time needed by the reasoner for completing the inference process and for creating the instances of the `Violation` class, where necessary. Concerning this validation, we recall that the reasoning is performed per-user (as mentioned in Sect. 5.2)

Secondly, we desired to analyze the throughput of the system with the aim of discovering which is the maximum load bearable by the system per unit of time. In particular, we focused our exploration on knowing how many users can be analyzed per minute by the system. Collected values may be indicative for estimating possible hardware requirements for making our platform deployable into different, larger scenarios, as planned in the "Trentino Salute 4.0" digital health initiative.

The configuration of the evaluation procedure deals with two parameters: (i) the number of days monitored for each user, and (ii) the number of monitoring rules contained in each profile. Concerning the number of days, we used values suggested by the domain experts as significant for monitoring purposes: 1, 7, 14, and 30 days. For the rules, we considered profiles with 1, 10, 50, and 100 rules, to simulate both `simple` profiles (1, 10 rules — e.g., typical of the dietary monitoring for healthy people) as well as `complex` ones (50, 100 rules — representative of more sensitive situations, with users to be monitored due to particular health conditions).

The graph shown on the left in Fig. 8 reports the observations of the time required for performing the reasoning activity.[5] On the x-axis, we reported the number of days for which user's data are provided, while on the y-axis, we reported the time necessary for completing the reasoning. There are two aspects that we may notice. Firstly, the timing necessary for reasoning in the more complex scenario (30 days and 100 rules) is acceptable for the implementation of the reasoner in a real-time system. Moreover, we can appreciate how the time necessary for completing the reasoning by considering simple profiles (1 or 10 rules) remains almost constant, independently of the number of the considered days. Secondly, we may notice a significant increment of the computational time when complex profiles are used.

The second aspect we wanted to highlight is the throughput of the system. The graph shown on the right in Fig. 8 reports the trend of the throughput of the system with an increasing number of rules. On the x-axis, we reported the number of days for which user's data are provided, while on the y-axis, we reported the number of users per-minute that the system is able to process. By performing reasoning operation on single-day data, the throughput is acceptable also for elaborating complex profiles. Indeed, on average the system is able to process approx. 456 user/minutes in the case that all of them are associated with profiles with 100 rules. On the contrary, by increasing the number of considered days, the support for a real-time reasoning starts to become unfeasible. Indeed, a throughput of less than 100 users per minute could be critical in a crowded scenario. However, this throughput can be arbitrarily increased using multiple machines to cope with large scale deployments.

7.3 Usability Evaluation of Experts Tool

The usability of the facilities used by the domain experts for building the monitoring rules has been evaluated through the System Usability Scale (SUS), analyzing the intuitiveness and simplicity of the interfaces. The evaluation protocol consisted in a multiple use sessions and followed the five steps below:

1. Training meetings with the experts involved in the evaluation for an introductory explanation of the functionalities available in the tool.
2. Definition of the first bunch of the Mediterranean Diet rules (95 rules).

[5] Testing machine: 8 CPUs Intel Core i7 870 2.93GHz, 16GB RAM, 1TB HDD.

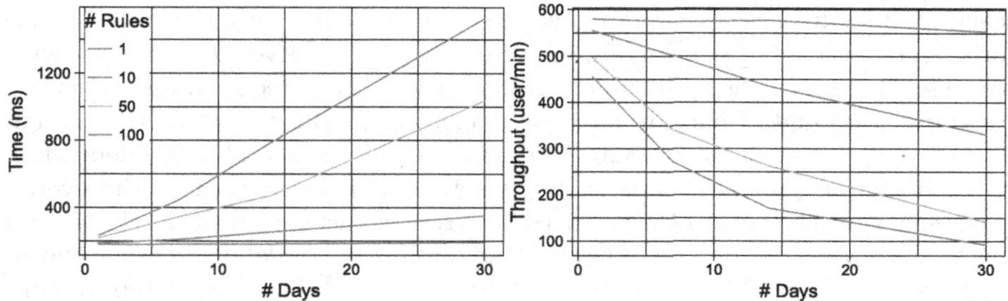

Fig. 8. Time (left) and throughput (right) of the reasoner on the scalability test.

3. Meetings with the experts for collecting questions about functionalities. Refinement of the interface integrating some improvements requested by the experts after the first four days of usage were deployed.
4. Definition of the second bunch of the Mediterranean Diet rules (126 rules)
5. Final meetings with the experts and distribution of the evaluation questionnaires.

The usability test of the tool involved six domain experts participated to the setup of the PerKApp platform for the *Key to Health* project. According to the usability test requirements provided by [17], the number of users involved in the test granted the discovery of 91% of the usability problems. The average score obtained from the SUS was 81.5, that, according to the adjective rating scale proposed by [18], corresponds to "excellent". Finally, the experts agreed about the capability of provided operators to cover the definition of rules necessary for supporting the overall monitoring activity of the Mediterranean Diet.

7.4 Lessons Learned

The design and development of the PerKApp platform provided interesting insights related to the challenges that we needed to address in our experience.

Concerning Challenge C1, we needed to address the lack of detailed structured data about both the food and physical activity domains. In our experience, we noticed that available resources for the food domain were not granular enough. While, concerning the physical activity domain, structured resource representing, not only an activities classification schema, but also other detailed information like the energy consumption and the effort level, were not available. Thus, despite the number of resources available in the Semantic Web community, our scenario demonstrated the necessity of further effort for the construction of resources with a level of detail useful for supporting the development of smart applications.

Regarding Challenge C2, we validated and discussed earlier in this section the effectiveness of the reasoning process implemented into the PerKApp platform. Our results derived from the optimization of rules design and rules evaluation schedule. In a first stage, we designed few complex rules for covering all monitoring activities. On the one hand, we were able to cover several constraints with one rule. But, on the other hand, the computational time required for evaluating these rules was high. Hence, in a second stage, we opted for splitting the rules in more simpler ones and, at the same time, to schedule their evaluation depending on their timing property. This strategy led to an improvement of the overall reasoning performance and allowed us to have an easier control on the overall reasoning process (exactness of the `Violation` instances, debugging operations, etc.). In the scenario addressed by the current deployment of PerKApp, reasoning operations are performed on sets of triple describing users' specific events. Part of the future work will be the investigation of stream reasoning strategies when it is necessary to monitor a continuous flow of information as well as to exploit learning strategies for suggesting, if possible, new rules or adaptations of existing ones. An example within the health domain is the real-time monitoring of the glycemic index.

Concerning Challenge C3, we observed the platform's behavior after the integration of an open source solution, namely RDF4J[6], for supporting data storage. The performance analysis highlighted that this solution was feasible for a real-world scenario. In particular, we observed the capability of managing the parallel data management operations in an efficient way. Thus, we may state that the engine implemented into RDF4J satisfies the scalability requirements of the PerKApp platform. Even if this finding sounds trivial, we believe that it is important to mention this result as a feasibility proof concerning the integration of open source solutions into real-world environments.

Finally, we mentioned in the Challenge C4 the necessity of supporting experts in specifying and revising users' prescriptions through the use of a collaborative tool. One of the potential issues is that the collaborative definition of these constraints may lead to possible disagreements between experts. In order to evaluate if the risk of having these disagreements truly exists, we built a set of 4 scenarios from the food consumption data collected in our triplestore, and we sent them to a group of 12 experts that were not involved in the setup of the PerKApp platform[7]. For each scenario, we proposed a list of possible violations based on the prescriptions modeled into the PerKApp platform and we asked them to confirm if such violations were correct or not. By analyzing the provided responses[8], we can observe that a full agreement was never reached. From this result, we may infer that a tool for supporting experts in the rule definition activity is necessary for trying to limit the disagreement between them. The necessity of reaching an agreement between experts is clear when they work for the same organization: if a healthcare organization has to monitor and support patients, the implemented guidelines should not be expert-dependent. Thus, possible disagreement between them has to be reconciled.

[6] http://rdf4j.org/.

[7] The selected experts are nutritionists of the Trento and Milan healthcare districts.

[8] https://perkapp.fbk.eu/downloads/ISWC2018_inuse_open_issues_dieta_mediterranea.csv.zip

8 Conclusions

In this paper, we presented the PerKApp platform and its key component, i.e., the rule-based reasoner adopted for monitoring users' behaviors in order to support the promotion of health lifestyles. We discussed the role of the knowledge base within the PerKApp platform and how it is equipped with monitoring rules inserted by domain experts through an easy-to-use interface integrated as an extension of the MoKi tool. Then, through a running example, we described how the reasoner operates on the data provided by the users, and we evaluated the reasoner performances by discussing: (i) the time required for processing users' data under different settings; (ii) the throughput of the system with the aim of inferring possible hardware requirements for deploying the PerKApp platform into different environments; and (iii) how the complexity of the users' profiles (i.e. the number of monitoring rules associated with each user) affects the overall efficiency of the system. Results demonstrated the possibility of adopting the system in real-world scenarios, and the reported lessons learned provide insights for future developments, in order to improve the overall efficiency, and thus allowing the deployment of the PerKApp platform in more challenging environments.

References

1. Intille, S.S.: Ubiquitous computing technology for just-in-time motivation of behavior change. In: UbiHealth 2003: 2nd International Workshop on Ubiquitous Computing for Pervasive Healthcare Applications (2003)
2. Dragoni, M., Bailoni, T., Eccher, C., Guerini, M., Maimone, R.: A semantic-enabled platform for supporting healthy lifestyles. In: Seffah, A., Penzenstadler, B., Alves, C., Peng, X., (eds.) Proceedings of the Symposium on Applied Computing, SAC 2017, Marrakech, Morocco, 3–7 April 2017, pp. 315–322. ACM (2017)
3. Izumi, S., Kuriyama, D., Itabashi, G., Togashi, A., Kato, Y., Takahashi, K.: An ontology-based advice system for health and exercise. In: IMSA, pp. 95–100 (2006)
4. Nassabi, M.H., op den Akker, H., Vollenbroek-Hutten, M.M.R.: An ontology-based recommender system to promote physical activity for pre-frail elderly. In: Mensch and Computer Workshopband, pp. 181–184 (2014)
5. Benmimoune, L., Hajjam, A., Ghodous, P., Andres, E., Talha, S., Hajjam, M.: Ontology-based medical decision support system to enhance chronic patients' lifestyle within e-care telemonitoring platform. Stud. Health Technol. Inform. **213**, 279–282 (2015)
6. Mikolajczak, S., Ruette, T., Tsiporkova, E., Angelova, M.,V.,B.: A semantic reasoning engine for lifestyle profiling in support of personalised coaching. In: International Conference on Global Health Challenges, pp. 79–83 (2015)
7. Jawaid, H., Latif, K., Mukhtar, H., Ahmad, F., Raza, S.A.: Healthcare data validation and conformance testing approach using rule-based reasoning. In: Yin, X., Ho, K., Zeng, D., Aickelin, U., Zhou, R., Wang, H. (eds.) HIS 2015. LNCS, vol. 9085, pp. 241–246. Springer, Cham (2015). https://doi.org/10.1007/978-3-319-19156-0_25
8. Hommersom, A., Lucas, P.J.F.: An Introduction to Knowledge Representation and Reasoning in Healthcare. In: Hommersom, A., Lucas, P.J.F. (eds.) Foundations of Biomedical Knowledge Representation. LNCS (LNAI), vol. 9521, pp. 9–32. Springer, Cham (2015). https://doi.org/10.1007/978-3-319-28007-3_2

9. Mamykina, L., Levine, M.E., Davidson, P.G., Smaldone, A.M., Elhadad, N., Albers, D.J.: Data-driven health management: reasoning about personally generated data in diabetes with information technologies. JAMIA **23**(3), 526–531 (2016)
10. Rospocher, M., Serafini, L.: An ontological framework for decision support. In: Takeda, H., Qu, Y., Mizoguchi, R., Kitamura, Y. (eds.) JIST 2012. LNCS, vol. 7774, pp. 239–254. Springer, Heidelberg (2013). https://doi.org/10.1007/978-3-642-37996-3_16
11. Bailoni, T., Dragoni, M., Eccher, C., Guerini, M., Maimone, R.: Healthy lifestyle support: the PerKApp ontology. In: Dragoni, M., Poveda-Villalón, M., Jimenez-Ruiz, E. (eds.) OWLED/ORE -2016. LNCS, vol. 10161, pp. 15–23. Springer, Cham (2017). https://doi.org/10.1007/978-3-319-54627-8_2
12. Dragoni, M., Bosca, A., Casu, M., Rexha, A.: Modeling, managing, exposing, and linking ontologies with a wiki-based tool. In: LREC, pp. 1668–1675 (2014)
13. Eccher, C., Ferro, A., Seyfang, A., Rospocher, M., Miksch, S.: Modeling clinical protocols using semantic MediaWiki: the case of the oncocure project. In: Riaño, D. (ed.) K4HelP 2008. LNCS (LNAI), vol. 5626, pp. 42–54. Springer, Heidelberg (2009). https://doi.org/10.1007/978-3-642-03262-2_4
14. Corcoglioniti, F., Rospocher, M., Mostarda, M., Amadori, M.: Processing billions of RDF triples on a single machine using streaming and sorting. In: ACM SAC, pp. 368–375 (2015)
15. Dragoni, M., Bailoni, T., Maimone, R., Eccher, C.: HeLiS: an ontology for supporting healthy lifestyles. In: The Semantic Web - ISWC 2018–17th International Semantic Web Conference. Lecture Notes in Computer Science, Monterey, California, USA, 8–12 October 2018. Springer (2018)
16. Di Francescomarino, C., Ghidini, C., Rospocher, M.: Evaluating wiki collaborative features in ontology authoring. IEEE Trans. Knowl. Data Eng. **26**(12), 2997–3011 (2014)
17. Nielsen, J., Landauer, T.K.: A mathematical model of the finding of usability problems. In: Proceedings of ACM INTERCHI 1993 Conference, pp. 24–29 (1993)
18. Bangor, A., Kortum, P., Miller, J.: Determining what individual sus scores mean: adding an adjective rating scale. J. Usability Stud. **4**, 114–123 (2009)

Reshaping the Knowledge Graph by Connecting Researchers, Data and Practices in ResearchSpace

Dominic Oldman(✉) and Diana Tanase(✉)

British Museum, London, UK
{doldman,dtanase}@britishmuseum.org

Abstract. ResearchSpace is an open source platform designed at the British Museum to help establish a community of researchers, where their underlying activities are framed by *data sharing*, *active engagement in formal arguments*, and *semantic publishing*. Using Semantic Web languages and technologies, the innovations of the system are shaped by a social conceptualisation of the graph-based representation of information. This is employed by integrated semantic components aimed at subject experts that offer mechanisms to create, annotate, assert, argue, search, cite, and justify data-driven research. This paper showcases a new onto-epistemological approach that supports researchers to contribute to a growing and sustainable corpus of knowledge that has history, not just provenance, built-in. It describes our considerations in designing for interdisciplinary collaboration, usability and trust in the digital space, highlighted by use cases in archaeology, art history, and history of science.

1 Introduction

The ResearchSpace open source project[1] is a vision of the Andrew W. Mellon Foundation, a major funder of humanities and cultural heritage research. This article describes the system developed based on this vision. Designed at the British Museum and developed in partnership with Metaphacts[2], Semantic Web technologies are combined to enable creation, sharing and linking of information in different configurations and levels of de-centralisation, between multiple confederated instances. The emphasis is on a representation of information that integrates, preserves multiple perspectives, and promotes collaborative research. The project's long term goal is to build a community of researchers that open and share their *data*, *knowledge*, *research practices*, and *arguments* with each other.

Museums, libraries, archives and other research institutions are often referred to as *knowledge* or *memory* institutions, but their contribution to the Web of

[1] http://github.com/researchspace/researchspace.

[2] https://metaphacts.com.

D. Vrandečić et al. (Eds.): ISWC 2018, LNCS 11137, pp. 325–340, 2018.
https://doi.org/10.1007/978-3-030-00668-6_20

Data is often disconnected from subject experts. The system presented here attempts to reconnect with those experts and focuses on knowledge representation of data for research, rather than just mapping traditional forms of digital documentation to Linked Data. The latter records substantive information about material things mainly for referential purposes (e.g. inventories, catalogues). This data, not originally designed for wider reuse, has become the basis for online publication and aggregation, but only represents a small part of overall institutional knowledge. Some types of administrative public data provide significant informational value particularly when transferred to formats like Resource Description Framework (RDF) [1]. The situation is less clear for complex subject areas that have wide historical significance with various and often contentious perspectives, and where new knowledge is constantly being identified. This is the case for art, cultural and historical knowledge, which is part of an ever changing understanding of the world and its development.

Considering these domain constraints, ResearchSpace was developed with an onto-epistemological approach to information representations aimed at research, relations discovery and dissemination using suitably expressive structures presented in an accessible form (see Sect. 2 for existing approaches). The current software implementation allows for:

- Assertion and argumentation models for tracing the multiple perspectives on history and the material world.
- Creation of semantic data and narratives, as well as expert-led refinements and expansions of existing data through a growing graph structure, namely the *knowledge graph*, that formally captures different worldviews and their provenance.
- Multi-level visual representation of resources (e.g. historical processes and entities), structured and comprehensive exploration of resources based on maps, timelines, charts, comparative image overlays, search and browsing across different heterogeneous datasets.
- Presentations of findings that explicitly record and describe researchers' views expressed in a graph that connects narrative, data, processes, and arguments.

We argue that such systems must be approached as a *psychological tool*[3], a mediation space, in which, despite recent trends in computing, and in particular machine learning, the focus should remain on the human as a builder of knowledge by reshaping the knowledge graph. This addresses the issue of how the computer deals with change and how thinking that we normally associate with writing narratives can be transferred and evolved within a digital environment. To this aim we propose a broader definition of the knowledge graph that encapsulates the unique nature of the relationship between humans and computers

[3] Lev Vygotsky used *psychological tools* in the context of sociocultural theory of cognitive development. They are described as "the form of mediation needed for the emergence of conceptualised thinking and the tools in question include, in addition to speaking and writing, gestures, sign systems such as maps and diagrams, and mnemonic techniques" [5].

necessary for a knowledge system to operate effectively (see Sect. 3 for a description of the system). In doing so, we put forward the following considerations for digital research systems:

- Computer systems should be designed to actively encourage human knowledge production.
- Knowledge is cumulative and should be built upon to reduce fragmentation.
- Research increases in effectiveness as a social and collaborative activity embracing formal models of argumentation to investigate contradictions.
- Computer systems that aim to reflect aspects of the world need to deal with changing knowledge and support questioning established history.
- Data should be contextualised to reflect complexity and presented in accessible ways to wider audiences, not just experts.

These key points constitute guiding principles for existing projects that use ResearchSpace in archaeology, art history, and history of science (see Sect. 4).

2 Challenges for a Digital Research Space

Throughout this project's development we identified four groups of challenges relating to data, infrastructure, publishing and the researchers' mindset. In this section we contrast our approach with existing work.

2.1 Creating, Integrating, Sharing and Reuse of Data

In the last decade several museums including the British Museum have opened their collection data to the World Wide Web. Yet, this is problematic since cultural heritage collection data systems were designed for internal administration by specialist users, where the shortfall in data specificity, ambiguities, or uncertainties are compensated by the knowledge of expert users who interpret it. The language and the knowledge required to understand the original meaning behind the data is not accessible to external users when this data is openly published in Linked Data format. It is here that we see the tension between those working with Linked Data focusing on reuse, and the needs of domain experts. Best practice[4] notes on data publication make no explicit recommendations based on source knowledge characteristics or consider the quality of the mapping process to Linked Data in terms of its usefulness for target applications.

We advocate for ontological representation of data from the cultural heritage domain, which provides a framework for a high degree of semantic meaning and contextual structure to be expressed, but also the ability to create inferred presentations for different audiences that communicate appropriately using the same underlying data. Therefore, the challenge goes beyond solving the technical tasks of creating, sharing, and reuse of data, but rather to consider its wider long term use and purpose.

[4] https://www.w3.org/TR/dwbp/.

2.2 An Infrastructure for Transferability of Knowledge

Just as data design can be affected by top down decisions, the same applies to infrastructure and this leads to concerns about the nature of a digital infrastructure and how this relates to its users. A common starting point is the National Science Foundation's Atkins report (2003) [2] which provided a general description as part of a case for investment in new physical infrastructures. It gave an overview of the technical components (hardware and software) and the multidisciplinary expertise required to operate it. Other studies, for example [19], emphasized that the components should be a conduit for interactions between technology and people, with the *researcher an integral component of the application execution*. The need for visualisation interfaces for human interaction was advocated as a crucial aspect.

In [17], a further demand is articulated from a research infrastructure, namely the ability to link to real research outcomes and prevent technology from distorting them. Hence, the importance of transparency at the level of data and process. A Digital Humanities model proposed by [9] sought to embed scholarly activities into the definition of infrastructure. These were refined by the Scholarly Domain Model (SDM) based on an initial, now expanded, set of scholarly primitives, namely discovering, annotating, comparing, referring, sampling, illustrating, and representing. Also [14,15] argued that a lack of attention to data design and data outputs compared to function, risked making research systems ineffectual by not having the required semantic elements for reliable knowledge building, also discussed by [3].

This literature review charts the continuous refinement of what is considered a good infrastructure for digital research projects. However, the value of having such an infrastructure in the cultural heritage and arts domain is to "explore theories, ways of perceiving, ways of knowing; to enter into other mindsets and world-views" [20]. This requires interpretative work that is not possible just with technologies, but needs human experts. The Semantic Web and Linked Data solutions address some types of data integration problem, but ignore the underlying need of experts to collaboratively grow information over time in a relevant way for their research. ResearchSpace makes this possible by using Semantic Web technologies in applications designed specifically for knowledge workers. The system presents the user with an interface to make research activities like creation, discovery, enrichment, argument formulation and publication - intuitive, while in the background it employs rich ontologies to record interactions as RDF. The new triples trace the connections between research activities thus adding the missing historical dimension.

2.3 Semantic Scholarly Publishing

The combination of narrative and data in semantic publishing is of increasing interest to the Semantic Web community. In [11], the authors provide a critique of current systems by pointing out that these solutions transform existing publications, rather than support direct semantic creation and publication. True semantic publishing should come directly from authors, and if possible as part of an integral knowledge generation activity not mediated by technologists which inevitably stifles the ability to use a digital environment as a place where modes of thinking actually take place.

Centralisation can lead to issues of data ownership, accessibility, bias and limitation of freedom of expression. Decentralisation of semantic publications and the existence of accessible tools that support it addresses these problems, but at the same time takes on aspects of traditional narratives and disciplinary fragmentation. These narratives host descriptive and analytical modes of communication. In descriptive narrative, such as those provided by Wikipedia, articles are restricted to *facts* about things (material and immaterial), events or people. The Wikipedia community is used to enforce a neutrality rather than express the different perspectives represented within it. Equally, traditionally authored analytical narratives use evaluation and comparison, inevitably selecting evidence that supports a particular hypothesis.

Decentralised semantic publication also draws upon this tradition using selected semantic elements that support certain perspectives without adequate comparative semantics to allow effective resolution of contradiction. Typically academic disciplines are silos of information and knowledge using their own narrative conventions that make interdisciplinary studies difficult. Narratives are highly heterogeneous, and have both linguistic complexity and ambiguity, but they have overlapping concepts, ideas and information that are important to building a history of interdisciplinary knowledge.

The descriptive narratives of Wikipedia have much in common with aggregated data services in that they tend towards an almost fixed structure and form. Aggregated data services, in attempting to deal with variable local resources, will force data to conform to a central model. In pursuit of being open to a range of audiences a dominant common denominator approach is developed based on a perception that general audiences require a reduction of complexity for reuse. These centralised services produce descriptive models with limited contextual richness and therefore narrow reuse value. These constraints are also apparent in cultural aggregations like Europeana, resulting in some researchers attempting to enrich the data using narratives, for example [13].

At the other extreme, completely decentralised publications also present problems for reuse and the ability to use the knowledge of different communities for progressive digital publication. Creating semantically enhanced publications, whether born semantic or whether semantically enriched existing publications, can use data to describe things difficult to describe precisely just through text. Equally, narratives can clarify and make accessible abstract data. Using structured data within narratives can potentially improve clarity and discovery, therefore support better assessments of a particular subject area. It also holds the possibility for computer inferences across publications. Decentralised articles which employ semantic enrichment, but embed raw data based on many different ontologies, place some limitations on both discovery and inference. A coherent framework of semantics used consistently across data and narratives allows greater scope to improve meaning within a particular narrative, but also across many narratives.

Applications like Dokieli [4] provide decentralised publishing environments in which the user can generate data enriched narratives and establish personal networks. However, reliance on completely decentralised modes of thinking risk continuing disciplinary fragmentation due to the lack of semantic integration and sometimes appropriateness (in terms of data representation) making analysis across genuine semantic publications no less problematic than with traditional publications.

2.4 Mindset and the Interpretation of the Knowledge Graph

While decentralised semantic publishing is still in its infancy, the whole notion of the Semantic Web remains opaque to many people. Many users have a narrow view of the role information systems have in their work, based on a dominant mindset around traditional database systems. Most institutions adopt information systems that fulfil an administrative, reference and operational mindset, rather than one which seeks to promote knowledge-based activity. There is a separation between what the computer provides and the intellectual processes and activities retained by the human operator, and used elsewhere. The aim of representing knowledge using the Semantic Web means that computer scientists and domain experts need to re-evaluate their relationship with these new types of information (knowledge base) system. This is required for a transition from traditional and individualistic methods of research, to collaborative and open research practices.

Domain experts need to be active in the design of the Semantic Web applications rather than just inform basic requirements gathering. In the same way that technologists cannot represent the knowledge of other domains, they are unable to implement appropriate knowledge systems without the direct involvement of source experts at the design level.

The situation is not helped by current technical definitions of a knowledge graph, not just because of the language used but also because they tend to be technology centered and fail to encapsulate the intellectual contribution of the user. The notion of knowledge graph has been discussed in [16] and further debated in [8]. One general definition views knowledge graphs as large networks of entities, their semantic types, properties, and relationships between them based on automatically derived and interlinked factual information from knowledge bases such as DBpedia[5] and others.

We propose a definition intended to influence the design of a knowledge oriented information system that recognises the wider role of the user in the creation of the graph-based representational structure at its core: *a knowledge graph is a continually changing informational structure that mediates between a human, the world and a computer. The graph itself is ontologically based and enhanced by human epistemology. These are closely linked in that the ontology provides*

[5] https://wiki.dbpedia.org/.

real world references and a structure of interrelated entities or processes, while the epistemology uses the graph to interpret and generate new knowledge. Growing the graph is based on both automated reasoning and crucially, collaborative human thinking and creativity.

This new conception of the knowledge graph, helps capture the nature of change in knowledge in a representation suitable for interdisciplinary scholarship enabling the ongoing process of bringing knowledge into being with historical provenance.

3 Connecting Researchers, Data and Practices

Lowestoft is a town on the most easterly point of Great Britain. It was featured in the 2015 British General Election as a coastal town economically depressed and mostly forgotten, but still hopeful and looking optimistically forward into the future. When we read the Wikipedia descriptive entry it is hard to understand where this optimism might come from, but when delving deeper into cultural heritage, and related resources, one finds a completely different impression from the encyclopaedic perspective. To the computer system a place is a static 'entity', but in our wider version of the knowledge graph definition, it is a changing 'process' with large numbers of relations to other processes, across time and space. The demonstration instance of the ResearchSpace system[6] uses an RDF graph based on a subset of the British Museum's catalogue data. This provides an example of semantic enhancement of an existing data resource achieved by mapping institutional data to an event-based ontology called CIDOC CRM (Conceptual Reference Model) [6]. ResearchSpace is setup to offer multiple paths of exploration and analysis. In a shared environment multiple researchers can investigate the history of Lowestoft afresh, or from a set of established resources (and perspectives) in a shared clipboard. This can contain predefined sets of semantic resources such as, charts, diagrams, arguments, and searches. Similarly, a Semantic Narrative can include the same resources taken from the clipboard and juxtaposed with text. Therefore, an existing narrative on Lowestoft might already hold defined searches or other resources that are relevant and act as an existing research object to build upon, or argue with. Typically, as a researcher finds relevant resources they are saved into a clipboard through a drag and drop mechanism and can be subsequently organised into different sets. Through the British Museum's data, a picture of Lowestoft's historical periods can be developed and knowledge from other sources added.

If a researcher needs to challenge existing British Museum data or assert completely new information then this can be added using the ontology rather than textual annotations. During this process of discovery and enrichment, the researcher will realise that Lowestoft has a long and rich history, and that it attracted many people, for example artists like, J.W.M. Turner, Muirhead Bone and Samuel Varsey. It produced fine porcelain exported around the world. Its position on the North Sea means that it has military associations, as well as

[6] https://demo.researchspace.org.

social and economic connections with other countries, particularly Scandinavia and Northern Europe. The Battle of Lowestoft in 1665, is documented within UK and Dutch museums. The military legacy and boat building tradition is reflected in objects in several cultural heritage institutions, including archival photographs, medals, newspaper cuttings, letters and so on. The ResearchSpace system is able to bring this data together semantically intact, and visualise it in multiple ways. The data found from direct connections to Lowestoft can have secondary semantic references to other parts of England, or the World, and the network of people with connections to Lowestoft becomes wider, including writers such as Joseph Conrad, Rudyard Kipling and Charles Dickens.

3.1 The Technology Stack

The ResearchSpace technology stack (see Fig. 1) builds on the metaphactory knowledge graph platform enabling customisation and extensibility of the interaction with the graph database (Blazegraph[7]) through the use of familiar open standards such as RDF and SPARQL, expressive ontologies for schema modeling based on CIDOC CRM[8], rules, constraints, and query specifications based on SPIN[9], W3C Web Components[10], W3C Open Annotation Data Model[11], and W3C Linked Data Platform Containers[12]. The platform is open source, integrating external tools including OntoDia[13], MIRADOR Image Viewer[14] with an IIIF Image Server[15]. Instantiating ResearchSpace for application projects involves creating templates, which are a mixture of HTML5[16], React Components[17] and Handlebars[18]. The custom HTML5/REACT Web components described in the next section are informed by domain experts, which operate on the result of SPARQL queries. They represent a selection of the ResearchSpace key features.

3.2 Semantic Components

Semantic Component: Knowledge Patterns They are predefined graph paths that express data creation, modification, deletion and visualisation. The use of the term *knowledge pattern* acknowledges their association with experts' needs for capturing, at various levels of detail, the contexts of processes involved in research. Technically, a knowledge pattern includes defining a set of SPARQL

[7] https://www.blazegraph.com.
[8] http://www.cidoc-crm.org.
[9] http://spinrdf.org.
[10] https://www.w3.org/TR/components-intro.
[11] https://www.w3.org/TR/annotation-model.
[12] https://www.w3.org/TR/ldp.
[13] http://www.ontodia.org.
[14] http://projectmirador.org.
[15] http://iiif.io.
[16] https://www.w3.org/TR/2010/WD-html5-20100624.
[17] https://reactjs.org.
[18] https://handlebarsjs.com.

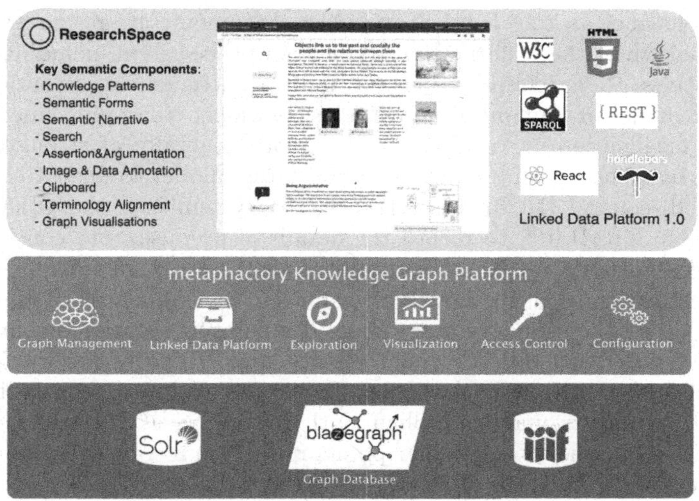

Fig. 1. ResearchSpace platform architecture

1.1 statements but additional metadata is used for their integration into other components such as data input, visualisations, arguments and custom search systems. A single knowledge pattern combines with others but is transferable as an LDP Resource between different instances of ResearchSpace. Knowledge patterns are defined on an ongoing basis and subject experts are encouraged to learn the main system ontology which is essential to the ResearchSpace design. CIDOC CRM is a rich ontology including a growing number of specialisations with adoption in cultural heritage and beyond. It provides a contextual framework under which diverse and variable information can be integrated without homogenisation. Carefully designed UI exposes the ontology (if requested) to non-technical users encouraging involvement in the design of new knowledge patterns covering different areas of interest. For example, the project, Late Hokusai (see below), models patterns that describe the condition states of a woodblock over time, and its relationship with impressions (prints) derived from it. They help answer specific art history inquiries, but also address wider societal questions.

Semantic Component: Assertions and Arguments Formalisation of argumentation is a process of reasoning in support of an observation, idea, action, theory, or interpretation of facts, and is increasingly implemented by systems that support decision-making through social interactions [18]. The implementation of argumentation in ResearchSpace enables the creation of structured data assertions and arguments. These challenge existing entities and relations or make new assertions based on direct observation, the adoption of belief from others, or inference based on premises that are resources in the system (data, images, narratives, other assertions or arguments, etc.). These assertions and arguments have clear knowledge provenance. Arguments are resources that can be combined with other information resources in the system including explanatory narratives. ResearchSpace does not limit a particular argumentation logic but uses a model of human argumentation in which, "reasoning may not only consist of falsification or verification, but more generally of strengthening or weakening hypotheses, and a way to connect this model to an ontology of the domain of discourse" [7]. In this instance, the system uses the CIDOC CRM specialisation, CRMInf[19], to record the details of premises and conclusions. In Fig. 2, we visualized a very simple argument that a water colour representation of a boat named *Hope*, in the Lowestoft shipyard, was influenced by the works of Cornelius Varley. The expert's notes about the painting mention a connection to the British painter without recording the facts supporting this observation. What is essential for researchers is the ability to see the evolution, composition, and revision of arguments making explicit both the processes of argument-making and the states of belief at particular points in time in a composite inference [7]. This is relevant to the needs of a trusted digital environment with an embedded history of arguments. Using these principles, highly complex arguments can be constructed.

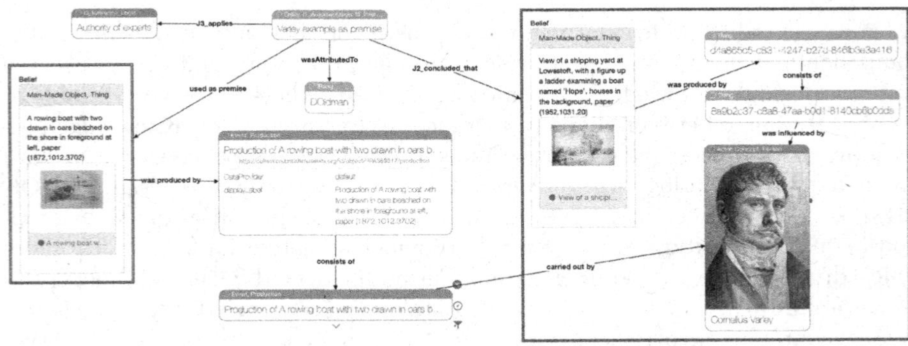

Fig. 2. Argument representation for expressing how Cornelius Varley influenced the production of a painting

Semantic Component: Data Enriched Narrative In the challenges section we outlined the problems with both centralised and decentralised publication. Centralisation imposes constraints on subject experts who should be directly responsible for their content. ResearchSpace provides the opportunity for semantic authoring, but within a particular community that agrees on an ontological framework. The objective is to ensure comparability, integration and a common purpose of building knowledge through different perspectives. These will almost certainly contain contradictions. Fully decentralised semantic publishing, even with data enrichment, has the same disadvantage as traditional analytical

[19] http://www.ics.forth.gr/isl/CRMext/CRMinf/docs/CRMinf-0.7.pdf.

narratives, in that contradictions and arguments are hard to identify, let alone resolve. In centralised systems contradictions create difficulties and are often avoided. The ability to integrate argument within the medium of narrative helps resolve analytical differences head on by fully understanding the nature of contradictions and providing an appreciation for the range of influences that cause change, and which individual researchers are unlikely to identify themselves. A third party non-technical visual editor, called Ory, was integrated to organise and visualise different RDF representations of content including ResearchSpace semantic resources into a narrative[20]. From the ResearchSpace clipboard, different resources are dragged into formatted blocks. For example, a Semantic Search can be saved as a resource, placed in the clipboard, and then dragged into a narrative. The Semantic Narrative component allows researchers to write text and use citations of contextualised data they, or others, have added, and visualise it appropriately. This means that embedded data can also be inspected by other readers. A particularly important example of this is the use of assertion and arguments providing the ability to combine the accessibility of narrative alongside the complex data it seeks to explain and to allow greater integration of information. It is important for the system to track and alert authors to new data, and particularly to arguments. They then can review and update their narratives, but also add and enrich entities to reflect new positions. The narrative acts as an accessible data reference point for performing these dynamic tasks that address the type of changes that computers are unable to interpret and respond to.

Semantic Component: Search Search is a classic task where the researcher formulates a question and the system answers with a set of relevant resources. ResearchSpace supports three different types of search scenarios. They differ in the way the system handles the formulation and transformation of a query into a set of resources. These are: (a) knowledge-graph driven search, (b) knowledge pattern-based search, and (c) text-based search.

Knowledge Graph Driven Search. Researchers expect to ask who, what, where, and when type of questions from computer information systems. ResearchSpace enables adding data and paths that support the formulation of 'why' questions based on arguments and semantic narratives. Using data from the British Museum, we built a semantic network of RDF triples that provides depth and detail to each entry. The solution to designing a UI based on a rich ontology, but with suitable recall and precision, uses an approach developed by the Foundation for Research and Technology Hellas (FORTH) [7]. The complexity of searching across a graph flows from using CIDOC CRM to capture sufficiently rich semantics. Exploring graph paths using exhaustive automatic reasoning would not be possible in real time. Therefore, a model for querying is employed based on six fundamental categories (FCs) that abstract the main entities in the graph: Thing (material and immaterial), Actor, Place, Time, Event, Concept. Relationships (properties) are similarly inferred creating a matrix of semantic shortcuts carving paths through the detailed informational space.

[20] https://github.com/ory.

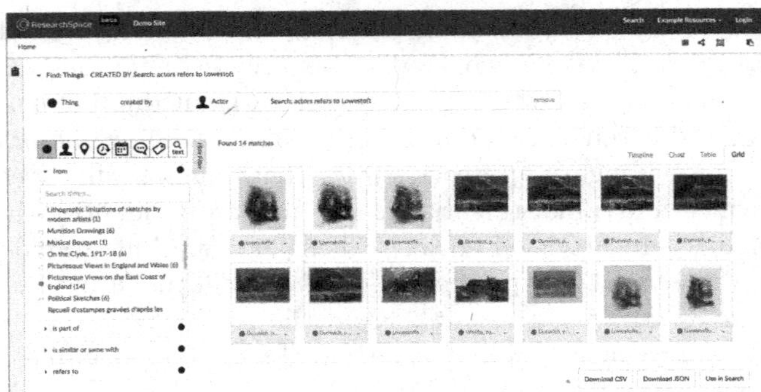

Fig. 3. A two part search that identifies first actors that refer to Lowestoft, and then the actual works created

```
@prefix rso:   <http://www.researchspace.org/ontology/>.
@prefix rdfs:  <http://www.w3.org/2000/01/rdf-schema#>.
rso:Actor_refers_to_Place rso:hasDomain rso:Actor;
rso:hasRange rso:Place;  rdfs:label "refers␣to" .
```

Listing 1.1. FR definition for Actor refers to Place

Figure 3 presents the query formulated by the researcher where Actor and Place are connected by a fundamental relationship (FR) *refers to* (see Listing 1.1). We exemplify in Listing 1.2 how that is specified in the system and the SPARQL construct to create *refers to* connections between actors and places. The current FCs and FRs are defined to cover the underlying data, but FCs and FRs can be adapted to reflect different onto-epistemological paths for particular datasets and according to different theoretical approaches. For example, archaeological data might be used to answer anthropological, economic and social questions and carefully designed FCs and FRs could encourage particular perspectives and approaches useful for different types of research. Therefore, this component is not simply a solution to the accessibility problem, but a component that incorporates expert rules for specialist data investigation.

```
PREFIX crm: <http://www.cidoc-crm.org/cidoc-crm/>
PREFIX rso: <http://www.researchspace.org/ontology/>
CONSTRUCT {?actor rso:Actor_refers_to_Place ?place .
               ?place rso:Place_is_referred_to_by_Actor ?actor}
WHERE {
        ?actor a rso:Actor .
        ?place a rso:Place .
        ?actor rso:Actor_is_creator_of_Thing
           / rso:Thing_refers_to_Place ?place}
```

Listing 1.2. SPARQL CONSTRUCT for inferring the FR Actor refers to Place

Structured Search with Knowledge Patterns In this case, search is about interrogating the structured data by using knowledge patterns i.e. SPARQL graph patterns rather than abstractions. However, it employs the same UI as graph driven searches using domain and range assignments avoiding the pre-processing of inferences. It allows researchers to apply specific knowledge patterns, useful in specialist projects.

Text-Based Search Complementary to the previous type of search, the textual data attached to the fundamental categories in the graph is indexed using Solr[21]. This complements the other search types and makes looking for known items in the graph a straightforward process.

4 Example Projects Using ResearchSpace

In this section, we highlight three very different projects in terms of research goals: archaeology, art history and history of science. The functionality incorporated by each has been determined by the scope of the projects, but they all share the same thinking in terms of digital scholarship. Explicitly, this refers to the sharing and reuse of data, the transferability of knowledge across contexts and over time, as well as its continuous enrichment, and to scholarly publications where data has the function to strengthen the argument. Early versions of ResearchSpace have already been deployed and used as the prototyping platforms for an archaeology and an art history project, while the third project has a live version of a customised instance of ResearchSpace. In each project the knowledge graphs are being grown to address different questions and accommodate different methods of research.

4.1 Archaeology: Geometric Reconstruction and Novel Semantic Reunification of Cultural Heritage Objects

This interdisciplinary project[22] aims to support reassociation of object fragments with shared characteristics (e.g. same school of production, style, age), the unification of object parts separated across collections, and, if possible, reassembly of fragments helped by algorithmic modeling. It employs a desktop application for 3D image analysis and annotations of 3D objects' representations and creates a knowledge graph in an integrated ResearchSpace instance pulling together data enriched by human agents and algorithmic analysis with explicit provenance.

The prototype application uses an existing body of information extracted from museums' inventories as a starting point for the knowledge graph. This is enriched with new information from 3D analysis and with assertions from the users. Using 3D scans of fragment edges, the algorithms developed within the project determine possible relations between fragments. The likely matches are assessed within the context of their historical data, and passed to humans for a

[21] https://demo.researchspace.org/resource/Help:SolrFullTextSearchSyntax.
[22] http://gravitate-project.eu.

final determination. The aim is to reduce the amount of time required for these operations overall, given a large number of dispersed fragments.

The organic expansion of the graph through continuous enrichment either from human or algorithmic agents, including structured argument, accompanied by knowledge provenance is paramount. It helps users understand the sources of the archaeological data and establish trust in the system. The prototype system has received positive and constructive feedback during a workshop session at the Computer Applications and Quantitative Methods in Archaeology (CAA) conference in March 2018.

4.2 Art History: Late Hokusai (British Museum/SOAS)

The focus of this project[23] was to bring together existing scholarly information about the work, life, and historical context of the Japanese artist Katsushika Hokusai into a searchable and explorable resource. Trained curators have undertaken the technical task of mapping to RDF using CIDOC CRM, custom scripts, and tools such as [12]. Hokusai and his works, namely his paintings, single-sheet impressions, illustrated books, drawings and letters from the British Museum (London) together with data from the Freer-Sackler Gallery (Washington DC), Metropolitan Museum of Art (New York) and the Art Research Center, Ritsumeikan University, have been semantically linked and enriched through transcriptions, translations and annotations. This enrichment applies to the primary material, and to secondary sources that document Hokusai's late period of activity in Japan. The Hokusai instance of ResearchSpace aims to be a multilingual research and knowledge platform, providing an alternative to current reference systems that restrict research input due to limited pre-defined categories. It provides new ways of investigating the archetypal representative of the ukiyo-e ('floating world') school with all its richness and complexity encoded in a specialist network of knowledge. Feedback on the development of the resource was received at workshops in Washington DC and New York (2018) with researchers expressing particular interest in the use of argumentation which is fundamental to various aspects of the project in particular connoisseurship.

4.3 History of Science: CorpusTracer (Max Planck)

This project uses ResearchSpace to investigates the knowledge networks and history surrounding one book: the Tractatus de sphaera of Johannes de Sacrobosco [10]. It is an example of a highly specialised digital monograph, but one which can be extended and integrated with other related digital resources and 'grown' into a network of interrelated knowledge. By applying methods from network analysis, it investigates how specific commentaries on the text circulated, which actors were responsible for them, and what factors supported

[23] http://gtr.rcuk.ac.uk/projects?ref=AH%2FN00440X%2F1.

or hindered the spread of specific kinds of knowledge. The core of this investigation is achieved through the construction of the Corpus Tracer, a graph-database that uses primarily the IFLA FRBRoo[24] standard, a specialisation of the CIDOC CRM developed by the International Federation of Library Associations and Institutions (IFLA) and the CIDOC CRM Special Interest Group. The structured search with knowledge patterns has proven valuable for exploratory searches of the specialist knowledge graph.

5 Conclusions

This paper describes the rationale and some of the in use capability of ResearchSpace, a Semantic Web knowledge oriented system that is designed to work in, or help transform, knowledge environments into collaborative, argumentative, digital scholarly spaces through the contextualisation of data using onto-epistemological processes for semantic modeling. It supports interdisciplinary research and is additionally underpinned by material culture representing world history through the products of social relations.

This type of knowledge, created directly by academics and subject experts, is not represented by existing modes of data dissemination, which although useful, are based on data created for a different rationale and purpose. This difference in design and purpose is directly linked to the benefits and value of the data. Its dynamic enables the creation of Semantic Web applications for people who are interested in collaborative research and knowledge building, or want better contextual engagement by placing things within historical and theoretical settings, not provided by raw Linked Data.

References

1. Alani, H.: Unlocking the potential of public sector information with semantic web technology. In: Aberer, K., et al. (eds.) ASWC/ISWC 2007. LNCS, vol. 4825, pp. 708–721. Springer, Heidelberg (2007). https://doi.org/10.1007/978-3-540-76298-0_51
2. Atkins, D.: Revolutionizing science and engineering through cyberinfrastructure: report of the national science foundation blue-ribbon advisory panel on cyberinfrastructure (2003). http://hdl.handle.net/10150/106224
3. Bechhofer, S., et al.: Why linked data is not enough for scientists. Future Gener. Comp. Syst. **29**(2), 599–611 (2013). https://doi.org/10.1016/j.future.2011.08.004
4. Capadisli, S., Guy, A., Verborgh, R., Lange, C., Auer, S., Berners-Lee, T.: Decentralised authoring, annotations and notifications for a read-write web with dokieli. In: Cabot, J., De Virgilio, R., Torlone, R. (eds.) ICWE 2017. LNCS, vol. 10360, pp. 469–481. Springer, Cham (2017). https://doi.org/10.1007/978-3-319-60131-1_33
5. Dean, K.: Towards a Historical-Materialist Science of Thinking, Chap. 18, pp. 53–73. Routledge, London (2014)

[24] https://www.ifla.org/node/10171.

6. Doerr, M.: The CIDOC conceptual reference module: an ontological approach to semantic interoperability of metadata. AI Mag. **24**(3), 75–92 (2003). http://dl.acm.org/citation.cfm?id=958671.958678
7. Doerr, M., Kritsotaki, A., Boutsika, K.: Factual argumentation - a core model for assertions making. JOCCH **3**(3), 8:1–8:34 (2011). http://doi.acm.org/10.1145/1921614.1921615
8. Ehrlinger, L., Wöß, W.: Towards a definition of knowledge graphs. In: SEMANTiCS (Posters, Demos, SuCCESS) (2016)
9. Gradmann, S., Hennicke, S., Tschumpel, G.: Modelling the scholarly domain beyond infrastructure. In: DHd 2016, p. 143 (2016)
10. Kräutli, F., Valleriani, M.: CorpusTracer: a CIDOC database for tracing knowledge networks. Digit. Scholarsh. Humanit. (2017). https://doi.org/10.1093/llc/fqx047
11. Kuhn, T., Dumontier, M.: Genuine semantic publishing. Data Sci. **1**(1–2), 139–154 (2017). https://content.iospress.com/articles/data-science/ds010
12. Marketakis, Y., et al.: X3ML mapping framework for information integration in cultural heritage and beyond. Int. J. Digit. Libr. **18**(4), 301–319 (2017). https://doi.org/10.1007/s00799-016-0179-1
13. Meghini, C., Bartalesi, V., Metilli, D., Benedetti, F.: Introducing narratives in Europeana: preliminary steps. In: Kirikova, M., et al. (eds.) ADBIS 2017. CCIS, vol. 767, pp. 333–342. Springer, Cham (2017). https://doi.org/10.1007/978-3-319-67162-8_33
14. Oldman, D., Doerr, M., Gradmann, S.: Zen and the Art of Linked Data, Chap. 18, pp. 251–273. Wiley-Blackwell (2015). https://onlinelibrary.wiley.com/doi/abs/10.1002/9781118680605.ch18
15. Oldman, D., Doerr, M., de Jong, G., Norton, B., Wikman, T.: Realizing Lessons of the last 20 years: a manifesto for data provisioning & aggregation services for the digital humanities (a position paper). D-Lib Mag. **20**(7/8) (2014). http://dblp.uni-trier.de/db/journals/dlib/dlib20.html#OldmanDJNW14
16. Paulheim, H.: Knowledge graph refinement: a survey of approaches and evaluation methods. Semant. Web **8**(3), 489–508 (2017). https://doi.org/10.3233/SW-160218
17. Rockwell, G.: As transparent as infrastructure: on the research of cyberinfrastructure in the humanities. In: Online Humanities Scholarship: The Shape of Things to Come, pp. 1–20 (2010)
18. Schneider, J., Groza, T., Passant, A.: A review of argumentation for the social semantic web. Semant. Web **4**, 159–218 (2013)
19. Stewart, C.A., Simms, S., Plale, B., Link, M., Hancock, D.Y., Fox, G.C.: What is cyberinfrastructure. In: Proceedings of the 38th Annual ACM SIGUCCS Fall Conference: Navigation and Discovery, SIGUCCS 2010, pp. 37–44. ACM, New York (2010). https://doi.org/10.1145/1878335.1878347
20. Veltman, K.H.: Towards a semantic web for culture. J. Digit. Inf. **4**(4) (2004). http://journals.tdl.org/jodi/article/view/113

Ontology-Based Recommendation of Editorial Products

Thiviyan Thanapalasingam[1(✉)], Francesco Osborne[1], Aliaksandr Birukou[2(✉)], and Enrico Motta[1]

[1] Knowledge Media Institute, The Open University, Milton Keynes MK7 6AA, UK
{thiviyan.thanapalasingam, francesco.osborne, enrico.motta}@open.ac.uk

[2] Springer-Verlag GmbH, Tiergartenstrasse 17, 69121 Heidelberg, Germany
aliaksandr.birukou@springer.com

Abstract. Major academic publishers need to be able to analyse their vast catalogue of products and select the best items to be marketed in scientific venues. This is a complex exercise that requires characterising with a high precision the topics of thousands of books and matching them with the interests of the relevant communities. In Springer Nature, this task has been traditionally handled manually by publishing editors. However, the rapid growth in the number of scientific publications and the dynamic nature of the Computer Science landscape has made this solution increasingly inefficient. We have addressed this issue by creating Smart Book Recommender (SBR), an ontology-based recommender system developed by The Open University (OU) in collaboration with Springer Nature, which supports their Computer Science editorial team in selecting the products to market at specific venues. SBR recommends books, journals, and conference proceedings relevant to a conference by taking advantage of a semantically enhanced representation of about 27K editorial products. This is based on the Computer Science Ontology, a very large-scale, automatically generated taxonomy of research areas. SBR also allows users to investigate why a certain publication was suggested by the system. It does so by means of an interactive graph view that displays the topic taxonomy of the recommended editorial product and compares it with the topic-centric characterization of the input conference. An evaluation carried out with seven Springer Nature editors and seven OU researchers has confirmed the effectiveness of the solution.

Keywords: Recommender systems · Ontology · User interface
Scholarly ontology · Scholarly data

1 Introduction

Major academic publishers need to be able to analyse their vast catalogue of editorial products and make data-driven decisions to ensure they are showcasing the right products to the right target market. This is a complex exercise that requires characterising with a high precision the topics of thousands of books and matching them with the interests of the relevant scientific communities.

D. Vrandečić et al. (Eds.): ISWC 2018, LNCS 11137, pp. 341–358, 2018.
https://doi.org/10.1007/978-3-030-00668-6_21

In Springer Nature, this task has traditionally been handled manually by publishing editors, who tend to rely on their domain knowledge and their personal experience for selecting the books to be marketed at scientific venues. In addition to this, they typically use Springer.com[1] for searching publications associated with keywords relevant to the conferences in question and find additional information by querying their internal database of editorial products. This approach lacks a user-friendly interface and can be very time-consuming, since it requires editors to manually browse a large and fast-growing catalogue of publications. For example, in order to select books for the International Semantic Web Conference one might want to search for all the publications produced in the last three years that have been authored by well-known researchers who are likely to attend the event. While the editorial products are tagged with product market codes characterizing their topics, these are only limited to high-level research fields, such as "Artificial Intelligence" and "Database Systems". The results of the editor queries may thus include hundreds of items. Another issue is that keyword-based queries do not take in consideration the relationships between topics and may miss pertinent publications that do not contain specific strings. For instance, searching all books about "ontology matching" may miss publications about "ontology alignment".

In this paper, we present Smart Book Recommender (SBR)[2], an ontology-based recommender system developed by The Open University (OU) in collaboration with Springer Nature (SN) for supporting their Computer Science editorial team in selecting products to market at specific venues. SBR recommends books, journals, and proceedings by taking advantage of a semantically enhanced representation of about 27K editorial products. In order to do so, we characterized all SN publications according to their associated research topics by exploiting the Computer Science Ontology (CSO), a large-scale automatically generated taxonomy of research areas [1]. Furthermore, SBR allows users to investigate why a certain publication was suggested by means of an interactive graph view that compares the topics of the suggested publication with those characterizing the input conference.

The rest of the paper is organized as follows. In Sect. 2, we discuss Smart Book Recommender in terms of its knowledge base, its architecture, and its user interface. In Sect. 3, we present the results of the user study. In Sect. 4, we discuss the steps required for large-scale deployment of the technology within the company. In Sect. 5, we review the state of the art and in Sect. 6 we conclude outlining future directions of research and development.

2 Smart Book Recommender

Smart Book Recommender takes as input a conference series and returns a list of editorial products that may be of interest for the attendees of the conference. This is achieved by representing SN books as a set of research topics drawn from a large-scale

---

[1] http://www.springer.com/.

[2] A demo of SBR is available at http://rexplore.kmi.open.ac.uk/SBR-demo.

Computer Science ontology, and ranking them according to their similarity with a topic-centric characterization of the conference. For instance, given the conference series "International Semantic Web Conference" (ISWC), SBR will return the books, journals, and conference proceedings that are characterized by a set of research topics similar to the one of ISWC, e.g., the "Handbook of Semantic Web Technologies" and "Proceedings of the European Semantic Web Conference". The primary purpose of SBR is to provide a concise and relevant list of publications that editors can quickly review to decide which books to market during a conference. However, it can also be used by researchers for finding publications relevant to a certain venue of interest.

SBR provides the web interface shown in Fig. 1. It works according to three main steps:

(1) It represents journals, books, and conferences according to the metadata of their chapters/articles and uses the Smart Topic API [2] to characterize each of them with a semantically enhanced topic vector.
(2) It computes the similarity between conferences and other editorial products and saves the results in a database.
(3) For a given input conference, it returns a list of relevant editorial products, ranked by their topic-centric similarity with the conference in question and filtered in accordance with a number of user preferences.

In order to make it easier for users to understand why a certain item was suggested, SBR offers also an interactive graph view that displays the topic taxonomy of the suggested editorial product and compares it with the input conference.

In the next sections, we will discuss the system in detail. In Sect. 3.1, we describe the knowledge bases used by SBR. In Sect. 3.2, we discuss the Smart Topic API, a service for tagging books with a set of relevant topics. In Sect. 3.3, we describe how we compute the similarity scores. Finally, in Sect. 3.4, we present the user interface.

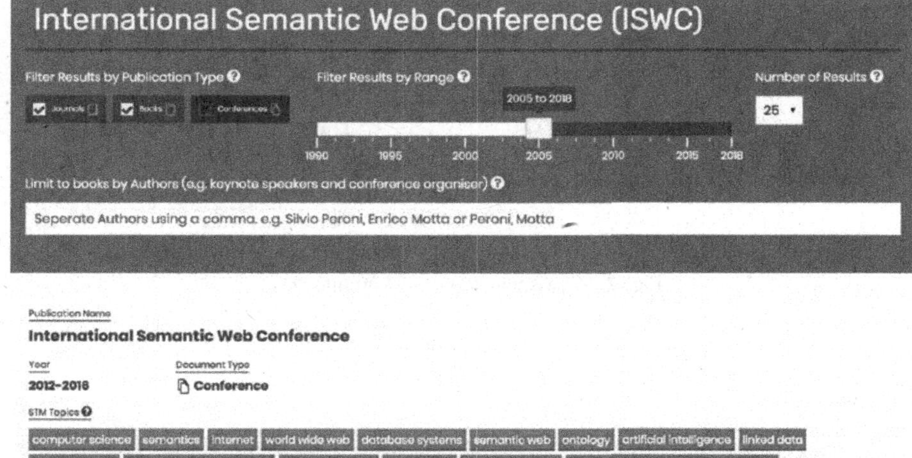

Fig. 1. The main interface of SBR.

2.1 Background Data

SBR relies on two background knowledge bases: a large database of metadata describing publications and the Computer Science Ontology[3].

The database of metadata contains titles, abstracts, keywords and other information describing the chapters of about **27K books and 320 journals** published by SN in the field of Computer Science. In the case of conference proceedings, journals, and edited books, each chapter is usually a research paper. Since we represent conferences according to their proceedings, SBR can only take as input conferences published by Springer Nature.

The Computer Science Ontology (CSO) [3] is a large-scale and granular ontology of research topics that was created automatically by running the Klink-2 algorithm [1] on the Rexplore dataset [4]. This consists of about 16 million publications, primarily in the field of Computer Science. The Klink-2 algorithm combines semantic technologies, machine learning, and background knowledge from a number of web sources, including DBpedia, calls for papers, and web pages, to identify research topics and their relationships from a given corpus of publications. CSO uses the Klink data model[4], which is an extension of the BIBO ontology[5], which in turn builds on SKOS[6]. This model includes three classes of semantic relations: *relatedEquivalent*, which

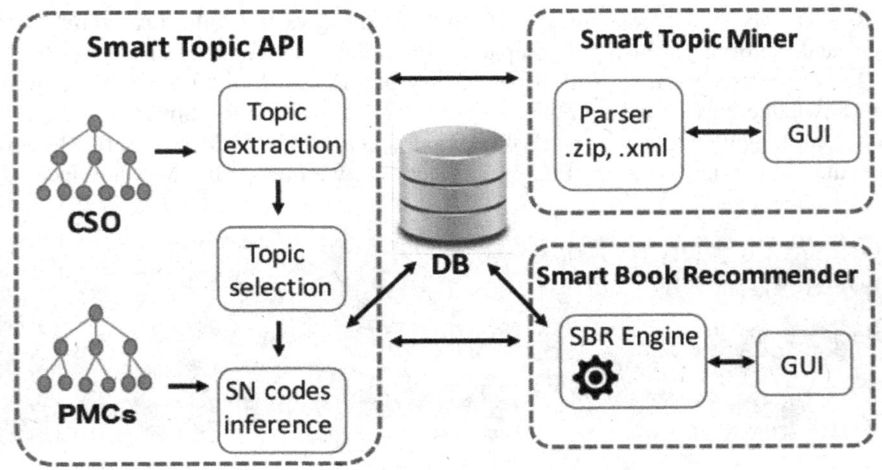

Fig. 2. The smart topic API architecture.

[3] http://skm.kmi.open.ac.uk/cso.

[4] http://technologies.kmi.open.ac.uk/rexplore/ontologies/BiboExtension.owl.

[5] http://purl.org/ontology/bibo/.

[6] http://www.w3.org/2004/02/skos/.

indicates that two topics can be treated as equivalent for the purpose of exploring research data; *skos:broaderGeneric/skos:narrowerGeneric*, which indicate that a topic is a super-area/sub-area of another one; and *contributesTo,* which indicates that the research outputs of one topic significantly contribute to the research work within another. The version of CSO used in the current prototype consists of approximately 15K semantic topics linked by 70K relationships.

2.2 The Smart Topic API

The ongoing collaboration between The Open University and Springer Nature has produced several semantic solutions for supporting the SN editorial team. These include the Smart Topic API [2, 5], an online service for automatically tagging publications with a set of relevant topics from CSO. This API supports a number of applications, including Smart Book Recommender, Smart Topic Miner [5], the Technology-Topic Framework [6], a system that forecasts the propagation of technologies across research communities, and the Pragmatic Ontology Evolution Framework [7], an approach to ontology evolution that is able to select new concepts on the basis of their contribution to specific computational tasks.

Figure 2 shows the architecture of the system. The Smart Topic API takes as input a JSON containing the metadata of a book and returns its description in terms of a taxonomy (or optionally a list) of topics, in which each topic is associated with the number of chapters in which it appeared. It works as following:

(1) For each topic in CSO (e.g., Semantic Web), it associates all the chapters that contain the label of the topic or the label of any *relatedEquivalent* or *skos:narrowerGeneric* (e.g., Linked Data) in the title, the abstract, or the keyword field.
(2) It reduces the list of topics associated with a book to a user-friendly number by means of set covering algorithms [5].
(3) It infers from the topics the product market codes (PMCs) used by SN as internal classification. It then returns a taxonomy of research topics and PMCs associated with the (number of) chapters in which they were detected.

The Smart Topic API powers Smart Topic Miner (STM) [5], a web interface that supports SN editors in classifying proceedings. STM allows editors to submit one or more proceedings, uses the API to annotate them, and then displays them as a taxonomy of research topics. It also offers a number of other options, such as the ability of explaining why a certain topic is relevant by showing the full set of sub-topics that were used to infer it. STM halves the time needed for classifying proceedings from 20–30 to 10–15 min and allows this task to be performed also by assistant editors, thus distributing the load and reducing costs [5].

```
Data: List of publications SN_dataset
Result: Pairwise Similarity Values pairwiseSimilarityScores
foreach conference in SN_dataset.getConferences() do
    /* Extract topic distributions the proceedings of the last 5 years with the Smart Topic API */
    topicsInConference = GetTopics(conference);
    foreach book in SN_dataset do
        pairwiseSimilarityScores=[];
        /* Extract topic distributions of the publication with the Smart Topic API */
        topicsInBook = GetTopics(book);
        /* Estimate the similarity using the Jaccard index */
        jaccard_heuristic = computeJaccardIndex(topicsInConference, topicsInBook);
        if jaccard_heuristic >= JaccardIndexThreshold then
            cosine_similarity = computeCosineSimilarity(topicsInConference, topicsInBook);
            if cosine_similarity >= SimilarityThreshold then
                pairwiseSimilarityScores[] = {conference, book, cosine_similarity};
            end if
        end if
    end foreach
    // Save set of similarity scores to SBR database
    storeInDatabase( pairwiseSimilarityScores );
end foreach
```

Algorithm 1. The SBR algorithm

2.3 Similarity Computation

In order to characterize specific journals, books, and conferences we group the publications as following: (1) for books, chapters are grouped by the book DOI; (2) for journals, the articles are grouped using the journal DOI and their publication year (e.g., Journal of Intelligent Information Systems in 2016), and (3) for conferences, papers are grouped using unique conference identifiers and considering only articles from the last five years. We use the persistent identifiers for conferences and conference series introduced in the Linked Open Data Conference Portal [8] and recently migrated to SciGraph [9]. Such identifiers make sure that the conference series links all relevant conferences, regardless of name changes (e.g., after a few years the "European Semantic Web Conference" became the "Extended Semantic Web Conference") and acronyms.

In an earlier version of SBR, we considered specific editions of conferences – e.g., ISWC 2013. However, on the basis of feedback from the editors, it was decided to consider full conferences series rather than individual editions. This solution simplifies the interface and allows us to reduce possible bias from specific conference editions, which may be affected by trendy topics exhibiting a transient burst of popularity.

We employ the Smart Topic API to associate each item with a vector in which the elements represent research topics and their value is the number of chapters/papers in which the topic was detected. Henceforth, this value will be referred to as *topic weight*.

We exploit this vector representation for computing the similarity between the conference series and the editorial products, as described in Algorithm 1. Since, the Smart Topic API associates publications containing topic T also with the *relatedEquivalent* and *skos:broaderGeneric* of T (see Sect. 2.2), the resulting vectors allow us to match publications that refer to the same concepts (e.g., "Deep Learning") with different key phrases ("Deep Neural Networks"), at different granularity levels ("Machine Learning", "AI").

We assess the similarity of two semantic vectors using the cosine similarity [10], since this measure relies on the orientation but not the magnitude of the topic weights in the vector space, allowing us to compare editorial products associated with a different number of chapters. The similarity computation is carried out offline.

Since it is computationally infeasible to calculate the cosine similarity between each book in the SN dataset, we first prune the number of candidate pairs by calculating their Jaccard index, which is a more lightweight similarity metric, and selecting only the ones that yield a value higher than a threshold. A data analysis revealed that by applying a threshold of 0.125 we halve the number of candidate pairs while still producing very good results. Finally, we save the cosine similarity of a pair in the database if it is greater than 0.5, since according to the editors, recommendations with similarity <0.5 are unhelpful.

2.4 The Web Interface

Figure 1 shows the user interface of SBR. The user can select a conference by typing either the conference name (e.g., "International Semantic Web Conference") or its abbreviated form (e.g., "ISWC"). In Fig. 1 the user has selected ISWC and SBR is showing the top fifteen topics that characterize this venue.

When the user selects a conference, the corresponding conference ID along with the other user preferences (e.g., publication type, year, maximum number of results) are sent as JSON file to the backend via a GET request. The backend is a REST API, which retrieves all relevant publications that meet the criteria and returns the results as a JSON file, which is then visualized by the web interface. The API was developed in Python and the data are pulled from a MariaDB database, while the frontend uses HTML5 and Javascript.

Here, we briefly describe the settings available to the users, to allow them to customise the behaviour of the system.

- *Types of publication* – Users can specify which types of editorial products should be included in the results. Currently, these include books, journals, and (other) conference proceedings.
- *Publication year* – Users can filter results to include only the ones published in a specified time interval. By default, this interval is set to the last three years.
- *Maximum number of results* – Users can set the maximum amount of results according to their needs. This functionality is provided as normally editors can only select a limited number of books to market during a conference.
- *Filter publications by authors and editors* – Users can narrow down the recommendations to books authored or edited by an individual or a group of academics using this free text field. This functionality is provided as editors often focus on marketing editorial products produced by key researchers with high visibility in the research fields relevant to a conference.
- *Exporting data* – Once a list of recommendations has been generated, it is possible for the user to export the results as a CSV or JSON file. These files are typically used by publishing editors to submit an order to the Exhibit Department, which takes care of dispatching the selected products to the conference.

Figure 3 shows the recommendation list that is loaded via an AJAX request after the user has selected a conference. The results are shown as *cards* and sorted in descending order of similarity. Each publication is summarized with respect to its key elements. These include title, publication year, the fifteen most significant topics, and the overall similarity score with the input conference. We display the authors of a book wherever there are less than five authors, otherwise we display editors.

The users can interact with each card by:

- *Examining the publication on SpringerLink*[7] – A hyperlink on the publication title redirects users to the relevant SpringerLink page. This enables editors to collect additional information regarding the publication, such as the authors of individual papers and the abstracts.
- *Providing feedback for a specific card* – We provide a binary feedback system that uses emoticon radio buttons to allow users to express their view on a recommendation. The feedback is used to improve the recommender engine.
- *Opening the graph view interface* – By clicking on the "visualize topic taxonomy" button, users can access a graph view, shown in Fig. 4, which makes it easier for them to make sense of the relationship between the selected output and the input conference.

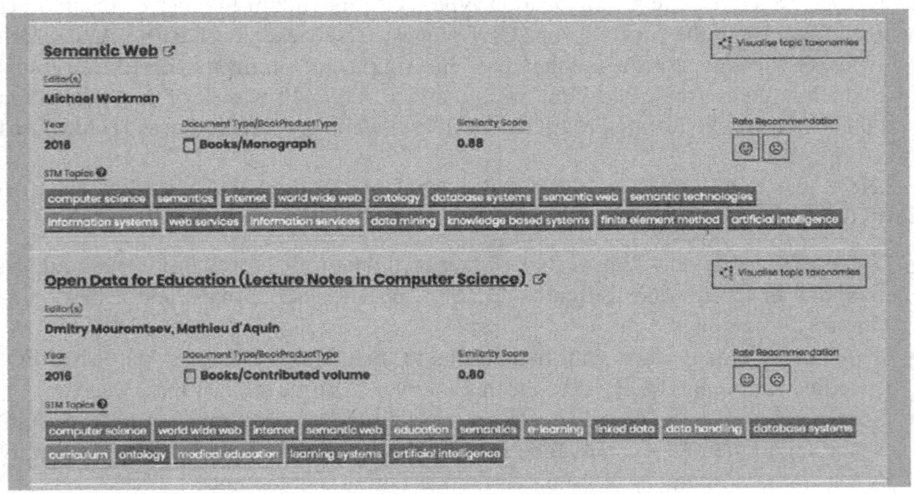

Fig. 3. Recommended SN books for ISWC.

The graph view[8] visualizes editorial products according to their taxonomy of research topics derived from the Computer Science Ontology. The purpose is allowing users to understand why a certain product was recommended and how its associated

7 https://link.springer.com/.

8 The graph view was realised in JavaScript, using the D3.js library (https://d3js.org/).

topics intersects with the ones characterizing the input conference. As an example, in Fig. 4 we show the comparison between the topic-centric characterization of the "*Handbook of Semantic Web Technologies*"[9] and the one of the International Semantic Web Conference. The user can choose whether to visualise only the topics of the conference, those of the recommended publication, their intersection, or all topics of the two items. Hovering over a topic shows the number of chapters/papers within the publication that are associated with the topic. A slider above the interface allows users to filter topics according to their weight.

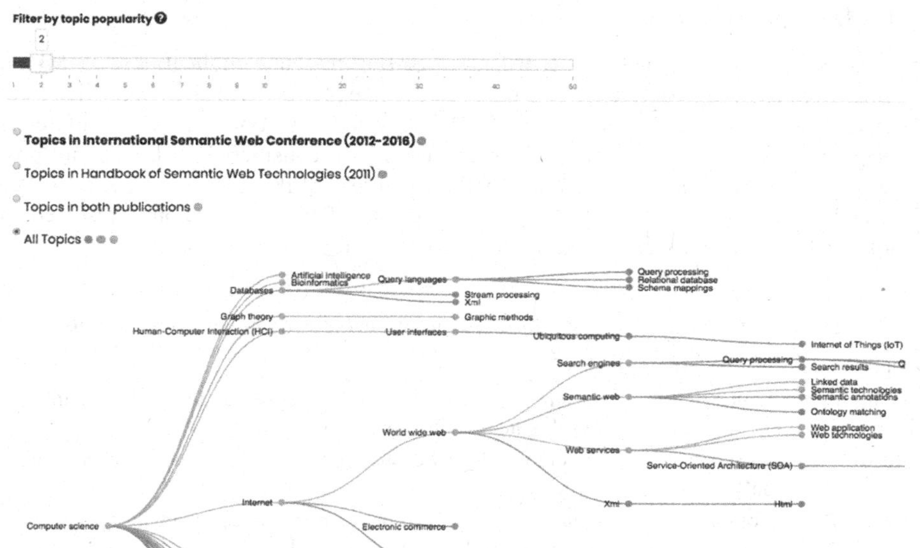

Fig. 4. Portion of the graph view showing the taxonomies of the topics associated with the input conference and one of the recommended editorial products.

3 Evaluation

We evaluated Smart Book Recommender by means of a user study involving seven SN editors and seven OU researchers. The goal of the study was to assess both the usability of SBR and also the quality of its recommendations. We structured the user study in three phases. First, we provided each subject with a 10 min introduction to SBR. Then we asked them to try the system for approximately 45 min and rate its recommendations. Finally, each subject filled a questionnaire about their experience with SBR.

While editors are the main users of the system, we also evaluated SBR with a number of researchers, given that the whole point of the application is to assist editors in selecting editorial products that researchers are likely to be interested in.

---

[9] https://link.springer.com/referencework/10.1007%2F978-3-540-92913-0.

The expertise of the evaluators covered a variety of Computer Science topics, including but not limited to Robotics, Semantic Web, Software Engineering, HCI, AI, Computational Biology, and Wireless Networks.

We assessed the quality of the results by considering the “bring it”, “read it”, and “relevant” books as relevant instances and computing the Precision @10, a standard metric for evaluating ranked lists of items.

The material produced for this evaluation is publically available at http://rexplore.kmi.open.ac.uk/SBR_eval_data and on FigShare[10].

3.1 Quantitative Analysis

We assessed the performance of SBR in suggesting relevant publications, by asking users to choose two conferences in their fields of expertise and then rate SBR recommendations. For each conference, SBR suggested 20 books and 10 conference proceedings. To keep recommendations consistent, we considered all books and proceedings published between 2005 and 2018, regardless of the authors and editors. We asked the users to rate each item by selecting one of the four options presented in Table 1. The sessions were video-recorded to allow further analysis.

Table 1. Options available to editors and researchers for rating recommendations.

Option	Applies to	Definition
Bring it	Editor only	The item is *relevant* to the conference and the editor would *bring* this item to the conference and market it
Read it	Researcher only	The item is *relevant* to the conference and the researcher would want to *read it*
Relevant	Both	The item is *relevant* to the conference, but the editor does not consider it suitable to be marketed or the researcher does not desire to read it. This could be for a variety of marketing or personal reasons
Debatable	Both	Whether the recommended item is relevant to the conference is open to discussion and different people may have different opinions
Irrelevant	Both	The recommended item is not relevant to the conference and should not be recommended

The Precision @10 is 76.8% for the books and 75.4% for the proceedings. It thus seems that there is not much difference in the quality of the recommendations regarding these two editorial products.

Figure 5 shows the percentage of recommendations that were tagged as “bring it/read it”, “relevant”, “debatable”, or “irrelevant” by the users. Editors rated “bring it” or “relevant” 72.9% of the recommendations while researchers rated “read it” or

[10] https://doi.org/10.6084/m9.figshare.6087032.v2.

"relevant" the 66.8% of them. In total, only 10.5% of the recommendations were rated as irrelevant by the editors and 7.2% by the researchers.

Editors would bring to the conference 31.9% of the recommended publications, considering the others not marketable for a variety of reasons, even when they were relevant. On the other hand, researchers would read 14.5% of it. This discrepancy may be explained by the fact that editors and researchers apply different decision-making strategies when choosing whether to "bring" or "read" a publication. Researchers are mainly interested in publications that address their specific needs and they consider also the time invested in reading it and the price. Conversely, editors take into account the preferences of a large group of people and consider a variety of other dimensions, such as how much the book sold in previous years, the popularity of the authors within the community, the potential audience size, and so on.

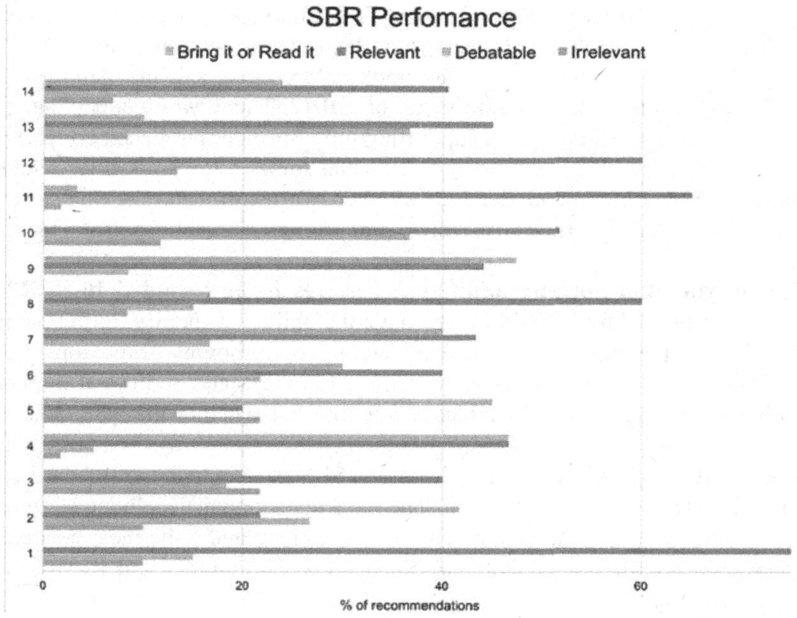

Fig. 5. SBR performance as rated by the evaluators (SN editors labelled 1-7 and OU researchers 8-14). The results of the 28 test cases were aggregated by user.

3.2 Qualitative Analysis

The questionnaire consisted of three sections: (i) an assessment of the evaluators' background and expertise, (ii) five open questions, and (iii) a standard System Usability Scale (SUS) questionnaire to assess the usability of SBR. On average, the editors had 15 years of experience in their role and extensive experience in selecting books for conferences. Three of them had more than 20 years of experience in their field. The OU researchers had an average seniority of about 5 years.

We will first summarize the answers to the open questions.

Q1. How do you find the interaction with the SBR interface? Both groups found the user interface very intuitive. Most attributed this to the "simple" and "well-organised" layout of SBR and the ability to perform queries with little user input. One researcher mentioned that there was a learning curve but it was "easy to pick up", and one editor suggested to make the text input field for searching conference series more noticeable.

Q2. How effectively does SBR support you in selecting relevant publications? Some editors placed the accuracy of the recommended conference proceedings higher than that of the books. One editor felt that some recommended titles were generic, possibly due to the "large margin of error associated with vast selections of conferences" and two pointed out that it would be beneficial to be able to select particular book types, such as handbooks, textbooks and monographs.

Q3. What are the most useful features of SBR? Five researchers found the visual analytics of taxonomies useful for understanding similarities. Three editors appreciated the hyperlinks to the Springer product page. Two researchers and one editor found particularly helpful the option of viewing books and conferences independently.

Q4. What are the main weaknesses of SBR? There was general agreement between editors and researchers that supporting only Springer published conferences is a significant drawback. Three editors indicated that some of the book titles relevant to their conference were not recommended. Another two mentioned that when searching for books, the system returned also some proceedings (i.e., books from the LNCS series).

Q5. Can you think of any additional features to be included in SBR? Two researchers and two editors would like to have the ability of modifying the automatic representation of the input conference by adding or removing some topics. Some editors would like to have direct links to conference pages and additional information about publications, e.g., the main subject discipline and whether they are open access or not.

The last part of the user survey consisted of a SUS questionnaire, a standard tool for assessing the usability of an application. The SUS questionnaire includes 10 questions on a 1 to 5 scale, where 1 is the most negative assessment and 5 the most positive. The average system is expected to score 68 out of 100. The editors and the researchers yielded respectively an average SUS score of 77.1 ± 15.2 and 80.3 ± 11.3, which converts in a percentile rank of about 75%.

Figure 6 shows the answers of the users to four SUS questions. The users believed that SBR was easy to use (with an average score of 4.4 ± 0.7) and its functions where well integrated (3.9 ± 0.6). They did not think that it was complex to use (1.4 ± 0.8) or that they would need the help of a technical person to use it in the future (1.5 ± 0.5).

3.3 Informal Feedback Beyond Computer Science Editorial Team

In addition to the formal evaluation reported in this section, we have also presented the SBR tool to a wider group of publishing editors and editorial assistants at Springer Nature. The fifteen participants (3 sessions with 5 participants each) first saw a short

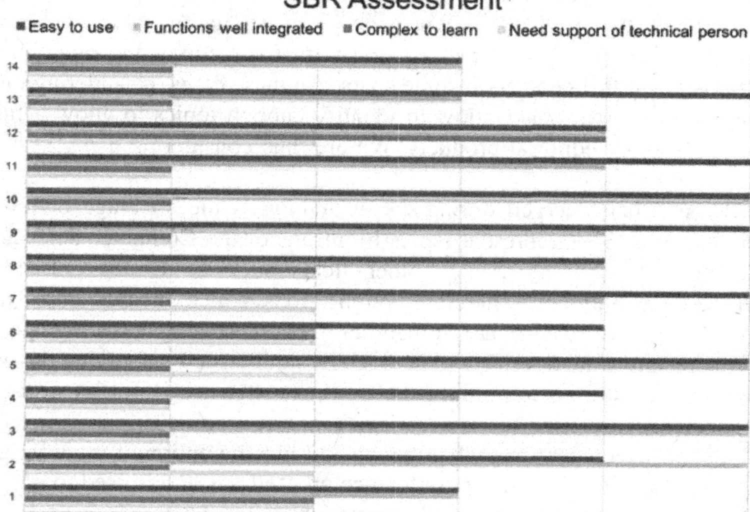

Fig. 6. SUS questionnaire results (editors labelled 1-7 and researchers 8-14).

3-min demo of the tool and then took part in a 10–12 min session where they were encouraged to ask questions and suggest improvements.

The participants saw strong potential for the SBR tool over current practices, which include "ask colleagues for relevant books and journals via e-mail, hoping they have time to reply and are in the office" and "compile a list of relevant titles using various systems, actually developed for other purposes". In particular, they appreciated the time range and type of product filters and the support for searching for books by keynote speakers. They also suggested areas for further improvements, such as the ability of (i) directly querying the system with a list of research topics; (ii) looking up people on the editorial board of a journal; (iii) expand the scope of the system to other disciplines (e.g., Mathematics).

3.4 Discussion and Limitations

SBR obtained a more than satisfactory performance in recommending relevant editorial products and received a high score in term of usability. Nonetheless, the evaluation highlighted some issues that we intend to address in future versions.

A first concern that was mentioned by a number of users is that SBR currently provides recommendations only for conferences which proceedings are published by Springer Nature[11], thus not providing support for marketing activities outside these conferences. In order to include more conferences, we need to also access to the

[11] Actually, since Springer Nature is one of the largest publishers of Computer Science conferences, its coverage of the conferences in this field is very extensive.

conference proceedings published by other editors. We are thus exploring the option of using datasets such as CrossRef[12], Dimensions[13], OpenCitations [11], and Core [12].

Another issue arising from the evaluation is that sometimes the topic characterization of books with few chapters is quite sparse. In these cases, considering only title, abstract and keywords may not allow to identify enough topics to allow a fair comparison with the other editorial products. We are thus considering using also the full text.

A third issue that emerged during the evaluation is the coverage of multidisciplinary publications. SBR represent topics by means of the Computer Science Ontology, and therefore scarcely covers other fields such as Biology, Engineering, Mathematics, or Economics. Therefore, publications which include other fields in addition to Computer Science are sometimes misrepresented, lowering the overall quality of the recommendations. We plan to address this issue by applying the ontology learning techniques utilized to produce the Computer Science Ontology also on other domains of science.

Finally, some users mentioned that they would like the option of modifying the set of topics that get extracted from the conference proceedings and is used to produce the recommendations. A further step in this direction would be to allow users to input directly a set of topics as a query. This would naturally require some significant changes to the backend, since currently all the similarity values are precomputed, but it would also allow for more flexibility. Indeed, this solution may also enable us to associate users with a representation of their research interests and automatically produce tailored recommendations.

4 Next Steps for Large Scale Deployment

SBR was well received by Springer Nature editors, but we must take some additional steps to fully integrate it into their workflow.

In the first instance, we intend to automatize the process for importing and processing the most recent editorial items. Currently, we renew our database every four months by importing a new dump of metadata and recalculating the similarity values. This solution suffers from two limitations: it requires human intervention and the system is updated only every four-months. We plan to fully automatize this process by developing a system for importing new metadata on a daily basis and recomputing seamlessly the relevant similarity values.

In the second instance, we plan to develop a new version of SBR that will address the most important requests that came up during the user study, as discussed in previous section.

Finally, we are exploring the ability of SBR to produce collections of documents relevant to certain topics, e.g., all recent publications in the field of Ontology Engineering. This has broader implications beyond selecting books for conferences, and can

[12] http://crossref.org.

[13] https://www.dimensions.ai.

help compiling ad-hoc packages for industry or academic institutions in the developing countries. Some initial experiments in this direction have already yielded promising results.

5 Related Work

Recommender systems are software tools and methods which provide suggestions for items to users, according to their preferences and needs [13]. They are typically classified as collaborative filtering approaches, content-based filtering approaches and hybrid approaches [14].

Content-based recommender systems [15] rely on a pre-existing domain knowledge to suggest items more similar to the ones that the user seems to like. They usually generate user models that describe user interests according to a set of features [16]. With the advent of the Semantic Web, several recommender systems started to adopt ontologies for representing both user interests and items [17]. Often these systems use an ontology so that, given user interest in an item represented in the ontology, they can then propagate such interest to relevant items and concepts. For example, given a positive feedback on "beagles", a system may infer (correctly or not) that a user is interested in "dogs", and more generally in "pets". SBR exploits a similar mechanism when it infers that a publication explicitly linked to a topic (e.g., Linked Data) is also about its *skos:broaderGeneric* concepts in CSO (e.g., Semantic Web). The main advantages of these solutions are (i) the ability to exploit the domain knowledge for improving the user modelling process, (ii) the ability to share and reuse system knowledge, and (iii) the alleviation of the cold-start and data sparsity problems [16, 18].

We will now discuss some of these ontology-based approaches. Sieg et al. [16] present an ontology-based recommender to improve personalised Web searching in which the user profiles are instances of a reference domain ontology and are incrementally updated based on the user interaction with the system. Middleton et al. [18] describe a hybrid recommender system that exploit ontologies for increasing the accuracy of the profiling process and hence the usefulness of the recommendations. Thiagarajan et al. [19] use a different strategy by representing user profiles as bags-of-words and weighing each term according to the user interests derived from a domain ontology. Razmerita et al. [20] describe OntobUM, an ontology-based recommender that integrates three ontologies: (i) the user ontology, which structures the characteristics of users and their relationships, (ii) the domain ontology, which defines the domain concepts and their relationships, and (iii) the log ontology, which defines the semantics of the user interactions with the system. Birukou et al. [21] present an agent-based system that learns the preferences of experienced researchers and provides specific suggestions to support search for scientific publications. Colombo-Mendoza et al. [22] propose RecomMetz, a context-aware mobile recommender system based on Semantic Web technologies. This system introduced some unique features, such as the composite structure of the items and the integration of temporal and crowd factors into a context-aware model. Finally, Cantador et al. [23] propose a hybrid recommendation model in which user preferences are described in terms of semantic concepts defined in domain ontologies. Similar to all these systems SBR builds a semantic representation of the items and exploits the ontology for inferring additional concepts. However, rather than creating a representation of a single user, it characterizes the overall interests of the research community associated with the proceedings of a conference series.

SBR builds on the Smart Topic API to represent publications as vectors of research topics. This is a useful representation that is used in a variety of systems for exploring the research landscape [4, 6]. In recent years, we have seen the emergence of several approaches to annotating research articles. For instance, DBpedia Spotlight [24] is often used for automatically annotating papers with DBpedia concepts. Gabor et al. [25] introduce an approach for annotating scientific corpora with domain-relevant concepts and semantic relations. The Dr. Inventor Framework [26] focuses instead on extracting structured textual contents, discursive characterization of sentences, and graph based representations of text excerpts.

6 Conclusions

In this paper, we presented Smart Book Recommender, a semantic recommender system developed in collaboration with Springer Nature which suggests editorial products to market at academic venues.

A user study involving seven SN editors and seven OU researchers showed that SBR was able to suggest relevant materials and scored high in usability. In particular, Springer Nature editors considered as relevant 72.9% of the SBR recommendations and assessed the system as very user friendly, yielding an average SUS score of 77.1.

We are now planning to further integrate the SBR tool into the process workflows at Springer Nature. To this purpose, we are going to develop a new version of the system, which will take into account a variety of suggestions which arose from the user study.

Acknowledgements. We would like to thank publishing editors at Springer Nature for assisting us in the evaluation of SBR and allowing us to access their large repositories of scholarly data.

References

1. Osborne, F., Motta, E.: Klink-2: integrating multiple web sources to generate semantic topic networks. In: Arenas, M., et al. (eds.) ISWC 2015, Part I. LNCS, vol. 9366, pp. 408–424. Springer, Cham (2015). https://doi.org/10.1007/978-3-319-25007-6_24
2. Osborne, F., Salatino, A., Birukou, A., Thanapalasingam, T., Motta, E.: Supporting Springer nature editors by means of semantic technologies. In: Nikitina, N., Song, D., Fokoue, A., Haase, P. (eds.) International Semantic Web Conference (Posters, Demos and Industry Tracks). CEUR Workshop Proceedings, vol. 1963. CEUR-WS.org (2017)
3. Salatino, A.A., Thanapalasingam, T., Mannocci, A., Osborne, F., Motta, E.: The computer science ontology : a large-scale taxonomy of research areas. In: Vrandečić, D., et al. (eds.) ISWC 2018, Part II. LNCS, vol. 11137, pp. 187–205. Springer, AG (2018)

4. Osborne, F., Motta, E., Mulholland, P.: Exploring scholarly data with rexplore. In: Alani, H., et al. (eds.) ISWC 2013, Part I. LNCS, vol. 8218, pp. 460–477. Springer, Heidelberg (2013). https://doi.org/10.1007/978-3-642-41335-3_29
5. Osborne, F., Salatino, A., Birukou, A., Motta, E.: Automatic classification of springer nature proceedings with smart topic miner. In: Groth, P., et al. (eds.) ISWC 2016, Part II. LNCS, vol. 9982, pp. 383–399. Springer, Cham (2016). https://doi.org/10.1007/978-3-319-46547-0_33
6. Osborne, F., Mannocci, A., Motta, E.: Forecasting the spreading of technologies in research communities. In: Proceedings of the Knowledge Capture Conference (2017)
7. Osborne, F., Motta, E.: Pragmatic ontology evolution: reconciling user requirements and application performance. In: Vrandečić, D., et al. (eds.) ISWC 2018, Part II. LNCS, vol. 11137, pp. 495–512. Springer, AG (2018)
8. Birukou, A., Bryl, V., Eckert, K., Gromyko, A., Kaindl, M.: Springer LOD conference portal. Demo paper. In: Nikitina, N., Song, D., Fokoue, A., Haase, P. (eds.) International Semantic Web Conference (Posters, Demos and Industry Tracks).CEUR Workshop Proceedings, vol. 1963. CEUR-WS.org (2017)
9. Hammond, T., Pasin, M., Theodoridis, E.: Data integration and disintegration: managing springer nature SciGraph with SHACL and OWL. In: Nikitina, N., Song, D., Fokoue, A., Haase, P. (eds.) International Semantic Web Conference (Posters, Demos and Industry Tracks). CEUR Workshop Proceedings, vol. 1963. CEUR-WS.org (2017)
10. Salton, G., Buckley, C.: Term-weighting approaches in automatic text retrieval. Inf. Process. Manag. **24**, 513–523 (1988)
11. Peroni, S., Dutton, A., Gray, T., Shotton, D.: Setting our bibliographic references free: towards open citation data. J. Doc. **71**, 253–277 (2015)
12. Knoth, P., Zdrahal, Z.: CORE: three access levels to underpin open access. D-Lib Mag. **18** (11–12), 4 (2012)
13. Ricci, F., Rokach, L., Shapira, B. (eds.): Recommender Systems Handbook. Springer, Boston (2015). https://doi.org/10.1007/978-1-4899-7637-6
14. Burke, R.: Hybrid web recommender systems. In: Brusilovsky, P., Kobsa, A., Nejdl, W. (eds.) The Adaptive Web. LNCS, vol. 4321, pp. 377–408. Springer, Heidelberg (2007). https://doi.org/10.1007/978-3-540-72079-9_12
15. Lops, P., de Gemmis, M., Semeraro, G.: Content-based recommender systems: state of the art and trends. In: Ricci, F., Rokach, L., Shapira, B., Kantor, P.B. (eds.) Recommender Systems Handbook, pp. 73–105. Springer, Boston, MA (2011). https://doi.org/10.1007/978-0-387-85820-3_3
16. Sieg, A., Mobasher, B., Burke, R.: Web search personalization with ontological user profiles. In: Conference on Information and Knowledge Management, p. 525. ACM Press, New York (2007)
17. de Gemmis, M., Lops, P., Musto, C., Narducci, F., Semeraro, G.: Semantics-aware content-based recommender systems. In: Ricci, F., Rokach, L., Shapira, B. (eds.) Recommender Systems Handbook, pp. 119–159. Springer, Boston, MA (2015). https://doi.org/10.1007/978-1-4899-7637-6_4
18. Middleton, S.E., Shadbolt, N.R., De Roure, D.C.: Ontological user profiling in recommender systems. ACM Trans. Inf. Syst. **22**, 54–88 (2004)
19. Thiagarajan, R., Manjunath, G., Stumptner, M.: Finding experts by semantic matching of user profiles. In: International Semantic Web Conference Personal Identification and Collaborations: Knowledge Mediation and Extraction (PICKME 2008), pp. 7–18 (2008)

20. Razmerita, L., Angehrn, A., Maedche, A.: Ontology-based user modeling for knowledge management systems. In: Brusilovsky, P., Corbett, A., de Rosis, F. (eds.) UM 2003. LNCS (LNAI), vol. 2702, pp. 213–217. Springer, Heidelberg (2003). https://doi.org/10.1007/3-540-44963-9_29
21. Birukou, A., Blanzieri, E., Giorgini, P.: A multi-agent system that facilitates scientific publications search. In: International Joint Conference on Autonomous Agents and Multiagent Systems, pp. 265–272. ACM Press (2006)
22. Colombo-Mendoza, L.O., Valencia-García, R., Rodríguez-González, A., Alor-Hernández, G., Samper-Zapater, J.J.: RecomMetz: a context-aware knowledge-based mobile recommender system for movie showtimes. Expert Syst. Appl. **42**, 1202–1222 (2015)
23. Cantador, I., Bellogín, A., Castells, P.: A multilayer ontology-based hybrid recommendation model. AI Commun. **21**, 203–210 (2008)
24. Mendes, P.N., Jakob, M., García-Silva, A., Bizer, C.: DBpedia spotlight: shedding light on the web of documents (2011)
25. Gábor, K., Zargayouna, H., Buscaldi, D., Tellier, I., Charnois, T.: Semantic annotation of the ACL anthology corpus for the automatic analysis of scientific literature (2016)
26. Ronzano, F., Saggion, H.: Dr. Inventor Framework: Extracting Structured Information from Scientific Publications (2015)

Synthesizing Knowledge Graphs from Web Sources with the MINTE$^+$ Framework

Diego Collarana[1,2(✉)], Mikhail Galkin[1,2,4], Christoph Lange[1,2], Simon Scerri[1,2], Sören Auer[3,6], and Maria-Esther Vidal[2,3,5]

[1] University of Bonn, Bonn, Germany
{collaran,galkin,langec,scerri}@cs.uni-bonn.de
[2] Fraunhofer Institute for Intelligent Analysis and Information Systems, Sankt Augustin, Germany
[3] TIB Leibniz Information Centre for Science and Technology, Hannover, Germany
{soeren.auer,maria.vidal}@tib.eu
[4] ITMO University, Saint Petersburg, Russia
[5] Universidad Simón Bolívar, Caracas, Venezuela
[6] L3S Research Centre, Leibniz University of Hannover, Hannover, Germany

Abstract. Institutions from different domains require the integration of data coming from heterogeneous Web sources. Typical use cases include Knowledge Search, Knowledge Building, and Knowledge Completion. We report on the implementation of the *RDF Molecule-Based Integration Framework MINTE*$^+$ in three domain-specific applications: Law Enforcement, Job Market Analysis, and Manufacturing. The use of RDF molecules as data representation and a core element in the framework gives MINTE$^+$ enough flexibility to synthesize knowledge graphs in different domains. We first describe the challenges in each domain-specific application, then the implementation and configuration of the framework to solve the particular problems of each domain. We show how the parameters defined in the framework allow to tune the integration process with the best values according to each domain. Finally, we present the main results, and the lessons learned from each application.

Keywords: Data integration · RDF · Knowledge graphs
RDF molecules

1 Introduction

We are living in the era of digitization. Today as never before in the history of mankind, we are producing a vast amount of information about different entities in all domains. The Web has become the ideal place to store and share this information. However, the information is spread across several web sources, with different accessibility mechanisms. The more the amount of information grows on the Web, the more important are efficient and cost-effective search, integration,

D. Vrandečić et al. (Eds.): ISWC 2018, LNCS 11137, pp. 359–375, 2018.
https://doi.org/10.1007/978-3-030-00668-6_22

and exploration of such information. Creating valuable knowledge out of this information is of interest not only to research institutions but to enterprises as well. Big companies such as Google or Microsoft spend a lot of resources in creating and maintaining so-called knowledge graphs. However, institutions such as law enforcement agencies, startups, or SMEs cannot spend comparable resources to collect, integrate, and create value out of such data.

In this paper, we present the use of MINTE$^+$, an RDF Molecule-Based Integration Framework, in three domain-specific applications. MINTE$^+$ is an integration framework that collects and integrates data from heterogenous web sources into a knowledge graph. MINTE$^+$ implements novel semantic integration techniques that rely on the concept of RDF molecules to represent the meaning of this data; it also provides fusion policies that enable *synthesis* of RDF molecules. We present the main results, showing a significant improvement of the task completion efficiency when the goal is to find specific information about an entity, and discuss the lessons learned from each application.

Although several approaches and tools have been proposed to integrate heterogeneous data, a complete and configurable framework specialized for web sources is still not easy to set up. The power of MINTE$^+$ comes with the parameters to tune the integration process according to the use case requirements and challenges. MINTE$^+$ builds on the main outcomes of the semantic research community such as semantic similarity measures [8], ontology-based information integration, RDF molecules [4], and semantic annotations [9] to identify relatedness between entities and integrate them into a knowledge graph.

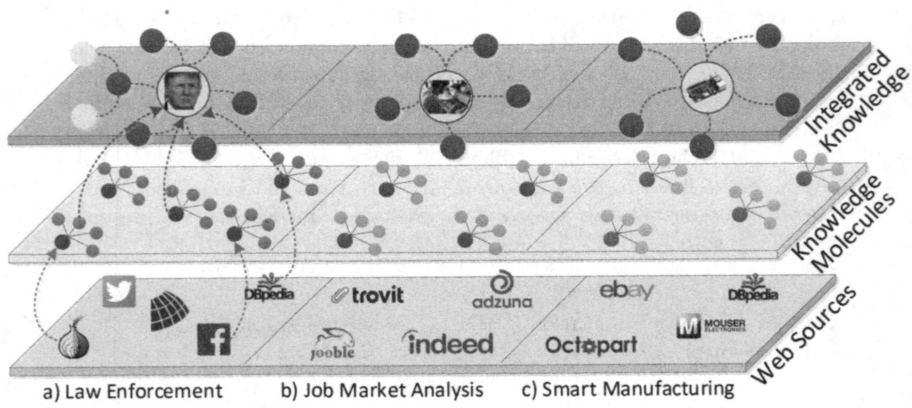

Fig. 1. Domain-specific applications. (a) Law Enforcement agencies need to synthesize knowledge about suspects. (b) For a Job Market analysis, the job offers from different job portals need to be synthesized. (c) A manufacturing company needs synthesized knowledge about providers from web sources.

Motivation: Law enforcement agencies need to find information about suspects or illegal products on *web sites*, *social networks*, or private web sources in the Deep Web such as OCCRP[1]. For a job market analysis, job offers from different web portals need to be integrated to gain a complete view of the market. Finally, manufacturing companies are interested in information about their *providers* available in knowledge graphs such as DBpedia, which can be used to complete the company's internal knowledge. Figure 1 illustrates the main problem and challenges of integrating pieces of knowledge from heterogeneous web sources. Although three different domain specific applications are presented, the core problem is shared: *synthesizing knowledge graphs from heterogeneous web sources*, involving, for example, knowledge about *suspects*, or *job postings*, or *providers* (top layer of Fig. 1). This knowledge is spread across different web sources such as *social networks*, *job portals*, or Open Knowledge Graphs (bottom layer of Fig. 1). However, the integration poses the following challenges:

- The lack of uniform representation of the pieces of knowledge.
- The need to identify semantically equivalent molecules.
- A flexible process for integrating these pieces of knowledge.

The remainder of the article is structured as follows: Sect. 2 describes MINTE$^+$. Then, the application of MINTE$^+$ in Law Enforcement (Sect. 3), Job Marked Analysis (Sect. 4), and Manufacturing (Sect. 5) is described. Finally, Sect. 6 presents our conclusions and outlines our future work.

2 The Synthesis of RDF Molecules Using MINTE$^+$

Grounded on the semantic data integration techniques proposed by Collarana et al. [3,4], we propose MINTE$^+$, an integration framework able to create, identify, and merge semantically equivalent RDF entities. Thus, a solution to the *problem of semantically integrating* RDF molecules is provided. Figure 2 depicts the main components of the MINTE$^+$ architecture. The pipeline receives a keyword-based query Q and a set of APIs of web sources (API_1, API_2, API_n) to run the query against. Additionally, the integration configuration parameters are provided as input. These parameters include: a semantic similarity measure Sim_f, a threshold γ, and an ontology O; they are used to determine when two RDF molecules are semantically equivalent. Furthermore, a set of fusion policies σ to integrate the RDF molecules is part of the configuration. MINTE$^+$ consists of three essential components: RDF molecule creation, identification, and integration. First, various RDF subgraphs coming from heterogeneous web sources are organized as RDF molecules, i.e., sets of triples that share the same subject. Second, the *identification component* discovers semantically equivalent RDF molecules, i.e., ones that refer to the same real-world entity; it performs two sub-steps, i.e., *partitioning* and *1-1 weighted perfect matching.* Third, having identified equivalent RDF molecules, MINTE$^+$'s semantic data integration techniques resemble the *chemical synthesis of molecules* [2], and the *integration component* integrates RDF molecules into complex RDF molecules in a knowledge graph.

[1] https://www.occrp.org/.

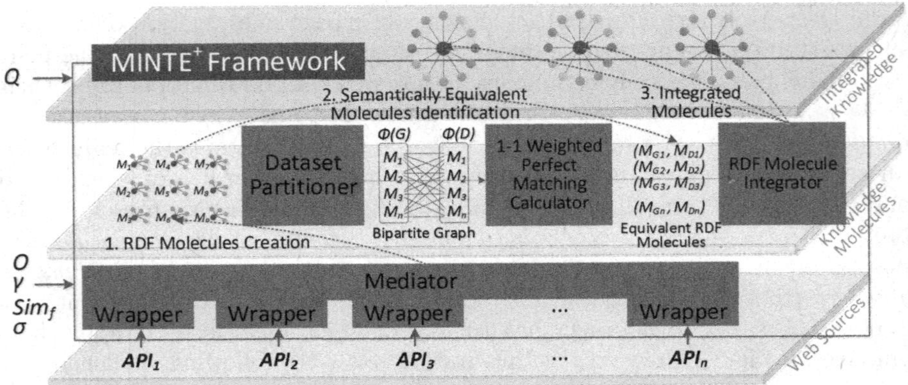

Fig. 2. The MINTE$^+$ Architecture. MINTE$^+$ receives a set of web APIs, a keyword query Q, a similarity function Sim_f, a threshold γ, an ontology O, and a fusion policy σ. The output is a semantically integrated RDF graph.

2.1 Creating RDF Molecules

The *RDF molecule creation component* relies on search API methods, e.g., the API for searching people on Google+[2], and transforms an initial keyword-based query Q into a set of API requests understandable by the given web sources. MINTE$^+$ implements the mediator-wrapper approach; wrappers are responsible for physical data extraction, while a mediator orchestrates transformation of the obtained data into a knowledge graph. An ontology O provides formal descriptions for RDF molecules, using which the API responses are transformed into RDF molecules using Silk Transformation Tasks[3]. All the available sources are queried, i.e., no source selection technique is applied. Nevertheless, the execution is performed in an asynchronous fashion, so that the process takes as much time as the slowest web API. Once a request is complete, wrappers transform the results into sets of RDF triples that share the same subject, i.e., RDF molecules. Then, the mediator aggregates RDF molecules into a knowledge graph, which is sent to the next component. These RDF molecule-based methods enable data transformation and aggregation tasks in a relatively simple way. Figure 3 depicts the interfaces implemented by a wrapper in order to be plugged into the pipeline.

2.2 Equivalent Molecules Identification

MINTE$^+$ employs a semantic similarity function Sim_f to determine whether two RDF molecules correspond to the same real-world entity, e.g., determining if two job postings are semantically equivalent. A similarity function has to

[2] https://developers.google.com/+/web/api/rest/latest/people/search.
[3] http://silkframework.org/.

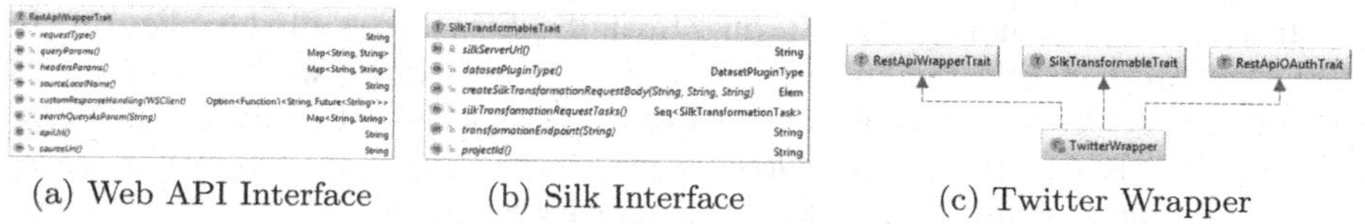

(a) Web API Interface (b) Silk Interface (c) Twitter Wrapper

Fig. 3. The MINTE$^+$ framework defines three basic interfaces for a Wrapper: WebApiTrait, SilkTransformationTrait, and OAuthTrait.

leverage semantics encoded in the ontology O. For instance, GADES [8] implementation[4] supports this requirement. Additional knowledge about class hierarchy (`rdfs:subClassOf`), equivalence of resources (`owl:sameAs`), and properties (`owl:equivalentProperty`) enables uncovering semantic relations at the molecule level instead of just comparing plain literals. The identification process involves two stages: (a) dataset partitioning and (b) finding a perfect matching between partitions.

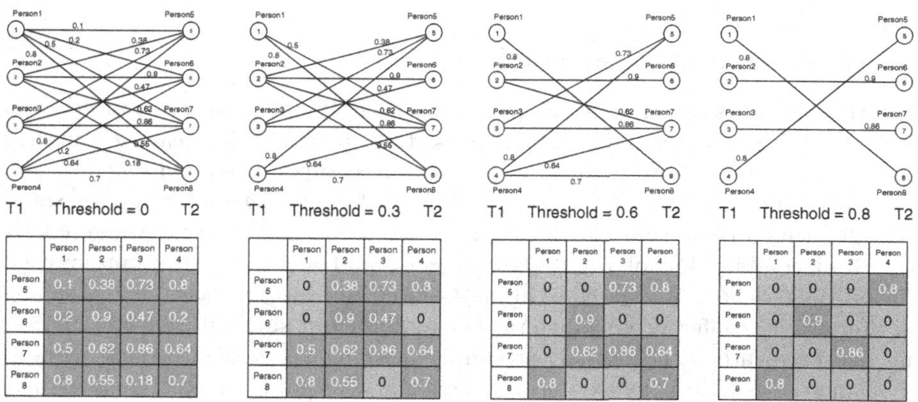

Threshold = 0

	Person 1	Person 2	Person 3	Person 4
Person 5	0.1	0.38	0.73	0.8
Person 6	0.2	0.9	0.47	0.2
Person 7	0.5	0.62	0.86	0.64
Person 8	0.8	0.55	0.18	0.7

Threshold = 0.3

	Person 1	Person 2	Person 3	Person 4
Person 5	0	0.38	0.73	0.8
Person 6	0	0.9	0.47	0
Person 7	0.5	0.62	0.86	0.64
Person 8	0.8	0.55	0	0.7

Threshold = 0.6

	Person 1	Person 2	Person 3	Person 4
Person 5	0	0	0.73	0.8
Person 6	0	0.9	0	0
Person 7	0	0.62	0.86	0.64
Person 8	0.8	0	0	0.7

Threshold = 0.8

	Person 1	Person 2	Person 3	Person 4
Person 5	0	0	0	0.8
Person 6	0	0.9	0	0
Person 7	0	0	0.86	0
Person 8	0.8	0	0	0

Fig. 4. Bipartite Graph Pruning. Various thresholds on a semantic similarity function and their impact on creating a bipartite graph between RDF molecules.

Dataset Partitioner. The partitioner component relies on a similarity measure Sim_f and an ontology O to determine relatedness between RDF molecules. Addressing flexibility, MINTE$^+$ allows for arbitrary, user-supplied similarity functions, e.g., simple string similarity and set similarity. We, however, advocate for semantic similarity measures as they achieve better results (as we show in [3]) by considering semantics encoded in RDF graphs. After computing similarity scores, the partitioner component constructs a bipartite graph between the sets of RDF molecules; it is used to match the RDF molecules.

Given a bipartite graph $G = (U, V, E)$, the set of vertices U and V are built from two collections of RDF molecules to be integrated (e.g., a wrapper

[4] https://github.com/RDF-Molecules/sim_service.

result and an in-memory RDF Graph). Initially, E includes all the edges in the Cartesian product of U and V, and each edge is annotated with the similarity value of the related RDF molecules. In case of a specified threshold, the edges annotated with a similarity value lower than the threshold are removed from E.

A threshold γ bounds the values of similarity when two RDF molecules cannot be considered similar. It is used to prune edges from the bipartite graph whose weights are lower than the threshold. Figure 4 illustrates how different threshold values affect the number of edges in a bipartite graph. Low threshold values, e.g., 0, result in graphs retaining almost all the edges. Contrarily, when setting a high threshold, e.g., 0.8, graphs are significantly pruned.

1-1 Weighted Perfect Matching. Having prepared a bipartite graph in the previous step, the *1-1 weighted perfect matching component* identifies the equivalent RDF molecules by matching them with the highest pairwise similarity score; a Hungarian algorithm is used to compute the matching. Figure 4 ($\gamma = 0.8$) illustrates the result of computing a 1-1 weighted perfect matching on the given bipartite graph. MINTE$^+$ demonstrates better accuracy when semantic similarity measures like GADES are applied when building a bipartite graph.

2.3 RDF Molecule Integration

The third component of MINTE$^+$, namely the RDF molecule *integration component*, leverages the identified equivalent RDF molecules in creating a unified knowledge graph. In order to retain knowledge completeness, consistency, and address duplication, MINTE$^+$ resorts to a set of *fusion policies* σ implemented by rules that operate on the RDF triple level. These rules are triggered by a certain combination of predicates, objects, and axioms in the ontology O. Fusion policies resemble flexible filters tailored for specific tasks, e.g., keeping all literals with different language tags or retaining an authoritative one, replacing one predicate with another, or simply merging all predicate-value pairs of given molecules. Ontology axioms are particularly useful when resolving conflicts and inequalities on different semantic levels. Types of fusion policies include the following: policies that process RDF resources such as dealing with URI naming conventions, are denoted as a subset $\sigma_r \in \sigma$. Policies that focus on properties are denoted as $\sigma_p \in \sigma$. Interacting with the ontology O, σ_p tackles property axioms, e.g., `rdfs:subPropertyOf`, `owl:equivalentProperty`, and `owl:FunctionalProperty`. Property-level fusion policies tackle sophisticated OWL restrictions on properties. That is, if a certain property can have only two values of some fixed type, σ_p has to guide the fusion process to ensure semantic consistency. Lastly, the policies dedicated to objects (both entities and literals) comprise a subset $\sigma_v \in \sigma$. On the literal level, the σ_v policies implement string processing techniques, such as recognition of language tags, e.g., *@en*, *@de*, to decide whether those literals are different. For object properties, the σ_v policies deal with semantics of the property values, e.g., objects of different properties are linked by `owl:sameAs`. In this application of MINTE$^+$, the following policies are utilized [4]:

Union Policy. The union policy creates a set of $(prop, val)$ pairs where duplicate pairs, i.e., pairs that are syntactically the same, are discarded retaining only one pair. In Fig. 5a the pair $(type, A)$ appears in both molecules. In Fig. 5b, only one pair is retained. The rest of the pairs are added directly.

Subproperty Policy. The policy tracks if a property of one RDF molecule is an `rdfs:subPropertyOf` of a property of another RDF molecule, i.e., $\{r_1, p_1, A\}$, $\{r_2, p_2, B\} + O + subPropertyOf(p_1, p_2) \models \{\sigma_r(r_1, r_2), p_2, \sigma_v(A, B)\}$. As a result of applying this policy, the property p_1 is replaced with a more general property p_2. The default σ_v object policy is to keep the property value of p_1 unless a custom policy is specified. In Fig. 5c, a property *brother* is generalized to *sibling* preserving the value C according to the subproperty ontology axiom in Fig. 5a.

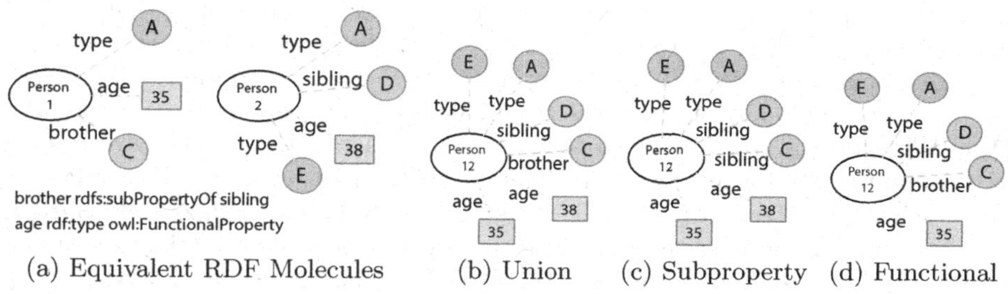

(a) Equivalent RDF Molecules (b) Union (c) Subproperty (d) Functional

Fig. 5. Merging Semantically Equivalent RDF Molecules. Applications of a fusion policy σ: (a) semantically equivalent molecules R_1 and R_2 with two ontology axioms; (b) simple union of all triples in R_1 and R_2 without tackling semantics; (c) p_3 is replaced as a subproperty of p_4; (d) p_2 is a functional property and R_1 belongs to the authoritative graph; therefore, literal C is discarded.

Authoritative Graph Policy. The policy selects one RDF graph as a major source when merging various configurations of $(prop, val)$ pairs:

- The **functional property policy** keeps track of the properties annotated as `owl:FunctionalProperty`, i.e., such properties may have only one value. The authoritative graph policy then retains the value from the primary graph: $\{r_1, p_1, B\}, \{r_2, p_1, C\} + O + functional(p_1) \models \{\sigma_r(r_1, r_2), p_1, \sigma_v(B, C)\}$. Annotated as a functional property in Fig. 5a, *age* has the value 35 in Fig. 5d, as the first graph has been marked as authoritative beforehand. The value 38 is therefore discarded.
- The **equivalent property policy** is triggered when two properties of two molecules are `owl:equivalentProperty`: $\{r_1, p_1, A\}, \{r_2, p_2, B\} + O + equivalent(p_1, p_2) \models \{\sigma_r(r_1, r_2), \sigma_p(p_1, p_2), \sigma_v(A, B)\}$. The authoritative policy selects a property from the authoritative graph, e.g., either p_1 or p_2. By default, the property value is taken from the chosen property. Custom σ_v policies may override thesecriteria.

- The **equivalent class or entity policy** contributes to the integration process when entities are annotated as `owl:equivalentClass` or `owl:sameAs`, i.e., two classes or individuals represent the same real-world entity, respectively: $\{r_1, p_1, A\}, \{r_2, p_2, B\} + O + \mathit{equivalent}(A, B) \models \{\sigma_r(r_1, r_2), \sigma_p(p_1, p_2), \sigma_v(A, B)\}$. Similarly to the equivalent property case, the value with its corresponding property is chosen from the primary graph. Again, custom σ_p policies may handle the merging of properties.

3 A Law Enforcement Application

3.1 Motivation and Challenges

Law enforcement agencies and other organizations with security responsibilities are struggling today to capture, manage and evaluate the amounts of data stored in countless heterogeneous web sources. As Fig. 1a shows, possible sources include the document-based Web (so-called "visible net"), usually indexed by search engines such as Google or Bing, the Social Web (e.g., Facebook or Twitter), the Deep Web and the Dark Web (so-called "invisible net"). Deep web sources, such as e-commerce platforms (e.g., Amazon or eBay), cannot be accessed directly, but only via web interfaces, e.g., REST APIs. The same holds for dark web sources, which are usually among the most relevant web sources for investigating online crime. Finally, open data catalogs in the Data Web, i.e., machine-understandable data from open sources such as Wikipedia, serve as sources of information for investigations. Law enforcement agencies spend a lot of time on searching, collecting, aggregating, and analyzing data from heterogeneous web sources. The main reason for such inefficient knowledge generation is that the agencies need different methods and tools to access this diversified information. If the investigators are not experts in a particular tool or technique, such as querying the Data Web using SPARQL, they may not find the information they need. Finally, most current search technology is based on simple keywords but neglects semantics and context. The latter is particularly important if you are looking for people with common names such as (in German) "Müller" or "Schmidt". Here, a context of related objects such as other people, places or organizations is needed to make a proper distinction. The main challenges of this application are the following:

C1. Heterogeneity of accessibility: Different access mechanisms need to be used to collect data from the web sources. Social networks require user-token authentication, deep web sources use access keys, and dark web sources require the use of the special software Tor Proxy[5].

C2. Provenance Management: Law enforcement institutions need to know the origin of the data, for a post-search veracity evaluation.

C3. Information Completeness: Although the process should be as automatic as possible, in this application no data should be lost, e.g., all aliases or names of a person should be kept.

[5] https://www.torproject.org/.

C4. Privacy by design: The system must be fully compliant with data protection laws, e.g., the strict ones that hold in the EU and especially in Germany. Citizens privacy is mainly protected by a fundamental design decision: No comprehensive data warehouse is built-up, but information is access on demand from the Web sources.

The LiDaKrA[6] project had as main goal the implementation of a Crime Analysis Platform to solve the challenges presented above. The platform concept should be offered as a platform-as-a-service intended to support end users, such as police departments, in the following use cases:

U1. Politically Exposed Persons: searching for politicians' activity in social networks, and possible relations with corruption cases and leaked documents detailing financial and client information of offshore entities. Relevant sources are Google+, Twitter, Facebook, DBpedia, OCCRP, Linked Leaks[7], etc.
U2. Fanaticism and terrorism: searching for advertising, accounts and posts on social networks. Relevant sources are Twitter, Google+, OCCRP, etc.
U3. Illegal medication: searching for web sites, posts, or video ads, with offers or links to darknet markets. Relevant sources are darknet markets, Tweets, Facebook posts, YouTube videos, ads, etc.

Table 1. MINTE$^+$ Configuration. The Law Enforcement Application

Parameter	Value	Description
Query	Free Text	Usually people, organizations, or products name or description
Ontology	LiDaKrA	The ontology describing the main concepts in the crime investigation domain
Web APIs	11	Facebook, Google+, VK, Twitter, Xing, ICIJ Offshore Leaks, DBpedia, eBay, darknet sites, crawled darknet markets, OCCRP reports
Simf	GADES [8]	A semantic similarity measure for entities in knowledge graphs
Threshold	0.9	Only highly similar molecules are synthesized
Fusion policy	Union	No information is lost, e.g., all alias names of a person are kept in the final molecule

[6] https://www.bdk.de/der-bdk/aktuelles/artikel/bdk-beteiligt-sich-im-forschungsprogramm-lidakra.
[7] http://data.ontotext.com/.

3.2 MINTE+ Configuration

To address the challenges of this application and support the use cases, we configured MINTE+ with the parameters shown in Table 1. As keyword Q, the users mainly provide people, organization, or product names, e.g., Donald Trump, Dokka Umarov, ISIS, or Fentanyl. Figure 7a shows the main RDF molecules described with the LiDaKrA domain-specific ontology O developed for this application. To address C1, thirteen wrappers were developed by implementing the interfaces described in Fig. 3. These interfaces were sufficient for the social network and deep web sources defined in the application. However, an extension to access dark web sources was needed. A new interface was defined to enable a wrapper to connect to the Darknet using a Tor Proxy. As the similarity function, we used GADES [8] with a threshold of 0.9. This high value guarantees that only very similar molecules are integrated.

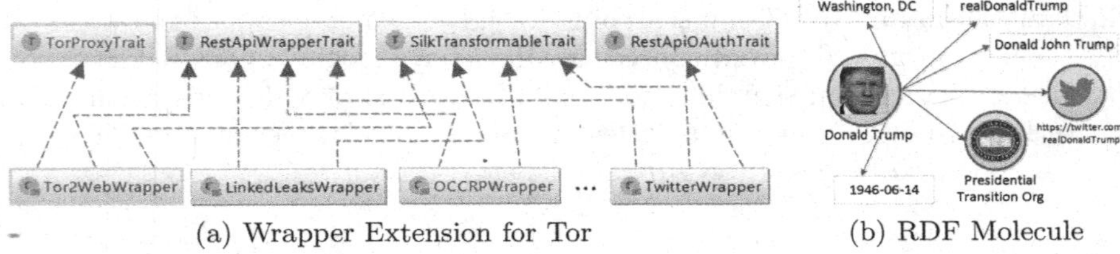

(a) Wrapper Extension for Tor (b) RDF Molecule

Fig. 6. MINTE+ in the Law Enforcement Application. (a) A new wrapper interface is implemented for querying the Dark Web. (b) An RDF molecule synthesized by the application; it synthesizes information about Donald Trump.

To address C2, each RDF molecule is annotated with its provenance at creation time using PROV-O[8], Fig. 6b shows an RDF molecule example. The fusion policy *Union* was selected to address C3; this guarantees no information is lost during the integration process, e.g., whenever a person has two aliases, both are kept in the final molecule. By design, MINTE+ does not persist any result in a triple store. All molecules are integrated on demand and displayed to the user. The on-demand approach addresses challenge C4.

To close the application cycle, a faceted browsing user interface exposes the integrated RDF graph to users. Figure 7b shows the UI; users *pose* keyword queries and *explore* results using a multi-faceted browsing user interface. We chose facets as a user-friendly mechanism for exploring and filtering a large number of search results [1]. In an earlier publication [5], we presented a demo of the user interface, comprising the following elements: a text box for the search query, a result list, entity summaries, and a faceted navigation component. Technically, MINTE+ provides a REST API to execute its pipeline on demand. JSON-LD is the messaging format between the UI and MINTE+ to avoid unnecessary data transformations for the UI components.

[8] https://www.w3.org/TR/prov-o/.

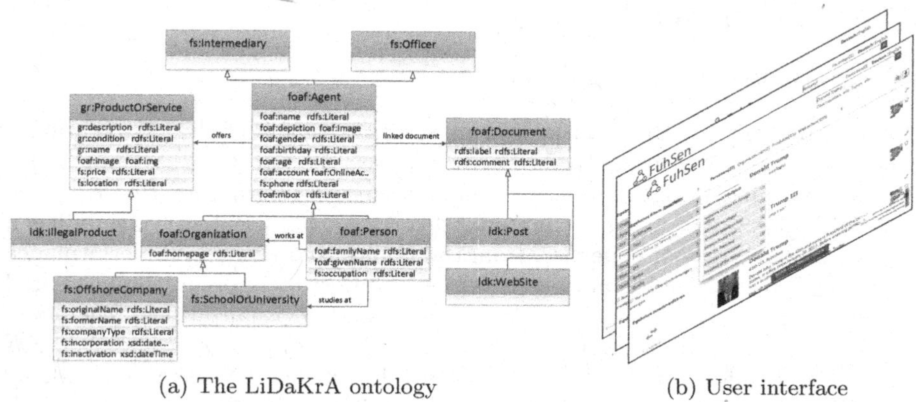

(a) The LiDaKrA ontology (b) User interface

Fig. 7. MINTE$^+$ in LiDaKrA. (a) LiDaKrA UML ontology profile view (cf. [6]) of the main RDF molecule types. (b) The faceted browsing user interface that allows the exploration of the synthesized RDF molecules.

3.3 Results and Lessons Learned

Currently, the application is installed in *four law enforcement agencies* in Germany for evaluation.[9] The user feedback is largely positive. The use of semantics in the integration process and as input for the faceted navigation gives the necessary context to facilitate the exploration and disambiguation of results, e.g., suspects with similar names. One main user concern about the application relates to the completeness of results, e.g., a person is not found by MINTE$^+$ but it is found via an interactive Facebook search. Since MINTE$^+$ is limited to the results returned by the *API*, completeness of results cannot be guaranteed.

Thanks to MINTE$^+$, law enforcement agencies can integrate new web sources into the system with low effort (1–2 person days). This dynamicity is important in this domain due to some web sources going online or offline frequently. The users furthermore emphasized the importance of the possibility to integrate internal data sources of the law enforcement agencies into the framework, which is possible thanks to the design of MINTE$^+$. The keyword search approach allows MINTE$^+$ to cope with all use cases defined for the system (e.g., U1, U2, and U3). In this application, we validate that the MINTE$^+$ framework works in an on-demand fashion. The main result of this application has become a product offered by Fraunhofer IAIS, which shows the maturity of MINTE$^+$'s approach.[10]

[9] For confidentiality we cannot state their names, nor gather usage data automatically.
[10] https://www.iais.fraunhofer.de/en/business-areas/enterprise-information-integration/federated-search.html.

4 A Job Market Application

4.1 Motivation and Challenges

Declared by Harvard Business Review as the "sexiest job of the 21st century", data scientists and their skills have become a key asset to many organizations. The big challenge for data scientists is making sense of information that comes in varieties and volumes never encountered before. A data scientist typically has a number of core areas of expertise, from the ability to operate high-performance computing clusters and cloud-based infrastructures, to applying sophisticated big data analysis techniques and producing powerful visualizations. Therefore, it is in the interest of all companies to understand the *job market* and the *skills* demand in this domain. The main goal of the European Data Science Academy (EDSA), which was established by an EU-funded research project and will continue to exist as an "Online Institute"[11], is to deliver learning tools that are crucially needed in order to close this problematic skills gap. One of these tools consists of a dashboard intended for the general public, such as students, training organizations, or talent acquisition institutions. Through this dashboard, users can monitor trends in the job market and fast evolving skill sets for data scientists. A key component of the dashboard is the demand analysis responsible for searching, collecting and integrating job postings from different job portals. The job postings need to be annotated with the skills defined in the SARO ontology [9] and enriched with geo-location information; it presents the following challenges:

C1. Complementary Information: A complete view of the European data science job market is needed by gathering job postings from all member states.
C2. Information Enrichment: The job posting description should be annotated with the required skills described in the text.
C3. Batch Processing: To get an updated status of the job market, job postings should be extracted at least every two weeks.

The EDSA dashboard uses the results of the MINTE$^+$ integration framework; it can address the following use cases:

U1. Searching for a job offer: Search for relevant data scientist jobs by EU country or based on specific skills (e.g., Python or Scala).
U2. Missing Skills Identification: it should be possible to identify what skills a person is missing on their learning path to becoming a data scientist.
U3. Analysis of Job Market By Country: analyze which EU country has more job offers, what is the average salary per country, etc.
U4. Top 5 Required Skills: identify the top 5 relevant skills for a data scientist at the time of search.

[11] http://edsa-project.eu/.

Table 2. MINTE$^+$ Configuration. The job market analysis application

Parameter	Value	Description
Query	Job Title + Country	List of 150 job titles, e.g., Machine Learning, and 28 EU Countries, e.g., IT (Italy)
Ontology	SARO [9]	The ontology describes data scientist job postings and skills
Web APIs	5	Adzuna, Trovit, Indeed, Jooble, and Xing
Simf	Silk [7]	Job title, description and hiring organization are used in the linking rules
Threshold	0.7	Best score to integrate the same job posting from different job portals
Fusion Policy	Authoritative	Adzuna was defined as the main source

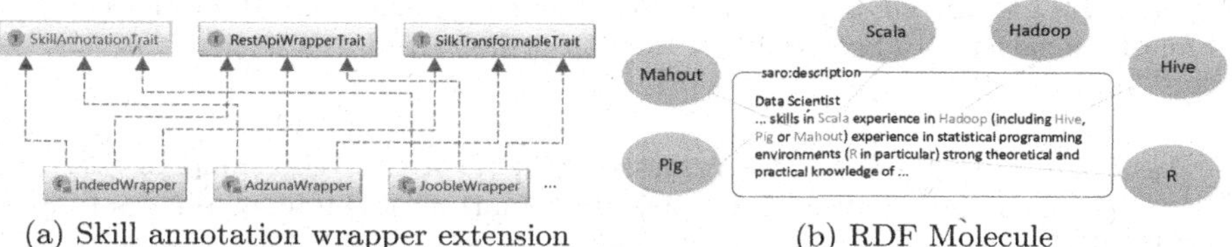

(a) Skill annotation wrapper extension (b) RDF Molecule

Fig. 8. MINTE$^+$ in the Job Market Application. (a) A new wrapper interface is implemented for annotating a job description with the corresponding skills defined in the SARO ontology. (b) An RDF molecule synthesized by the application; it synthesizes an annotated job description.

4.2 MINTE$^+$ Configuration

To address the stated challenges and to support the use cases, we configured MINTE$^+$ with the parameters shown in Table 2. A query Q is constructed from a list of 150 job titles and 28 countries. The combination of both is used as a keyword, e.g., "Machine Learning IT", yielding a total of 4,200 results. Figure 9a depicts the RDF molecule described with the SARO ontology O [9]. To address C1, five wrappers (Adzuna, Trovit, Indeed, Jooble, and Xing) were developed by implementing the interfaces described in Fig. 3. The data sources were selected covering as many countries as possible, e.g., Adzuna provides insights on the DE, FR, UK, IT markets. Indeed complements with data from NL, PL, ES. To address C2, a new interface *SkillAnnotationTrait* was defined. Figure 8a shows how the wrappers implement this new interface in addition to the standard ones defined in the framework. Technically, we employ GATE Embedded[12] with a custom REST service[13] to do the annotation using the SARO ontology.

[12] https://gate.ac.uk/family/embedded.html.
[13] https://github.com/EDSA-DataAcquisition/skill-annotation.

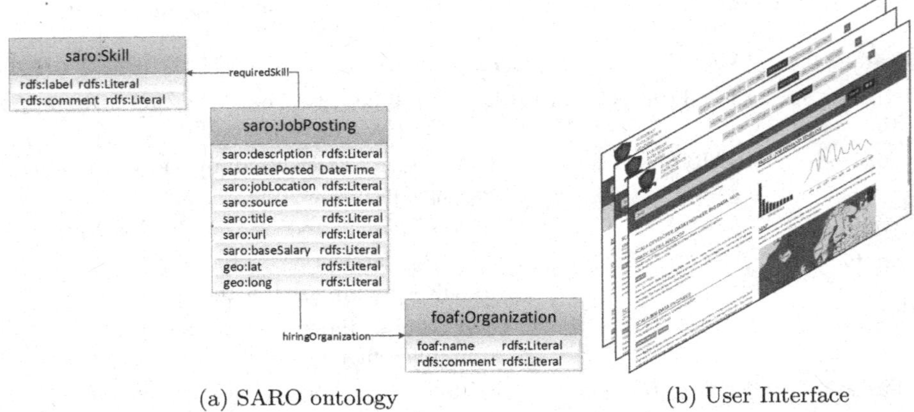

(a) SARO ontology (b) User Interface

Fig. 9. MINTE$^+$ in EDSA. (a) The SARO ontology defines the RDF molecules for job market analysis. (b) Screenshot of the EDSA dashboard.

As a similarity function, we resort to Silk [7] with a threshold of 0.7. The threshold was assigned after an empirical evaluation of the linkage rules in Silk. The RDF molecules created from job postings are similar in terms of properties. The Authoritative fusion policy was configured in this scenario, as only one property is required for fusion. Adzuna was defined as a main source. To periodically extract and integrate the job postings, a script was developed. The script reads the file containing the list of job titles and countries, calls MINTE$^+$ through its API, and saves the results in a triple store. Thus, batch processing (challenge C3) is addressed. The EDSA dashboard is then able to use this data to show integrated information about the EU job market.

4.3 Results and Lessons Learned

The EDSA dashboard[14] is running and open to the general public. Thanks to the flexibility of the wrappers, the skills annotation behavior was easy to implement. The integrated job posting knowledge graph serves as the information source to address the defined use cases (U1, U3) by using the dashboard. Using a semantic representation of job postings, it was feasible to link the job market analysis with the supply analysis (i.e., the analysis of learning material) and the learning path identified in use cases U2 and U4. The main conclusion on this application is that MINTE$^+$ is able to support an intense integration process (batch mode). Overall, it takes one day to execute all the query combinations and update the status of the job market.

[14] http://edsa-project.eu/resources/dashboard/.

5 Smart Manufacturing Application

5.1 Motivation and Challenges

The application is motivated by a global manufacturing company[15], which needs to complement their internal knowledge about parts providers with external web sources. The final usage of this external knowledge is to improve the user experience of some applications the company has been running already. The main challenges are:

C1. Entity Matching: identify the internal provider information with the external data sources. No matching entities should be discarded.
C2. Context Validation: we have to validate whether the external provider's data belongs to the manufacturing domain.

The use case (U1) is: based on the internal metadata of the providers, the company wants to complete their knowledge about them from external sources.

5.2 MINTE+ Configuration

To address the challenges of this application and support the use case, MINTE+ was configured with the parameters shown in Table 3. As query Q, metadata about the providers, e.g., the provider's name, is sent to MINTE+. As the ontology O, schema.org was configured, in particular, the subset that describes the Organization concept[16] was extended: `theCompany:PartsProvider` (a subclass of `schema:Organization`), having the property `theCompany:industry` with values such as "Semiconductors". Four wrappers were developed for this application. For confidentiality reasons, we can mention just DBpedia and Google Knowledge Graph. To address challenge C1, Silk was configured to provide values of similarity, i.e., it is used in MINTE+ as a similarity function. In this application, only one rule was configured in Silk to measure the similarity between a Google Knowledge Graph molecule with a DBpedia molecule. Only if the organization Wikipedia page[17] in both molecules refers to the same URL, they are considered the same. This is the reason for a threshold of 1.0. DBpedia is selected as major source in the authoritative fusion policy configured for this application. To provide the necessary interface for other systems on top of the MINTE+ API, a new REST method returning just JSON was designed with the company. To address the C2 challenge, a SPARQL Construct query filters the manufacturing context of the molecules (`theCompany:industry = Semiconductors`).

5.3 Results and Lessons Learned

The application is in production state. The company has more than 300 providers in their internal catalog. We evaluated the accuracy of knowledge completion

[15] For confidentiality reasons we cannot mention the name.
[16] http://schema.org/Organization.
[17] http://schema.org/ContactPage.

Table 3. MINTE$^+$ Configuration. The manufacturing application

Parameter	Value	Description
Query	Provider metadata	Includes company name, address, web site
Ontology	Schema.org	An extension of organization concept is used to describe the providers
Web APIs	4	DBpedia, Google Knowledge Graph, plus further confidential sources
Simf	Silk	Wikipedia page is used in the linking rule
Threshold	1.0	Providers with same Wikipedia page are integrated
Fusion Policy	Authoritative	DBpedia is defined as the main source

(U1) by randomly selecting 100 molecules and manually creating a gold standard, then compared the results produced by MINTE$^+$ to the gold standard. We obtained 85% accuracy, which means 85 times out of 100 MINTE$^+$ was able to complete the internal knowledge about providers with molecules coming from DBpedia and Google Knowledge Graph. Matching failures are explained mostly by outdated information from the providers, e.g., when the name of a subcontractor has changed. Although the percentage is not high, it still impacts user experience in the company's control system. Thanks to the good results regarding providers, the next step is to apply MINTE$^+$ to other entities handled by the company, such as "Components".

6 Conclusions, Lessons Learned, and Future Work

We described the MINTE$^+$ and discussed its implementation in three domain-specific applications to synthesize RDF molecules into a knowledge graph. The three applications are either under evaluation or in production. As global lessons learned we may emphasize that the role of semantic web technology is central to the success of the framework. MINTE$^+$ is able to generate knowledge graphs from remote Web API sources. However, the framework still depends on the quality, consistency, and completeness of the given data. That is, the better the source data, the better is the resulting knowledge graph. We showed the benefits of MINTE$^+$ framework in terms of the configurability and extensibility of its components. The effort to configure, extend and adapt MINTE$^+$ framework is relatively low (new fusion policies, similarity functions, wrappers may be developed and plugged into the framework); state-of-art approaches can be easily integrated. MINTE$^+$ is started to be used in biomedical applications to integrate and transform big data into actionable knowledge. Therefore, MINTE$^+$ is being extended to scale up to large volumes of diverse data. Moreover, we are developing machine learning techniques to automatically configure MINTE$^+$ according to the characteristics of the data sources and application domain.

Acknowledgements. Work supported by the European Commission (project SlideWiki, grant no. 688095) and the German Ministry of Education and Research (BMBF) in the context of the projects LiDaKrA ("Linked-Data-basierte Kriminalanalyse", grant no. 13N13627) and InDaSpacePlus (grant no. 01IS17031).

References

1. Arenas, M., Grau, B.C., Kharlamov, E., Marciuska, S., Zheleznyakov, D.: Faceted search over RDF-based knowledge graphs. J. Web Seman. **37**, 55–74 (2016)
2. Ball, P.: Chemistry: why synthesize? Nature **528**(7582), 327 (2015)
3. Collarana, D., Galkin, M., Ribón, I.T., Lange, C., Vidal, M., Auer, S.: Semantic data integration for knowledge graph construction at query time. In: 11th IEEE International Conference on Semantic Computing, ICSC 2017, pp. 109–116 (2017)
4. Collarana, D., Galkin, M., Ribón, I.T., Vidal, M., Lange, C., Auer, S.: MINTE: semantically integrating RDF graphs. In: Proceedings of the 7th International Conference on Web Intelligence, Mining and Semantics, WIMS 2017, pp. 22:1–22:11 (2017)
5. Collarana, D., Lange, C., Auer, S.: FuhSen: a platform for federated, RDF-based hybrid search. In: Proceedings of the 25th International Conference on World Wide Web, pp. 171–174 (2016)
6. Gasevic, D., Djuric, D., Devedzic, V.: Model Driven Engineering and Ontology Development, 2nd edn. Springer, Heidelberg (2009). https://doi.org/10.1007/978-3-642-00282-3
7. Isele, R., Bizer, C.: Active learning of expressive linkage rules using genetic programming. J. Web Seman. **23**, 2–15 (2013)
8. Ribón, I.T., Vidal, M., Kämpgen, B., Sure-Vetter, Y.: GADES: a graph-based semantic similarity measure. In: Proceedings of the 12th International Conference on Semantic Systems, SEMANTICS 2016, pp. 101–104 (2016)
9. Sibarani, E.M., Scerri, S., Morales, C., Auer, S., Collarana, D.: Ontology-guided job market demand analysis: a cross-sectional study for the data science field. In: Proceedings of the 13th International Conference on Semantic Systems, SEMANTICS 2017, pp. 25–32 (2017)

Getting the Most Out of Wikidata: Semantic Technology Usage in Wikipedia's Knowledge Graph

Stanislav Malyshev[1(✉)], Markus Krötzsch[2(✉)], Larry González[2(✉)], Julius Gonsior[2(✉)], and Adrian Bielefeldt[2(✉)]

[1] Wikimedia Foundation, San Francisco, U.S.A.
smalyshev@wikimedia.org
[2] cfaed, TU Dresden, Dresden, Germany
{markus.krotzsch,larry.gonzalez,julius.gonsior,
adrian.bielefeldt}@tu-dresden.de

Abstract. Wikidata is the collaboratively curated knowledge graph of the Wikimedia Foundation (WMF), and the core project of Wikimedia's data management strategy. A major challenge for bringing Wikidata to its full potential was to provide reliable and powerful services for data sharing and query, and the WMF has chosen to rely on semantic technologies for this purpose. A live SPARQL endpoint, regular RDF dumps, and linked data APIs are now forming the backbone of many uses of Wikidata. We describe this influential use case and its underlying infrastructure, analyse current usage, and share our lessons learned and future plans.

1 Introduction

Since its inception in late 2012, Wikidata [19] has become one of the largest and most prominent collections of open data on the web. Its success was facilitated by the proximity to its big sister Wikipedia, which has supported Wikidata both socially and technically, e.g., with reliable server infrastructure and global user management. This has helped Wikidata to engage editors, and paved its way of becoming one of the most active projects of the Wikimedia Foundation in terms of contributors, surpassing most Wikipedia editions.[1] Many organisations now donate data and labour to help Wikidata's volunteers in selecting and integrating relevant pieces of information. A widely noted case was Google's migration of Freebase to Wikidata [17], but most organisational contributions naturally are coming from non-profits and research projects, such as Europeana.

Wikidata thereby has grown into one of the largest public collections of general knowledge, consisting of more than 400 million statements about more than

[1] Wikidata has attracted contributions from over >200 K registered editors (>37 K in Jan 2018).

D. Vrandečić et al. (Eds.): ISWC 2018, LNCS 11137, pp. 376–394, 2018.
https://doi.org/10.1007/978-3-030-00668-6_23

45 million entities. These figures still exclude over 60 million links from Wikidata entities to Wikipedia articles (familiar from Wikipedia's *Languages* toolbar), over 200 million labels and aliases, and over 1.2 billion short descriptions in several hundred languages. Wikidata thus has become the central integration point for data from all Wikipedia editions and many external sources, an authoritative reference for numerous data curation activities, and a widely used information provider. Applications range from user-facing tools such as Apple's Siri or EuroWings' in-flight app to research activities, e.g., in the life sciences [4] and in social science [20].

However, this success crucially depends on the ability of the underlying system to serve the needs of its growing community. Many thriving community projects have turned into deserted digital wastelands due to a failure to adopt to changing requirements. Unfortunately, the core of Wikidata is not well-adapted to the needs of data analysts and ontology engineers. It is built upon the well-tested ideas and methods of Wikipedia, which were, however, developed for encyclopedic texts. Internally, Wikidata's content likewise consists of strings, stored and versioned as character blobs in a MySQL database.

The underlying MediaWiki software that ensured the initial adoption and stability of Wikidata is therefore hardly suitable for advanced data analysis or query mechanisms. This was an acceptable trade-off for the first phases of the project, but it was clear that additional functionality would soon become a requirement for moving on. Most critically, the lack of reliable query features is a burden for the volunteer editors who are trying to detect errors, find omission or biases, and compare Wikidata to external sources.

The Wikimedia Foundation has considered many possible solutions for addressing this challenge, including own custom-built software and the use of existing NoSQL databases of several data models. The final conclusion was to build the core functionality on semantic technologies, in particular on RDF and SPARQL, as a mature technology which can address Wikidata's need to share, query, and analyse data in a uniform way. The main reason for this choice was the availability of well-supported free and open source tools for the main tasks, especially for SPARQL query answering.

Three years into the project, semantic technologies have become a central component in Wikidata's operations. The heart of the new functionality is a live SPARQL endpoint, which answers over a hundred million queries each month, and which is also used as a back-end for several features integrated with Wikidata and Wikipedia. Linked data web services and frequent RDF dumps provide further channels for sharing Wikidata content.

In this paper, we first discuss this influential application in detail. We present Wikidata and its RDF encoding and export services (Sects. 2 and 3), introduce the Wikidata SPARQL service *WDQS* (Sect. 4), and discuss its technical characteristics and current usage (Sects. 5 and 6). We conclude with some lessons we learned and a brief outlook.

2 Encoding Wikidata in RDF

The Wikidata knowledge graph is internally stored in JSON format, and edited by users through custom interfaces. We therefore start by explaining how this content is represented in RDF. Conceptually, the graph-like structure of Wikidata is already rather close to RDF: it consists of items and data values connected by properties, where both entities and properties are first-class objects that can be used globally. For example, Wikidata states that the item *Germany* for property *speed limit* has a value of *100* km/h. All items and properties are created by users, governed not by technical restrictions but by community processes and jointly developed guidelines. Properties are assigned a fixed datatype that determines which values they may take. Possible types include numbers, strings, other items or properties, coordinates, URLs, and several others. Unique identifiers are used to address entities in a language-agnostic way, e.g., *Germany* is Q183 and *speed limit* is P3086. It is natural to use these identifiers as local names for suitable IRIs in RDF.

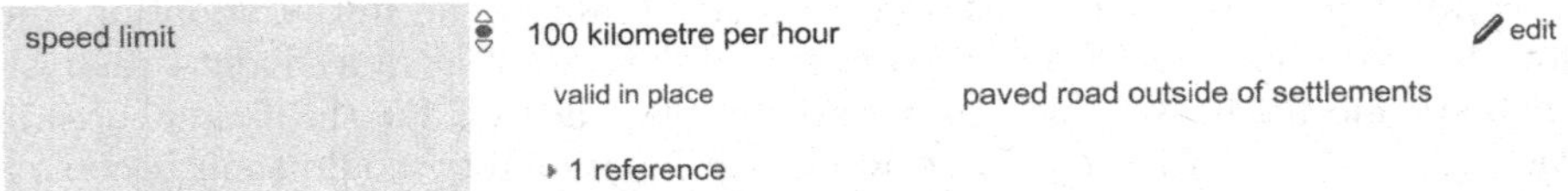

Fig. 1. Wikidata statement on the page about Germany (https://www.wikidata.org/wiki/Q183)

What distinguishes Wikidata from RDF, however, is that many components of the knowledge graph carry more information than plain RDF would allow a single property or value to carry. There are essentially two cases where this occurs:

- Data values are often compound objects that do not correspond to literals of any standard RDF type. For example, the value *100* km/h has a numeric value of 100 and a unit of measurement *km/h* (Q180154).
- Statements (i.e., subject-property-object connections) may themselves be the subject of auxiliary property-value pairs, called *qualifiers* in Wikidata. For example, the Wikidata statement for a speed limit of Germany is actually as shown in Fig. 1, where the additional qualifier *valid in place* (P3005) clarifies the context of this limit (*paved road outside of settlements*, Q23011975). Nesting of annotations is not possible, i.e., qualifiers cannot be qualified further.

Qualifiers used in statements can be arbitrary Wikidata properties and values, but Wikidata also has some built-in annotations of special significance. Most importantly, each statement can have one or more *references*, which in turn are complex values characterised by many property-value pairs. This can be a list of bibliographical attributes (title, author, publisher, ...), or something as simple as a reference URL.

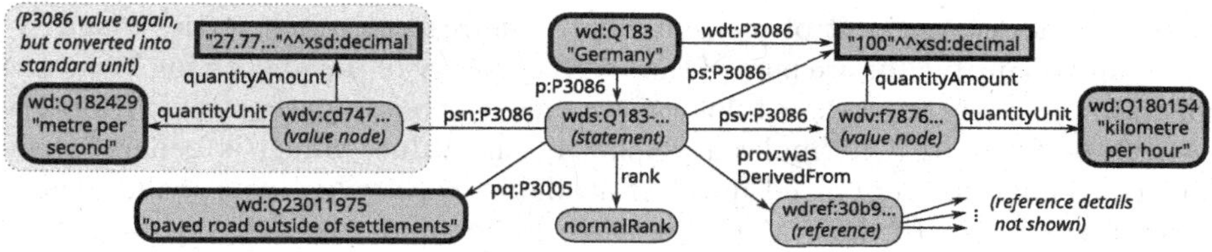

Fig. 2. RDF graph for the statement from Fig. 1, with added annotations and highlights for readability; properties with P3086 are facets of *speed limit*; pq:P3005 is the *valid in place* qualifier

Another type of built-in annotation on statements is the *statement rank*, which can be *normal* (default), *preferred*, or *deprecated*. Ranks are a simple filtering mechanism when there are many statements for one property. For example, cities may have many population numbers for different times, the most recent one being *preferred*. Such ranks simplify query answering, since it could require complex SPARQL queries to find the most recent population without such a hint. The rank *deprecated* is used for recording wrong or inappropriate statements (e.g., the statement that Pluto is a planet is deprecated).

We have developed a vocabulary and mapping for representing Wikidata in RDF. Details are found online,[2] so we focus on the main aspects here. The basic approach follows the encoding by Erxleben et al. [5], who had created early RDF dumps before the official adoption of RDF by Wikimedia. Complex values and annotated statements both are represented by auxiliary nodes, which are the subject of further RDF triples to characterise the value or statement annotations, respectively.

The statement in Fig. 1 therefore turns into an RDF graph as shown in Fig. 2, which we explain next. For readability, we have added labels for Wikidata items, and hints on the type of auxiliary nodes (whose lengthy IRIs are shortened). The two Wikidata properties P3086 (*speed limit*) and P3005 (*valid in place*) are associated with several RDF properties to disambiguate the use of properties in different contexts.

The simplest case is wdt:P3086, which relates a simplified version of the value (the number 100) directly to the subject. This triple is only generated for statements with the highest rank among all statements for the same Wikidata property and subject. As shown by our example, the simple encoding often captures only part of the available information, but this suffices for many applications, especially since values that are strings, URLs, or Wikidata entities can be represented in full detail as a single RDF entity.

Quantifiers and references can be accessed by following p:P3086 to the statement node. From the statement node, one can access again the simplified RDF value (via ps:P3086), or the complete value (via psv:P3086), which specifies the

[2] https://www.mediawiki.org/wiki/Wikibase/Indexing/RDF_Dump_Format, which also defines the namespace prefixes we use herein (all URLs in this paper were retrieved on 15 June 2018).

numeric amount and unit of measurement. The respective properties quantityValue and quantityUnit are part of the OWL ontology of the underlying Wikibase software (http://wikiba.se/ontology#; we omit the namespace prefix). Property psn:P3086 leads to a normalised version of the value, which is generated by unit conversion for supported units. If available, each value node is linked to a normalised version of that value, which we do not show in the figure.

Qualifier values are also accessed from the statement node. The only qualifier here is P3005, represented by qualifier property pq:P3005 here. The statement further has a rank (*normal* in this case), and a reference, connected through RDF property prov:wasDerivedFrom from the W3C provenance vocabulary [8]. References also have their own RDF property IRIs for any Wikidata property used (not shown here).

Overall, this encoding of statements leads to graphs with many, sometimes redundant triples. This design is meant to simplify query answering, since users can easily ignore unwanted parts of this encoding, allowing queries to be as simple as possible and as complicated as needed. The remainder of the data of Wikidata, including labels, aliases, descriptions, and links to Wikipedia articles, is exported as described by Erxleben et al. [5], mostly by direct triples that use standard properties (e.g., rdfs:label).

3 Wikidata RDF Exports

Based on the above RDF mapping, the Wikimedia Foundation generates real-time linked data and weekly RDF dumps. Dumps are generated in two formats: a complete dump of all triples in Turtle, and a smaller dump of only the simplified triples for wdt: properties in N-Triples.[3] Dumps and linked data exports both are generated by Wikibase, the software used to run Wikidata, with RDF generated by the PHP library Purtle[4]. Both tools are developed by Wikimedia Germany, and published as free and open source software.

As of April 2018, the RDF encoding of Wikidata comprises over 4.9 billion triples (32 GB in gzipped Turtle). The dataset declares over 47,900 OWL properties, of which about 71% are of type owl:ObjectProperty (linking to IRIs) while the rest are of type owl:DatatypeProperty (linking to data literals). Most of these properties are used to encode different uses of over 4400 Wikidata properties, which simplifies data processing and filtering even where not required for avoiding ambiguity.

Besides the RDF encodings of over 415 million statements, there are several common types of triples that make up major parts of the data. Labels, descriptions, and aliases become triples for rdfs:label (206 M), schema:description (1.3 B), and skos:altLabel (22 M), respectively. Over 63 M Wikipedia articles and pages of other Wikimedia projects are linked to Wikidata, each using four triples for (schema:about, schema:inLanguage, schema:name, and schema:isPartOf).

[3] Current dumps are found at https://dumps.wikimedia.org/wikidatawiki/entities/.
[4] See http://wikiba.se/ and https://www.mediawiki.org/wiki/Purtle.

Finally, rdf:type is not used to encode the conceptually related Wikidata property *instance of* (P31) but to classify the type of different resources in RDF. There is a total of over 480 M triples of this kind.

Linked data exports for each Wikidata entity can be accessed in two ways: clients that support content negotiation will simply receive the expected data from the IRIs of Wikdiata entities. Users who wish to view a particular data export in a browser may do so using URLs of the form http://www.wikidata.org/wiki/Special:EntityData/Q183.nt. Other recognised file endings include ttl (Turtle) and rdf (RDF/XML).

During March 2018, the complete Turtle dump has been downloaded less than 100 times,[5] which is small as compared to the over 7,000 downloads of the similarly sized JSON dump. In contrast, more than 270 M requests for linked data in Turtle format have been made during that time, making this format more popular than JSON (16 M requests), RDF/XML (1.3 M requests), and NT (76 K requests) for per-page requests.[6] Part of this traffic can be attributed to our SPARQL endpoint, which fetches linked data for live updates, but there seem to be other crawlers that prefer linked data over dumps. The limited popularity of RDF dumps might also be due to the availability of a SPARQL query service that can directly answer many questions about the data.

4 The Wikidata Query Service

Since mid 2015, Wikimedia provides an official public Wikidata SPARQL query service (WDQS) at http://query.wikidata.org/,[7] built on top of the BlazeGraph RDF store and graph database.[8] This backend was chosen after a long discussion, mainly due to its very good personal support, well-documented code base, high-availability features, and the standard query language.[9] Data served by this endpoint is "live" with a typical update latency of less than 60 s. The service further provides several custom extensions and features beyond basic SPARQL support, which we describe in this section. Extensive user documentation with further details is available through the Wikidata Query Service help portal.[10] All software that is used to run WDQS is available online.[11]

4.1 User Interface

The most obvious extension of WDQS is the web user interface, shown in Fig. 3. It provides a form-based query editor (left), a SPARQL input with syntax highlighting and code completion (right), as well as several useful links (around the

[5] https://grafana.wikimedia.org/dashboard/db/wikidata-dump-downloads.
[6] https://grafana.wikimedia.org/dashboard/db/wikidata-special-entitydata.
[7] This is the user interface; the raw SPARQL endpoint is at https://query.wikidata.org/sparql.
[8] https://www.blazegraph.com/.
[9] https://lists.wikimedia.org/pipermail/wikidata-tech/2015-March/000740.html.
[10] https://www.wikidata.org/wiki/Help:SPARQL.
[11] https://github.com/wikimedia/wikidata-query-rdf.

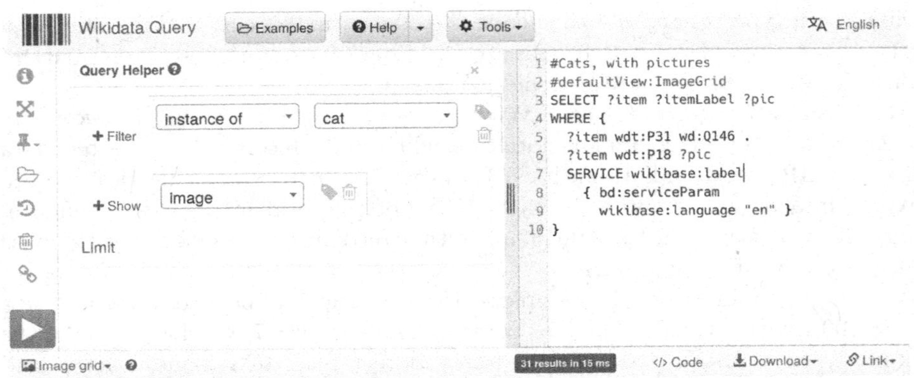

Fig. 3. Wikidata SPARQL query service UI with an example query

edges). As opposed to general-purpose SPARQL editors such as YASGUI [15], the interface has been customised for Wikidata to improve its functionality. For example, hovering the mouse pointer over any Wikidata entity in the SPARQL source displays a tooltip with a label in the user's browser language (or, if missing, a fallback) together with a short textual description from Wikidata (e.g., hovering on Q146 shows "cat (Q146): domesticated species of feline"). Likewise, auto-completion of Wikidata entities uses main labels and aliases, and ranks proposals by popularity in Wikipedia. These features are important to work with the opaque technical identifiers of Wikidata in an efficient way. Indeed, the interface is mostly targeted at developers and power users who need to design queries for their applications. A community-curated gallery of several hundred example queries (linked at the top) shows the basic and advanced features in practical queries.

Besides the usual tabular view, the front-end supports a variety of result visualisations, which can be linked to directly or embedded into other web pages. The most popular views that were embedded from January through March 2018 were timeline, map, bar chart, graph, image grid, and bubble chart.[12] The front-end further generates code snippets for obtaining the result of a given query in a variety of programming languages and tools, so as to help developers in integrating SPARQL with their applications.

4.2 Custom SPARQL Extensions

Besides the front-end, the SPARQL service as such also includes some custom extensions that are specific to Wikidata. In order to conform to the W3C SPARQL 1.1 standard, extensions were added using the SERVICE keyword, which SPARQL provides for federated queries to remote endpoints. Custom extensions then appear as sub-queries to special service URIs, which may, however, produce variable bindings in arbitrary ways. The query in Fig. 3 shows a

[12] https://grafana.wikimedia.org/dashboard/db/wikidata-query-service-ui.

call to the service wikibase:label, which takes as input variables from the outer query (e.g., ?item) and returns labels for the Wikidata entities they are mapped to as bindings to derived variable names (e.g., ?itemLabel). The labelling service is very popular, and it is the reason why the familiar query pattern

```
OPTIONAL { ?item rdfs:label ?itemLabel } FILTER (lang(?itemLabel) = 'en')
```

is rarely encountered in SPARQL queries. Besides efficiency, the main reason for introducing this service was to support user-specified language fallback chains that can be used to obtain labels even if there is no label in the preferred language.

Further custom services support searches for coordinates around a given point (wikibase:around) and within a given bounding box (wikibase:box), respectively. They are complemented by a custom function (used in SPARQL filter and bind clauses) to compute the distance between points on a globe, which one can use, e.g., to query for entities that are closest to a given location.

The special service wikibase:mwapi provides access to selected results from the MediaWiki Web API. For example, one can perform a full text search on English Wikipedia or access the (relevance-ordered) list of text-based suggestions produced by the search box on Wikidata when typing article names.[13]

Further services include several native services of BlazeGraph, such as graph traversal services that can retrieve nodes at up to a certain distance from a starting point, which is otherwise difficult to achieve in SPARQL.[14]

The service-based extension mechanism, which is already introduced by BlazeGraph, adds some procedural aspects to SPARQL query answering, but retains full syntactic conformance. A disadvantage of the current design is that the patterns that are used inside service calls may not contain enough information to compute results without access to the rest of the query, which one would not normally expect from a federated query.

Besides the custom services, the SPARQL endpoint also supports real federated queries to external endpoints. A white-list of supported endpoints is maintained to avoid untrusted sources that could be used, e.g., to inject problematic content into result views or applications that trust Wikidata.[15]

The only other customisations to BlazeGraph are some added functions, e.g., for URL-decoding, a greatly extended value space for xsd:dateTime (using 64bit Unix timestamps, for being able to accurately handle time points across the history of the universe), and support for a number of default namespace declarations.

[13] https://www.mediawiki.org/wiki/Wikidata_query_service/User_Manual/MWAPI.

[14] https://wiki.blazegraph.com/wiki/index.php/RDF_GAS_API.

[15] The current list of supported endpoints is at https://www.mediawiki.org/wiki/Wikidata_query_service/User_Manual/SPARQL_Federation_endpoints.

5 Realisation and Performance

In this section, we give technical details on the practical realisation of the Wikidata SPARQL query service, and performance figures obtained under current usage conditions.

The current release of Wikidata's query tools (v0.3.0) uses BlazeGraph v2.1.5. The graph database, including the live updater, is deployed on several servers of the Wikimedia Foundation, which share incoming traffic. As of April 2018, six machines in two U.S.-based data centres ("eqiad" in Ashburn and"codfw" in Carrollton) are utilised (CPU: dual Intel(R) Xeon(R) CPU E5-2620 v4 8 core; mem: 128 G; disk: Dual RAID 800 G SSD). The three servers in Carrollton are mostly for geographic redundancy and are currently not running under full load. The servers are not coordinated and each performs updates independently, so that the exact data may differ slightly between them.

Load balancing is managed by the Linux Virtual Server (LVS) software that is used throughout Wikimedia.[16] LVS is part of the operating system and does not require integration with BlazeGraph in any way. A web accelerator operates on top of the Web service endpoints by caching outputs. We use the Varnish system that is used for all Wikimedia page and API content. All responses (including error responses) are cached for 5 min, during which the exact same request will not be forwarded again to the endpoints.

The detailed configuration of BlazeGraph is part of the public source code of WDQS.[17] For the current content of almost 5 billion triples, the database needs more than 400 GB of storage space. The system is configured to use a query answering timeout of 60 s – queries running longer than this will be aborted. In addition, each client is allowed only up to five concurrent queries, and a limited amount of processing time and query errors. The latter is implemented with a *token bucket* protocol that allows clients an initial 120 s of processing time, and 60 query errors that they may use up. These buckets are refilled by 60 s/30 errors each minute, allowing clients to work at a steady rate. Clients that use up their allowed resources receive HTTP error code 429 (too many requests) with details on when they may resume querying. Importantly, WDQS does not impose limits on result size, and indeed one can effectively obtain partial data dumps of over a million results (subject to current traffic conditions).

The updater that synchronises the graph database with Wikidata is a separate process that polls Wikidata for recent changes, aggregates the modified entities, and retrieves RDF for all modified entities from the linked data service. Updates are performed using SPARQL update queries that delete all statements that are directly linked from modified entities and not retrieved with the new dataset. This may leave some auxiliary nodes (used for representing complex data values) orphaned, and another update query is used to remove such nodes

[16] See https://wikitech.wikimedia.org/wiki/LVS for technical details.

[17] https://github.com/wikimedia/wikidata-query-deploy/blob/master/RWStore.properties.

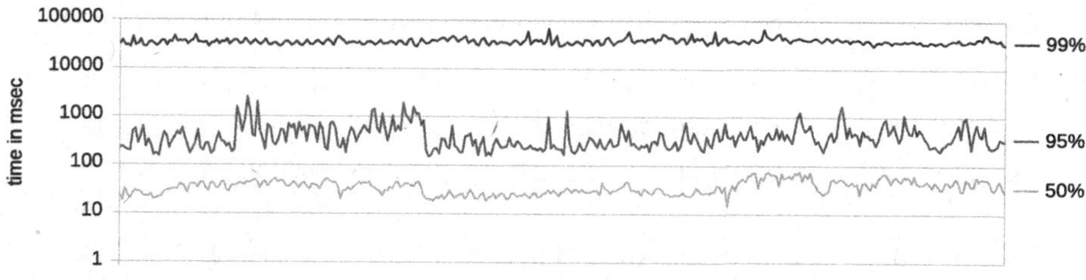

Fig. 4. Average query answering times for top 50%, 95%, and 99% of queries on a logarithmic scale; Ashburn cluster ("eqiad"); January through March 2018

(restricted to nodes used in the updated items' data before). Updates happen every 500 changes, or every few seconds if no changes occur.

Since its launch in 2015, WDQS has experienced a continued increase in query traffic. In 2018, there have been over 321 million requests within the twelve weeks from 1st January to 25th March, for a rate of 3.8 million requests per day or 44 per second. We have created several online dashboards with live usage metrics. Core load and performance figures are found in the Wikidata Query Service dashboard.[18] Among other metrics, the dashboard shows, for each server, the average number of requests and errors per second, the total number of triples stored, CPU load, and available memory. "Varnish latency" measures the time that the cache had to wait for content to be produced by the service, i.e., it is an upper bound for BlazeGraph's internal query answering times. Figure 4 shows the resulting average times that it took to answer various percentiles of queries on the (busier) "eqiad" cluster. The top 50% of the queries are answered in less than 40ms on average, while the top 95% finish in an average of 440 ms. The top 99% can reach peak latencies that are close to the 60 s timeout, but on average also stay below 40 s. At the same time, the total number of timeouts generated by BlazeGraph on all servers during this interval was below 100,000, which is less than 0.05% of the requests received.

Hernández et al. have studied the performance of query answering over Wikidata using several graph databases, including BlazeGraph and Virtuoso, and several ways of encoding statements [7]. They concluded that best performance might be achieved using Virtuoso and named graphs, which might seem at odds with our positive experience with BlazeGraph and statement reification. However, it is hard to apply their findings to our case, since they used BlazeGraph on spinning disks rather than SSD, which we discovered to have a critical impact on performance. Moreover, they used a plain version of BlazeGraph without our customisations, and focused on hypothetical query loads that heavily rely on accessing statements in full detail. It is therefore hard to tell if Virtuoso could retain a performance advantage under realistic conditions, making it an interesting topic for future investigations.

---

[18] https://grafana.wikimedia.org/dashboard/db/wikidata-query-service.

Table 1. Query dataset sizes with contribution of robotic and organic queries

	Start – End	Total	Valid	Robotic	Organic
I1	2017-06-12 – 2017-07-09	79,082,916	59,555,701	59,364,020	191,681
I2	2017-07-10 – 2017-08-06	82,110,141	70,397,955	70,199,977	197,978
I3	2017-08-07 – 2017-09-03	90,733,013	78,393,731	78,142,971	250,760
I4	2018-01-01 – 2018-01-28	106,074,877	92,100,077	91,504,428	595,649
I5	2018-01-29 – 2018-02-25	109,617,007	96,407,008	95,526,402	880,606
I6	2018-02-26 – 2018-03-25	100,133,104	84,861,808	83,998,328	863,480

Metrics on the usage of the Web query user interface are collected in a second dashboard.[19] A detailed usage analysis of this interface is not in scope for this paper, but we can note that the overall volume of queries that are posed through this interface is much lower, with several hundred to several thousand page loads per day. At the same time, the queries posed trigger many errors (over 50% on average since January), which are mostly due to malformed queries and timeouts. This suggests that the interface is indeed used to design and experiment with new queries, as intended.

6 SPARQL Service Usage Analysis

In this section, we focus on the current practical usage of WDQS, considering actual use cases and usage patterns. To this end, we analyse server-side request logs (Apache Access Log Files) of WDQS. As logs contain sensitive information (especially IP addresses), this data is not publicly available, and detailed records are deleted after a period of three months. For this research, we therefore created less sensitive (but still internal) snapshots that contain only SPARQL queries, request times, and user agents, but no IPs. We plan to release anonymised logs; see https://kbs.inf.tu-dresden.de/WikidataSPARQL.

6.1 Evaluation Data

We extend our recent preliminary analysis, which considered series of WDQS logs from 2017 [1] by including another set of logs recorded with some months distance, giving us a better impression of usage change over time. We consider the complete query traffic in six intervals, each spanning exactly 28 days, as shown in Table 1, which also shows the total number of requests for each interval. Further parts of this table are explained below.

We process all queries with the Java module of Wikidata's BlazeGraph instance, which is based on OpenRDF Sesame with minimal modifications in the parsing process re-implemented to match those in BlazeGraph. In particular,

[19] https://grafana.wikimedia.org/dashboard/db/wikidata-query-service-ui.

BIND clauses are moved to the start of their respective sub-query after the first parsing stage. This resulted in a total of 481,716,280 valid queries (208,347,387 from 2017 and 273,368,893 from 2018).

A major challenge in analysing SPARQL query logs is that a large part of the requests is generated systematically by software tools. Due to the huge impact that a single developer's script can have, such effects do not average out, at least not at the query volumes that we are observing in Wikidata. Traditional statistical approaches to the analysis of (user-generated) traffic therefore are heavily biased by the activities of individual users. There is no causal relation between the number of queries generated by a tool and the relevance or utility of these queries to the Wikidata community: in extreme cases, we have witnessed one and the same malformed SPARQL query being issued several million times, evidently generated by some malfunctioning script.

To address this issue, we have introduced an approach of classifying queries into *robotic* and *organic* requests [1]. Robotic traffic generally comprises all high-volume, single-source components of the total query load, whereas organic traffic comprises low-volume, multi-source queries. Our hypothesis is that organic traffic is dominated by queries that indicate an immediate information need of many users, and therefore is more interesting for usage analysis. Robotic traffic is still interesting from a technical perspective, since strongly it dominates the overall query load and therefore governs server performance. Notably, we do not want to single out hand-written queries of power users, as done, e.g., by Rietfeld and Hoekstra [14], but we let organic traffic include queries that users issue by interacting with software or websites, possibly without being aware of the underlying technology. The boundary between the two categories cannot be sharp, e.g., since some advanced browser applications may generate significant numbers of queries, and since some interactive applications may pre-load query results to prepare for a potential user request that may never happen. We believe that our classification is robust to the handling of such corner cases, since they do not dominate query load. As we will see below, the different traffic types exhibit distinctive characteristics.

We have classified queries using the following steps:

(1) All queries issued by non-browser user agents were considered robotic. These were mostly self-identified bots and generic programming language names (e.g., "Java").
(2) We scanned the remaining queries with browser-like user agent strings for comments of the form "`#TOOL:` *toolname*," which are conventionally used in the community to identify software clients in the query string. Of the tools that occurred, we considered *WikiShootMe* (nearby sites to photograph), *SQID* (data browser), and *Histropedia* (interactive timelines) as sources of organic traffic, and all others as robotic.
(3) We manually inspected the remaining browser-based queries to identify high-volume, single-source query loads. For this, we abstracted queries into query patterns by disregarding concrete values used for subjects, objects, LIMIT and OFFSET. We consider each agent that issues more than 2,000 queries

of the same type during one interval, and classify it as robotic if its temporal query distribution is highly irregular (typically in the form of peaks with thousands of queries for one or more hours, followed by inactivity during other days). It is often hard to determine the source of an identified traffic peak, but based on the number of distinct query types encountered, we estimate that at most 300 sources were filtered in this way (probably less, since some those query types are very similar and likely stem from one tool).

The outcome of our query type classification is shown in Table 1. The total of 2,980,154 organic queries (640,419 in 2017 and 2,339,735 in 2018) accounts for 0.6% of the traffic, and would therefore be overlooked completely in any non-discriminative analysis. We observe a general increase in query traffic over time, with a faster relative increase of organic traffic. This seems to be caused by the appearance of more and more user-space applications that use SPARQL to access Wikidata. The slight decline in overall traffic in I6 can be attributed to a single extremely active query source (the data-integration bot "auxiliary matcher"), which issued around 20.4 M queries in I5 but only 5.4 M in I6. This case illustrates how fragile query traffic statistics can be if such sources are not identified.

6.2 SPARQL Feature Analysis

We now report on several structural features of SPARQL queries throughout the datasets, which allow us to observe important differences between organic and robotic queries, but also between Wikidata and other SPARQL services. Our previous work includes an extended discussion of the different characteristics of organic and robotic traffic, including temporal distribution and language usage, which we do not repeat here [1].

Table 2 shows the prevalence of most common solution set modifiers, graph matching patterns, and aggregation functions. *Join* refers to the (often implicit) SPARQL join operator; *Filter* includes FILTER expressions using NOT EXIST; and SERVICE calls are split between the Wikidata labelling service (*lang*) and others. Queries with REDUCED, EXISTS, GRAPH, + in property paths, and other aggregation functions generally remained below 1%. Any other feature not shown in the table has not been counted.

The differences between organic and robotic traffic are clearly visible. Features that are much more common in organic queries include LIMIT, DISTINCT, OPTIONAL, ORDER BY, subqueries, services, and all aggregates. Robotic queries more often include BIND, UNION, and VALUES. FILTER and path queries occur in both traffic types to varying (high) degrees. In general, organic queries are more varied and use more analytical features, which is consistent with our expectations.

We can also see changes over time. Measures for robotic traffic are generally less stable, even from one month to the next, but especially when comparing I1–I3 with I4–I6, we can also see major changes for organic traffic. This includes strong increases in property paths, DISTINCT, LIMIT, and ORDER BY, and

Table 2. Prevalence of SPARQL features among valid queries in percent of total queries per dataset

Feature	Organic						Robotic					
	I1	I2	I3	I4	I5	I6	I1	I2	I3	I4	I5	I6
Limit	31.21	39.42	46.44	51.46	50.44	36.12	21.50	16.90	17.28	20.44	11.51	15.27
Distinct	26.71	31.37	18.67	59.19	60.12	64.22	14.96	5.53	4.21	4.29	7.54	10.84
Order By	17.18	14.27	12.80	46.51	46.16	34.15	12.08	8.17	6.77	8.81	7.74	17.82
Offset	0.38	2.75	0.35	0.08	0.07	0.04	7.94	6.04	6.29	0.10	0.07	0.10
Join	87.41	87.74	89.76	82.45	91.67	86.99	88.32	79.09	67.47	73.21	61.18	69.62
Optional	42.82	46.69	56.50	50.32	40.37	40.86	24.25	11.69	11.31	12.74	15.41	29.69
Filter	25.21	28.58	21.66	12.30	11.40	11.38	20.73	17.98	13.64	14.44	16.56	29.44
Path with *	14.91	15.21	12.55	40.20	31.24	29.69	16.89	19.58	14.81	20.65	17.37	23.71
Subquery	13.29	0.00	0.00	6.35	4.99	5.32	0.34	0.00	0.00	0.09	0.13	0.11
Bind	9.87	9.26	8.66	4.81	3.98	4.21	16.70	12.28	9.57	11.85	13.61	23.01
Union	5.14	5.75	12.78	2.46	1.90	2.68	11.56	8.79	7.61	14.03	13.16	18.93
Values	4.33	3.05	11.02	3.18	3.04	3.23	36.70	31.35	28.97	29.97	24.00	11.91
Not Exists	3.28	3.27	2.44	1.08	0.71	0.66	0.19	0.22	0.19	0.26	0.29	0.35
Minus	2.00	2.81	1.55	0.81	0.56	0.70	0.52	0.92	1.06	1.45	1.24	1.79
Service lang	44.74	41.82	54.98	50.19	40.51	42.05	9.41	6.25	4.27	7.13	7.85	8.95
Service other	11.55	10.66	10.38	7.50	13.51	2.30	4.62	0.18	1.15	0.16	0.15	0.39
Group By	17.18	19.86	12.88	6.89	5.19	5.06	0.41	0.37	0.46	0.20	0.21	0.36
Sample	8.97	11.15	4.68	1.61	1.64	1.64	0.03	0.03	0.05	0.04	0.03	0.09
Count	7.50	7.26	7.98	5.27	3.79	3.71	1.17	4.38	0.30	1.53	0.65	0.90
GroupConcat	1.80	2.79	1.14	0.87	0.85	0.74	0.06	0.09	0.02	0.03	0.02	0.28
Having	1.16	1.12	0.70	0.65	0.26	0.33	0.01	0.01	0.00	0.00	0.00	0.04

decreases in FILTER, UNION, Subquery, and aggregates. We should note that numbers are relative to each dataset: a decreasing value might be caused by an actual decline in queries using a feature, but also by an increase in queries with other features.

Our results show significant differences from previously reported findings. The most systematic recent overview across several datasets is given by Bonifati et al., who have analysed SPARQL logs from BioPortal, the British Museum, DBpedia, LinkedGeoData, OpenBioMed, and Semantic Web Dog Food [3]. They found SERVICE, VALUES, BIND, and property paths to occur in less than 1% of queries, while they have great relevance in all parts of our data. For SERVICE, this can be explained by the custom SPARQL extensions of WDQS discussed in Sect. 4, and the results seem to confirm their utility. VALUES and BIND are useful for building easy to modify query templates, with VALUES being particularly useful for (typically robotic) batch requests. Property paths express reachability queries, which are particularly useful for navigating (class) hierarchies and can in fact be used to emulate some amount of ontological reasoning in SPARQL [2]. The reduced occurrence of these features in other logs may be

Table 3. Co-occurrence of SPARQL features in percent of total queries per dataset (Join, Filter, Optional, Union, Path, Values, Subquery)

							organic		robotic	
J	**F**	**O**	**U**	**P**	**V**	**S**	I1–I3	I4–I6	I1–I3	I4–I6
			(none)				8.18	9.35	19.67	27.96
J							15.23	32.31	10.81	10.10
	F						1.09	0.94	1.95	1.27
J	F						8.87	2.37	2.61	1.50
J				P			2.93	1.63	13.72	14.09
J	F			P			2.49	0.58	0.39	0.06
J					V		0.41	2.04	30.91	17.65
		O					1.28	1.61	0.12	0.64
J		O					25.97	7.16	1.88	1.95
J		O		P			2.10	28.41	0.36	0.05

							organic		robotic	
J	**F**	**O**	**U**	**P**	**V**	**S**	I1–I3	I4–I6	I1–I3	I4–I6
J	F	O					3.30	1.30	1.78	1.19
J		O	U				3.57	0.25	0.02	0.00
J		O			V		3.45	0.40	0.11	0.43
J		O		P	V		1.01	0.06	0.16	0.04
J						S	0.86	1.43	0.06	0.01
J		O				S	1.64	0.63	0.00	0.01
J	F					S	0.64	2.17	0.02	0.01
J	F	O		P			0.87	0.31	0.65	1.60
J			U	P	V		0.01	0.01	0.05	1.94
			All cases shown				**83.90**	**92.96**	**85.27**	**80.50**

due to the fact that they are new in SPARQL 1.1, and therefore might have a lower acceptance in communities that have worked with SPARQL 1.0 before.

Low prevalence was also reported for subqueries, SAMPLE, and GroupConcat. Our robotic traffic is similar, but our organic traffic paints a different picture, and especially contains many subqueries, which are often needed for complex analytical queries.

We also analysed how features are combined in queries. Table 2 suggests that join, OPTIONAL, UNION, FILTER, subqueries, property paths, and VALUES should be taken into account here. SERVICE is also common, but mostly for labelling, where it has little impact on query expressivity. We therefore ignore the labelling service entirely. We do not count queries that use any other type of service, or any other feature not mentioned explicitly (the most common such feature is BIND). Solution set modifiers (LIMIT etc.) and aggregates (COUNT etc.) are ignored: they can add to the complexity of query answering only when used in subqueries, so this feature can be considered an overestimation of the amount of such complex queries.

Results are shown in Table 3 for all operator combinations found in more than 1% of queries in some dataset. Therein, path expressions include all queries with * or +. We separate the 2017 and 2018 intervals to show change. The shown combinations account for most queries (see last line of tables). Most remaining queries use some uncounted feature, especially BIND. As before, we can see significant variations across columns. Regarding organic traffic, the largest changes from 2017 to 2018 are the increase in join-only queries and the decrease of JO queries in favour of JOP queries. For robotic traffic, we see an increase of queries without any feature and a decrease of JV queries.

Other studies of SPARQL usage noted a strong dominance of conjunctive-filter-pattern (CFP) queries, typically above 65% [3,13]. The corresponding combinations *none*, J, F, and JF only account for 33%–45% in our datasets. We can cover many more robotic queries (58%–66%) by also allowing VALUES. However, filters are actually not very relevant in our queries. A much more prominent fragment are *conjunctive 2-way regular path queries* (C2PRQs), which correspond to combinations of J, P, and (by a slight extension) V [6]. They cover 70%–75% or robotic traffic. To capture organic traffic, we need optional: queries with J, O, P, and V cover 60%–83% there. This is expected, since users are often interested in as much (optional) information as possible, while automated processing tends to consider specific patterns with predictable results.

We see very little use of UNION, whereas previous studies found UNION alone to account for as much as 7.5% of queries [3]. Since especially robotic traffic contains many queries with UNION (Table 2, we conclude that this feature typically co-occurs with features not counted in Table 3, most likely BIND.

7 Discussion and Lessons Learned

The decision to rely upon semantic technologies for essential functionality of Wikidata has taken Wikimedia into unknown territory, with many open questions regarding technical maintainability as well as user acceptance. Three years later, the work has surpassed expectations, in terms of reliability and maintainability, as well as community adoption.

The early "beta" query service has developed into a stable part of Wikimedia's infrastructure, which is used as a back-end for core functions in Wikimedia itself. For example, Wikidata displays warnings to editors if a statement violates community-defined constraints, and this (often non-local) check is performed using SPARQL. WDQS is also used heavily by Wikidata editors to analyse content and display progress, and to match relevant external datasets to Wikidata.[20] Results of SPARQL queries can even be embedded in Wikipedia (in English and other languages) using, e.g, the graph extension.

Wikimedia could not follow any established path for reaching this goal. Public SPARQL endpoints were known for some time, but are rarely faced with hard requirements on availability and robustness [18]. The use of continuously updated live endpoints was pioneered in the context of DBpedia [12], albeit at smaller scales: as of April 2018, the DBpedia Live endpoint reports 618,279,889 triples across all graphs (less than 13% of the size of Wikidata in RDF).[21] Moreover, the update mechanisms for DBpedia and Wikidata are naturally very different, due to the different ways in which data is obtained.

Many Wikidata-specific extensions and optimisations have been developed and released as free software to further improve the service. This includes custom query language extensions to integrate with Wikimedia (e.g., many-language support and Web API federation), a Wikidata-specific query interface, and low-level optimisations to improve BlazeGraph performance (e.g., a custom dictionary implementation reduces storage space for our prefixes). Some of our most important lessons learnt are as follows:

BlazeGraph Works for Wikimedia. The choice of graph database followed a long discussion of many systems, and some of the top-ranked choices did not use

[20] See, e.g., https://tools.wmflabs.org/mix-n-match/.
[21] http://live.dbpedia.org/sparql.

RDF at all. While there could have been other successful approaches, BlazeGraph has proven to be an excellent choice, regarding stability but also (open source) extensibility.

Customisation Matters. We are convinced that our own extensions to the query service played an important role in its success. The labelling, support for Wikidata's special value types (e.g., coordinates on other planets), and Wikidata-aware assistance in the Web UI improve utility and acceptance of the service.

SPARQL is Usable. Most of the Wikidata community, including developers, had no prior contact with SPARQL. An impressive amount of SPARQL-literacy has developed very quickly. There is extensive documentation and support now, including a community project *Request a Query*[22] where experts design queries for novices.

SPARQL is Affordable. Even for such a large site with a custom data model and a large global user base, the total cost of development and administrative effort, and of hardware needed for a highly reliable system is affordable.

Merging Views in RDF. Our RDF encoding merges several representations of the same data, ranging from very simple to very detailed. Combining these views in one graph proved extremely useful, since it allows queries to use just the right amount of complexity. Almost 10% of the queries we analysed for 2018 are using an RDF property that is part of the more complex, reified statement encoding.

Tangible Benefits. The effort of providing these services to the public has clearly paid off. Wikidata is benefiting hugely from the improved capabilities of data analysis and integration, used, e.g., for quality improvements. We are also seeing much use of the services in a wider public, e.g., in mobile apps and in journalism.

Slow Take-Up Among Semantic Web Researchers. We received surprisingly little input from this community so far. This is problematic since Wikidata seems to differ greatly from use cases studied before, as evident, e.g., from our analysis of SPARQL usage. Wikidata also faces hard ontology engineering challenges [16] and is a major integration point of open data.[23] It is one of the very few large knowledge graphs of a major organisation that is completely public, including its full history and all related software. We can only guess what might be holding up semantic web researchers from applying their findings in this context, and would like to invite comments from the community.

We will continue working on WDQS for the foreseeable future. At the time of this writing, ongoing development work includes, e.g., a better distribution of data updates by sharing infrastructure among the servers. The RDF encoding is also continuously extended, especially to integrate new features of Wikidata, such as the upcoming major extensions of Wikidata to support the management

[22] https://www.wikidata.org/wiki/Wikidata:Request_a_query.

[23] Wikidata links to over 2,500 external data collections though properties of type "External ID".

of media meta-data and lexicographic information.[24] We also plan to publish anonymised query logs based on the data we analysed, pending approval by the Wikimedia Foundation.[25] On the research side, we are exploring how to provide further ontological modelling support for the annotated graph structure of Wikidata [9,10], but we can see many other areas where Wikidata would benefit from increased research activities, and we hope that the semantic web community can still make many more contributions to Wikidata in the future.

Acknowledgements. This work was partly supported by the German Research Foundation (DFG) in CRC 912 (HAEC) and in Emmy Noether grant KR 4381/1-1 (DIAMOND).

References

1. Bielefeldt, A., Gonsior, J., Krötzsch, M.: Practical linked data access via SPARQL: the case of wikidata. In: Proceedings of WWW2018 Workshop on Linked Data on the Web (LDOW-18). CEUR Workshop Proceedings, CEUR-WS.org (2018)
2. Bischof, S., Krötzsch, M., Polleres, A., Rudolph, S.: Schema-Agnostic Query Rewriting in SPARQL 1.1. In: Mika, P., et al. (eds.) ISWC 2014, Part I. LNCS, vol. 8796, pp. 584–600. Springer, Cham (2014). https://doi.org/10.1007/978-3-319-11964-9_37
3. Bonifati, A., Martens, W., Timm, T.: An analytical study of large SPARQL query logs. Proc. VLDB Endow. **11**, 149–161 (2017)
4. Burgstaller-Muehlbacher, S., Waagmeester, A., Mitraka, E., Turner, J., Putman, T., Leong, J., Naik, C., Pavlidis, P., Schriml, L., Good, B.M., sSu, A.I.: Wikidata as a semantic framework for the Gene Wiki initiative. Database 2016, baw015 (2016)
5. Erxleben, F., Günther, M., Krötzsch, M., Mendez, J., Vrandečić, D.: Introducing wikidata to the linked data web. In: Mika, P. et al. [11], pp, 50–65
6. Florescu, D., Levy, A., Suciu, D.: Query containment for conjunctive queries with regular expressions. In: Mendelzon, A.O., Paredaens, J. (eds.) Proceedings of 17th Symposium on Principles of Database Systems (PODS 1998), pp. 139–148. ACM (1998)
7. Hernández, D., Hogan, A., Riveros, C., Rojas, C., Zerega, E.: Querying wikidata: comparing SPARQL, relational and graph databases. In: Groth, P., et al. (eds.) ISWC 2016, Part II. LNCS, vol. 9982, pp. 88–103. Springer, Cham (2016). https://doi.org/10.1007/978-3-319-46547-0_10
8. Lebo, T., Sahoo, S., McGuinness, D. (eds.): PROV-O: The PROV Ontology. W3C Recommendation, 30 April 2013. http://www.w3.org/TR/prov-o
9. Marx, M., Krötzsch, M.: SQID: Towards ontological reasoning for Wikidata. In: Nikitina, N., Song, D. (eds.) Proceedings of the ISWC 2017 Posters & Demonstrations Track. CEUR Workshop Proceedings, CEUR-WS.org, October 2017
10. Marx, M., Krötzsch, M., Thost, V.: Logic on MARS: Ontologies for generalised property graphs. In: Proceedings of 26th International Joint Conference on Artificial Intelligence (IJCAI 2017), pp. 1188–1194 (2017)

[24] See https://www.wikidata.org/wiki/Wikidata:WikiProject_Commons and https://www.wikidata.org/wiki/Wikidata:Lexicographical_data.

[25] See https://kbs.inf.tu-dresden.de/WikidataSPARQL for information on data availability.

11. Mika, P., et al.: ISWC 2014, Part I. LNCS, vol. 8796. Springer, Cham (2014). https://doi.org/10.1007/978-3-319-11964-9
12. Morsey, M., Lehmann, J., Auer, S., Stadler, C., Hellmann, S.: DBpedia and the live extraction of structured data from Wikipedia. Program: Electron. Libr. Inf. Syst. **46**(2), 157–181 (2012)
13. Picalausa, F., Vansummeren, S.: What are real SPARQL queries like? In: Virgilio, R.D., Giunchiglia, F., Tanca, L. (eds.) Proceedings of International Workshop on Semantic Web Information Management (SWIM 2011), p. 6. ACM (2011)
14. Rietveld, L., Hoekstra, R.: Man vs. machine: Differences in SPARQL queries. In: Proceedings of 4th USEWOD Workshop on Usage Analysis and the Web of Data. usewod.org (2014)
15. Rietveld, L., Hoekstra, R.: The YASGUI family of SPARQL clients. Seman. Web **8**(3), 373–383 (2017)
16. Spitz, A., Dixit, V., Richter, L., Gertz, M., Geiß, J.: State of the union: A data consumer's perspective on Wikidata and its properties for the classification and resolution of entities. In: Proceedings of ICWSM 2016 Wiki Workshop. AAAI Workshops, vol. WS-16-17. AAAI Press (2016)
17. Tanon, T.P., Vrandecic, D., Schaffert, S., Steiner, T., Pintscher, L.: From Freebase to Wikidata: The great migration. In: Bourdeau, J., Hendler, J., Nkambou, R., Horrocks, I., Zhao, B.Y. (eds.) Proceedings of 25th International Conference on World Wide Web (WWW 2016), pp. 1419–1428. ACM (2016)
18. Vandenbussche, P., Umbrich, J., Matteis, L., Hogan, A., Buil Aranda, C.: SPARQLES: monitoring public SPARQL endpoints. Seman. Web **8**(6), 1049–1065 (2017)
19. Vrandečić, D., Krötzsch, M.: Wikidata: A free collaborative knowledgebase. Commun. ACM **57**(10), 78–85 (2014)
20. Wagner, C., Graells-Garrido, E., Garcia, D., Menczer, F.: Women through the glass ceiling: gender asymmetries in wikipedia. EPJ Data Sci. **5**(1), 5 (2016)

Author Index

Printed in the United States
by Baker & Taylor Publisher Services